Springer Proceedings in Mathematics & Statistics

Volume 71

For further volumes:
http://www.springer.com/series/10533

Springer Proceedings in Mathematics & Statistics

This book series features volumes composed of selected contributions from workshops and conferences in all areas of current research in mathematics and statistics, including OR and optimization. In addition to an overall evaluation of the interest, scientific quality, and timeliness of each proposal at the hands of the publisher, individual contributions are all refereed to the high quality standards of leading journals in the field. Thus, this series provides the research community with well-edited, authoritative reports on developments in the most exciting areas of mathematical and statistical research today.

Anne Frühbis-Krüger •
Remke Nanne Kloosterman • Matthias Schütt
Editors

Algebraic and Complex Geometry

In Honour of Klaus Hulek's 60th Birthday

Springer

Editors
Anne Frühbis-Krüger
Leibniz Universität Hannover
Hannover, Germany

Remke Nanne Kloosterman
Humboldt-University Berlin
Berlin, Germany

Matthias Schütt
Leibniz Universität Hannover
Hannover, Germany

ISSN 2194-1009 ISSN 2194-1017 (electronic)
ISBN 978-3-319-05403-2 ISBN 978-3-319-05404-9 (eBook)
DOI 10.1007/978-3-319-05404-9
Springer Cham Heidelberg New York Dordrecht London

Library of Congress Control Number: 2014950053

Mathematics Subject Classification (2010): 14J10, 14E30, 14J28, 14J32, 14J50, 14T05

Printed on acid-free paper

Springer is part of Springer Science+Business Media (www.springer.com)

Preface

This volume of PROMS grew out of the international conference on "Algebraic and Complex Geometry", which took place at Leibniz Universität Hannover during the week of September 10 through 14, 2012. The event was organised on the occasion of Klaus Hulek's 60th birthday; it is with great pleasure that we dedicate this volume to him.

We would like to thank the members of the Scientific Advisory Board of the conference, David Eisenbud, Nigel Hitchin and Thomas Peternell, for their crucial input in setting up the program. The conference would have been impossible without the generous support from the following institutions:

- Deutsche Forschungsgemeinschaft
- MWK Niedersachsen
- Foundation Compositio
- GRK 1463 "Analysis, Geometry and String Theory"
- Many other institutions at Leibniz Universität Hannover

Especially, we would like to express our gratitude to all members of the Institute of Algebraic Geometry at LUH for all the helping hands before and during the conference, in particular to our secretaries Nicole Rottländer and Simone Reimann.

Our thanks go to all the authors contributing to this volume, and to the referees for the thorough job they have done. Matthias Freise took care of the compilation of this volume, and we thank him very much.

Hannover, Germany — Anne Frühbis-Krüger
Berlin, Germany — Remke Kloosterman
Hannover, Germany — Matthias Schütt

Contents

Introduction

The following paragraphs are meant to give a brief outline of the topics of the conference "Algebraic and Complex Geometry" and the content of this PROMS volume. Algebraic and complex geometry are exceptionally active areas of research in pure mathematics which have seen many novel developments in recent years, influencing numerous other areas such as differential geometry, number theory, representation theory, and mathematical physics. Many of these interesting aspects will be reflected in what follows.

1 Topic of the Conference

The program of the conference was designed to review the latest achievements and innovations in algebraic and complex geometry. The program featured 23 lectures from various subfields, allowing a broad scope, but putting specific emphasis on two subjects of spectacular recent and ongoing progress: *geometry of moduli spaces* and *irreducible symplectic manifolds* (Hyperkähler manifolds).

Geometry of Moduli Spaces

Moduli spaces are a key object of study in algebraic and complex geometry. Originally introduced by Riemann in the case of curves, moduli spaces turned out to be interesting both for their own sake and for the numerous implications to other fields such as e.g. number theory (arithmetic geometry) and mathematical physics (string theory).

Recently, there has been a particular interest in establishing the geometric and topological properties of moduli spaces. In particular, newly developed techniques yield results on the Kodaira dimension and on the cohomology of several moduli spaces. Most of the recent results are for moduli spaces of curves, of abelian varieties and of K3 surfaces.

K3 surfaces are a special case of holomorphic symplectic manifolds, which brings us to the second central topic of the conference.

Irreducible Symplectic Manifolds (Hyperkähler Manifolds)

Irreducible symplectic manifolds (or Hyperkähler manifolds, defined by the existence of an everywhere non-degenerate holomorphic 2-form) behave in many ways similar to abelian varieties and K3 surfaces. Yet they remain quite mysterious objects.

As an illustration, there are only a few known constructions of irreducible symplectic manifolds due to Beauville, Huybrechts, Beauville-Donagi, Debarre-Voisin, and O'Grady. It is still unclear whether there might be any more. The moduli spaces of irreducible symplectic manifolds are conjectured to be locally symmetric varieties, as in the case of K3 surfaces.

The conference program highlighted several important aspects of moduli spaces and irreducible holomorphic symplectic manifolds.

For the reader's information, we decided to include a full list of speakers with titles and abstracts in the Appendix "Complete List of Talks".

At the same time, this volume reflects the broad diversity of lectures at the conference beyond the above focal topics. We continue with a short tour of the content of this book.

2 Tour of Content of This Volume

This volume comprises 11 papers on current research from different areas of algebraic and complex geometry. Reflecting the diversity of topics at the conference, we sorted the article in alphabetic order by the first author instead of grouping them by topic. Below we give a brief survey of the content.

The general topic of the paper by Barja and Stoppino concerns the relation between stability conditions and positivity in algebraic geometry. Given a 1-parameter family of polarized varieties, the authors study three different methods, all of them involving stability conditions, to prove the positivity of a natural numerical invariant associated to the family.

Beauville studies a problem related with the algebraic topology of algebraic varieties. The author expresses the second quotient of the lower central series of the fundamental group of a topological space X in terms of the homology and cohomology of X. As an application, the author considers the Fano surface parametrizing lines in a cubic threefold, where he recovers a result due to Collino.

Blume's contribution extends the classical McKay correspondence for finite subgroups G of $\mathrm{SL}(2, \mathbb{C})$ to non-algebraically closed fields. More precisely,

Blume constructs for arbitrary fields K of characteristic zero a bijection between isomorphism classes of nontrivial irreducible representations of $G \subset \mathrm{SL}(2, K)$ and the irreducible components of the exceptional divisor in the minimal resolution of the quotient singularity $\mathbb{A}^2_K/G$.

The paper by Caporaso studies the interplay between the theory of linear series on algebraic curves and on graphs. To this end, the author introduces the notion of d-gonality for weighted graphs using harmonic indexed morphisms. Then a combinatorial locus of the moduli space of curves contains a d-gonal curve if the corresponding graph is d-gonal and of Hurwitz type. Conversely the dual graph of a d-gonal stable curve is equivalent to a d-gonal graph of Hurwitz type. A detailed study of the hyperelliptic case is included.

A classical problem is the subject of Catanese's considerations. He proves for a plane curve C that the map from C to its caustic is a birational map and he concludes with similar results for matrix projections.

The starting point for the extensive work of Ciliberto and Dedieu are degenerations of complex K3 surfaces. Given a degeneration of complex K3 surfaces, they investigate the limits of the corresponding Severi varieties parametrizing irreducible δ-nodal plane sections of the K3 surfaces. Applications include counting plane nodal curves through base points in special position, the irreducibility of Severi varieties of a general quartic surface, and the monodromy of the universal family of rational curves on quartic K3 surfaces.

Fujino and Gongyo consider the behaviour of divisors under smooth morphisms between smooth complex projective varieties with a special view towards nefness. Their arguments lead to a Hodge theoretic proof of the fact that nefness of the anti-canonical divisor of the source space implies the same for the target space. Previous proof of these results had been derived using positive characteristic arguments. The present work relies on a generalization of Viehweg's weak positivity theorem due to Campana.

Haydys introduced the notion of the hyperholomorphic line bundle on a hyperkähler manifold with an S^1-action of a certain type. Previous descriptions involved twistor spaces and gauge theory, illustrating the relevance for physics. The paper by Hitchin gives examples and more general results with a more geometrical flavour.

Hollborn and Müller-Stach start from a local system $\mathbb{V}$ induced by a family of Calabi-Yau threefolds over a smooth quasi-projective curve S. Using Higgs cohomology, they determine the Hodge numbers of the cohomology group $H^1_{L^2}(S, \mathbb{V}) = H^1(\bar{S}, j_*\mathbb{V})$. This generalizes previous work to the case of quasi-unipotent local monodromies at infinity and has applications to Rohde's families of Calabi-Yau 3-folds without maximally unipotent degenerations.

Compact Kähler holomorphic-symplectic manifolds, which are deformation equivalent to the Hilbert scheme of length n subschemes of a K3 surface, are the subject of Markman's contribution. Motivated by the K3 case, Markman investigates criteria when the linear system associated with a nef line-bundle is base point free and when this linear system induces a Lagrangian fibration.

The concluding paper by Peternell and Schrack studies complex compact Kähler manifolds X carrying a contact structure (which is in some sense the opposite of a foliation). If X is almost homogeneous and $b_2(X) \geq 2$, then they show that X is a projectivised tangent bundle. Moreover, any global projective deformation of the projectivised tangent bundle over a projective space is again of this type unless it is the projectivisation of a special unstable bundle over a projective space.

Stability Conditions and Positivity of Invariants of Fibrations

M.A. Barja and L. Stoppino

Abstract We study three methods that prove the positivity of a natural numerical invariant associated to 1-parameter families of polarized varieties. All these methods involve different stability conditions. In dimension 2 we prove that there is a natural connection between them, related to a yet another stability condition, the linear stability. Finally we make some speculations and prove new results in higher dimension.

1 Introduction

The general topic of this paper regards how stability conditions in algebraic geometry imply positivity. One of the first results in this direction is due to Hartshorne [25]: a μ-semistable vector bundle of positive degree over a curve is ample. Other seminal results are Bogomolov Instability Theorem [15] and Miyaoka's Theorem on the nef cone of projective bundles over a curve [37]. These theorems – not accidentally – are recalled and used in this paper (Theorems 8 and 4).

M.A. Barja (✉)
Departament de Matemàtica Aplicada I, Universitat Politècnica de Catalunya, ETSEIB Avda. Diagonal, 08028 Barcelona, Spain

Partially supported by MICINN-MTM2009-14163-C02-02/FEDER, MINECO-MTM2012-38122-C03-01 and by Generalitat de Catalunya 2005SGR00557
e-mail: Miguel.Angel.Barja@upc.edu

L. Stoppino
Dipartimento di Scienza ed Alta Tecnologia, Università dell'Insubria, Via Valleggio 11, Como, Italy

Partially supported by PRIN 2009 "Spazi di moduli e Teoria di Lie", by FAR 2011 (Uninsubria) and by G.N.S.A.G.A.–I.N.d.A.M
e-mail: lidia.stoppino@uninsubria.it

A. Frühbis-Krüger et al. (eds.), *Algebraic and Complex Geometry*, Springer Proceedings in Mathematics & Statistics 71, DOI 10.1007/978-3-319-05404-9_1,

An important example of this kind of result is provided by the various proofs of the so-called *slope inequality* for a non-locally trivial relatively minimal fibred surface $f: S \longrightarrow B$, with general fibre F of genus $g \geq 2$:

$$K_f^2 \geq 4\frac{g-1}{g}\chi_f.$$

There are at least three different proofs of this result. One is due to Cornalba and Harris for the Deligne-Mumford non-hyperelliptic stable case [18] (generalized to the general case by the second author [50]), and uses the Hilbert stability of the canonical morphism of the general fibre of f. In [16] Bost proves a similar result assuming Chow stability. Although the proofs of Cornalba-Harris and Bost are different, the results are almost identical, being Chow and Hilbert stability very close (Remark 15). Another proof of the slope inequality, due to Xiao [52], uses the Clifford Theorem on the canonical system of the general fibre combined with the Harder-Narashiman filtration of the vector bundle $f_*\omega_f$. A third approach has been introduced more recently by Moriwaki in [39]; this method uses the μ-stability of the kernel of the relative evaluation map $f^* f_*\omega_f \longrightarrow \omega_f$ restricted on the general fibres. In [3] there is a good account of the last two proofs. Miyaoka's Theorem is a key tool in the proof of Xiao, and Bogomolov Theorem is the main ingredient of Moriwaki's approach. So we see at least two stabilities conditions involved in the proof of the slope inequality for fibred surfaces: Hilbert (or Chow) stability and μ-stability.

In this paper we study these three methods in a general setting. Firstly we present them with arbitrary line bundles – instead of the relative canonical one – and in arbitrary dimension, when possible. Then we make a comparison between them, finding that in dimension 2 there is a yet another stability condition, the *linear stability*, that connects them. Finally we make some speculations about the higher dimensional case, and we prove a couple of new applications.

Let us describe in more detail the contents of the paper. We consider the following setting. Let $f: X \longrightarrow B$ a fibred variety, $\mathcal{L}$ a line bundle on X, and let $\mathcal{G} \subseteq f_*\mathcal{L}$ be a subsheaf of rank r. A great deal of the results presented in the paper are in a more general setting, but let us assume here for the sake of simplicity that the general fibre of $\mathcal{G}$ is generating and that $\mathcal{L}$ is nef. Following [18], we consider the number $e(\mathcal{L}, \mathcal{G}) = rL^n - n \deg \mathcal{G}(L_{|F})^{n-1}$, which is an invariant of the fibration (Remark 1). We introduce the following notation (Definition 3): we say that $(\mathcal{L}, \mathcal{G})$ is *f-positive* when $e(\mathcal{L}, \mathcal{G}) \geq 0$. In the case $n = 2$, choosing $\mathcal{L} = \omega_f$, the slope inequality is equivalent to f-positivity of $(\omega_f, f_*\omega_f)$.

The structure of the paper is the following. In Sect. 2, after giving the first definitions, we make some useful computations via the Grothendieck-Riemann-Roch Theorem (Theorem 2 and Propositions 2 and 3): the number $e(\mathcal{L}, \mathcal{G})$ appears as the leading term of a polynomial expression associated to the relative Noether morphism

$$\gamma_h : \mathrm{Sym}^h \mathscr{G} \longrightarrow f_* \mathscr{L}^{\otimes h}, \text{ for } h \gg 0.$$

We then give a new elementary proof of a consequence of Miyaoka's result (Theorem 3): if $\mathscr{L}$ is nef and $\mathscr{G}$ is sheaf semistable, then $(\mathscr{L}, \mathscr{G})$ is f-positive. This is the first case we see where a stability condition implies f-positivity.

In Sect. 3 we describe the three methods, adding here and there some new contribution. As an illustration we re-prove along the way the slope inequality for fibred surfaces via the three methods (Examples 2, 3, and 5). The neat idea would be to extend them so that they all give as an output f-positivity of the couple $(\mathscr{L}, \mathscr{G})$, under some suitable assumptions. The Cornalba-Harris and Bost methods are originally stated in the general setting; we present them providing a slight generalization of the first one. They prove f-stability with the assumption that the fibre over general $t \in B$ is Hilbert or Chow semistable together with the morphism defined by the fibre $G_t := \mathscr{G} \otimes \mathbb{C}(t)$ (Theorems 6 and 7).

After discussing these methods, we make in Sect. 3.2 a digression on some applications that are specific to the Cornalba-Harris method. In particular we give in Proposition 4 a bound on the canonical slope of the fibred surfaces such that the k-th Hilbert point of (F, ω_F) is semistable for *fixed* k. This suggests a possible meaningful stratification of the moduli space of curves $\mathscr{M}_g$.

The method of Xiao was extended in higher dimensions by Konno [30] and Ohno [45]. We give a general compact version (Proposition 5). Xiao's method does not provide in general f-positivity; it gives an inequality between the invariants L^n and $\deg \mathscr{G}$ that has to be interpreted case by case.

Moriwaki's method is described in Sect. 3.4. It only works in dimension 2, and it gives f-positivity if the restriction of the kernel sheaf $\ker(f^*\mathscr{G} \longrightarrow \mathscr{L})$ is μ-semistable on the general fibres. We also provide a new condition for f-positivity, independent from the one of the theorem of Moriwaki (Theorem 10).

It is natural to try and make a comparison between these results, and between their assumptions: in particular, in the case of fibred surfaces all the three methods work because the canonical system enjoys many different properties or is there a red thread binding the three approaches? In Sect. 4 we study the 2-dimensional case. It turns out that there is a yet another stability concept, the *linear stability*, playing a central role in all three methods. Indeed, we observe the following:

- Section 4.1: linear (semi-)stability can be assumed as hypothesis in the Cornalba Harris method, as it implies Chow (semi-)stability (Mumford and others).
- Section 4.2: linear (semi-)stability is the key assumptions that assures that the method of Xiao produces f-positivity.
- Section 4.3: linear (semi-)stability is implied by the stability assumption needed in Moriwaki's method and in a large class of cases is equivalent to it (Mistretta-Stoppino).

So the picture goes as follows:

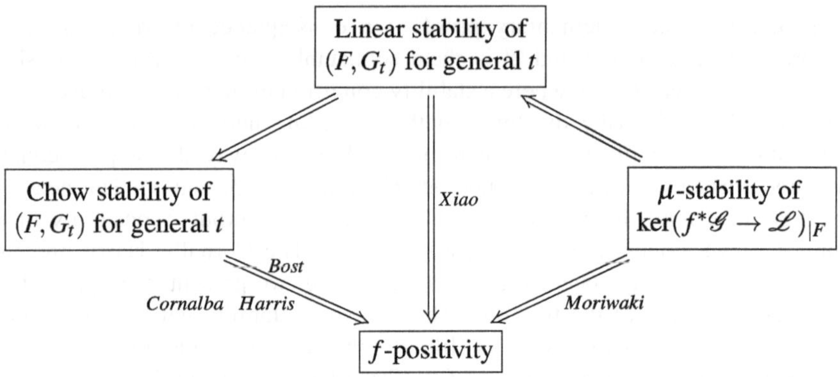

In Sect. 4.2 we also prove some positivity results that can be proved via Xiao's method with weaker assumptions.

Finally in Sect. 5 we consider the higher dimensional case. At this state of art, there is no hope to reproduce in higher dimension the beautiful connection between the three methods described for dimension 2. First of all, the method of Moriwaki seemingly can not even be extended to dimension higher than 2 (Remark 21). However, we provide some results regarding the other two methods. Firstly we prove that the hypothesis of linear stability still implies a positivity result via Xiao's method (Proposition 11). In Sect. 5.2, using known stability results, we can prove new inequalities for families of abelian varieties and of K3 surfaces via the Cornalba-Harris and Bost methods. Moreover, we conjecture a higher-dimensional slope inequality to hold for fibred varieties whose relative canonical sheaf is relatively nef and ample (Conjecture 1). We end the paper with an application of the (conjectured) slope inequality in higher dimension: using the techniques of Pardini [47] it is possible to derive from the slope inequality a sharp Severi inequality $K_X^n \geq 2n!\chi(\omega_X)$ for n-dimensional varieties with maximal Albanese dimension (Proposition 14). It is worth noticing that in [4] the first author proves this Severi inequality, and Severi type inequalities for any nef line bundle, independently of such conjectured slope inequality.

2 First Results

2.1 First Definitions and Motivation

We work over the complex field. All varieties, unless differently specified, will be normal and projective. Given a line bundle $\mathscr{L}$ on a variety X, we call L any (Cartier) divisor associated. It is possible to develop the major part of the theory for reflexive sheaves associated to Weil $\mathbb{Q}$-Cartier divisors, but in order to avoid cumbersome arguments, we will stitch to this setting.

Let X be a variety of dimension n, and B a smooth projective curve. Let $f: X \longrightarrow B$ be a flat proper morphism with connected fibres. Throughout the paper we shall call this data $f: X \longrightarrow B$ a *fibred variety.*

Let $\mathscr{L}$ be a line bundle on X. The pushforward $f_*\mathscr{L}$ is a torsion free coherent sheaf on the base B, hence it is locally free because B is smooth 1-dimensional. Let $\mathscr{G} \subseteq f_*\mathscr{L}$ be a subsheaf of rank r. The sheaf $\mathscr{G}$ defines a family of r-dimensional linear systems on the fibres of f,

$$G_t := \mathscr{G} \otimes \mathbb{C}(t) \subseteq H^0(F, \mathscr{L}_{|F}),$$

where $t \in B$ and $F = f^*(t)$. Let us recall that the evaluation morphism

$$ev: f^*\mathscr{G} \longrightarrow \mathscr{L}$$

is surjective at every point of X if and only if it induces a morphism φ from X to the relative projective bundle $\mathbb{P} := \mathbb{P}_B(\mathscr{G})$ over B

$$\begin{array}{ccc} X & \xrightarrow{\ \varphi\ } & \mathbb{P}_B(\mathscr{G}) := \mathbb{P} \\ {\scriptstyle f}\big\downarrow & \swarrow{\scriptstyle \pi} & \\ B & & \end{array}$$

such that $\mathscr{L} = \varphi^*(\mathscr{O}_{\mathbb{P}}(1))$. We will denote the surjectivity condition for *ev* by saying that the sheaf $\mathscr{G}$ is *generating* for $\mathscr{L}$. If *ev* is only generically surjective, it defines a rational map $\varphi: X \dashrightarrow \mathbb{P}$. In this case, let D be the unique effective divisor such that $f^*\mathscr{G} \longrightarrow \mathscr{L}(-D)$ is surjective in codimension 1. The divisor D is called the *fixed locus* of $\mathscr{G}$ in X. Clearly the evaluation morphism $f^*\mathscr{G} \longrightarrow \mathscr{L}(-D)$ is surjective in codimension 1.

Moreover, by Hironaka's Theorem, there exist a desingularization $\nu: \tilde{X} \longrightarrow X$ and a morphism $\tilde{\varphi}: \tilde{X} \longrightarrow \mathbb{P}$ such that $\tilde{\varphi} = \varphi \circ \nu$, and an effective ν-exceptional divisor E on $\tilde{X}$ such that

$$\tilde{\varphi}^*(\mathscr{O}_{\mathbb{P}}(1)) \cong \nu^*(\mathscr{L}(-D)) \otimes \mathscr{O}_{\tilde{X}}(-E).$$

See [45, Lemma 1.1] for a detailed proof of these facts. Define $\mathscr{M} := \tilde{\varphi}^*\mathscr{O}_{\mathbb{P}}(1) \subseteq \nu^*\mathscr{L}$; following [45] we call this the *moving part* of the couple $(\mathscr{L}, \mathscr{G})$, and we define the *fixed part* of $(\mathscr{L}, \mathscr{G})$ on $\tilde{X}$ to be $Z := \nu^*(D) + E$. Call $\tilde{f} := f \circ \nu$ the induced fibration. Clearly the evaluation homomorphism $\tilde{f}^*\mathscr{G} \longrightarrow \mathscr{M}$ is surjective at every point of $\tilde{X}$, i.e. $\mathscr{G}$ is generating for $\mathscr{M}$ on $\tilde{X}$.

Example 1. Let $f: S \longrightarrow B$ be a fibred surface, assuming for simplicity that S is smooth. Let $\omega_f = \omega_S \otimes f^*\omega_B^{-1}$ be the relative dualizing sheaf of f. Let g be the (arithmetic) genus of the fibres. The general fibres are smooth curves of genus g.

Let us assume that $g \geq 2$: then the restriction of ω_f on the general fibres is ample. Hence the base divisor D is vertical with respect to f. Moreover, the line bundle ω_f has negative degree only on the (-1)–curves contained in the fibres. So, all the vertical (-1)–curves of S are contained in D. It is possible to contract these curves preserving the fibration, and obtaining a unique *relatively minimal* fibration associated whose relative dualizing sheaf is f-nef. However, there could still be a divisorial fixed locus, as we see now for the case of nodal fibrations.

Let us suppose that f is a nodal fibration, i.e. that any fibre of f is a reduced curve with only nodes as singularities. We now describe explicitly the moving and the fixed part of $(\omega_f, f_*\omega_f)$. Let us first recall the following simple result, that can be found in [39, Prop. 2.1.3]. If C is a nodal curve, the base locus of ω_C is given by all the disconnecting nodes and all the smooth rational components of C that are attached to the rest of the fibre only by disconnecting nodes; following [39] we call these components of *socket type*.

The fixed locus of $(\omega_f, f_*\omega_f)$ is the union D of all components of socket type. Indeed, by what observed above the evaluation homomorphism $ev\colon f^*f_*\omega_f \longrightarrow \omega_f$ factors through $\omega_f(-D)$. On the other hand, it is easy to verify that the restriction of $\omega_f(-D)$ on any fibre is well defined except that on the disconnecting nodes not lying on components of socket type, so $f^*f_*\omega_f(-D) \longrightarrow \omega_f(-D)$ is surjective in codimension one.

Let $\nu\colon \tilde{S} \longrightarrow S$ be the blow up of all the base points of the map induced by $f_*\omega_f(-D)$; call E the exceptional divisor, and $\tilde{f} = f \circ \nu$ the induced fibration on $\tilde{S}$. Then we have that all the components of E are of socket type for the corresponding fibre, and that the union of all the components of socket type of the fibres of $\tilde{f}$ is $\tilde{D} + E$, where $\tilde{D}$ is the inverse image of D. Thus $\tilde{D} + E$ is the fixed part of $(\omega_{\tilde{f}}, \tilde{f}_*\omega_{\tilde{f}})$, and the evaluation homomorphism

$$\tilde{f}^*\tilde{f}_*\omega_{\tilde{f}}(-\tilde{D} - E) \longrightarrow \omega_{\tilde{f}}(-\tilde{D} - E)$$

is surjective at every point. Noting that $\omega_{\tilde{f}} \cong \nu^*(\omega_f) \otimes \mathcal{O}_{\tilde{S}}(E)$ (see for instance [10, Chap. 1, Theorem 9.1]), we have that

$$\omega_{\tilde{f}}(-\tilde{D} - E) \cong \nu^*(\omega_f) \otimes \mathcal{O}_{\tilde{S}}(-\tilde{D}) \cong \nu^*(\omega_f(-D)) \otimes \mathcal{O}_{\tilde{S}}(-E).$$

So the moving part of $(\omega_f, f_*\omega_f)$ is $\mathcal{M} \cong \nu^*(\omega_f(-D)) \otimes \mathcal{O}_{\tilde{S}}(-E)$.

Let us now come to the definition of the main characters of the play.

Definition 1. With the above notation, define the *Cornalba-Harris invariant*

$$e(\mathcal{L}, \mathcal{G}) := rL^n - n \deg \mathcal{G}(L_{|F})^{n-1},$$

where L is a divisor such that $\mathcal{L} \cong \mathcal{O}_X(L)$, and F is a general fibre.

Remark 1. The number $e(\mathcal{L}, \mathcal{G})$ is indeed invariant by twists of line bundles from the base curve B. Indeed, if $\mathcal{A}$ is a line bundle on B we have

$$\operatorname{rank}(\mathcal{G} \otimes \mathcal{A}) = \operatorname{rank}\mathcal{G} = r, \quad \deg(\mathcal{G} \otimes \mathcal{A}) = \deg \mathcal{G} + r \deg \mathcal{A},$$

$$(L + f^*A)^n = L^n + n \deg \mathcal{A} \, L_{|F}^{n-1}, \quad (\mathcal{L} \otimes f^*\mathcal{A})_{|F} \cong \mathcal{L}_{|F}.$$

It is therefore immediate to verify that $e(\mathcal{L} \otimes f^*\mathcal{A}, \mathcal{G} \otimes \mathcal{A}) = e(\mathcal{L}, \mathcal{G})$.

Remark 2. There is another significant incarnation of the C-H invariant: the number $r^{n-1}e(\mathcal{L}, \mathcal{G})$ is the top self-intersection of the divisor $rL - \deg \mathcal{G} F$.

Let us now consider again a fibred surface $f : S \longrightarrow B$ as in Example 1. Let $g \geq 2$ be the genus of the fibres and b the genus of the base curve B. The main relative invariants for f are $K_f^2 = K_S^2 - 8(b-1)(g-1)$ and $\chi_f = \chi(\mathcal{O}_S) - \chi(\mathcal{O}_B)\chi(\mathcal{O}_F) = \chi(\mathcal{O}_S) - (g-1)(b-1)$. By Leray's spectral sequence and Riemann-Roch one sees that $\chi_f = \deg f_*\omega_f$. The *canonical slope* s_f of the fibration is defined as the ratio between K_f^2 and χ_f. The slope s_f have been extensively studied in the literature (see [3, 6, 52]).

In a more general setting, given a line bundle $\mathcal{L}$ on X and a subsheaf $\mathcal{G} \subseteq f_*\mathcal{L}$, one can consider, when possible, the ratio between L^n and $\deg \mathcal{G}$, as follows.

Definition 2. With the same notation as above, let us suppose moreover that $\deg \mathcal{G} > 0$. We define the *slope of the couple* $(\mathcal{L}, \mathcal{G})$ as

$$s_f(\mathcal{L}, \mathcal{G}) := \frac{L^n}{\deg \mathcal{G}}.$$

When $\mathcal{G} = f_*\mathcal{L}$, we shall use the notation $s_f(\mathcal{L})$.

There is a rich literature about the search of lower bounds for the slope, in particular about the canonical one. The most general result is the following (see [5]).

Proposition 1. *Assume that $\mathcal{L}$ and $f_*\mathcal{L}$ are nef. Then $s_f(\mathcal{L}) \geq 1$.*

This bound is attained by a projective bundle on B and its tautological line bundle.

Remark 3. The slope is *not* invariant by twists of line bundles. Indeed, let $F = f^*(t)$ be a general fibre, and $G_t := \mathcal{G} \otimes \mathbb{C}(t) \subseteq H^0(F, \mathcal{L}_{|F})$. Attached to the triple $(f, \mathcal{G}, \mathcal{L})$ a *natural* ratio appears, which depends on the geometry of the triple $(F, G_t, \mathcal{L}_{|F})$. Indeed, consider the line bundle $\mathcal{L}(kF)$ obtained by "perturbing" $\mathcal{L}$ with kF for $k \in \mathbb{N}$, and the corresponding perturbed sheaf $\mathcal{G} \otimes \mathcal{O}_B(kt) \subseteq f_*(\mathcal{L}(kF)) \cong f_*\mathcal{L} \otimes \mathcal{O}_B(kt)$. Then we have that

$$s_f(\mathcal{L}, \mathcal{G})(k) := s_f(\mathcal{L}(kF), \mathcal{G} \otimes \mathcal{O}_B(kt)) = \frac{(L + kF)^n}{\deg \mathcal{G} \otimes \mathcal{O}_B(kt)} = \frac{L^n + kn(L_{|F})^{n-1}}{\deg(\mathcal{G}) + k \operatorname{rank}\mathcal{G}}.$$

Hence

$$\lim_{k \to \infty} s_f(\mathcal{L}, \mathcal{G})(k) = n \frac{(L_{|F})^{n-1}}{\operatorname{rank}\mathcal{G}}.$$

This asymptotic ratio is related to $e(\mathscr{L},\mathscr{G})$ as follows; we have that

$$s_f(\mathscr{L},\mathscr{G}) \geq n\frac{(L_{|F})^{n-1}}{\operatorname{rank}\mathscr{G}} \iff e(\mathscr{L},\mathscr{G}) \geq 0. \tag{1}$$

The positivity of the Cornalba-Harris invariant thus coincides with this natural bound on $s_f(\mathscr{L},\mathscr{G})$.

Remark 4. Let us consider inequality (1) in the case of a fibred surface $f: S \longrightarrow B$ of genus $g \geq 1$. It becomes

$$K_f^2 \geq 2\frac{\deg \omega_F}{\operatorname{rank} f_*\omega_f} \deg f_*\omega_f = 4\frac{g-1}{g}\chi_f.$$

This bound is the famous *slope inequality* for fibred surfaces mentioned in the introduction. It holds true for non locally trivial relatively minimal fibred surfaces of genus $g \geq 2$ ([18] and [39, 50, 52]).

The case of surfaces allows us to single out some positivity conditions on the family that seem to be necessary in general.

- The genus g of the fibration is $\geq 2 \iff \omega_f$ is ample on the general fibres of f;
- f is non-locally trivial $\iff \chi_f > 0$;
- f is relatively minimal $\iff$ the divisor K_f is nef (Arakelov).

In particular, if the fibration is not relatively minimal, the slope inequality is easily seen to be false. We see that indeed in order to prove the positivity of $e(\mathscr{L},\mathscr{G})$ we will often need similar conditions, in particular the relative nefness of $\mathscr{L}$. In Sect. 5.2 we conjecture and discuss a natural slope inequality in higher dimension.

By now we have seen how the condition of positivity of $e(\mathscr{L},\mathscr{G})$ is very natural and produces significant bounds for the geometry of the fibration. We shall thus give a name to this phenomenon:

Definition 3. The couple $(\mathscr{L},\mathscr{G})$ is said to be *f-positive* (resp. *strictly f-positive*) if $e(\mathscr{L},\mathscr{G}) \geq 0$ (resp. > 0).

2.2 Some Intersection Theoretic Computations

As above, let $f: X \longrightarrow B$ be a fibred variety over a curve B. Let $\mathscr{L}$ be a line bundle on X and $\mathscr{G} \subseteq f_*\mathscr{L}$ a subsheaf of rank r. Consider the natural morphism of sheaves

$$\gamma_h: \operatorname{Sym}^h\mathscr{G} \longrightarrow f_*\mathscr{L}^{\otimes h},$$

for $h \geq 1$. The fibres of this morphism on general $t \in B$ are just the multiplication maps

$$\gamma_h \otimes \mathbb{C}(t)\colon \mathrm{Sym}^h G_t = H^0(\mathbb{P}^{r-1}, \mathscr{O}_{\mathbb{P}^{r-1}}(h)) \longrightarrow H^0(F, \mathscr{L}_{|F}^{\otimes h}),$$

where $F = f^*(t)$ and $G_t = \mathscr{G} \otimes \mathbb{C}(t) \subseteq H^0(F, \mathscr{L}_{|F})$. Call $\mathscr{G}_h$ the image sheaf, and $\mathscr{K}_h$ the kernel of γ_h. If $\mathscr{G}$ is relatively ample then for $h \gg 0$ we have that $\mathscr{G}_h = f_* \mathscr{L}^{\otimes h}$ and that $\mathscr{K}_h$ is just $\mathscr{I}_{X/\mathbb{P}}(h)$, the ideal sheaf of the image of X in the relative projective space $\mathbb{P}$ twisted by $\mathscr{O}_{\mathbb{P}}(h)$.

Remark 5. Suppose now that $\mathscr{G}$ is generating. Let $\overline{X} := \varphi(X) \overset{j}{\hookrightarrow} \mathbb{P}$ be the image of X, let $\overline{f}\colon \overline{X} \longrightarrow B$ the induced fibration, and let $\overline{\mathscr{L}} = j^*(\mathscr{O}_{\mathbb{P}}(1))$. Then if $\alpha\colon X \to \overline{X}$ is the restriction of φ, we have that $\mathscr{L} = \alpha^* \overline{\mathscr{L}}$. Clearly, for $h \gg 0$ the sheaf $\mathscr{G}_h$ coincides with $\overline{f}_* \overline{\mathscr{L}}^{\otimes h}$, and $\mathscr{K}_h$ with $\mathscr{I}_{\overline{X}/\mathbb{P}}(h)$.

Let us recall that the *slope*[1] of a vector bundle $\mathscr{F}$ on a smooth curve C is the following rational number $\mu(\mathscr{F}) = \deg \mathscr{F} / \mathrm{rank}(\mathscr{F})$.

Remark 6. Note that f-positivity is equivalent to an upper bound on the slope of the sheaf $\mathscr{G}$, namely

$$\mu(\mathscr{G}) \leq \frac{L^n}{n(L_{|F})^{n-1}}.$$

We can now prove a simple condition for f-positivity.

Theorem 1. *Suppose that there exists an integer $m \geq 1$ such that*

(i) The couple $(\mathscr{L}^{\otimes m}, \mathscr{G}_m)$ is f-positive;
(ii) $m\mu(\mathscr{G}) \leq \mu(\mathscr{G}_m)$.

Then $(\mathscr{L}, \mathscr{G})$ is f-positive.

Proof. Assumption (i) tells us that

$$\mu(\mathscr{G}_m) \leq \frac{mL^n}{nL_{|F}^{n-1}}.$$

which, combined with (ii), gives the desired inequality.

We see below that the C-H class appears naturally as the leading term of the expression

$$r \deg \mathscr{G}_h - h \deg \mathscr{G} \mathrm{rank} \mathscr{G}_h$$

when computed as a polynomial in h. This produces the following condition for f-positivity in terms of the slope of $\mathscr{G}$ and of the one of $\mathscr{G}_h$.

---

[1]Unfortunately this crash of terminology seems unavoidable, as both the notations are well established.

Theorem 2. *With the above notation, suppose that the sheaf $\mathscr{G} \subseteq f_*\mathscr{L}$ is generating and such that the morphism φ it induces is generically finite on its image.*

Then the following implications hold

1. *If $\mu(\mathscr{G}_h) \geq h\mu(\mathscr{G})$ for infinitely many $h > 0$, then $(\mathscr{L}, \mathscr{G})$ is f-positive.*
2. *If $(\mathscr{L}, \mathscr{G})$ is strictly f-positive, then $\mu(\mathscr{G}_h) \geq h\mu(\mathscr{G})$ for $h \gg 0$.*

Proof. As in Remark 5, let $\overline{X} := \varphi(X) \stackrel{j}{\hookrightarrow} \mathbb{P}$ be the image of X, let $\overline{f}: \overline{X} \longrightarrow B$ be the induced fibration, and let $\overline{\mathscr{L}} = j^*(\mathscr{O}_{\mathbb{P}}(1))$. As observed in the remark, the sheaf $\mathscr{G}_h$ coincides with $\overline{f}_*\overline{\mathscr{L}}^{\otimes h}$ for $h \gg 0$. By Grothendieck-Riemann-Roch theorem we have that

$$\deg \mathscr{G}_h = \deg \overline{f}_*\overline{\mathscr{L}}^{\otimes h} = h^n \frac{(\overline{L})^n}{n!} + \sum_{i \geq 1} (-1)^{i+1} \deg R^i \overline{f}_*\overline{\mathscr{L}}^{\otimes h} + \mathscr{O}(h^{n-1}),$$

and that

$$\operatorname{rank} \mathscr{G}_h = \operatorname{rank} \overline{f}_*\overline{\mathscr{L}}^{\otimes h} = h^0(F, \overline{\mathscr{L}}_{|F}^{\otimes h}) =$$
$$= h^{n-1} \frac{(\overline{L}_{|F})^{n-1}}{(n-1)!} - \sum_{i \geq 1} (-1)^i h^i(F, \overline{\mathscr{L}}_{|F}^{\otimes h}) + \mathscr{O}(h^{n-2}).$$

Moreover, $\mathscr{G}$ is relatively very ample as a subsheaf of $\overline{f}_*\overline{\mathscr{L}}$, and so by Serre's vanishing theorem $\deg R^i \overline{f}_*\overline{\mathscr{L}}^{\otimes h} = 0$ and $h^i(F, \overline{\mathscr{L}}_{|F}^{\otimes h}) = 0$ for $h \gg 0$, and $i \geq 1$. By the assumption, the map $\alpha: X \longrightarrow \overline{X}$ is generically finite of degree say d. Hence

$$L^n = (\alpha^*\overline{L})^n = d(\overline{L})^n \quad \text{and} \quad (L_{|F})^{n-1} = (\alpha^*\overline{L}_{|F})^{n-1} = d(\overline{L}_{|F})^{n-1}.$$

Putting all together, we have

$$\begin{aligned} \operatorname{rank} \mathscr{G} \deg \mathscr{G}_h - h \deg \mathscr{G} \operatorname{rank} \mathscr{G}_h &= \frac{h^n}{d(n!)} \left(\operatorname{rank} \mathscr{G} L^n - n \deg \mathscr{G} L_{|F}^{n-1} \right) + \mathscr{O}(h^{n-1}) = \\ &= \frac{h^n}{d(n!)} e(\mathscr{L}, \mathscr{G}) + \mathscr{O}(h^{n-1}). \end{aligned} \tag{2}$$

So, if we have that $\mu(\mathscr{G}_h) \geq h\mu(\mathscr{G})$ for infinitely many $h > 0$, then the leading term of $\operatorname{rank} \mathscr{G} \deg \mathscr{G}_h - h \deg \mathscr{G} \operatorname{rank} \mathscr{G}_h$ as a polynomial in h must be non-negative (in particular inequality $\mu(\mathscr{G}_h) \geq \mu(\mathscr{G})$ is satisfied for $h \gg 0$). Vice-versa, if the leading term is strictly positive, then $\mu(\mathscr{G}_h) \geq h\mu(\mathscr{G})$ for $h \gg 0$.

Remark 7. If we have that $e(\mathscr{L}, \mathscr{G})$ is zero, then of course we cannot conclude that

$$\operatorname{rank} \mathscr{G} \deg \mathscr{G}_h - h \deg \mathscr{G} \operatorname{rank} \mathscr{G}_h \geq 0 \quad \text{for} \quad h \gg 0.$$

However, we can in this case consider the term in h^{n-1}, which is

$$\frac{h^{n-1}}{(n-1)!}\left((n-1)\deg\mathscr{G}L_{|F}^{n-2}K_F - L^{n-1}K_f\mathrm{rank}\mathscr{G}\right).$$

Using the equality $\mathrm{rank}\mathscr{G}L^n = n\deg\mathscr{G}L_{|F}^{n-1}$, this term becomes

$$\frac{rh^{n-1}}{(n-1)!}\left(\frac{n-1}{n}\frac{L_{|F}^{n-2}K_F}{L_{|F}^{n-1}}L^n - L^{n-1}K_f\right).$$

Note that in case $\mathscr{L} = \omega_f$ we obtain $-\frac{1}{n}K_f^n$, so that we can observe that if $K_f^n > 0$ and ω_f is ample on the general fibres, then if $\mu(\mathscr{G}_h) \geq h\mu(\mathscr{G})$ for infinitely many $h > 0$, $(\omega_f, f_*\omega_f)$ is strictly f-positive.

Remark 8. We can observe the following. Consider the function $\psi(h) := \mu(\mathscr{G}_h)/h$, and assume the same hypothesis as Theorem 2. Then, by the very same computations contained in the proof of Theorem 2, we see that

$$\lim_{h\to\infty}\psi(h) = \frac{L^n}{n(L_{|F}^{n-1})}.$$

Moreover observe that, for any $h \geq 1$

$$(\mathscr{L}^{\otimes h}, \mathscr{G}_h) \text{ is } f\text{-positive} \iff \psi(h) \leq \frac{L^n}{n(L_{|F}^{n-1})}.$$

Theorem 2 can thus be rephrased as the following behavior of the function ψ.

(1) If $\psi(h) \geq \psi(1)$ for infinitely many h, then $\psi(1) \leq L^n/(nL_{|F}^{n-1})$.
(2) If $\psi(1) < L^n/(nL_{|F}^{n-1})$, then $\psi(h) \geq \psi(1)$ for $h \gg 0$.

We state now a couple of results along the lines of Theorem 2, when we weaken as much as possible the assumptions needed in order to obtain f-positivity.

Proposition 2. *With the same notation as above, suppose that the line bundle $\mathscr{L}$ is nef on X and that the base locus of $\mathscr{G}$ is concentrated on fibres.*
If $\mu(\mathscr{G}_h) \geq h\mu(\mathscr{G})$ for infinitely many h, then $(\mathscr{L}, \mathscr{G})$ is f-positive.

Proof. If the map φ induced by $\mathscr{G}$ is not generically finite on its image then $e(\mathscr{L}, \mathscr{G}) = 0$, hence f-positivity is trivially satisfied. If on the contrary φ is finite on its image, we can apply Theorem 2 using, instead of $\mathscr{L}$, the moving part of $(\mathscr{L}, \mathscr{G})$

$$\mathscr{M} = \nu^*(\mathscr{L}(-D)) \otimes \mathscr{O}_{\tilde{X}}(-E),$$

where we follow the notation of Sect. 2. Let M be a divisor associated to $\mathcal{M}$. By Theorem 2, we have that the assumption $\mu(\mathcal{G}_h) \geq h\mu(\mathcal{G})$ for $h \gg 0$ implies that $(\mathcal{M}, \mathcal{G})$ is f-positive, so that $M^n \geq n\mu(\mathcal{G})(M_{|F})^{n-1}$. By the assumption on the base locus of $\mathcal{G}$, we have that $M_{|F} \sim L_{|F}$. Moreover, as $\mathcal{L}$ and $\mathcal{M}$ are nef and $\mathcal{M}$ is $\mathcal{L}$ minus an effective divisor, we have that $L^n \geq M^n$. Summing up, we have $L^n - n\mu(\mathcal{G})L_{|F}^{n-1} \geq M^n - n\mu(\mathcal{G})(M_{|F})^{n-1} \geq 0$, and so we are done.

Remark 9. It is worth noticing that in the statement of Proposition 2 above, we could replace the assumption of $\mathcal{L}$ being nef with $\mathcal{L}$ being *relatively* nef. Indeed, as $e(\mathcal{L}, \mathcal{G})$ is invariant by twists with pullback of line bundles on the base (Remark 1), we can always replace a relatively nef line bundle with a nef one, by twisting with the pullback of a sufficiently ample line bundle on B.

Proposition 3. *With the same notation as above, suppose that*

($\star$) *for* $h \gg 0$ *and* $i \geq 1$ $\deg R^i f_*\mathcal{L}^{\otimes h} = \mathcal{O}(h^{n-1})$ *and* $h^i(F, \mathcal{L}_{|F}^{\otimes h}) = \mathcal{O}(h^{n-2})$.

Suppose moreover that one of the following conditions hold

(*a*) *The sheaf* $\mathcal{G} \subseteq f_*\mathcal{L}$ *is normally generated for general* $t \in B$*;*
(*b*) *The sheaf* $f_*\mathcal{L}^{\otimes h}$ *is nef for* $h \gg 0$.

Then if $\mu(\mathcal{G}_h) \geq h\mu(\mathcal{G})$ *for infinitely many* $h > 0$, *then* $(\mathcal{L}, \mathcal{G})$ *is* f*-positive.*

Proof. Suppose that condition (a) holds: then for $h \gg 0$ the sheaf $\mathcal{G}_h$ generically coincides with (and is contained in) $f_*\mathcal{L}^{\otimes h}$. Hence, as we are on a smooth curve, $\deg \mathcal{G}_h \leq \deg f_*\mathcal{L}^{\otimes h}$ for $h \gg 0$. The same inequality holds true if condition (b) is satisfied.

By Grothendieck-Riemann-Roch theorem as in Theorem 2 we have that

$$\deg f_*\mathcal{L}^{\otimes h} = h^n \frac{L^n}{n!} + \sum_{i\geq 1}(-1)^{i+1} \deg R^i f_*\mathcal{L}^{\otimes h} + \mathcal{O}(h^{n-1}),$$

$$\operatorname{rank} f_*\mathcal{L}^{\otimes h} = h^0(F, \mathcal{L}_{|F}^{\otimes h}) = h^{n-1}\frac{(L_{|F})^{n-1}}{(n-1)!} + \sum_{i\geq 1}(-1)^i h^i(F, \mathcal{L}_{|F}^{\otimes h}) + \mathcal{O}(h^{n-2}).$$

Putting all together and using assumption ($\star$) we have

$$\begin{aligned} \operatorname{rank}\mathcal{G} \deg \mathcal{G}_h - h \deg \mathcal{G} \operatorname{rank}\mathcal{G}_h &\geq \frac{h^n}{n!}\left(\operatorname{rank}\mathcal{G} L^n - \deg \mathcal{G} L_{|F}^{n-1}\right) + \mathcal{O}(h^{n-1}) = \\ &= \frac{h^n}{n!} e(\mathcal{L}, \mathcal{G}) + \mathcal{O}(h^{n-1}), \end{aligned} \tag{3}$$

and the conclusion follows as in the above theorem.

Remark 10. Note that if we drop assumption ($\star$), we still obtain an inequality, involving a correction term due to the higher direct image sheaves.

The results above are generalizations of a computation contained in the proof of the main theorem of [18] (see also [50] and [7, sec.2]), where it is treated the case where the general fibre of $\mathscr{G}$ is very ample.

2.3 Stability and f-Positivity: First Results

Let us recall that a vector bundle $\mathscr{F}$ over a smooth curve B is said to be *μ-stable (resp. μ-semistable)* if for any proper subbundle $\mathscr{S} \subset \mathscr{F}$ we have $\mu(\mathscr{S}) < \mu(\mathscr{F})$ (resp. $\leq$). This is equivalent to asking that for any quotient bundle $\mathscr{F} \twoheadrightarrow \mathscr{Q}$ we have $\mu(\mathscr{Q}) > \mu(\mathscr{F})$ (resp. $\geq$).

Let us now consider as usual a fibred variety $f: X \longrightarrow B$ over a curve B. Let $\mathscr{L}$ be a line bundle on X and $\mathscr{G} \subseteq f_*\mathscr{L}$ a generating subsheaf of rank r. We see here that μ-semistability of $\mathscr{G}$ implies f-positivity. This is the first case we encounter where a stability condition implies the positivity of the C-H invariant. However, μ-semistability on the base is quite a restrictive condition to ask (see Remark 12). In Sect. 3.3, we will see a method due to Xiao that uses vector bundle techniques on $\mathscr{G}$ to prove some positivity results, but does not need to assume μ-semistability. However, we will see in Sect. 4.2 that, in order to give f-positivity as a result, Xiao's method needs another stability condition on the general fibres, the so-called linear stability.

We will need the following simple remark.

Remark 11. Let $\mathscr{F}$ be a vector bundle of rank r on a smooth curve B. Observe that, if h is any integer ≥ 1, we have the following equalities:

$$\deg(\mathrm{Sym}^h \mathscr{F}) = \binom{h+r-1}{r} \deg \mathscr{F}, \quad \mathrm{rank}(\mathrm{Sym}^h \mathscr{F}) = \binom{h+r-1}{r-1}.$$

We thus easily deduce the following.

$$\mu(\mathrm{Sym}^h \mathscr{F}) = h\mu(\mathscr{F}). \tag{4}$$

Theorem 3. *With the notation above, let us suppose that $\mathscr{G}$ is generating, or that the assumptions of Proposition 2 or of Proposition 3 hold. Then the following holds: if the sheaf $\mathscr{G}$ is μ-semistable, then $(\mathscr{L}, \mathscr{G})$ is f-positive.*

Proof. If $\mathscr{G}$ is μ-semistable then $\mathrm{Sym}^h\mathscr{G}$ is μ-semistable for any h, so that we have that the inequality $\mu(\mathrm{Sym}^h\mathscr{G}) \leq \mu(\mathscr{G}_h)$ is satisfied. But $\mu(\mathrm{Sym}^h\mathscr{G}) = h\mu(\mathscr{G})$ by formula (4) above.

Then if the conditions in Proposition 2 or in Proposition 3 are satisfied, we are done.

Let us now suppose that $\mathcal{G}$ is generating. If the morphism φ it induces is not generically finite on its image then $e(\mathcal{L},\mathcal{G})=0$. If on the contrary φ is generically finite on its image, by what we have seen above, we are in the conditions to apply Theorem 2.

Remark 12. From the above argument, we see that the μ-stability of $\mathcal{G}$ is much more than we need to prove f-positivity: indeed, in order to assure f-positivity, we just need that for infinitely many $h>0$ the sheaf $\mathcal{G}_h$ is not destabilizing for $\mathrm{Sym}^h\mathcal{G}$, and this condition is almost necessary (Theorem 2). The condition of μ-stability of $\mathrm{Sym}^h\mathcal{G}$ implies instead that this sheaf does not have *any* destabilizing quotient.

Indeed, it seems that the μ-stability of $\mathcal{G}$ is an extremely restrictive condition to ask. In order to illustrate this, consider any variety fibred over $\mathbb{P}^1$, and consider the relative canonical sheaf ω_f. If the sheaf $f_*\omega_f$ is μ-semistable, then necessarily its rank has to divide its degree, so that $h^0(F,K_F)$ necessarily divides $\deg f_*\omega_f$. Any fibred variety violating this numerical condition cannot have $f_*\omega_f$ μ-semistable. Moreover, let us recall Fujita's decomposition theorem for the pushforward of the relative canonical sheaf. Given a fibration $f\colon X\longrightarrow B$, we have that

$$f_*\omega_f=\mathcal{A}\oplus(\oplus^{q_f}\mathcal{O}_B), \tag{5}$$

where $q_f:=h^1(B,f_*\omega_X)$, and $H^0(B,\mathcal{A}^*)=0$. From this result we see that $f_*\omega_f$ fails to be semistable as soon as $q_f>0$. For instance, for any fibred surface $f\colon S\longrightarrow B$ with $q(S)>b$, the pushforward of the relative canonical sheaf needs to be μ-unstable. See [52] (in particular Theorem 3) for some related results.

A weaker version of Theorem 3 can be proved as a corollary of a beautiful result due to Miyaoka, as we see below.

Let us first define the setting of Miyaoka's Theorem. Let $\mathcal{F}$ be a vector bundle over a smooth curve B. Let $\pi\colon\mathbb{P}:=\mathbb{P}_B(\mathcal{F})\longrightarrow B$ be the relative projective bundle, and let H be a tautological divisor on $\mathbb{P}$, i.e. $\mathcal{O}_{\mathbb{P}}(H)\cong\mathcal{O}_{\mathbb{P}}(1)$, and let Σ be a general fibre of π.

Theorem 4 (Miyaoka [37]). *Using the above notations, the sheaf $\mathcal{F}$ is μ-semistable if and only if the $\mathbb{Q}$-divisor*

$$H-\mu(\mathcal{F})\Sigma$$

is nef.

Applying Theorem 4 to our situation we can deduce the following

Corollary 1. *Let $\mathcal{L}$ be a nef line bundle, $f\colon X\longrightarrow B$ a fibration and $\mathcal{G}\subseteq f_*\mathcal{L}$ has base locus vertical with respect to f. If the sheaf $\mathcal{G}$ is μ-semistable, then the couple $(\mathcal{L},\mathcal{G})$ is f-positive.*

Proof. With the notations of Sect. 2, let us observe that

$$\tilde{\varphi}^*(H-\mu(\mathcal{G})\Sigma)=\nu^*(L-D)-E-\mu(\mathcal{G})F.$$

Recalling that $\tilde{\varphi}$ is a morphism, by Theorem 4 the divisor $\nu^*(L-D)-E-\mu(\mathscr{G})F$ is nef. This divisor therefore has non-negative top self-intersection, and so the result follows using the same computations of Proposition 2 and Remark 2.

3 The Three Methods

3.1 Cornalba-Harris and Bost: Hilbert and Chow Stability

We now present the method of Cornalba and Harris [18], in the generalized setting introduced in [50]. Let us start with a definition. Let X be a variety, with a linear system $V \subseteq H^0(X, \mathscr{D})$, for some line bundle $\mathscr{D}$ on X. Fix $h \geq 1$ and call G_h the image of the natural homomorphism

$$\mathrm{Sym}^h V \xrightarrow{\varphi_h} H^0(X, \mathscr{D}^{\otimes h}).$$

Set $N_h = \dim G_h$ and take exterior powers

$$\bigwedge^{N_h} \mathrm{Sym}^h V \xrightarrow{\wedge^{N_h}\varphi_h} \bigwedge^{N_h} G_h = \det G_h. \tag{6}$$

The map $\wedge^{N_h}\varphi_h$ defines uniquely an element $\left[\wedge^{N_h}\varphi_h\right] \in \mathbb{P}(\wedge^{N_h}\mathrm{Sym}^h V^\vee)$ which we call the *generalized h-th Hilbert point associated to the couple* (X, V).

Definition 4. With the above notation, we say that the couple (X, V) is *Hilbert (semi)stable* if its generalized h-th Hilbert points are GIT (semi)stable for infinite $h \in \mathbb{N}$.

Remark 13. Let (X, V) be as above. Consider the factorization of the induced map through the image

$$X \dashrightarrow \overline{X} \overset{j}{\hookrightarrow} \mathbb{P}^r.$$

Set $\overline{\mathscr{D}} = j^*(\mathscr{O}_{\mathbb{P}^r}(1))$ and let $\overline{V} \subseteq H^0(\overline{X}, \overline{\mathscr{D}})$ be the linear systems associated to j. The homomorphism (6) factors as follows:

$$\mathrm{Sym}^h V \cong \mathrm{Sym}^h \overline{V} \xrightarrow{\overline{\varphi}_h} H^0(\overline{X}, \overline{\mathscr{D}}^{\otimes h}) \hookrightarrow H^0(X, \mathscr{D}^{\otimes h}),$$

where the homomorphism $\overline{\varphi}_h$ is the h-th Hilbert point of the embedding j; notice that, by Serre's vanishing theorem, this homomorphism is onto (and, in particular, $G_h = H^0(\overline{X}, \overline{\mathscr{D}}^{\otimes h})$) for large enough h. The generalized h-th Hilbert point of (X, V) is therefore naturally identified with the h-th Hilbert point of $(\overline{X}, \overline{V})$, and the generalized Hilbert stability of (X, V) coincides with the classical Hilbert stability of the embedding j.

Now consider a fibred variety $f: X \longrightarrow Y$, where the base Y is smooth but not necessarily of dimension 1. Let $\mathscr{L}$ be a line bundle on X, and let $\mathscr{G} \subseteq f_*\mathscr{L}$ be a subsheaf of rank r. Consider the homomorphism of sheaves $\mathrm{Sym}^h\mathscr{G} \longrightarrow f_*\mathscr{L}^{\otimes h}$ and, as usual, call $\mathscr{G}_h$ its image.

Theorem 5 (Cornalba-Harris). *With the above notation, suppose that for general $y \in Y$ the h-th generalized Hilbert point of the fibre $G_y := \mathscr{G} \otimes \mathbb{C}(y) \subseteq H^0(F, \mathscr{L}_{|F})$ is semistable.*

Then the line bundle

$$\mathscr{L}_h := \det(\mathscr{G}_h)^{\otimes r} \otimes (\det \mathscr{G})^{-\otimes h \mathrm{rank} \mathscr{G}_h}$$

is pseudo-effective.

The above result is the key point of the proof of [18, Theorem 1.1]. In particular, when the base Y is a smooth curve, we obtain the following inequality

$$\mathrm{rank}\mathscr{G} \deg \mathscr{G}_h - h \deg \mathscr{G} \mathrm{rank}\mathscr{G}_h \geq 0, \tag{7}$$

In the general case with base of arbitrary dimension it is possible, under some assumptions, to compute the first Chern class of $\mathscr{L}_h$ as a polynomial in h with coefficients in $CH_1(Y)_{\mathbb{Q}}$ and to conclude that its leading term is a pseudoeffective class ([18, Theorem 1.1] and [50, Corollary 1.6]).

Applying the results of Sect. 2, we obtain the following condition for f-positivity, which provides an improvement of Theorem 1.1 of [18] in the case of 1-dimensional base.

Theorem 6. *With the notation above, suppose that the base $Y = B$ is a curve. Suppose that the sheaf $\mathscr{G}$ is either generating, or it satisfies the conditions of Proposition 2 or 3. Suppose moreover that for general $t \in B$ the fibre $G_t \subseteq H^0(F, \mathscr{L}_{|F})$ is Hilbert semistable. Then $(\mathscr{L}, \mathscr{G})$ is f-positive.*

Proof. Apply Theorem 5 above, and Theorem 2 and Propositions 2 and 3.

Bost's Result: Chow Stability

We now describe a result of Bost, which is almost equivalent to the one of Cornalba-Harris, except that it uses as assumption the Chow stability on the general fibres. Moreover it has to be mentioned that Bost's result holds in positive characteristic.

Let us first recall some definitions. Let X be an n-dimensional variety together with a finite morphism of degree a in the projective space $\varphi: X \longrightarrow \mathbb{P}^r$ associated to a linear system $V \subseteq H^0(X, \mathscr{D})$. Consider

$$Z(X) := \{n - \text{spaces } \pi \text{ of } V \mid Ann(\pi) \cap \varphi(X) \neq \emptyset\} \subset Gr(n, V).$$

The set $Z(X)$ is an hypersurface of degree $d = \deg\varphi/a$, in the grassmanian $Gr(n,V)$. The homogeneous polynomial $F_X \in H^0(Gr(n,V), \mathcal{O}_{Gr(n,V)}(d))$ representing $Z(X)$ is the *Chow form* of (X,V) and the *Chow point* of (X,V) is the class of F_X in $\mathbb{P}(H^0(Gr(n,V), \mathcal{O}_{Gr(n,V)}(d)))$.

The couple (X,V) is *Chow (semi)stable* if its Chow point is GIT (semi)stable with respect to the natural $SL(V)$ action.

Remark 14. Note that X is Chow (semi)stable if and only if the cycle mX is, for any integer m: see for instance [16], proof of Proposition 4.2. So, in particular, the Chow (semi)stability of (X,V) as above coincides with the Chow (semi)stability of the cycle image $\varphi_*(X)$ together with the linear system of the immersion induced by φ. This fact should be compared with the behavior of the Hilbert stability described in Remark 13.

Remark 15. In [40, Corollary 3.5], it is proven that Chow stability implies Hilbert stability, while for *semi*stability, the arrows are reversed. In [13] some examples are given of curves Hilbert unstable and Chow (strictly) semistable. However, the two concepts asymptotically coincide (see Sect. 3.2 below).

Hence, we can apply Theorem 6 if we replace the assumption of Hilbert semistability with Chow stability, but we can not assume Chow semistability.

In [16], Bost has proven an arithmetic analogue to the theorem of Cornalba and Harris, assuming the Chow semistability of the maps on the general fibres. The geometric counterpart of Bost's result in the case when the base is 1-dimensional is the following. Consider as usual a fibred variety $f: X \longrightarrow B$. Let $\mathcal{L}$ be a line bundle on X, and let $\mathcal{G} \subseteq f_*\mathcal{L}$ be a subsheaf of rank r.

Theorem 7 ([16] Theorem 3.3). *With the above notation, suppose that*

1. *For $t \in B$ general, the fibre $G_t := \mathcal{G} \otimes \mathbb{C}(t) \subseteq H^0(F, \mathcal{L}_{|F})$ is base-point free;*
2. *If $\alpha: F \longrightarrow \mathbb{P}^r$ is the morphism induced, the cycle $\alpha_*(F) \in Z_p(\mathbb{P}^r)$ is Chow semi-stable;*
3. *The line bundle $\mathcal{L}$ is relatively nef.*

Then the couple $(\mathcal{L}, \mathcal{G})$ is f-positive.

Example 2. Let us prove the slope inequality for fibred surfaces via these methods. Let $f: S \longrightarrow B$ be a relatively minimal fibred surface of genus $g \geq 2$. Recall that the relative dualizing sheaf ω_f is nef [11]. The slope inequality for relatively minimal fibred surfaces now follows right away from Proposition 2, using the fact that the restriction of ω_f to the general smooth fibre is Hilbert and Chow semistable (Sect. 3.2 above), and base-point free. An alternative proof can be obtained using Proposition 3, by proving, as in [18], that condition ($\star$) holds.

Let us now refine the computation in the case of a relatively minimal *nodal* fibred surface. In this case we have given in Example 1 an explicit description of the moving and the fixed part of $(\omega_f, f_*\omega_f)$. Recall that the moving part is $\mathcal{M} \cong \nu^*\omega_f(-D) \otimes \mathcal{O}_{\tilde{S}}(-E)$, where D is the union of all socket type components, $\nu: \tilde{S} \longrightarrow S$ is the blow up of S in the disconnecting nodes of the fibres of f that do not belong to a socket type component and E is the exceptional divisor of ν. Let $\tilde{f} = \nu \circ f$ be the induced fibration. From the proof of Proposition 2, we can derive the following inequality:

$$0 \leq M^2 - 2\mu(\tilde{f}_*\omega_{\tilde{f}}) \deg \omega_{|\tilde{F}} = K_{\tilde{f}}^2 + D^2 + E^2 - 2K_f D - 4\frac{(g-1)}{g} \deg f_*\omega_f.$$

Let us compute explicitly the term $D^2 + E^2 - 2K_f D$. Let n be the total number of disconnecting nodes contained in the fibres, k the number of nodes lying on a socket type component and $l = n - k = -E^2$. Let r be the number of connected components of socket type in the fibres, so that $D = D_1 + \ldots + D_r$ with the D_i's connected and disjoint. Then we have that $K_f D = -2r + k$, so that $D^2 + E^2 - 2K_f D = 3k - 4r + l$. Note that the condition of relative minimality is equivalent to $2r \leq k$, so we obtain inequality

$$K_{\tilde{f}}^2 \geq 4\frac{g-1}{g}\chi_f + n. \tag{8}$$ [2]

In particular any fibred surface satisfying the slope *equality* necessarily has all fibres free from disconnecting nodes. It is interesting to compare this result with the inequalities obtained via Xiao's method (Example 3) and with Moriwaki's method (Example 5).

3.2 Some Remarks on GIT Stabilities and Applications

It comes out the interest in understanding when a variety, endowed with a map in a projective space, is Hilbert or Chow semistable. The following is a (without any doubt non-complete) list of cases where Hilbert (or Chow) semistability is known. In this list any time we use the term "stability" without specification, we mean that both the Hilbert and the Chow (semi)stabilities are known to coincide.

- Homogeneous spaces embedded by complete linear systems are semistable; abelian varieties embedded by complete linear systems are semistable [28].
- Linear systems on curves: if C is a smooth curve of genus $g \geq 2$, the canonical embedding is Chow semistable, and it is Chow stable as soon as C is non-hyperelliptic. Any line bundle of degree $d \geq 2g + 1$ induces a Chow stable embedding [42]. Deligne-Mumford stable curves are semistable for the linear system induced by the m-th power of the dualizing sheaf for $m \geq 5$ ([22, 42], [23, Chap.4, Sec.C]). See [26, 48] for curves Chow stable with respect to lower powers of the dualizing sheaf.

[2] For the reader familiar with the moduli space of curves $\overline{\mathcal{M}}_g$, this inequality means that the divisor $g\kappa_1 - 4(g-1)\lambda - g\sum_{i>0}\delta_i \sim (8g+4)\lambda - g\delta_0 - 2g\sum_{i>0}\delta_i$ is nef outside the boundary $\partial\overline{\mathcal{M}}_g$.

- Morrison in [40] studies the Chow stability of ruled surfaces in connection with the μ-stability of the associated rank 2 vector bundle: he proves that if $\mathscr{E}$ is a stable rank 2 bundle on a smooth curve C then the ruled surface $\pi\colon \mathbb{P}(\mathscr{E}) \longrightarrow C$ is Chow stable with respect to the polarisation $\mathscr{O}_{\mathbb{P}(\mathscr{E})}(1) \otimes \pi_* \mathscr{O}_C(k)$ for $k \gg 0$. Seyyedali in [49] extends the results of Morrison to higher rank vector bundles and to higher dimensional bases. See also [27] for another generalization.
- General $K3$ surfaces: a $K3$ surface with Picard number 1 and degree at least 12 is Hilbert semistable [41].
- Hypersurfaces: in [43, Prop. 4.2] it is proven that smooth hypersurfaces of $\mathbb{P}^n$ of degree ≥ 3 are stable. In [42] it is studied the stability of (singular) plane curves and surfaces in $\mathbb{P}^3$. A hypersurface $F \subset \mathbb{P}^r$ of degree $d \geq r+2$ and only log terminal singularities is Hilbert semistable [51].
- Higher codimensional varieties: Lee [32] proved that a subvariety $F \subset \mathbb{P}^r$ of degree d is Chow semistable as far as the log canonical threshold of its Chow form is greater or equal to $\frac{r+1}{d}$ (resp. > for stability). In [13] both the Chow and the Hilbert stability of curves of degree d and arithmetic genus g in $\mathbb{P}^{d-g}$ are studied.

A lot of remarkable results – due to Gieseker, Viehweg and many others – are known regarding *asymptotic* stability: given a line bundle $\mathscr{D}$ and a linear subsystem $V \subseteq H^0(X, \mathscr{D})$, this is the stability of the couple $(\mathscr{D}^{\otimes h}, V_h)$, for high enough h, where

$$V_h := \operatorname{Im}(\operatorname{Sym}^h V \longrightarrow H^0(X, \mathscr{D}^{\otimes h})).$$

In this case Hilbert and Chow stability have been proved to be equivalent by Fogarty [21] and Mabuchi [33]. There are beautiful results due to Donaldson, Ross, Thomas and many others relating asymptotic Chow stability to differential geometry properties, such that the existence of a constant scalar curvature metric. Unfortunately, if a bound is not known on the power of the line bundle, the Cornalba-Harris theorem does not give interesting consequences: if a couple $(\mathscr{G}, \mathscr{L})$ is asymptotically semistable on a general fibre, then the Cornalba-Harris theorem implies that $L^n \geq 0$.

On the other hand, it has come out recently, also in relation with the minimal model program for the moduli space of curves initiated in [26], the interest in the stability of the h-th Hilbert point for *fixed h*. The main result obtained in this topic is that general canonical and bicanonical curves have the h-th Hilbert point semistable for $h \geq 2$ [1].

The Cornalba-Harris method can be applied with this kind of assumption. For instance we can prove the following result (cf. [20] for $h = 2$).

Proposition 4. *Let $f\colon S \longrightarrow B$ be a relatively minimal non-hyperelliptic fibred surface of genus $g \geq 2$. Suppose that the h-th Hilbert point of a general fibre F with its canonical sheaf is semistable (with $h \geq 2$). Then the following inequality holds*

$$K_f^2 \geq 2\frac{2(g-1)h^2+(1-g)h-g}{gh(h-1)}\chi_f. \tag{9}$$

Proof. With the usual notation, we choose $\mathscr{L} = \omega_f$ and $\mathscr{G} = f_*\omega_f$. Then by the assumption, using Theorem 5, we have that $\operatorname{rank}\mathscr{G} \deg \mathscr{G}_h - h \deg \mathscr{G} \operatorname{rank}\mathscr{G}_h \geq 0$. By Riemann-Roch, $\operatorname{rank}\mathscr{G}_h = (2h-1)(g-1)$, and $\deg \mathscr{G}_h = \frac{h(h-1)}{2}K_f^2 + \chi_f$, and the computation is immediate.

The computations with higher powers of the relative canonical sheaf gives worse inequalities than the slope one.

Remark 16. By a result of Fedorchuck and Jensen [20] (that improves the result in [1]), the best inequality in Eq. (9), reached for $h = 2$, holds for relatively minimal fibred surfaces whose general fibres are non-hyperelliptic curves of genus g whose canonical image does not lie on a quadric of rank 3 or less. In particular this is the case for fibred surfaces of even genus whose general fibres are trigonal with Maroni invariant 0 (ibidem. and [7]). It is quite interesting to notice that this very same bound is obtained by Konno in [29, Lemma 2.5] under the assumption that the pushforward sheaf $f_*\omega_f$ is μ-semistable.

From the above proposition we can derive a new proof of the following result (cf. [18, Theorem. 4.12] and [50, Prop. 2.4]). The same result follows from the computation contained in Remark 7.

Corollary 2. *If a relatively minimal non-locally trivial fibred surface of genus $g \geq 2$ reaches the slope inequality, then it is hyperelliptic.*

Proof. Observe that the function of h appearing in inequality (9) is strictly decreasing and – of course – it tends to the ratio of the slope inequality $4(g-1)/g$ for $h \mapsto \infty$. So, for any non-hyperelliptic fibration in the conditions of the theorem, a strictly stronger bound than the slope one is satisfied.

A New Stratification of $\mathscr{M}_g$

It is widely believed (see for instance [7, 31]) that there should exist a lower bound for the slope of fibred surfaces increasing with the gonality of the general fibres (under some genericity assumption). This conjecture, however, is only proved for some step: hyperelliptic fibrations (the slope inequality), trigonal fibrations [7, 19] and fibrations with general gonality [24, 31]. Recently Beorchia and Zucconi [12] have proved some results also on fourgonal fibred surfaces.

Let us consider the following open subsets of $\mathscr{M}_g$

$$\mathscr{S}_h := \{[C] \in \mathscr{M}_g \text{ such that the } k\text{-th Hilbert point is semistable for } k \geq h\}.$$

Clearly $\mathscr{S}_i \subseteq \mathscr{S}_j$ for $i \leq j$, and for some $m \in \mathbb{N}$ the sequence becomes stationary, i.e. $\mathscr{S}_i = \mathscr{S}_j$ for every $i, j \geq m$ (cf. [22]). If we consider the subsets

$\mathcal{S}_2, \mathcal{S}_3 \setminus \mathcal{S}_2, \ldots, \mathcal{S}_m \setminus \mathcal{S}_{m-1}$, it seems possible that these provide an alternative stratification of $\mathcal{M}_g$ minus the hyperelliptic locus. For such a stratification, a lower bound for the slope increasing with the dimension of the strata would be provided by Proposition 4. However, it does not seem clear, at least to the authors, to give a geometrical characterization of the curves lying in $\mathcal{S}_i \setminus \mathcal{S}_{i-1}$, and an estimate on the codimensions of these strata.

3.3 Xiao's Method: The Harder-Narashiman Filtration

As we have seen in the previous section, μ-semistability of $\mathcal{G}$ implies f-positivity. What about the case when the sheaf $\mathcal{G}$ is not semistable as a vector bundle? We describe here a method based on Miyaoka's Theorem 4, which exploits the Harder-Narashimann filtration of the sheaf $\mathcal{G}$.

The main idea is given by Xiao in [52], where he uses the method in the case of fibred surfaces. Later on, Ohno [45] and Konno [30] extended the method to higher dimensional fibred varieties over curves. We present here a compact version of the general formula (see Proposition 5 below).

We need to recall the definition of the *Harder-Narashimann filtration* of a vector bundle $\mathcal{G}$ over a curve B: it is the unique filtration of subbundles

$$0 = \mathcal{G}_0 \subset \mathcal{G}_1 \subset \ldots \subset \mathcal{G}_l = \mathcal{G}$$

satisfying the following assumptions

- For any $i = 0, \ldots l$ the sheaf $\mathcal{G}_i/\mathcal{G}_{i-1}$ is μ-semistable;
- If we set $\mu_i := \mu(\mathcal{G}_i/\mathcal{G}_{i-1})$, we have that $\mu_i > \mu_{i-1}$.

Note that $\mu_1 > \mu(\mathcal{E}) > \mu_l$, unless $\mathcal{G}$ is μ-semistable, in which case $1 = l$ and these numbers are equal. If H is a divisor associated to the tautological line bundle of $\mathbb{P}(\mathcal{G})$ and Σ is a general fibre then an $\mathbb{R}$-line bundle $H - x\Sigma$ is pseudoeffective if and only if $x \leq \mu_1$ [44, Cor. 3.7] and it is nef if and only if $x \leq \mu_l$ [37].

As usual, consider an n-dimensional fibred variety $f : X \longrightarrow B$ and be a line bundle $\mathcal{L}$ on X. Let F be a general smooth fibre of f. Consider $\mathcal{G} \subseteq f_*\mathcal{L}$ a subbundle, and its corresponding Harder-Narashiman filtration as above. Set $r_i = \operatorname{rank}\mathcal{G}_i$.

For each $i = 1, \ldots, l$, we consider the pair $(\mathcal{L}, \mathcal{G}_i)$ as in Sect. 2.1 and a common resolution of indeterminacies $\nu : \tilde{X} \longrightarrow X$. Let M_i be the moving part of $(\mathcal{L}, \mathcal{G}_i)$, and let $N_i = M_i - \mu_i F$. By Miyaoka's theorem 4 we have that N_i is a nef $\mathbb{Q}$–divisor (not necessarily effective). The linear system $P_i := N_{i|\tilde{F}}$ is free from base points and induces a map $\phi_i : \tilde{F} \longrightarrow \mathbb{P}^{r_i-1}$. By construction we have $P_l \geq P_{l-1} \geq \ldots \geq P_2 \geq P_1$. Define $a_{l+1} = 0$ and $N_{l+1} = N_l$. Then, we can state the generalized Xiao's inequality as follows. We refer to [30] for proofs.

For any set of indexes $I = \{i_1, \ldots, i_m\} \subseteq \{1, 2, \ldots, l\}$, define $i_{m+1} = l + 1$ and consider the partition of I given by

$$I_s = \{i_k \mid k = 1, \ldots, m \text{ such that } \dim\phi_{i_k}(\hat{F}) = s\}.$$

Define now $b_n = l + 1$ and decreasingly

$$b_s = \begin{cases} \min I_s & \text{if } I_s \neq \emptyset \\ b_{s+1} & \text{otherwise.} \end{cases}$$

Proposition 5 (Xiao, Konno). *With the above notation, assume the $\mathcal{L}$ and $\mathcal{G}$ are nef. Then the following inequality holds*

$$L^n = (\nu^* L)^n \geq N_{l+1}^n \geq \left(\sum_{s=n-1}^{1} \Big(\prod_{n-1 \geq k > s} P_{b_k} \Big) \sum_{j \in I_s} \Big(\sum_{r=0}^{s} P_j^{s-r} P_{j+1}^r \Big) \right) (\mu_j - \mu_{j+1}). \tag{10}$$

Remark 17. As we see Xiao's method does not give as a result f-positivity, but an inequality for the top self-intersection L^n that has to be interpreted case by case. On the other hand, it basically only has one hypothesis: the nefness of $\mathcal{L}$ and of $\mathcal{G}$. However, as we will see in Sect. 4.2 we can derive results on $\mathcal{G} = f_*\mathcal{L}$ even if it is not a nef vector bundle. One of the contributions of this article is to frame Xiao's result in a more general setting, and to prove that with the right stability condition in the couple $(F, \mathcal{G}_{|F})$, for F general, Xiao's method produces f-positivity, at least in the case of dimension 2.

Example 3. Let us describe how inequality (10) implies the slope inequality in the case of fibred surfaces. We use the above formula for $n = 2$, $\mathcal{L} = \omega_f$, $\mathcal{G} = f_*\omega_f$ and the sets of indexes $I = \{1, \ldots, l\}$ and $I' = \{1, l\}$. If we call $d_i = \deg P_i$ inequality (10) becomes, respectively

$$K_f^2 \geq \sum_{i=1}^{l} (d_i + d_{i+1})(\mu_i - \mu_{i+1}),$$

$$K_f^2 \geq (d_1 + d_l)(\mu_1 - \mu_l) + 2d_l\mu_l \geq d_l(\mu_1 + \mu_l) = (2g - 2)(\mu_1 + \mu_l).$$

Let us note that by Cifford's theorem we have inequality $d_i \geq 2r_i - 2$. Observing now that $r_{i+1} \geq r_i + 1$, and that $\deg f_*\omega_f = \sum_{i=1}^{l} r_i(\mu_i - \mu_{i+1})$, we obtain straight away the slope inequality

$$K_f^2 \geq 4\frac{g-1}{g} \deg f_*\omega_f.$$

In fact, the above proof gives an inequality for N_l^2. In the case of nodal fibrations, using the same notations as in Example 2, since $N_l = \nu^*(K_f(-D))(-E)$, we obtain the inequality $N_l^2 \leq K_f^2 - n$, which gives the very same inequality (8) obtained via the Cornalba-Harris method.

Example 4. It could be interesting to have explicitly written the case $n = 3$ for the complete set of indexes $\{1, \ldots, l\}$. Assume that N_l induces a generically finite map on the surface F. Hence we have $I_2 \neq \emptyset$ and so

$$L^3 \geq 3P_l^2\mu_l + (P_l^2 + P_lP_{l-1} + P_{l-1}^2)(\mu_{l-1} - \mu_l) + \ldots$$

$$\ldots + (P_{b_2+1}^2 + P_{b_2+1}P_{b_2} + P_{b_2}^2)(\mu_{b_2} - \mu_{b_2+1}) +$$

$$+P_{b_2}[(P_{b_2} + P_{b_2-1})(\mu_{b_2-1} - \mu_{b_2}) + \ldots + (P_{b_1+1} + P_{b_1})(\mu_{b_1+1} - \mu_{b_1})].$$

Observe that $b_1 = 1$ except for the case $r_1 = 1$ where $b_1 = 2$.

Since the linear systems induced by P_i for $i = b_1, \ldots, b_2 - 1$ map F onto curves C_i, we have a chain of projections between these curves in such a way that the fibration part of the Stein factorization of the maps $F \longrightarrow C_i$ are the same. Hence we have a fibration

$$\pi : F \longrightarrow C.$$

Call D the general fibre, and let Q_i be base point free linear systems on C such that $P_i = \pi^* Q_i$ of rank $h^0(C, Q_i) \geq r_i = \mathrm{rank}\mathcal{G}_i$ and degrees which we call d_i. Writing this information and using that for all j

$$P_{j+1}^k P_j^l \geq P_{j+1}^{k-1} P_j^{l+1},$$

since $P_j \leq P_{j+1}$ and they are nef, we obtain a simplified (and weaker) version of the previous inequality:

$$\begin{aligned} L^3 \geq\ & 3(P_l^2\mu_l + P_{l-1}^2(\mu_{l-1} - \mu_l) + \ldots + P_{b_2}^2(\mu_{b_2} - \mu_{b_2+1})) + \\ & +2\lambda(d_{b_2-1}(\mu_{b_2-1} - \mu_{b_2}) + \ldots + d_{b_1}(\mu_{b_1} - \mu_{b_1+1})), \end{aligned} \tag{11}$$

where $\lambda = DP_{b_2}$.

3.4 Moriwaki's Method: μ-Stability on the Fibres

In this paragraph we shall restrict ourselves to the case $n = 2$; see Remark 21 below for a discussion on higher-dimensional results. Let $X = S$ be a smooth surface. We need the following fundamental result due to Bogomolov, which can be found in [15].

Definition 5. Let $\mathscr{E}$ be a torsion free sheaf over S. The class

$$\Delta(\mathscr{E}) := 2\,\mathrm{rank}\mathscr{E}\, c_2(\mathscr{E}) - (\mathrm{rank}\mathscr{E} - 1)c_1^2(\mathscr{E}) \in A^2_{\mathbb{Q}}(S)$$

is the *discriminant* of the vector bundle $\mathscr{E}$. Let $\delta(\mathscr{E})$ denote its degree.

Theorem 8 (Bogomolov Instability Theorem). *With the above notation, if $\delta(\mathscr{E}) < 0$ then there exists a saturated subsheaf $\mathscr{F} \subseteq \mathscr{E}$ such that the class*

$$D = -\mathrm{rank}\mathscr{F}\, c_1(\mathscr{E}) + \mathrm{rank}\mathscr{E}\, c_1(\mathscr{F})$$

belongs to the positive cone $K^+(S)$ of $\mathrm{Pic}_{\mathbb{Q}}(S)$.

Recall that the positive cone K^+ is defined as follows: consider the (double) cone

$$K(S) = \{A \in N^1(S)_{\mathbb{Q}} \mid A^2 > 0\} \subset N^1(S)_{\mathbb{Q}}.$$

The cone $K^+(S)$ is the connected component of $K(S)$ containing the ample cone.

Remark 18. Recall the definition of semistable sheaf in higher dimension: if X is a variety of dimension n and $\mathscr{F}$ a locally free sheaf on X, let $\mathscr{H}$ be an ample line bundle on X. We say that $\mathscr{F}$ is $\mathscr{H}$-(semi)stable if for any proper subsheaf $0 \neq \mathscr{R} \subset \mathscr{F}$

$$\frac{c_1(\mathscr{R}) \cdot H^{n-1}}{\mathrm{rank}\mathscr{R}} \leq \frac{c_1(\mathscr{F}) \cdot H^{n-1}}{\mathrm{rank}\mathscr{F}} \qquad \text{(resp. } <\text{)},$$

where H is the class of $\mathscr{H}$. In particular from the strong instability condition provided by the theorem above, we have that if $\mathscr{E}$ is $\mathscr{H}$-semistable with respect to *any* ample line bundle $\mathscr{H}$ on S, then $\delta(\mathscr{E}) \geq 0$.

The argument of Moriwaki relies on two key observations. The first is the following: if the surface S carries a fibration, then, in order to ensure the non-negativity of $\delta(\mathscr{E})$ for a vector bundle $\mathscr{E}$, one can assume that $\mathscr{E}$ is semistable *on the general fibres* of f.

Proposition 6 ([39] Theorem 2.2.1). *Let us consider a fibred surface $f: S \longrightarrow B$. Let $\mathscr{E}$ be a sheaf on S such that the restriction of $\mathscr{E}$ on a general fibre of f is a μ-semistable sheaf. Then $\delta(\mathscr{E}) \geq 0$.*

Proof. Suppose by contradiction that $\delta(\mathscr{E}) < 0$. Then by the Bogomolov Instability Theorem there exists a saturated subsheaf $\mathscr{F} \subseteq \mathscr{E}$ such that the divisor $D = \mathrm{rank}\mathscr{F}\, c_1(\mathscr{E}) - \mathrm{rank}\mathscr{E}\, c_1(\mathscr{F})$ satisfies that $D^2 > 0$. As a fibre F is nef, and $F^2 = 0$, by the Hodge Index Theorem [10, sec.IV, Cor. 2.16], we have that

$$0 < D \cdot F = \mathrm{rank}\mathscr{E}\, \deg \mathscr{F}_{|F} - \mathrm{rank}\mathscr{F}\, \deg \mathscr{E}_{|F}.$$

So $\mathscr{F}_{|F}$ is a destabilizing subsheaf of $\mathscr{E}_{|F}$, against the assumption.

Remark 19. It is worth noticing that the hypothesis of the above proposition that the restriction $\mathscr{E}_{|F}$ is μ-semistable on the general fibres F does not imply, neither is implied, by some semistability of the sheaf $\mathscr{E}$ on S. Indeed, we can say only the following:

- Let $\mathscr{H}$ be an ample line bundle on S and C be a general curve in $|\mathscr{H}^{\otimes d}|$ with $d \geq 1$. Suppose that $\mathscr{E}_{|C}$ is μ-semistable, then $\mathscr{E}$ is $\mathscr{H}$-semistable. Indeed if $\mathscr{F}$ would be an $\mathscr{H}$-destabilizing subsheaf of $\mathscr{E}$, then $\mathscr{F}_{|C}$ would be destabilizing for $\mathscr{E}_{|C}$, because

$$\mu(\mathscr{F}_{|C}) = \frac{\deg \mathscr{F}_{|C}}{\mathrm{rank}\mathscr{F}_{|C}} = d\frac{\deg(c_1(\mathscr{F}) \cdot H)}{\mathrm{rank}\mathscr{F}} > d\frac{\deg(c_1(\mathscr{E}) \cdot H)}{\mathrm{rank}\mathscr{E}}$$
$$= \frac{\deg \mathscr{E}_{|C}}{\mathrm{rank}\mathscr{E}_{|C}} = \mu(\mathscr{E}_{|C})$$

- If $\mathscr{E}$ is $\mathscr{H}$-semistable with respect to some ample line bundle $\mathscr{H}$, and C is a general curve in $|\mathscr{H}^{\otimes m}|$, for sufficiently large m, then $\mathscr{E}_{|C}$ is μ-semistable [34].

Note that as a fibre F of any fibration $f: S \longrightarrow B$ satisfies $F^2 = 0$, it cannot be ample. However, if the fibration is rational (i.e. $B \cong \mathbb{P}^1$), the conditions above can hold true after some blow down of sections of the fibration.

Let us consider now a fibred surface $f: S \longrightarrow B$, a line bundle $\mathscr{L}$ and a rank r subsheaf $\mathscr{G} \subseteq f_*\mathscr{L}$. The second point of Moriwaki's argument, using our terminology, relates $\delta(\mathscr{E})$ to $e(\mathscr{L}, \mathscr{G})$, for a suitably chosen vector bundle $\mathscr{E}$, as follows.

Let $\mathscr{M}$ be the kernel of the evaluation morphism $f^*\mathscr{G} \subset f^*f_*\mathscr{L} \longrightarrow \mathscr{L}$. The following is a generalization of a computation contained in [39].

Proposition 7. *With the above notation, if either*

(a) The sheaf $\mathscr{G}$ is generating in codimension 2, or
(b) The line bundle $\mathscr{L}$ is f-nef, and $\mathscr{G}$ has base locus vertical with respect to f,

then

$$\delta(\mathscr{M}) \leq e(\mathscr{L}, \mathscr{G}).$$

Proof. Let us call $\mathscr{K}$ the image of the evaluation morphism, so that we have the following exact sequence

$$0 \longrightarrow \mathscr{M} \longrightarrow f^*\mathscr{G} \stackrel{\varphi}{\longrightarrow} \mathscr{K} \longrightarrow 0.$$

Note that, with the notations of Sect. 2.1, $c_1(\mathscr{K}) = c_1(\mathscr{L}(-D))$, where D is the fixed locus of $\mathscr{G}$, and $c := c_2(\mathscr{K}) = -E^2 \geq 0$, where E is as in Sect. 2.1. Indeed,

$c_2(\mathcal{K})$ is the length of the isolated base points of the variable part (with natural scheme structure) [3]. So we have

- $c_1(\mathcal{M}) = f^*c_1(\mathcal{G}) - c_1(\mathcal{L}(-D))$;
- $c_2(\mathcal{M}) = -c_1(\mathcal{M})c_1(\mathcal{L}(-D)) - c$.

Hence $\deg c_2(\mathcal{M}) = (L-D)^2 - \deg\mathcal{G}(L-D)F - c$. Let $r = \mathrm{rank}\mathcal{G}$, so that $\mathrm{rank}\mathcal{M} = r-1$. We can make the following computation

$$\begin{aligned}\delta(\mathcal{M}) &= 2(r-1)\big[(L-D)^2 - \deg\mathcal{G}(L-D)F - c\big] + \\ &-(r-2)\big[(L-D)^2 - 2\deg\mathcal{G}(L-D)F\big] = \\ &= r(L-D)^2 - 2\deg\mathcal{G}(L-D)F - 2(r-1)c.\end{aligned}$$

In case (a) $D=0$ and we thus obtain $\delta(\mathcal{M}) = e(\mathcal{L},\mathcal{G}) - 2(r-1)c \leq e(\mathcal{L},\mathcal{G})$. In case assumption (b) holds, observe that $(L-D)^2 = L^2 - 2LD + D^2 \leq L^2$; indeed being D effective and vertical, we have $D^2 \leq 0$ by Zariski's Lemma, and $LD \geq 0$ because L is supposed to be f-nef. On the other hand, $(L-D)F = LF$ again because D is vertical. Hence, we still obtain the desired inequality, and the proof is concluded.

Combining Propositions 6 and 7 we get immediately the following result

Theorem 9. *With the notation above, suppose that the restriction $\mathcal{M}_{|F}$ to a general fibre F is a semistable sheaf on it. Suppose moreover that one of the following conditions holds.*

- *The sheaf $\mathcal{G}$ is generating in codimension 2;*
- *The line bundle $\mathcal{L}$ is f-nef, and $\mathcal{G}$ has base locus vertical with respect to f.*

Then the couple $(\mathcal{L},\mathcal{G})$ is f-positive.

Example 5. Let $f\colon S \longrightarrow B$ be a relatively minimal non-locally trivial fibred surface. Let us prove the slope inequality via Moriwaki's method. Let us consider the couple $(\omega_f, f_*\omega_f)$, and let $\mathcal{M}$ be the kernel sheaf of the evaluation morphism $f^*f_*\omega_f \longrightarrow \omega_f$. The hypotheses of the above theorem are satisfied. Indeed, the assumption that $\mathcal{M}_{|F}$ is a semistable sheaf has been proved by Pranjape and Ramanan in [46]. The evaluation morphism $f^*f_*\omega_f \longrightarrow \omega_f$ can fail to be surjective on some vertical divisor, so that Theorem 9 can be applied with assumption (b) holding, and leads to the slope inequality (see also [3]).

In case f is a nodal fibration, we can obtain a finer inequality, as follows. From the computations of Proposition 7, using the results contained in Example 1, we have that

$$0 \leq \delta(\mathcal{E}) = gK_f^2 - 4(g-1)\chi_f - g(3k-4r) - 2(g-1)l,$$

where, as in Example 2, k is the number of disconnecting nodes lying on components of socket type of the fibres, l is the number of the others disconnecting nodes, and r is the number of components of socket type. So, we get

$$K_f^2 \geq 4\frac{g-1}{g}\chi_f + k + 2\frac{g-1}{g}l.$$

Note that this inequality is slightly better than the one obtained in Examples 2 and 3 using respectively the Cornalba-Harris and the Xiao methods.

Remark 20. Moriwaki in [39] uses, for nodal fibred surfaces $S \longrightarrow B$, as line bundle $\mathcal{L}$ on S an ad hoc modification of the relative canonical bundle ω_f on the singular fibres, and as $\mathcal{G}$ the whole $f_*\mathcal{L}$. The result is an inequality, involving some contributions due to the singular fibres, stronger than the one obtained in Example 5.

From Proposition 7, combining it with Remark 18, we can straightforwardly deduce the following condition for f-positivity.

Theorem 10. *With the notations above, if the kernel of the evaluation morphism*

$$f^*\mathcal{G} \longrightarrow \mathcal{L}$$

is $\mathcal{H}$-semistable with respect to an ample line bundle $\mathcal{H}$ on S, then $(\mathcal{L}, \mathcal{G})$ is f-positive.

This result is not implied by Moriwaki's Theorem 9, by what observed in Remark 19.

Remark 21. It would be nice to be able to extend Moriwaki's method to higher dimensions. Thanks to Mumford-Metha-Ramanathan's restriction theorem [34], it is possible to obtain the following Bogomolov-type result. Let X be a variety of dimension n, and $\mathcal{E}$ a vector bundle on X. Let $\mathcal{H}$ be an ample line bundle on S. Define

$$\delta(\mathcal{E}) := \deg\left(2\,\mathrm{rank}\mathcal{E}\, c_2(\mathcal{E})H^{n-2} - (\mathrm{rank}\mathcal{E} - 1)c_1^2(\mathcal{E})H^{n-2}\right).$$

Then, if $\mathcal{E}$ is $\mathcal{H}$-semistable, then $\delta(\mathcal{E}) \geq 0$. Unfortunately, this beautiful result does not imply f-positivity in dimension greater than 2. One should consult also the paper [38], of Moriwaki himself, for other inequalities along the same lines.

4 Linear Stability: A Thread Binding the Methods for $n = 2$

In this section we introduce the *linear stability*, for curves together with a linear series. We see that in the case of fibred surfaces linear stability represents a link between the three methods described in the previous section, that are from all other aspects extremely different.

Let us start by recalling the notion of linear stability for a curve and a linear series on it [50]. This is a straightforward generalization in the case of curves of the one given by Mumford in [42].

Let C be a smooth curve, and let $\varphi: C \longrightarrow \mathbb{P}^{r-1}$ be a non-degenerate morphism. This corresponds to a globally generated line bundle $\mathscr{L}$ on C, and a base-point free linear subsystem $V \subseteq H^0(C, \mathscr{L})$ of dimension r such that φ is induced from the linear series $|V|$. Let d be the degree of $\mathscr{L}$ (i.e. $|V|$ is a g_d^{r-1} on C). Linear stability gives a lower bound on the slope between the degree and the dimension of any projections, depending on the degree and dimension of the given linear series as follows.

Definition 6. With the above notation, we say that the couple (C, V), is *linearly semistable (resp. stable)* if any linear series of degree d' and dimension $r' - 1$ contained in $|V|$ satisfies

$$\frac{d'}{r'-1} \geq \frac{d}{r-1} \quad \text{(resp. } >\text{)}$$

In case $V = H^0(\mathscr{L})$, we shall talk of the stability of the couple $(C, \mathscr{L})$. It is easy to see that it is sufficient to verify that the inequality of the definition holds for any *complete* linear series in $|V|$.

Example 6. Some of the known results are the following.

1. The canonical system on a curve of genus ≥ 2 is linearly semistable and it is stable if and only if the curve is non-hyperelliptic. This follows from Clifford's Theorem and Riemann-Roch Theorem (see [2, chap.14, sec.3]).
2. It is immediate to check that a plane curve of degree d is linearly semistable (with respect to its immersion in $\mathbb{P}^2$) if and only if it has points of multiplicity at most $d/2$.
3. Using Riemann-Roch Theorem, it is easy to check that the morphism induced on a curve of genus g by a line bundle of degree $\geq 2g + 1$ is linearly stable (see [42]).
4. For a non-hyperelliptic curve of genus ≥ 2, generic projections of low codimension from the canonical embedding are linearly stable [6].
5. Given a base-point free linear system $V \subseteq H^0(C, \mathscr{L})$ on a curve C of genus g, if $\deg \mathscr{L} \geq 2g$, and the codimension of V in $H^0(C, \mathscr{L})$ is less or equal than $(\deg \mathscr{L} - 2g)/2$, then (C, V) is linearly semistable [35].

4.1 Linear Stability and the Cornalba-Harris Method

Mumford introduced the concept of linear stability in order to find a more treatable notion that the ones of GIT stability. The importance of linear stability from this point of view lies indeed in the following result [42, Theorem 4.12]

Theorem 11 (Mumford). *If $(C, \mathscr{L})$ is linearly (semi-)stable and $\mathscr{L}$ is very ample, then $(C, \mathscr{L})$ is Chow (semi-)stable.*

In [2] it is proved the following result ([2, Theorem (2.2)])

Theorem 12. *If $(C, \mathscr{L})$ is linearly stable and $\mathscr{L}$ is very ample, then $(C, \mathscr{L})$ is Hilbert stable.*

By Morrison's result [40, Corollary 3.5], we see that Theorem 11 implies Theorem 12. The arguments of [2, Theorem (2.2)] cannot be pushed through to the semistable case, so at present it is not known if linear strict semistability implies Hilbert strict semistability (through the authors would be surprised if it doesn't).

It is easy to extend the proof of both Mumford's Theorem 11, and [2, Theorem (2.2)] to the case of a very ample non necessarily complete linear system (see e.g. the second author's Ph.D. Thesis).

Remark 22. Let V be a base-point free linear system on C inducing a morphism $\varphi: C \longrightarrow \mathbb{P}^r$ of positive degree on the image. It is immediate to see that the linear (semi)stability of V is equivalent to the linear stability of the image $\varphi(C)$ with its embedding in $\mathbb{P}^r$ (compare with Remarks 13 and 14).

Using the results of Sect. 3.1 we can thus state the following results.

Theorem 13. *Let $f: S \longrightarrow B$ be a fibred surface, $\mathscr{L}$ a line bundle on S and $\mathscr{G} \subseteq f_*\mathscr{L}$ a subsheaf. Suppose that for general $t \in B$ the couple (F, G_t) is linearly semistable. Suppose moreover that we are in one of the following situations:*

(i) The couple (F, G_t) is strictly linearly stable, and the sheaf $\mathscr{G}$ is either generating, or it satisfies the conditions of Proposition 2 or 3;
(ii) For $t \in B$ general, the fibre $G_t \subseteq H^0(F, \mathscr{L}_{|F})$ is base-point free and the line bundle $\mathscr{L}$ is relatively nef.

Then the couple $(\mathscr{L}, \mathscr{G})$ is f-positive (via the Cornalba-Harris method).

4.2 Linear Stability and Xiao's Method

We verify here that the method of Xiao gives as a result the f-positivity under the assumption of linear stability.

Theorem 14. *Let $f: S \longrightarrow B$ a fibred surface, F a general fibre, $\mathscr{L}$ a nef line bundle on S and $\mathscr{G} \subseteq f_*\mathscr{L}$ a nef rank r subsheaf. Assume that the linear system on F induced by $\mathscr{G}$ is linearly semistable. Then $(\mathscr{L}, \mathscr{G})$ is f-positive (via Xiao's method).*

Proof. Following the description of Xiao's method given in Sect. 3.3, consider the linear systems P_i induced on F by the pieces of the Harder-Narashiman filtration of $\mathscr{G}$, of rank r_i. Let $d_i = \deg P_i$, and observe that $d_l = \deg L_{|F} =: d$. Linear stability condition implies

$$\frac{d_i}{r_i - 1} \geq \frac{d_l}{r_l - 1} = \frac{d_l}{r - 1} =: a \qquad \text{for any } i = 1, \ldots, l.$$

Observe that if $r_1 = 1$ then $d_1 = 0$ and the above inequality should be read as $d_1 \geq ar_1$ and still holds. Consider now the sets of indexes $I = \{1, \ldots, l\}$ and $I' = \{1, l\}$. Then we have

$$L^2 \geq \sum_{i=1}^{l} (d_i + d_{i+1})(\mu_i - \mu_{i+1})$$

and

$$L^2 \geq (d_1 + d_l)(\mu_1 - \mu_l) + 2d_l\mu_l \geq d_l(\mu_1 + \mu_l).$$

Use now that $d_i \geq a(r_i - 1)$ for $i = 1, \ldots, l$ ($d_{l+1} = d_l$) and that $r_{i+1} \geq r_i + 1$. Observe that $\deg\mathcal{G} = \sum_{i=1}^{l} r_i(\mu_i - \mu_{i+1})$ to get

$$L^2 \geq 2a\deg\mathcal{G} - a(\mu_1 + \mu_l),$$

which finally proves

$$L^2 \geq \frac{2ad_l}{a + d_l}\deg\mathcal{G} = 2\frac{d}{r}\deg\mathcal{G}.$$

Remark 23. The fact that we used Clifford's theorem in the proof of the slope inequality via Xiao's method in Example 3 can thus be rephrased in the following way: Clifford's theorem implies the linear semistability of the general fibres of f together with their canonical systems.

We can make the following improvement for the complete case.

Proposition 8. *With the notations above, assume that $\mathcal{L}$ is nef and that $\mathcal{L}_{|F}$ is linearly semistable. Then $(\mathcal{L}, f_*\mathcal{L})$ if f-positive, i.e.*

$$L^2 \geq 2\frac{d}{r}\deg f_*\mathcal{L}.$$

Proof. Take $\mathcal{G}$ to be the biggest piece of the Harder-Narashiman filtration of $f_*\mathcal{L}$ such that $\mu_i \geq 0$. It is nef and we have that $\frac{d_i}{r_i-1} \geq \frac{d}{r-1}$ by linear semistability and that $\deg\mathcal{G} \geq \deg f_*\mathcal{L}$. Then apply the same method as in the proof of Theorem 14.

Remark 24. From the proof of the Theorem 14 we get an inequality even if we do not assume linear semistability condition on fibres. Indeed, observe that, if the linear subsystems of P, the one induced by $\mathcal{G}$, verify

$$\frac{d_i}{r_i - 1} \geq a \qquad \text{for any } i = 1, \ldots, l$$

for some constant a, then we obtain the following inequality for the slope

$$L^2 \geq \frac{2ad}{a+d}\deg \mathscr{G}.$$

where $d = \deg P$.

Take $\mathscr{G}$ as in the proof of the previous proposition. Observe that $\frac{2ad_1}{a+d_1} \geq \frac{2ad_2}{a+d_2}$ if $d_1 \geq d_2$. Hence we can conclude that if $\mathscr{L}$ is nef and induces a base point free linear system on F of degree d, such that all its linear subsystems verify

$$\frac{d_i}{r_i - 1} \geq a,$$

then

$$L^2 \geq \frac{2ad}{a+d}\deg f_*\mathscr{L}.$$

This remark allows us to give a general result for a nef line bundle $\mathscr{L}$ depending of its *degree of subcanonicity* (compare with [4]).

Proposition 9. *Let $f : S \longrightarrow B$ be a fibred surface with general fibre F of genus $g \geq 2$ and let $\mathscr{L}$ be a nef line bundle on S. Let d be the degree of the moving part of $\mathscr{L}_{|F}$. Then*

(i) If $\mathscr{L}_{|F}$ is subcanonical then

$$L^2 \geq \frac{4d}{d+2}\deg f_*\mathscr{L}.$$

(ii) If $d \geq 2g+1$ then

$$L^2 \geq \frac{2d}{d-g+2}\deg f_*\mathscr{L}.$$

Proof. (i) Just take $a = 2$ in the previous remark using Clifford's theorem.
(ii) If $d \geq 2g+1$ then the linear system $\mathscr{L}_{|L}$ is linearly semistable and hence we can take, by Riemann-Roch theorem on F,

$$a = \frac{d}{r-1} = \frac{d}{d+1-g}.$$

4.3 Linear Stability and Moriwaki's Method

Let C be a curve, $\mathscr{L}$ a line bundle on C, and $V \subseteq H^0(C, \mathscr{L})$ a linear subsystem of degree d and dimension r. We now compare the concept of linear stability for a couple (C, V) with the stability needed for the application of Moriwaki's method. We call

$$M_{\mathscr{L},V} := \ker(V \otimes \mathscr{O}_C \longrightarrow \mathscr{L}),$$

the *dual span bundle* (DSB) of the line bundle $\mathscr{L}$ with respect to the generating subspace $V \subseteq H^0(C, \mathscr{L})$. This is a vector bundle of rank $r - 1$ and degree $-d$. When $V = H^0(C, \mathscr{L})$ we denote it $M_{\mathscr{L}}$.[3]

Remark 25. An interesting geometric interpretation of this sheaf is the following. Consider the Euler sequence on $\mathbb{P}^n$:

$$0 \longrightarrow \Omega^1_{\mathbb{P}^n}(1) \longrightarrow \mathscr{O}_{\mathbb{P}^n}^{\oplus(n+1)} \longrightarrow \mathscr{O}_{\mathbb{P}^n}(1) \longrightarrow 0.$$

Applying the pullback of φ we obtain

$$0 \longrightarrow \varphi^*(\Omega^1_{\mathbb{P}^n}(1)) \longrightarrow V \otimes \mathscr{O}_C \longrightarrow \mathscr{L} \longrightarrow 0.$$

Hence the kernel of the evaluation morphism coincides with the restriction of the cotangent bundle of the projective space $\mathbb{P}^n$ to the curve C.

The μ-stability of the DSB is the stability condition assumed to hold on the general fibres for the method of Moriwaki.

Proposition 10. *With the above notation, if the DSB sheaf $M_{\mathscr{L},V}$ is μ-(semi)stable, then the couple (C, V) is linearly (semi)stable.*

Proof. Let us consider any $g^{r'-1}_{d'}$ in $|V|$. Let V' be the associated subspace of V. Consider the evaluation morphism $V' \otimes \mathscr{O}_C \longrightarrow \mathscr{L}$, which is not surjective unless $d' = d$, and let $\mathscr{G}$ be its kernel. Then $\mathscr{G}$ is a vector subbundle of $M_{\mathscr{L},V}$ with $\deg \mathscr{G} = -d'$, $\operatorname{rank} \mathscr{G} = r' - 1$. So from the stability condition on $M_{\mathscr{L},V}$ we obtain that $d'/(r'-1) \geq d/(r-1)$.

Remark 26. Note that any $\mathscr{G} \subseteq M_{\mathscr{L},V}$ as in the above theorem fits into the diagram

$$\begin{array}{ccccccccc}
0 & \longrightarrow & \mathscr{G} & \longrightarrow & V' \otimes \mathscr{O}_C & \longrightarrow & \overline{\mathscr{L}} & \longrightarrow & 0 \\
 & & \downarrow & & \downarrow & & \downarrow & & \\
0 & \longrightarrow & M_{\mathscr{L},V} & \longrightarrow & V \otimes \mathscr{O}_C & \longrightarrow & \mathscr{L} & \longrightarrow & 0
\end{array}$$

where $\overline{\mathscr{L}} \subseteq \mathscr{L}$ is the image of the evaluation morphism $V' \otimes \mathscr{O}_C \longrightarrow \mathscr{L}$.

[3]Note that we make here, as in [36], an abuse of notation: properly speaking the dual span bundle is the dual bundle of $M_{V,\mathscr{L}}$, which is indeed spanned by V^*.

The converse implication is studied in [36]. Let us now briefly discuss the question. Consider a linearly semistable couple (C, V). Let us consider a proper saturated subsheaf $\mathcal{G} \subseteq M_{\mathcal{L},V}$. We have that $\mathcal{G}^*$ is generated by its global sections. Consider the image of the natural morphism $W^* := \mathrm{im}(V^* \to H^0(C, \mathcal{G}^*))$. The evaluation morphism $W^* \otimes \mathcal{O}_C \longrightarrow \mathcal{G}^*$ is surjective. We thus have the following commutative diagram (cf. [17])

$$\begin{array}{ccccccccc} 0 & \longrightarrow & \mathcal{G} & \longrightarrow & W \otimes \mathcal{O}_C & \longrightarrow & \mathcal{F} & \longrightarrow & 0 \\ & & \downarrow & & \downarrow & & \downarrow{\scriptstyle \alpha} & & \\ 0 & \longrightarrow & M_{\mathcal{L},V} & \longrightarrow & V \otimes \mathcal{O}_C & \longrightarrow & \mathcal{L} & \longrightarrow & 0 \end{array} \qquad (12)$$

where $\mathcal{F}$ is a vector bundle without trivial summands (this follows from the choice of W). Note that the morphism α is non-zero, because $W \otimes \mathcal{O}_C$ is not contained in the image of $M_{\mathcal{L},V}$ in $V \otimes \mathcal{O}_C$. Let us suppose that $\mathcal{G}$ is destabilizing: if the sheaf $\mathcal{F}$ is a line bundle, then we would be in the situation described in Remark 26, and we could easily deduce from the linear (semi)stability of (C, V) the μ-(semi)stability of $M_{\mathcal{L},V}$. But we cannot exclude that a destabilizing subsheaf exists which is not the transform of a line bundle contained in $\mathcal{L}$.

In [36] the second author and E. C. Mistretta prove that indeed this is the case in the following cases, which depend on the Clifford index $\mathrm{Cliff}(C)$ of the curve C.

Theorem 15 ([36] Theorem 1.1). *Let $\mathcal{L}$ be a globally generated line bundle on C, and $V \subseteq H^0(C, \mathcal{L})$ a generating space of global sections such that*

$$\deg \mathcal{L} - 2(\dim V - 1) \leqslant \mathit{Cliff}(C).$$

Then linear (semi)stability of (C, V) is equivalent to μ-(semi)stability of $M_{V,\mathcal{L}}$ in the following cases:

1. $V = H^0(\mathcal{L})$;
2. $\deg \mathcal{L} \leqslant 2g - \mathit{Cliff}(C) + 1$;
3. $\mathrm{codim}_{H^0(\mathcal{L})} V < h^1(\mathcal{L}) + g/(\dim V - 2)$;
4. $\deg \mathcal{L} \geqslant 2g$, *and* $\mathrm{codim}_{H^0(\mathcal{L})} V \leqslant (\deg \mathcal{L} - 2g)/2$.

5 Results in Higher Dimensions

5.1 Linear Stability in Higher Dimensions and Xiao's Method

Mumford's original definition of linear stability is in any dimension, as follows.

Definition 7 ([42], Definition 2.16). An m-dimensional variety of degree d in $\mathbb{P}^{r-1}$ is linearly semistable (resp. linearly stable) if for any projection $\pi: \mathbb{P}^{r-1} \dashrightarrow \mathbb{P}^{s-1}$ such that the image of X is still of dimension m, the following inequality holds:

$$\frac{\deg(\pi_*(X))}{s-m} \geq \frac{d}{r-m} \qquad (resp. >),$$

where $\pi_*(X)$ denotes the image cycle of X in $\mathbb{P}^s$.

For example, it easy to verify that a K3 surface with Picard number 1 is linearly semistable. However, as Mumford himself remarks some lines after the definition, this condition in dimension higher than 1 seems to be difficult to handle. Moreover, it does not imply anymore Hilbert or Chow stability.

It seems that there is no sensible connection between linear stability and the method of Cornalba-Harris.

The relation of linear semistability with f-positivity via Xiao's method appears clear enough when all the induced maps of the Harder-Narashiman pieces are generically finite onto its image. More concretely, we obtain an inequality very close to f-positivity (it is possible to get something slightly better with much more effort).

Proposition 11. *Let $f : X \longrightarrow B$ be a fibration with general fibre F, $n = \dim X$ and $\mathcal{L}$ a nef line bundle. Let $\mathcal{G} \subseteq f_*\mathcal{L}$ be a nef subbundle. Assume that all the induced maps on F by the Harder-Narashimann pieces of $\mathcal{G}$ are generically finite and that the one induced by $\mathcal{G}$ is (Mumford-)linearly semistable. Then*

$$L^n \geq n\frac{d}{r+(n-1)^2}\deg\mathcal{G}.$$

Proof. A similar argument as in Theorem 14 applies. Let $a = d/(r-n+1)$, where d and r are the degree and rank of the base-point free linear map induced on (a suitable blow-up of) F by $\mathcal{G}$. By Xiao's inequality, using that all the induced maps on fibres are generically finite onto their images and that $P_{i+1}^k P_i^{r-1-k} \geq P_i^{r-1}$for all i, we obtain

$$\begin{aligned} L^n &\geq n(\sum_i P_i^{n-1}(\mu_i - \mu_{i+1})) \geq \sum_i (nar_i - n(n-1)a)(\mu_i - \mu_{i+1}) \\ &= na\deg\mathcal{G} - n(n-1)a\mu_1. \end{aligned}$$

Since $L - \mu_1 F$ is pseudoeffective and L is nef we have that $L^{n-1}(L - \mu_1 F) \geq 0$ and so

$$L^n \geq \mu_1 d,$$

which finally gives

$$L^n \geq n\frac{d}{r+(n-1)^2}\deg\mathcal{G}.$$

Clearly, the argument above does not work in general for $\dim X \geq 3$, due to the presence of induced map on fibres which are not generically finite. In some situations, however, it is possible to control such maps and conclude again f-positivity. In [9] we do this analysis for families of $K3$ surfaces, obtaining a significant generalization of Proposition 13 below.

5.2 *New Inequalities and Conjectures via the C-H Method*

We now state a couple of new results obtained via the CH method, using known stability results in dimension ≥ 2, and make some speculation and natural conjectures.

Families of Abelian Varieties

Let us consider a fibred variety $f\colon X \longrightarrow B$ of dimension n such that the general fibre is an abelian variety. Suppose that $\mathscr{L}$ is a line bundle on X such that $f_*\mathscr{L}$ is either generating, or it satisfies the conditions of Proposition 2 or 3.

Proposition 12. *Under the above assumption, suppose that $\mathscr{L}$ is very ample on the general fibre. Then $(\mathscr{L}, \mathscr{G})$ is f-positive, i.e.*

$$L^n \geq n! \deg f_*\mathscr{L}. \tag{13}$$

Proof. We can apply Theorem 6 because the immersion induced by $\mathscr{L}_{|F}$ on the general fibre F is Hilbert semistable by Kempf's result [28]. Observe then that as F is abelian, and $\mathscr{L}$ is very ample, we have that $h^0(F, \mathscr{L}_{|F}) = \chi(\mathscr{L}_{|F}) = L_{|F}^{n-1}/(n-1)!$, and so f-positivity translates in formula (13).

Families of K3 Surfaces

Let $f\colon T \longrightarrow B$ be a fibred threefold such that the general fibre is a $K3$ surface of genus g. Let $\mathscr{L}$ be a line bundle on T such that $f_*\mathscr{L}$ is either generating, or it satisfies the conditions of Proposition 2 or 3. The following result follows right away from Theorem 6 applying Morrison's result [41].

Proposition 13. *In the above situation, suppose that the general fibres F have Picard number* 1, *that $\mathscr{L}_{|F}$ is the primitive divisor class, and that its degree is at least* 12. *Then $\mathscr{L}$ is f-positive, i.e. the following inequality holds:*

$$L^3 \geq 6\frac{g-1}{g+1} \deg f_*\mathscr{L}.$$

Remark 27. It is interesting to notice that the bound $6(g-1)/(g+1)$ appearing in the inequality of Proposition 13 coincides with the one obtained in [31] and in [24] for the canonical slope of fibred surfaces of odd degree g whose general fibre is of maximal gonality.

Moreover, it is easy to prove that the canonical slope of a family of curves contained in a *fixed* $K3$ surface is indeed bounded from below by $6(g-1)/(g+1)$.

A Conjecture on the Slope Inequality in Higher Dimension

Let $f\colon X \longrightarrow B$ be an n-dimensional fibred variety. It is natural to define as a possible canonical slope the ratio between K_f^n and $\deg f_*\omega_f$, another possibility being to use the relative characteristic χ_f: in higher dimension the values $\deg f_*\omega_f$ and χ_f are not equal, but it holds an inequality between them [5, 45].

A natural slope inequality in higher dimension would be the following

$$K_f^n \geq n\frac{K_F^{n-1}}{h^0(F,\omega_F)}\deg f_*\omega_f, \tag{14}$$

which is equivalent to the f-positivity of ω_f. From the Cornalba-Harris and Bost method we can derive inequality (14) any time we have a Hilbert-Chow semistable canonical map on the general fibres. Although there are not much general results, it seems natural in the framework of GIT to conjecture that the stability of a variety has a connection with its singularities: a stable or asymptotically stable variety has mild singularities and it seems that also a vice-versa to this statement should hold. In consideration of this fact, and in analogy with the case of curves, it seems natural to state the following conjecture. See also Remark 4 for an account the natural positivity conditions on ω_f.

Conjecture 1. Let $f\colon X \longrightarrow B$ be a fibred n-dimensional variety whose relative canonical sheaf ω_f is relatively nef and ample on the general fibres, and whose general fibres have sufficiently mild singularities (e.g. they are log canonical, or semi-log-canonical). Then the fibration satisfies the slope inequality (14).

Almost nothing is known about this conjecture in dimension higher than 2. In [8] we prove this inequality for families of hypersurfaces whose general fibres satisfy a very weak singularity condition expressed in terms of its log canonical threshold and depending upon the degree of the hypersurfaces (see [32]).

Remark 28. Recall that the Severi inequality for surfaces S of maximal Albanese dimension $K_X^2 \geq 4\chi(\mathcal{O}_X)$ has been proved in full generality by Pardini in [47]. In [4] the first author proves that higher dimensional Severi inequalities of the form $L^n \geq 2n!\chi(\mathcal{L})$ hold n arbitrary dimensions for any nef line bundle $\mathcal{L}$. The classical proof of Severi inequality for surfaces and $\mathcal{L} = \omega_S$ given by Pardini makes use of the slope inequality for fibred surfaces. We prove now that her argument can be generalized, assuming that Conjecture 1 holds.

Proposition 14. *Let $m > 0$ be an integer. Suppose that slope inequality (14) holds for all varieties of dimension $\leq m$ that have maximal Albanese dimension and are* *fibred over $\mathbb{P}^1$. Then for any variety X of dimension $n \leq m$ with maximal Albanese dimension it holds the following sharp Severi inequality:*

$$K_X^n \geq 2n!\chi(\omega_X). \tag{15}$$

Proof. We proceed by induction on $n = \dim X$. For $n = 1$ inequality (15) is trivially true. Take now $n \geq 2$. First of all observe that from the slope inequality we can deduce a stronger result for maximal Albanese dimensional varieties with fibrations $f\colon X \longrightarrow \mathbb{P}^1$. Indeed, consider an étale Galois cover of X of degree r, and the induced fibration $\tilde{f}$. Then, applying inequality (14) we obtain

$$rK_f^n = K_{\tilde{f}}^n \geq n\frac{rK_F^{n-1}}{r\chi(\omega_F) + \epsilon_1}(r\chi_f + \epsilon_2),$$

where $\epsilon_1 = (h^1(F, \omega_F) - \ldots + (-1)^{n-2}h^{n-1}(F, \omega_F))$ and $\epsilon_2 = (\deg R^1 f_*\omega_f - \ldots \ldots + (-1)^{n-2}R^{n-1}f_*\omega_f)$. Since the inequality holds for all r we obtain

$$K_F^n + 2nK_F^{n-1} = K_f^n \geq n\frac{K_F^{n-1}}{\chi(\omega_F)}(\chi(\omega_X) + 2\chi(\omega_F)).$$

Applying induction hypothesis for F (which is clearly of maximal Albanese dimension), we deduce the inequality

$$K_F^n + 2nK_F^{n-1} \geq 2n!(\chi(\omega_X) + 2\chi(\omega_F)). \tag{16}$$

Now we can "eliminate the contribution due to F" just mimetizing Pardini's argument in [47], which we sketch here.

Consider the following cartesian diagram

$$\begin{array}{ccc} \tilde{X} & \xrightarrow{\mu} & X \\ \downarrow{\scriptstyle \tilde{a}} & & \downarrow{\scriptstyle a} \\ A & \xrightarrow{\mu} & A \end{array}$$

where $a\colon X \longrightarrow A$ is the Albanese map, and the maps μ are multiplication by d in A and so are Galois étale maps of degree d^{2q}. Fix a very ample line bundle $\mathcal{H}$ on A and let $\mathcal{M} = a^*(\mathcal{H})$ and $\tilde{\mathcal{M}} = \tilde{a}^*(\mathcal{H})$. By [14, Ch2. Prop.3.5] we have that

$$\tilde{M} \equiv \frac{1}{d^2}\mu^*(M) \quad \text{(numerical equivalence).}$$

Take general elements $F, F' \in |\tilde{M}|$ and perform a blow-up $Y \longrightarrow X$ to obtain a fibration $f : Y \longrightarrow \mathbb{P}^1$. Then we apply (16) to f and obtain

$$K_Y^n + 2nK_F^{n-1} \geq 2n!(\chi(\omega_Y) + 2\chi(\omega_F)).$$

Now an easy computation through the blow-up and the étale cover μ shows that

- $K_Y^n = d^{2q} K_X^n + \mathcal{O}(d^{2q-4})$.
- $K_F^{n-1} = K_{\tilde{X}}^{n-1}\tilde{M} + (n-1)K_{\tilde{X}}^{n-2}\tilde{M} = \mathcal{O}(d^{2q-2})$.
- $\chi(\omega_Y) = \chi(\omega_{\tilde{X}}) = d^{2q}\chi(\omega_X)$.
- $\chi(\omega_F) = \mathcal{O}(d^{2q-2})$ by Riemann-Roch theorem on $\tilde{X}$.

Since these equalities holds for any d we conclude that

$$K_X^n \geq 2n!\chi(\omega_X).$$

References

1. J. Alper, M. Fedorchuk, D. Smyth, Finite Hilbert stability of (bi)canonical curves. Invent. Math. **191**(3), 671–718 (2013)
2. E. Arbarello, M. Cornalba, P.A. Griffiths, J. Harris, *Geometry of Algebraic Curves*. Grundlehren Der Mathematischen Wissenschaften, vol. II (Springer, New York, 2011), p. 268
3. T. Ashikaga, K. Konno, Global and local properties of pencils of algebraic curves, in *Algebraic Geometry 2000*, Azumino. Advanced Studies in Pure Mathematics, vol. 36 (Mathematical Society of Japan, Tokyo, 2002), pp. 1–49
4. M.A. Barja, Generalized clifford-Severi inequality (preprint). arXiv:1303.3045 [math.AG]
5. M.A. Barja, On the slope of fibred threefolds. Int. J. Math. **11**(4), 461–491 (2000)
6. M.A. Barja, L. Stoppino, Linear stability of projected canonical curves, with applications to the slope of fibred surfaces. J. Math. Soc. Jpn. **60**(1), 171–192 (2008)
7. M.A. Barja, L. Stoppino, Slopes of trigonal fibred surfaces and of higher dimensional fibrations. Ann. Scuola Norm. Sup. Pisa Cl. Sci. **VIII**(5), 647–658 (2009)
8. M.A. Barja, L. Stoppino, Stability and singularities of relative hypersurfaces (in preparation)
9. M.A. Barja, L. Stoppino, Invariants of families of K3 surfaces (in preparation)
10. W.P. Barth, K. Hulek, C.A.M. Peters, A. Van de Ven, *Compact Complex Surfaces*, 2nd edn. (Springer, Berlin/New York, 2004)
11. A. Beauville, L'inégalité $p_g \geq 2q - 4$ pour les surfaces de type général. Bull. Soc. Math. Fr. **110**, 343–346 (1982)
12. V. Beorchia, F. Zucconi, On the slope conjecture for the fourgonal locus in $\overline{\mathcal{M}}_g$. arXiv:1209.3571 [math.AG]
13. G. Bini, M. Melo, F. Viviani, GIT for polarized curves, arXiv: 1109.6908
14. C. Birkenhake, H. Lange, *Complex Abelian Varieties*. Grundlehren der mathematischen Wissenschaften, vol. 302 (Springer, Berlin/New York, 2004)
15. F. Bogomolov, Holomorphic tensors and vector bundles on projective manifolds. Izv. Akad. Nauk SSSR Ser. Mat. **42**(6), 1227–1287 (1978)
16. J. Bost, Semi-stability and heights of cycles. Invent. Math. **118**(2), 223–253 (1994)
17. D.C. Butler, Normal generation of vector bundles over a curve (1997). arXiv:9705009 [math.AG]
18. M. Cornalba, J. Harris, Divisor classes associated to families of stable varieties, with applications to the moduli space of curves. Ann. Sc. Ec. Norm. Sup. **21**(4), 455–475 (1988)
19. A. Deopurkar, A. Patel, Sharp slope bounds for sweeping families of trigonal curves. arXiv:1211.2827 [math.AG]
20. M. Fedorchuk, D. Jensen, Stability of 2nd Hilbert points of canonical curves. Int. Math. Res. Not. (2012, to appear). doi:10.1093/imrn/rns204

21. J. Fogarty, Truncated Hilbert functors. J. Reine Angew. Math. **234**, 65–88 (1969)
22. D. Gieseker, *Lectures on Moduli of Curves*. Tata Lectures Notes, vol. 69 (Bombay, Published for the Tata Institute of Fundamental Research, 1982)
23. J. Harris, I. Morrison, *Moduli of Curves*. Graduate Texts in Mathematics, vol. 187 (Springer, New York, 1998)
24. J. Harris, D. Mumford, On the Kodaira dimension of the moduli space of curves. Invent. Math. **67**(1), 23–88 (1982)
25. R. Hartshorne, Ample vector bundles on curves. Nagoya Math. J. **43**, 73–89 (1971)
26. B. Hassett, D. Hyeon, Log canonical model for the moduli space of curves: first divisorial contraction. Trans. Am. Math. Soc. **361**(8), 4471–4489 (2009)
27. J. Keller, J. Ross, A note on chow stability of the projectivisation of Gieseker stable bundles. J. Geom. Anal. doi:10.1007/s12220-012-9384-3
28. G.R. Kempf, Instability in invariant theory. Ann. Math. **108**, 299–316 (1978)
29. K. Konno, Non-hyperelliptic fibrations of small genus and certain irregular canonical surfaces. Ann. Sc. Norm. Sup. Pisa Ser. IV **20**, 575–595 (1993)
30. K. Konno, A lower bound of the slope of trigonal fibrations. Int. J. Math. **7**(1), 19–27 (1996)
31. K. Konno, Clifford index and the slope of fibered surfaces. J. Algebr. Geom. **8**(2), 207–220 (1999)
32. Y. Lee, Chow stability criterion in terms of log canonical threshold. J. Korean Math. Soc. **45**(2), 467–477 (2008)
33. T. Mabuchi, Chow-stability and Hilbert-stability in Mumford's geometric invariant theory. Osaka J. Math. **45**(3), 833–846 (2008)
34. V.B. Mehta, A. Ramanathan, Restriction of stable sheaves and representations of the fundamental group. Invent. Math. **77**(1), 163–172 (1984)
35. E.C. Mistretta, Stability of line bundle transforms on curves with respect to low codimensional subspaces. J. Lond. Math. Soc. (2) **78**1, 172–182 (2008)
36. E.C. Mistretta, L. Stoppino, Linear series on curves: stability and clifford index. Int. J. Math. **23**(12) (2012). doi:10.1142/S0129167X12501212
37. Y. Miyaoka, The Chern classes and Kodaira dimension of a minimal variety, in *Algebraic Geometry*, Sendai, 1985. Advanced Studies in Pure Mathematics, vol. 10 (Kinokuniya/North-Holland/Elsevier, Tokyo/Amsterdam/New York, 1987), pp. 449–476
38. A. Moriwaki, Semi-stably polarized fiber spaces and Bogomolov-Gieseker type inequality. J. Math. Kyoto Univ. **32**(4), 843–872 (1992)
39. A. Moriwaki, A sharp slope inequality for general stable fibrations of curves. J. Reine Angew. Math. **480**, 177–195 (1996)
40. I. Morrison, Projective stability of ruled surfaces. Invent. Math. **56**(3), 269–304 (1980)
41. I. Morrison, *Stability of Hilbert Points of Generic $K3$ Surfaces*, vol. 401 (Centre de Recerca Matemática, Bellaterra, 1999)
42. D. Mumford, Stability of projective varieties. L'Ens. Math. **23**, 39–110 (1977)
43. D. Mumford, J. Fogarty, F. Kirwan, *Geometric Invariant Theory*. Ergebnisse der Mathematik und ihrer Grenzgebiete, vol. 34(2), 3rd edn. (Springer, Berlin, 1994)
44. N. Nakayama, *Zariski-Decomposition and Abundance*. MSJ Memoires, vol. 14 (Mathematical Society of Japan, Tokyo, 2004)
45. K. Ohno, Some inequalities for minimal fibrations of surfaces of general type over curves. J. Math. Soc. Jpn. **44**(4), 643–666 (1992)
46. K. Paranjape, S. Ramanan, *On the Canonical Ring of a Curve*. Algebraic Geometry and Commutative Algebra, vol. II (Kinokuniya, Tokyo, 1988), pp. 503–516
47. R. Pardini, The Severi inequality $K^2 \geq 4\chi$ for surfaces of maximal Albanese dimension. Invent. Math. **159**(3), 669–672 (2005)
48. D. Schubert, A new compactification of the moduli space of curves. Compos. Math. **78**(3), 297–313 (1991)
49. R. Seyyedali, Balanced metrics and Chow stability of projective bundles over Käler manifolds. Duke Math. J. **153**(3), 573–605 (2010)

50. L. Stoppino, Slope inequalities for fibered surfaces via GIT. Osaka J. Math. **45**(4), 1027–1041 (2008)
51. G. Tian, The k-energy on hypersurfaces and stability. Commun. Anal. Geom. **2**(2), 239–265 (1994)
52. G. Xiao, Fibred algebraic surfaces with low slope. Math. Ann. **276**, 449–466 (1987)

On the Second Lower Quotient of the Fundamental Group

Arnaud Beauville

Dedicated to Klaus Hulek on his 60th birthday

Abstract Let X be a topological space, $G = \pi_1(X)$ and $D = (G, G)$. We express the second quotient $D/(D, G)$ of the lower central series of G in terms of the homology and cohomology of X. As an example, we recover the isomorphism $D/(D, G) \cong \mathbb{Z}/2$ (due to Collino) when X is the Fano surface parametrizing lines in a cubic threefold.

1 Introduction

Let X be a connected topological space. The group $G := \pi_1(X)$ admits a lower central series

$$G \supseteq D := (G, G) \supseteq (D, G) \supseteq \dots$$

The first quotient G/D is the homology group $H_1(X, \mathbb{Z})$. We consider in this note the second quotient $D/(D, G)$. In particular when $H_1(X, \mathbb{Z})$ is torsion free, we obtain a description of $D/(D, G)$ in terms of the homology and cohomology of X (see Corollary 2 below).

As an example, we recover in the last section the isomorphism $D/(D, G) \cong \mathbb{Z}/2$ (due to Collino) for the Fano surface parametrizing the lines contained in a cubic threefold.

A. Beauville (✉)
Laboratoire J.-A. Dieudonné, UMR 7351 du CNRS, Université de Nice Sophia Antipolis, Parc Valrose, F-06108 Nice cedex 2, France
e-mail: arnaud.beauville@unice.fr

A. Frühbis-Krüger et al. (eds.), *Algebraic and Complex Geometry*, Springer Proceedings in Mathematics & Statistics 71, DOI 10.1007/978-3-319-05404-9_2,

2 The Main Result

Proposition 1. *Let X be a connected space homotopic to a CW-complex, with $H_1(X,\mathbb{Z})$ finitely generated. Let $G = \pi_1(X)$, $D = (G,G)$ its derived subgroup, $\tilde{D}$ the subgroup of elements of G which are torsion in G/D. The group $D/(\tilde{D},G)$ is canonically isomorphic to the cokernel of the map*

$$\mu : H_2(X,\mathbb{Z}) \to \mathrm{Alt}^2(H^1(X,\mathbb{Z})) \quad \textit{given by } \mu(\sigma)(\alpha,\beta) = \sigma \frown (\alpha \wedge \beta)\,,$$

where $\mathrm{Alt}^2(H^1(X,\mathbb{Z}))$ is the group of skew-symmetric integral bilinear forms on $H^1(X,\mathbb{Z})$.

Proof. Let H be the quotient of $H_1(X,\mathbb{Z})$ by its torsion subgroup; we put $V := H \otimes_{\mathbb{Z}} \mathbb{R}$ and $T := V/H$. The quotient map $\pi : V \to T$ is the universal covering of the real torus T.

Consider the surjective homomorphism $\alpha : \pi_1(X) \to H$. Since T is a $K(H,1)$, there is a continuous map $a : X \to T$, well defined up to homotopy, inducing α on the fundamental groups. Let $\rho : X' \to X$ be the pull back by a of the étale covering $\pi : V \to T$, so that $X' := X \times_T V$ and ρ is the covering associated to the homomorphism α.

Our key ingredient will be the map $f : X \times V \to T$ defined by $f(x,v) = a(x) - \pi(v)$. It is a locally trivial fibration, with fibers isomorphic to X'. Indeed the diagram

$$\begin{array}{ccc} X' \times V & \xrightarrow{\ g\ } & X \times V \\ {\scriptstyle \mathrm{pr}_2}\downarrow & & \downarrow{\scriptstyle f} \\ V & \xrightarrow{\ \pi\ } & T \end{array}$$

where $g((x,v),w) = (x, v - w)$, is cartesian.

It follows from this diagram that the monodromy action of $\pi_1(T) = H$ on $H_1(X',\mathbb{Z})$ is induced by the action of H on X'; it is deduced from the action of $\pi_1(X)$ on $\pi_1(X')$ by conjugation in the exact sequence

$$1 \to \pi_1(X') \xrightarrow{\rho_*} \pi_1(X) \to H \to 1\,. \tag{1}$$

The homology spectral sequence of the fibration f (see for instance [5]) gives rise in low degree to a five terms exact sequence

$$H_2(X,\mathbb{Z}) \xrightarrow{a_*} H_2(T,\mathbb{Z}) \longrightarrow H_1(X',\mathbb{Z})_H \xrightarrow{\rho_*} H_1(X,\mathbb{Z}) \longrightarrow H_1(T,\mathbb{Z}) \longrightarrow 0\,, \tag{2}$$

where $H_1(X',\mathbb{Z})_H$ denote the coinvariants of $H_1(X',\mathbb{Z})$ under the action of H.

The exact sequence (1) identifies $\pi_1(X')$ with $\tilde{D}$, hence $H_1(X', \mathbb{Z})$ with $\tilde{D}/(\tilde{D}, \tilde{D})$, the action of H being deduced from the action of G by conjugation. The group of coinvariants is the largest quotient of this group on which G acts trivially, that is, the quotient $\tilde{D}/(\tilde{D}, G)$.

The exact sequence (2) gives an isomorphism $\operatorname{Ker}\rho_* \xrightarrow{\sim} \operatorname{Coker} a_*$. The map $\rho_* : H_1(X', \mathbb{Z})_H \to H_1(X, \mathbb{Z})$ is identified with the natural map $\tilde{D}/(\tilde{D}, G) \to G/D$ deduced from the inclusions $\tilde{D} \subset G$ and $(\tilde{D}, G) \subset D$. Therefore its kernel is $D/(\tilde{D}, G)$. On the other hand since T is a torus we have canonical isomorphisms

$$H_2(T, \mathbb{Z}) \xrightarrow{\sim} \operatorname{Hom}(H^2(T, \mathbb{Z}), \mathbb{Z}) \xrightarrow{\sim} \operatorname{Alt}^2(H^1(T, \mathbb{Z})) \xrightarrow{\sim} \operatorname{Alt}^2(H^1(X, \mathbb{Z})),$$

through which a_* corresponds to μ, hence the Proposition. □

Corollary 1. 1. *There is a canonical surjective map* $D/(D, G) \to \operatorname{Coker}\mu$ *with finite kernel.*

2. *There are canonical exact sequences*

$$H_2(X, \mathbb{Q}) \xrightarrow{\mu_{\mathbb{Q}}} \operatorname{Alt}^2(H^1(X, \mathbb{Q})) \longrightarrow D/(D, G) \otimes \mathbb{Q} \to 0$$

$$0 \to \operatorname{Hom}(D/(D, G), \mathbb{Q}) \longrightarrow \wedge^2 H^1(X, \mathbb{Q}) \xrightarrow{c_{\mathbb{Q}}} H^2(X, \mathbb{Q}),$$

where $c_{\mathbb{Q}}$ *is the cup-product map.*

Proof. (2) follows from (1), and from the fact that the transpose of $\mu_{\mathbb{Q}}$ is $c_{\mathbb{Q}}$. Therefore in view of the Proposition, it suffices to prove that the kernel of the natural map $D/(D, G) \to D/(\tilde{D}, G)$, that is, $(\tilde{D}, G)/(D, G)$, is finite. Consider the surjective homomorphism

$$G/D \otimes G/D \to D/(D, G)$$

deduced from $(x, y) \mapsto xyx^{-1}y^{-1}$. It maps $\tilde{D}/D \otimes G/D$ onto $(\tilde{D}, G)/(D, G)$; since $\tilde{D}/D$ is finite and G/D finitely generated, the result follows. □

Corollary 2. *Assume that* $H_1(X, \mathbb{Z})$ *is torsion free.*

1. *The second quotient* $D/(D, G)$ *of the lower central series of* G *is canonically isomorphic to* $\operatorname{Coker}\mu$.
2. *For every ring* R *the group* $\operatorname{Hom}(D/(D, G), R)$ *is canonically isomorphic to the kernel of the cup-product map* $c_R : \wedge^2 H^1(X, R) \to H^2(X, R)$.

Proof. We have $\tilde{D} = D$ in that case, so (1) follows immediately from the Proposition. Since $H_1(X, \mathbb{Z})$ is torsion free, the universal coefficient theorem provides an isomorphism $H^2(X, R) \xrightarrow{\sim} \operatorname{Hom}(H_2(X, \mathbb{Z}), R)$, hence applying $\operatorname{Hom}(-, R)$ to the exact sequence

$$H_2(X, \mathbb{Z}) \to \operatorname{Alt}^2(H^1(X, \mathbb{Z})) \to D/(D, G) \to 0$$

gives (2). □

Remark 1. The Proposition and its Corollaries hold (with the same proofs) under weaker assumptions on X, for instance for a connected space X which is paracompact, admits a universal cover and is such that $H_1(X,\mathbb{Z})$ is finitely generated. We leave the details to the reader.

Remark 2. For compact Kähler manifolds, the isomorphism $\mathrm{Hom}(D/(D,G),\mathbb{Q}) \cong \mathrm{Ker}\, c_{\mathbb{Q}}$ (Corollary 1) is usually deduced from Sullivan's theory of minimal models (see [1], ch. 3); it can be used to prove that certain manifolds, for instance Lagrangian submanifolds of an abelian variety, have a non-abelian fundamental group.

3 Example: The Fano Surface

Let $V \subset \mathbb{P}^4$ be a smooth cubic threefold. The Fano surface F of V parametrizes the lines contained in V. It is a smooth connected surface, which has been thoroughly studied in [2]. Its Albanese variety A is canonically isomorphic to the intermediate Jacobian JV of V, and the Albanese map $a : F \to A$ is an embedding. Recall that $A = JV$ carries a principal polarization $\theta \in H^2(A,\mathbb{Z})$; for each integer k the class $\dfrac{\theta^k}{k!}$ belongs to $H^{2k}(A,\mathbb{Z})$. The class of F in $H^6(A,\mathbb{Z})$ is $\dfrac{\theta^3}{3!}$ ([2], Proposition 13.1).

Proposition 2. *The maps $a^* : H^2(A,\mathbb{Z}) \to H^2(F,\mathbb{Z})$ and $a_* : H_2(F,\mathbb{Z}) \to H_2(A,\mathbb{Z})$ are injective and their images have index 2.*

Proof. We first recall that if $u : M \to N$ is a homomorphism between two free $\mathbb{Z}$-modules of the same rank, the integer $|\det u|$ is well-defined: it is equal to the absolute value of the determinant of the matrix of u for any choice of bases for M and N. If it is nonzero, it is equal to the index of $\mathrm{Im}\, u$ in N.

Poincaré duality identifies a_* with the Gysin map $a_* : H^2(F,\mathbb{Z}) \to H^8(A,\mathbb{Z})$, and also to the transpose of a^*. The composition

$$f : H^2(A,\mathbb{Z}) \xrightarrow{a^*} H^2(F,\mathbb{Z}) \xrightarrow{a_*} H^8(A,\mathbb{Z})$$

is the cup-product with the class $[F] = \dfrac{\theta^3}{3!}\cdot$ We have $|\det a^*| = |\det a_*| \neq 0$ ([2], 10.14), so it suffices to show that $|\det f| = 4$.

The principal polarization defines a unimodular skew-symmetric form on $H^1(A,\mathbb{Z})$; we choose a symplectic basis $(\varepsilon_i, \delta_j)$ of $H^1(A,\mathbb{Z})$. Then

$$\theta = \sum_i \varepsilon_i \wedge \delta_i \qquad \text{and} \qquad \frac{\theta^3}{3!} = \sum_{i<j<k} (\varepsilon_i \wedge \delta_i) \wedge (\varepsilon_j \wedge \delta_j) \wedge (\varepsilon_k \wedge \delta_k)\,.$$

If we identify by Poincaré duality $H^8(A,\mathbb{Z})$ with the dual of $H^2(A,\mathbb{Z})$, and $H^{10}(A,\mathbb{Z})$ with $\mathbb{Z}$, f is the homomorphism associated to the bilinear symmetric form $b : (\alpha, \beta) \mapsto \alpha \wedge \beta \wedge \frac{\theta^3}{3!}$, hence $|\det f|$ is the absolute value of the discriminant of b. Let us write $H^2(A, \mathbb{Z}) = M \oplus N$, where M is spanned by the vectors $\varepsilon_i \wedge \varepsilon_j$, $\delta_i \wedge \delta_j$ and $\varepsilon_i \wedge \delta_j$ for $i \neq j$, and N by the vectors $\varepsilon_i \wedge \delta_i$. The decomposition is orthogonal with respect to b; the restriction of b to M is unimodular, because the dual basis of $(\varepsilon_i \wedge \varepsilon_j, \delta_i \wedge \delta_j, \varepsilon_i \wedge \delta_j)$ is $(-\delta_i \wedge \delta_j, -\varepsilon_i \wedge \varepsilon_j, -\varepsilon_j \wedge \delta_i)$. On N the matrix of b with respect to the basis $(\varepsilon_i \wedge \delta_i)$ is $E - I$, where E is the 5-by-5 matrix with all entries equal to 1. Since E has rank 1 we have $\wedge^k E = 0$ for $k \geq 2$, hence

$$\det(E - I) = -\det(I - E) = -I + \operatorname{Tr} E = 4 ;$$

hence $|\det f| = 4$. □

Corollary 3. *Set $G = \pi_1(F)$ and $D = (G, G)$. The group $D/(D, G)$ is cyclic of order 2.*

Indeed $H_1(F, \mathbb{Z})$ is torsion free [3], hence the result follows from Corollary 2. □

Remark 3. The deeper topological study of [3] gives actually the stronger result that D is generated as a normal subgroup by an element σ of order 2 (see [3], and the correction in [4], Remark 4.1). Since every conjugate of σ is equivalent to σ modulo (D, G), this implies Corollary 3.

Remark 4. Choose a line $\ell \in F$, and let $C \subset F$ be the curve of lines incident to ℓ. Let $d : H^2(F, \mathbb{Z}) \to \mathbb{Z}/2$ be the homomorphism given by $d(\alpha) = (\alpha \cdot [C])$ (mod. 2). We claim that the image of $a^* : H^2(A, \mathbb{Z}) \to H^2(F, \mathbb{Z})$ is $\operatorname{Ker} d$. Indeed we have $(C^2) = 5$ (the number of lines incident to two given skew lines on a cubic surface), hence $d([C]) = 1$, so that $\operatorname{Ker} d$ has index 2; thus it suffices to prove $d \circ a^* = 0$. For $\alpha \in H^2(A, \mathbb{Z})$, we have $d(a^*\alpha) = (a^*\alpha \cdot [C]) = (\alpha \cdot a_*[C])$ mod. 2; this is 0 because the class $a_*[C] \in H^8(A, \mathbb{Z})$ is equal to $2\frac{\theta^4}{4!}$ ([2], Lemma 11.5), hence is divisible by 2.

We can identify a^* with the cup-product map c; thus we have an exact sequence

$$0 \to \wedge^2 H^1(F, \mathbb{Z}) \xrightarrow{c} H^2(F, \mathbb{Z}) \xrightarrow{d} \mathbb{Z}/2 \to 0 \quad \text{with } d(\alpha) = (\alpha \cdot [C]) \text{ (mod. 2)}.$$

References

1. J. Amorós, M. Burger, K. Corlette, D. Kotschick, D. Toledo, *Fundamental Groups of Compact Kähler Manifolds*. Mathematical Surveys and Monographs, vol. 44 (AMS, Providence, 1996)
2. H. Clemens, P. Griffiths, The intermediate Jacobian of the cubic threefold. Ann. Math. **95**(2), 281–356 (1972)
3. A. Collino, The fundamental group of the Fano surface I, II, in *Algebraic Threefolds (Varenna, 1981)*. Lecture Notes in Mathematics, vol. 947 (Springer, Berlin/New York, 1982), pp. 209–220
4. A. Collino, Remarks on the topology of the Fano surface. Preprint arXiv:1211.2621
5. S.-T. Hu, *Homotopy Theory* (Academic, New York/London, 1959)

McKay Correspondence over Non Algebraically Closed Fields

Mark Blume

Abstract The classical McKay correspondence for finite subgroups G of $\mathrm{SL}(2,\mathbb{C})$ gives a bijection between isomorphism classes of nontrivial irreducible representations of G and irreducible components of the exceptional divisor in the minimal resolution of the quotient singularity $\mathbb{A}^2_{\mathbb{C}}/G$. Over non algebraically closed fields K there may exist representations irreducible over K which split over $\overline{K}$. The same is true for irreducible components of the exceptional divisor. In this paper we show that these two phenomena are related and that there is a bijection between nontrivial irreducible representations and irreducible components of the exceptional divisor over non algebraically closed fields K of characteristic 0 as well.

1 Introduction

Let G be a finite group operating on a smooth variety M over $\mathbb{C}$, e.g. $M = \mathbb{A}^n_{\mathbb{C}}$ and a linear operation of a finite subgroup $G \subset \mathrm{SL}(n,\mathbb{C})$. Usually the quotient M/G is singular and one considers resolutions of singularities $Y \to M/G$ with some minimality property. A method to construct resolutions of quotient singularities is the G-Hilbert scheme G-Hilb$M$ introduced in [10, 11]. Under some conditions the G-Hilbert scheme is irreducible, nonsingular and G-Hilb$M \to M/G$ a crepant resolution [6]. In particular, this applies to the operation of finite subgroups $G \subset \mathrm{SL}(n,\mathbb{C})$ on $\mathbb{A}^n_{\mathbb{C}}$ for $n \leq 3$. For $G \subset \mathrm{SL}(2,\mathbb{C})$ there are also other methods to show that the G-Hilbert scheme is the minimal resolution, see [10, 11].

The McKay correspondence in general describes the resolution Y in terms of the representation theory of the group G, see [16, 17] for expositions of this subject. Part of the correspondence for $G \subset \mathrm{SL}(2,\mathbb{C})$ is a bijection between irreducible

M. Blume (✉)
Mathematisches Institut, Universität Münster, Einsteinstrasse 62, 48149 Münster, Germany
e-mail: mark.blume@uni-muenster.de

A. Frühbis-Krüger et al. (eds.), *Algebraic and Complex Geometry*, Springer Proceedings in Mathematics & Statistics 71, DOI 10.1007/978-3-319-05404-9_3,

components of the exceptional divisor E and isomorphism classes of nontrivial irreducible representations of the group G and moreover an isomorphism of graphs between the intersection graph of components of E and the representation graph of G, both being graphs of ADE type. This was the observation of McKay [14].

The new contribution in this paper is to consider McKay correspondence over non algebraically closed fields. We will work over a field K that is not assumed to be algebraically closed but always of characteristic 0 and extend the McKay correspondence to this slightly more general situation. Over non algebraically closed K it is natural to consider finite group schemes instead of simply finite groups. In comparison with the situation over algebraically closed fields there may exist both representations of G and components of E that are irreducible over K but split over the algebraic closure. We will see that these two kinds of splitting that arise by extending the ground field are related by investigating the operation of the Galois group. For this we introduce Galois-conjugate representations and consider the Galois operation on the G-Hilbert scheme. The following McKay correspondence over arbitrary fields K of characteristic 0 will be consequence of more detailed theorems in Sect. 5.

Theorem 1. *Let K be any field of characteristic 0 and $G \subset \mathrm{SL}(2, K)$ a finite subgroup scheme. Then there is a bijection between the set of irreducible components of the exceptional divisor E and the set of isomorphism classes of nontrivial irreducible representations of G and moreover an isomorphism between the intersection graph of the irreducible components of E_{red} and the representation graph of G.*

Examples are discussed in Sect. 5.5, the possible graphs can be found in Sect. 4.4. As already observed in [13], considering the rational double points over non algebraically closed fields one finds the remaining Dynkin diagrams of types (B_n), (C_n), (F_4), (G_2). The methods of this paper should also apply to other situations, in particular to the McKay correspondence for finite small subgroups of $\mathrm{GL}(2, \mathbb{C})$ and give a similar generalisation as in the SL-case.

This paper is organised as follows. Section 2 shortly summarises some techniques used in this paper, namely G-sheaves for group schemes G and G-Hilbert schemes. Section 3 is concerned with the relations between Galois operations and decompositions into irreducible components both of schemes and representations. We introduce the notion of Galois-conjugate representations and G-sheaves and we describe the Galois operation on G-Hilbert schemes. In Sect. 4 we collect some data of the finite subgroup schemes of $\mathrm{SL}(2, K)$ and list possible representation graphs. In addition we investigate under what conditions a finite subgroup of $\mathrm{SL}(2, C)$, C the algebraic closure of K, is realisable as a subgroup of $\mathrm{SL}(2, K)$. Section 5 contains the theorems of McKay correspondence over non algebraically closed fields. We consider two constructions, the stratification of the G-Hilbert scheme and the tautological sheaves, originating from [10] and [8] respectively, that are known to give a McKay correspondence over $\mathbb{C}$ and formulate them for not necessarily algebraically closed K.

Notations. In general we write a lower index for base extensions, for example if X, T are S-schemes then X_T denotes the T-scheme $X \times_S T$ or if V is a representation over a field K then V_L denotes the representation $V \otimes_K L$ over the extension field L. Likewise, if $\varphi: X \to Y$ is a morphism of S-schemes, we write $\varphi_T: X_T \to Y_T$ for its base extension with respect to $T \to S$.

2 Preliminaries

2.1 G-Sheaves

Let K be a field. Let G be a group scheme over K with $p: G \to \operatorname{Spec} K$ the projection, $e: \operatorname{Spec} K \to G$ the unit, and $m: G \times_K G \to G$ the multiplication. For affine $G = \operatorname{Spec} A$, A has the structure of a Hopf algebra over K, the coalgebra structure being equivalent to the group structure of G.

Let X be a G-scheme over K, that is a K-scheme with an operation $s_X: G \times_K X \to X$ of the group scheme G over K. We have to use a more general notion of a G-sheaf than in [6], we adopt the definition of [15]: a (quasicoherent, coherent) G-sheaf on X is a (quasicoherent, coherent) $\mathcal{O}_X$-module $\mathscr{F}$ with an isomorphism $\lambda^{\mathscr{F}}: s_X^*\mathscr{F} \xrightarrow{\sim} p_X^*\mathscr{F}$ of $\mathcal{O}_{G\times_K X}$-modules satisfying the conditions (i) the restriction of $\lambda^{\mathscr{F}}$ to the unit in G_X is the identity, i.e. $e_X^*\lambda^{\mathscr{F}}: e_X^* s_X^*\mathscr{F} \to e_X^* p_X^*\mathscr{F}$ identifies with $id_{\mathscr{F}}: \mathscr{F} \to \mathscr{F}$, and (ii) $(m \times id_X)^*\lambda^{\mathscr{F}} = p_{23}^*\lambda^{\mathscr{F}} \circ (id_G \times s_X)^*\lambda^{\mathscr{F}}$, where $p_{23}: G \times_K G \times_K X \to G \times_K X$ is the projection to the factors 2 and 3.

Remark 1. We summarise relevant properties of G-sheaves.

1. There is the canonical notion of G-equivariant homomorphisms between G-sheaves $\mathscr{F}, \mathscr{G}$ on X, the set of these is denoted by $\operatorname{Hom}_X^G(\mathscr{F}, \mathscr{G})$. Kernels and co-kernels of G-equivariant homomorphisms have natural G-sheaf structures.
2. Assume $G = \operatorname{Spec} A$ affine and let X be a G-scheme with trivial G-operation, i.e. $s_X = p_X$. Then the G-sheaf structure of a G-sheaf $\mathscr{F}$ is equivalent to a homomorphism of $\mathcal{O}_X$-modules $\varrho: \mathscr{F} \to A \otimes_K \mathscr{F}$ satisfying the usual conditions of a comodule. This relation can be constructed using the adjunction $(p_X{}^*, p_{X*})$. Further, notions such as "subcomodule", "homomorphism of comodules", etc. correspond to "G-subsheaf", "equivariant homomorphism", etc. The G-invariant part $\mathscr{F}^G \subseteq \mathscr{F}$ is defined by $\mathscr{F}^G(U) := \{f \in \mathscr{F}(U) \mid \varrho(f) = 1 \otimes f\}$ for open $U \subseteq X$.
3. For an A-comodule $\mathscr{F}$ on X a decomposition of A into a direct sum $A = \bigoplus_i A_i$ of subcoalgebras A_i determines a direct sum decomposition $\mathscr{F} = \bigoplus_i \mathscr{F}_i$ into subcomodules (take preimages $\varrho^{-1}(A_i \otimes_K \mathscr{F})$), where the comodule structure of $\mathscr{F}_i$ reduces to an A_i-comodule structure.
4. A G-sheaf on $X = \operatorname{Spec} K$ (or an extension field of K) we also call a representation. Dualisation of an A-comodule V over K leads to a KG-module $V^\vee$, where $KG = A^\vee = \operatorname{Hom}_K(A, K)$ with algebra structure dual to the coalgebra structure of A.

5. For quasicoherent G-sheaves $\mathscr{F},\mathscr{G}$ with $\mathscr{F}$ finitely presented the sheaf $\mathscr{H}om_{\mathscr{O}_X}(\mathscr{F},\mathscr{G})$ carries a natural G-sheaf structure. For locally free $\mathscr{F}$ one defines the dual G-sheaf by $\mathscr{F}^\vee = \mathscr{H}om_{\mathscr{O}_X}(\mathscr{F},\mathscr{O}_X)$. In the case of trivial G-operation on X there is the component $\mathscr{H}om^G_{\mathscr{O}_X}(\mathscr{F},\mathscr{G})$ of $\mathscr{H}om_{\mathscr{O}_X}(\mathscr{F},\mathscr{G})$, the sheaf of equivariant homomorphisms, that can either be described as G-invariant part $(\mathscr{H}om_{\mathscr{O}_X}(\mathscr{F},\mathscr{G}))^G \subseteq \mathscr{H}om_{\mathscr{O}_X}(\mathscr{F},\mathscr{G})$ or by $\mathscr{H}om^G_{\mathscr{O}_X}(\mathscr{F},\mathscr{G})(U) = \mathrm{Hom}^G_U(\mathscr{F}|_U,\mathscr{G}|_U)$ for open $U \subseteq X$.
6. Functors for sheaves like $\otimes$, $f^*,\ldots$ as well have analogues for G-sheaves, e.g. for equivariant $f: Y \to X$ and a G-sheaf $\mathscr{F}$ on X the sheaf $f^*\mathscr{F}$ has a natural G-sheaf structure.
7. Natural isomorphisms for sheaves lead to isomorphisms for G-sheaves, e.g. under some conditions there is an isomorphism $f^*\mathscr{H}om_{\mathscr{O}_X}(\mathscr{F},\mathscr{G}) \cong \mathscr{H}om_{\mathscr{O}_Y}(f^*\mathscr{F}, f^*\mathscr{G})$ and this isomorphism becomes an isomorphism of G-sheaves provided that f is equivariant and $\mathscr{F},\mathscr{G}$ are G-sheaves. Other examples are $f^*(\mathscr{F} \otimes_{\mathscr{O}_X} \mathscr{G}) \cong f^*\mathscr{F} \otimes_{\mathscr{O}_Y} f^*\mathscr{G}$, $\mathscr{H}om_{\mathscr{O}_X}(\mathscr{F} \otimes_{\mathscr{O}_X} \mathscr{E},\mathscr{G}) \cong \mathscr{H}om_{\mathscr{O}_X}(\mathscr{F},\mathscr{E}^\vee \otimes_{\mathscr{O}_X} \mathscr{G})$.
8. Base extension $K \to L$ makes out of a G-scheme X over K a scheme X_L with a G-scheme or a G_L-scheme structure, the operation given by $s_{X_L} = (s_X)_L$. A G-sheaf $\mathscr{F}$ on a G-scheme X gives rise to a G-sheaf $\mathscr{F}_L = \mathscr{F} \otimes_K L = f^*\mathscr{F}$ on X_L, where $f: X_L \to X$. $\mathscr{F}_L$ can be considered as a G_L-sheaf on the G_L-scheme X_L over L.

2.2 G-Hilbert Schemes

Let $G = \mathrm{Spec} A$ be a finite group scheme over a field K, assume that its Hopf algebra A is cosemisimple (that is, A is sum of its simple subcoalgebras, see [20, Ch. XIV] and Sect. 3.1 below).

For us the G-Hilbert scheme $\mathrm{G\text{-}Hilb}_K X$ of a G-scheme X over K will be by definition the moduli space of G-clusters, i.e. parametrising G-stable finite closed subschemes $Z \subseteq X_L$, L an extension field of K, with $H^0(Z,\mathscr{O}_Z)$ isomorphic to the regular representation of G over L. We recall its construction (a variation of the Quot scheme construction of [9]), for a detailed discussion including the generalisation to finite group schemes with cosemisimple Hopf algebra over arbitrary base fields see [1].

Let X be a G-scheme algebraic over K, assume that a geometric quotient $\pi: X \to X/G$, π affine, exists. Then the G-Hilbert functor $\underline{\mathrm{G\text{-}Hilb}}_K X$: $(K\text{-}schemes)^\circ \to (sets)$, given by

$$\underline{\mathrm{G\text{-}Hilb}}_K X(T) := \left\{ \begin{array}{l} \textit{Quotient } G\textit{-sheaves } [0 \to \mathscr{I} \to \mathscr{O}_{X_T} \to \mathscr{O}_Z \to 0] \textit{ on } X_T, \\ Z \textit{ finite flat over } T,\ \textit{for } t \in T\textit{: } H^0(Z_t,\mathscr{O}_{Z_t}) \textit{ isomorphic} \\ \textit{to the regular representation} \end{array} \right\}$$

for K-schemes T, is represented by an algebraic K-scheme G-Hilb$_K X$. Here we write $[0 \to \mathscr{I} \to \mathscr{O}_{X_T} \to \mathscr{O}_Z \to 0]$ for an exact sequence $0 \to \mathscr{I} \to \mathscr{O}_{X_T} \to \mathscr{O}_Z \to 0$ of quasicoherent G-sheaves on X_T with $\mathscr{I}, \mathscr{O}_Z$ specified up to isomorphism, that is either a quasicoherent G-subsheaf $\mathscr{I} \subseteq \mathscr{O}_{X_T}$ or an equivalence class $[\mathscr{O}_{X_T} \to \mathscr{O}_Z]$ of surjective equivariant homomorphisms of quasicoherent G-sheaves with two of them equivalent if their kernels coincide.

There is the natural morphism $\tau\colon \text{G-Hilb}_K X \to X/G$, which is projective and as a map of points takes G-clusters to the corresponding orbits.

In this paper we are interested in the case $G \subset \mathrm{SL}(2, K)$ operating on $X = \mathbb{A}^2_K$ over fields K of characteristic 0.

Proposition 1. *The G-Hilbert scheme* G-Hilb$_K \mathbb{A}^2_K$ *is irreducible and nonsingular. The morphism* $\tau\colon \text{G-Hilb}_K \mathbb{A}^2_K \to \mathbb{A}^2_K/G$ *is birational and the minimal resolution of* $\mathbb{A}^2_K/G$.

Proof. This is known for algebraically closed fields of characteristic 0 [6, 10, 11]. From this the statements about irreducibility and nonsingularity for not necessarily algebraically closed K follow, use that for C the algebraic closure $(\text{G-Hilb}_K \mathbb{A}^2_K)_C \cong G_C\text{-Hilb}_C \mathbb{A}^2_C$ (see [1]). The morphism $\tau\colon \text{G-Hilb}_K \mathbb{A}^2_K \to \mathbb{A}^2_K/G$ is known to be birational. The base extension $(\text{G-Hilb}_K \mathbb{A}^2_K)_C \to (\mathbb{A}^2_K/G)_C$ identifies with the natural morphism $G_C\text{-Hilb}_C \mathbb{A}^2_C \to \mathbb{A}^2_C/G_C$ (follows directly from the functorial definition of τ, see e.g. [1]). So the statement about minimality as well follows from the same statement for algebraically closed fields. □

3 Galois Operation and Irreducibility

3.1 (Co)semisimple (Co)algebras and Galois Extensions

Let K be a field and $K \to L$ a Galois extension, $\Gamma := \mathrm{Aut}_K(L)$. As reference for simple and semisimple algebras we use [3, Algébre, Ch. VIII], for coalgebras and comodules [20]. Note that for a K-vector space V (maybe with some additional structure) Γ operates on the base extension $V_L = V \otimes_K L$ via the second factor.

Proposition 2. *Let F be a simple K-algebra. Assume that F_L is semisimple, let $F_L = \bigoplus_{i=1}^r F_{L,i}$ be its decomposition into simple components. Then Γ permutes the simple summands $F_{L,i}$ and the operation on the set $\{F_{L,1}, \dots, F_{L,r}\}$ is transitive.*

Proof. The $F_{L,i}$ are the minimal two-sided ideals of F_L. Since any $\gamma \in \Gamma$ is an automorphism of F_L as a K-algebra or ring, the $F_{L,i}$ are permuted by Γ.

Let $U = \sum_{\gamma \in \Gamma} \gamma F_{L,1}$ and V the sum over the remaining $F_{L,i}$. Then $F_L = U \oplus V$, U and V are Γ-stable and thus $U = U'_L$, $V = V'_L$ for K-subspaces $U', V' \subseteq F$ by [2, Algebra II, Ch. V, § 10.4], since $K \to L$ is a Galois extension. It follows that $F = U' \oplus V'$ with U', V' two-sided ideals of F. Since F is simple, $V' = 0$, $U = F_L$ and the operation is transitive. □

A coalgebra $C \neq 0$ is called simple, if it has no subcoalgebras except $\{0\}$ and C. A coalgebra is called cosemisimple, if it is the sum of its simple subcoalgebras, in which case this sum is direct. For cosemisimple C the simple subcoalgebras are the isotypic components of C as a C-comodule (left or right), so they correspond to the isomorphism classes of simple representations of G over K.

Proposition 3. *Let C be a finite dimensional coalgebra over K. Then C is cosemisimple if and only if C_L is cosemisimple.*

Proof. This is equivalent to the dual statement for finite dimensional semisimple K-algebras [3, Algébre, Ch. VIII, § 7.6, Thm. 3, Cor. 4]. □

For simple coalgebras there is a result similar to Proposition 2 and proven analogously, note that simple coalgebras are finite dimensional.

Proposition 4. *Let C be a simple coalgebra over K. Then C_L is cosemisimple, and if $C_L = \bigoplus_i C_{L,i}$ is its decomposition into simple components, then Γ transitively permutes the simple summands $C_{L,i}$.*

Corollary 1. *Let C be a cosemisimple coalgebra over K. Then C_L is cosemisimple, and if $C = \bigoplus_j C_j$ resp. $C_L = \bigoplus_i C_{L,i}$ are the decompositions into simple subcoalgebras, then:*

(i) *The decomposition $C_L \cong \bigoplus_i C_{L,i}$ is a refinement of the decomposition $C_L \cong \bigoplus_j (C_j)_L$.*
(ii) *Γ transitively permutes the summands $C_{L,i}$ of $(C_j)_L$ for any j.*
Therefore $(C_j)_L = \sum_{\gamma \in \Gamma} \gamma C_{L,i}$, if $C_{L,i}$ is a summand of $(C_j)_L$.

This applies to the situation considered in this paper. Assume that the field K is of characteristic 0 and let $G = \operatorname{Spec} A$ be a finite group scheme over K, $|G| := \dim_K A$. Define KG to be the K-vector space $A^\vee = \operatorname{Hom}_K(A, K)$ with algebra structure dual to the coalgebra structure of A. In this situation the algebra A is always reduced and for a suitable algebraic extension field L of K the group scheme G_L is discrete. Then $G(L)$ is a finite group of order $|G|$ and the algebra $LG = (KG)_L$ is isomorphic to the group algebra of the group $G(L)$ over L. By semisimplicity of group algebras for finite groups over fields of characteristic 0 and Proposition 3 one obtains:

Proposition 5. *Let $G = \operatorname{Spec} A$ be a finite group scheme over a field K of characteristic 0. Then the Hopf algebra A is cosemisimple and so are its base extensions A_L with respect to field extensions $K \to L$.*

3.2 Irreducible Components of Schemes and Galois Extensions

Let X be a K-scheme. For an extension field L of K the group $\Gamma = \operatorname{Aut}_K(L)$ operates on $X_L = X \times_K \operatorname{Spec} L$ by automorphisms of K-schemes via the second factor. For simplicity we denote the morphisms $\mathrm{Spec}L \to \mathrm{Spec}L$, $X_L \to X_L$ coming from $\gamma\colon L \to L$ by γ as well.

A point of X may decompose over L, this way a point $x \in X$ corresponds to a set of points of X_L, the preimage of x with respect to the projection $X_L \to X$. In particular this applies to closed points and to irreducible components. These sets are known to be exactly the Γ-orbits.

Proposition 6. *Let X be an algebraic K-scheme and $K \to L$ be a Galois extension, $\Gamma := \mathrm{Aut}_K(L)$. Then points of X correspond to Γ-orbits of points of X_L, the Γ-orbits are finite.*

Proof. Taking fibers, the proposition reduces to the following statement:

Let F be the quotient field of a commutative integral K-algebra of finite type. Then $F_L = F \otimes_K L$ has only finitely many prime ideals and they are Γ-conjugate.

Proof. F_L is integral over F because this property is stable under base extension [5, Commutative Algebra, Ch. V, § 1.1, Prop. 5]. It is clear that every prime ideal of F_L lies above the prime ideal (0) of F. There are no inclusions between the prime ideals of F_L [5, Commutative Algebra, Ch. V, § 2.1, Proposition 1, Corollary 1]. Since every prime ideal of F_L is a maximal ideal and F_L is noetherian (a localisation of an L-algebra of finite type), F_L is artinian, it has only finitely many prime ideals $Q_1, \ldots, Q_r$.

F_L has trivial radical [3, Algébre, Ch. VIII, § 7.3, Thm. 1, also § 7.5 and § 7.6, Cor. 3]. Being an artinian ring without radical, i.e. semisimple [3, Algébre, Ch. VIII, § 6.4, Thm. 4, Cor. 2 and Prop. 9], F_L decomposes as a L-algebra into a direct sum

$$F_L \cong \bigoplus_{i=1}^{r} F_{L,i}$$

of fields $F_{L,i} \cong F_L/Q_i$ (this can easily be seen directly, however, it is part of the general theory of semisimple algebras developed in [3, Algébre, Ch. VIII] that contains the representation theory of finite groups schemes with cosemisimple Hopf algebra as another special case).

Γ operates on F_L, it permutes the Q_i and the simple components $F_{L,i}$ of F_L transitively by Proposition 2. □

3.3 Galois Operation on G-Hilbert Schemes

Let Y be a K-scheme, L an extension field of K and $\Gamma = \mathrm{Aut}_K(L)$.

For an L-scheme $f\colon T \to \mathrm{Spec}L$ and $\gamma \in \Gamma$ define the L-scheme $\gamma_* T$ to be the scheme T with structure morphism $\gamma \circ f$. For a morphism $\alpha\colon T' \to T$ of L-schemes let $\gamma_*\alpha$ be the same morphism α considered as an L-morphism $\gamma_* T' \to \gamma_* T$.

For a morphism $\alpha\colon Y_L \to Y'_L$ of L-schemes and $\gamma \in \Gamma$ define the conjugate morphism α^γ by $\alpha^\gamma := \gamma \circ (\gamma_*\alpha) \circ \gamma^{-1}$, which again is a morphism of L-schemes. Here $\gamma\colon \gamma_* Y_L \to Y_L$ is a morphism over L.

Let T be an L-scheme defined over K, that is $T = T'_L$ for some K-scheme T'. The group Γ operates on the set $Y_L(T)$ of morphisms $T \to Y_L$ over L by

$$\begin{aligned} \gamma : Y_L(T) &\to Y_L(T) \\ \alpha &\mapsto \alpha^\gamma = \gamma \circ (\gamma_* \alpha) \circ \gamma^{-1} \end{aligned}$$

Consider the case of G-Hilbert schemes. Let G be a finite group scheme over K, X be a G-scheme over K and assume that the G-Hilbert functor is represented by a K-scheme G-Hilb$_K X$. There is the canonical isomorphism of L-schemes $(\text{G-Hilb}_K X)_L \cong G_L\text{-Hilb}_L X_L$ (see [1]), obtained by identifying $X \times_K T = X_L \times_L T$ for L-schemes T.

Proposition 7. *Let T be an L-scheme defined over K. Then, for a morphism $\alpha: T \to G_L\text{-Hilb}_L X_L$ of L-schemes corresponding to a quotient $[0 \to \mathscr{I} \to \mathscr{O}_{X_T} \to \mathscr{O}_Z \to 0]$ and for $\gamma \in \Gamma$, the γ-conjugate morphism α^γ corresponds to the quotient $[0 \to \gamma_* \mathscr{I} \to \mathscr{O}_{X_T} \to \mathscr{O}_{\gamma Z} \to 0]$.*

Proof. For a morphism of L-schemes $\alpha: T \to G_L\text{-Hilb}_L X_L \cong (\text{G-Hilb}_K X) \times_K \operatorname{Spec} L$ consider the commutative diagram of L-morphisms

$$\begin{array}{ccc} \gamma_* T & \xrightarrow{\gamma_* \alpha} & (\text{G-Hilb}_K X) \times_K (\gamma_* \operatorname{Spec} L) \\ {\scriptstyle \gamma}\downarrow & \overset{\gamma \circ (\gamma_* \alpha)}{\searrow} & \downarrow {\scriptstyle id \times \gamma} \\ T & \xrightarrow[\alpha^\gamma]{} & (\text{G-Hilb}_K X) \times_K \operatorname{Spec} L \end{array}$$

The morphism α is given by a quotient $[0 \to \mathscr{I} \to \mathscr{O}_{X_T} \to \mathscr{O}_Z \to 0]$ on $X_T = X_L \times_L T$. Under the identification $G_L\text{-Hilb}_L X_L = (\text{G-Hilb}_K X_K)_L$ the T-valued point α corresponds to a morphism $T \to \text{G-Hilb}_K X$ of K-schemes, that is a quotient $[0 \to \mathscr{I} \to \mathscr{O}_{X \times_K T} \to \mathscr{O}_Z \to 0]$ on $X \times_K T$, and the structure morphism $f: T \to \operatorname{Spec} L$. We have the correspondences

$$\begin{aligned} \alpha &\longleftrightarrow \begin{cases} [0 \to \mathscr{I} \to \mathscr{O}_{X \times_K T} \to \mathscr{O}_Z \to 0] \\ f: T \to \operatorname{Spec} L \end{cases} \\ \gamma \circ (\gamma_* \alpha) &\longleftrightarrow \begin{cases} [0 \to \mathscr{I} \to \mathscr{O}_{X \times_K \gamma_* T} \to \mathscr{O}_Z \to 0] \\ \gamma \circ (\gamma_* f): \gamma_* T \to \operatorname{Spec} L \end{cases} \\ \alpha^\gamma = \gamma \circ (\gamma_* \alpha) \circ \gamma^{-1} &\longleftrightarrow \begin{cases} [0 \to \gamma^{-1*} \mathscr{I} \to \mathscr{O}_{X \times_K T} \to \gamma^{-1*} \mathscr{O}_Z \to 0] \\ f = \gamma \circ (\gamma_* f) \circ \gamma^{-1}: T \to \operatorname{Spec} L \end{cases} \\ &\longleftrightarrow \begin{cases} [0 \to \gamma_* \mathscr{I} \to \mathscr{O}_{X \times_K T} \to \mathscr{O}_{\gamma Z} \to 0] \\ f = \gamma \circ (\gamma_* f) \circ \gamma^{-1}: T \to \operatorname{Spec} L \end{cases} \end{aligned}$$

Under the identification $(G\text{-Hilb}_K X)_L = G_L\text{-Hilb}_L X_L$ the last morphism corresponds to the quotient $[0 \to \gamma_* \mathscr{I} \to \mathscr{O}_{X_T} \to \mathscr{O}_{\gamma Z} \to 0]$ on $X_T = X_L \times_L T$. □

In particular, in the case $X = \mathbb{A}^2_K$ the γ-conjugate of an L-valued point given by an ideal $I \subseteq L[x_1, x_2]$ or a G_L-cluster $Z \subset \mathbb{A}^2_L$ is given by the γ-conjugate ideal $\gamma^{-1} I \subset L[x_1, x_2]$ or the γ-conjugate G_L-cluster $\gamma Z \subset \mathbb{A}^2_L$.

Every point x of the L-scheme $G_L\text{-Hilb}_L \mathbb{A}^2_L$ such that $\kappa(x) = L$ corresponds to a unique L-valued point $\alpha\colon \operatorname{Spec} L \to G_L\text{-Hilb}_L \mathbb{A}^2_L$. The γ-conjugate point γx corresponds to the γ-conjugate L-valued point $\alpha^\gamma\colon \operatorname{Spec} L \to G_L\text{-Hilb}_L \mathbb{A}^2_L$.

Corollary 2. *Let x be a closed point of $G_L\text{-Hilb}_L \mathbb{A}^2_L$ such that $\kappa(x) = L, \alpha\colon \operatorname{Spec} L \to G_L\text{-Hilb}_L \mathbb{A}^2_L$ the corresponding L-valued point given by an ideal $I \subset L[x_1, x_2]$. Then for $\gamma \in \Gamma$ the conjugate point γx corresponds to the γ-conjugate L-valued point $\alpha^\gamma\colon \operatorname{Spec} L \to G_L\text{-Hilb}_L \mathbb{A}^2_L$, which is given by the ideal $\gamma^{-1} I \subset L[x_1, x_2]$.*

3.4 Conjugate G-Sheaves

Let $G = \operatorname{Spec} A$ be a group scheme over a field K, X be a G-scheme over K, let $K \to L$ be a field extension and $\Gamma = \operatorname{Aut}_K(L)$. Again, Γ operates on X_L by automorphisms $\gamma\colon X_L \to X_L$ over K, these are equivariant with respect to the G-scheme structure of X_L defined in Remark 1(8).

Proposition–Definition 1 *Let $\mathscr{F}$ be a G_L-sheaf on X_L. For $\gamma \in \Gamma$ the $\mathscr{O}_{X_L}$-module $\gamma_* \mathscr{F}$ has a natural G_L-sheaf structure given by*

$$
\begin{array}{ccc}
\gamma_* s^*_{X_L} \mathscr{F} & \xrightarrow{\gamma_* \lambda^{\mathscr{F}}} & \gamma_* p^*_{X_L} \mathscr{F} \\
\uparrow \wr & & \uparrow \wr \\
s^*_{X_L} \gamma_* \mathscr{F} & \xrightarrow{\lambda^{\gamma_* \mathscr{F}}} & p^*_{X_L} \gamma_* \mathscr{F}
\end{array}
\qquad (1)
$$

This G_L-sheaf $\gamma_* \mathscr{F}$ is called the γ-conjugate G_L-sheaf of $\mathscr{F}$. For a morphism of G_L-sheaves $\varphi\colon \mathscr{F} \to \mathscr{G}$ the morphism $\gamma_* \varphi\colon \gamma_* \mathscr{F} \to \gamma_* \mathscr{G}$ is a morphism of G_L-sheaves between the sheaves $\gamma_* \mathscr{F}$ and $\gamma_* \mathscr{G}$ with γ-conjugate G_L-sheaf structure.

Remark 2. This way functors γ_* are defined, similarly one may define functors γ^*, then γ_* and $(\gamma^{-1})^*$ are isomorphic. In the case of trivial operation they preserve trivial G-sheaf structures.

The functors γ_* commute with functors f_L^*, f_{L*} for equivariant morphisms f and with bifunctors like $\mathscr{H}om$ and $\otimes$:

Lemma 1. *There are the following natural isomorphisms of G_L-sheaves:*

(i) *For G_L-sheaves $\mathscr{F}, \mathscr{G}$ on X_L : $\gamma_*(\mathscr{F} \otimes_{\mathscr{O}_{X_L}} \mathscr{G}) \cong \gamma_*\mathscr{F} \otimes_{\mathscr{O}_{X_L}} \gamma_*\mathscr{G}$.*

(ii) *Let $f: Y \to X$ be an equivariant morphism of G-schemes over K and $\mathscr{F}$ a G_L-sheaf on X_L. Then $\gamma_* f_L^* \mathscr{F} \cong f_L^* \gamma_* \mathscr{F}$.*

(iii) *For quasicoherent G_L-sheaves $\mathscr{F}, \mathscr{G}$ on X_L with $\mathscr{F}$ finitely presented:$\gamma_* \mathscr{H} om_{\mathscr{O}_{X_L}}(\mathscr{F}, \mathscr{G}) \cong \mathscr{H}om_{\mathscr{O}_{X_L}}(\gamma_*\mathscr{F}, \gamma_*\mathscr{G})$. If the G-operation on X is trivial, it follows that $\gamma_*(\mathscr{H}om^{G_L}_{\mathscr{O}_{X_L}}(\mathscr{F}, \mathscr{G})) \cong \mathscr{H}om^{G_L}_{\mathscr{O}_{X_L}}(\gamma_*\mathscr{F}, \gamma_*\mathscr{G})$.*

Remark 3. If $\mathscr{F} \cong \mathscr{F}'_L$ for some G-sheaf $\mathscr{F}'$ on X, then there are maps (not L-linear) $\gamma: \mathscr{F} \to \mathscr{F}$ resp. isomorphisms of G_L-sheaves $\gamma: \mathscr{F} \to \gamma_*\mathscr{F}$ on X_L. For a subsheaf $\mathscr{G} \subseteq \mathscr{F}$ the above isomorphisms of G_L-sheaves restrict to isomorphisms of G_L-sheaves $\gamma: \gamma^{-1}\mathscr{G} \to \gamma_*\mathscr{G}$.

3.5 Conjugate Comodules and Representations

Let $G = \operatorname{Spec} A$ be an affine group scheme over a field K, X be a G-scheme over K, let $K \to L$ be a field extension and $\Gamma = \operatorname{Aut}_K(L)$.

Remark 4. For $\gamma \in \Gamma$ there are maps $\gamma: A_L \to A_L$. Taking the canonically defined conjugate Hopf algebra structure on the target, these maps become isomorphisms $\gamma: A_L \to \gamma_* A_L$ of Hopf algebras over L. They correspond to isomorphisms $\gamma: \gamma_* G_L \to G_L$ of group schemes over L.

Proposition 8. *Let $\mathscr{F}$ be a G_L-sheaf on X_L, X with trivial G-operation, the G_L-sheaf structure equivalent to an A_L-comodule structure $\varrho^{\mathscr{F}}: \mathscr{F} \to A_L \otimes_L \mathscr{F}$. Then for $\gamma \in \Gamma$ the G_L-sheaf structure of the γ-conjugate G_L-sheaf $\gamma_*\mathscr{F}$ is equivalent to the comodule structure $\varrho^{\gamma_*\mathscr{F}}: \gamma_*\mathscr{F} \to A_L \otimes_L \gamma_*\mathscr{F}$ determined by commutativity of the diagram*

$$\begin{array}{ccc} \gamma_*\mathscr{F} & \xrightarrow{\gamma_*\rho^{\mathscr{F}}} & \gamma_*A_L \otimes_L \gamma_*\mathscr{F} \\ {\scriptstyle id}\uparrow & & \uparrow{\scriptstyle \gamma\otimes id} \\ \gamma_*\mathscr{F} & \xrightarrow{\rho^{\gamma_*\mathscr{F}}} & A_L \otimes_L \gamma_*\mathscr{F} \end{array}$$

Proof. Apply the construction mentioned in Remark 1(2) to diagram (1). □

In the special case of representations the definition of conjugate G-sheaves leads to the notion of a conjugate representation: Instead of a sheaf $\gamma_*\mathscr{F}$ one has an L-vector space $\gamma_* V$, the vector space structure given by $(l, v) \mapsto \gamma(l)v$ using the original structure. The choice of a K-structure $V = V'_L$ gives an isomorphism $\gamma: V \to \gamma_* V$ of L-vector spaces and leads to the diagram

$$\begin{array}{ccc}
\gamma_*V & \xrightarrow{\gamma_*\rho^V} & \gamma_*A_L \otimes_L \gamma_*V \\
{\scriptstyle id}\uparrow & & \uparrow{\scriptstyle \gamma\otimes id} \\
\gamma_*V & \xrightarrow{\rho^{\gamma_*V}} & A_L \otimes_L \gamma_*V \\
{\scriptstyle \gamma}\uparrow & & \uparrow{\scriptstyle id\otimes\gamma} \\
V & \xrightarrow{(\rho^V)^\gamma} & A_L \otimes_L V
\end{array}$$

for definition of the γ-conjugate A_L-comodule structure $(\varrho^V)^\gamma$ on V—this definition is made, such that $\gamma\colon (V, (\varrho^V)^\gamma) \to (\gamma_* V, \varrho^{\gamma_* V})$ is an isomorphism of A_L-comodules. We write V^γ for V with the conjugate A_L-comodule structure.

Remark 5. Let V' be an A-comodule over K and $V = V'_L$. Then as a special case of Remark 3 there are maps $\gamma\colon V \to V$ resp. isomorphisms of A_L-comodules $\gamma\colon V \to \gamma_* V$. For any A_L-subcomodule $U \subseteq V$ these restrict to isomorphisms of A_L-comodules $\gamma^{-1}U \xrightarrow{\sim} \gamma_* U \cong U^\gamma$.

3.6 Decomposition into Isotypic Components and Galois Extensions

Let $G = \mathrm{Spec} A$ be an affine group scheme over a field K, let $K \to L$ be a Galois extension, $\Gamma = \mathrm{Aut}_K(L)$. Assume that A, A_L are cosemisimple.

Recall the relations between the Galois operation on A_L given by maps $\gamma\colon A_L \to A_L$ resp. isomorphisms $\gamma\colon A_L \to \gamma_* A_L$ of Hopf algebras or of A_L-comodules (see Remarks 4 or 5) and the decompositions $A = \bigoplus_{j\in J} A_j$ and $A_L = \bigoplus_{i\in I} A_{L,i}$ into simple subcoalgebras described in Corollary 1. We relate this to conjugation of representations. The subcoalgebras $A_{L,i}$ are the isotypic components of A_L as a left-(or right-)comodule, let V_i be the isomorphism class of simple A_L-comodules corresponding to $A_{L,i}$. Define an operation of Γ on the index set I by $V_{\gamma(i)} = V_i^\gamma$. Using Remark 5 one obtains:

Lemma 2. $\gamma^{-1} A_{L,i} = A_{L,\gamma(i)}$.

The decomposition of A into simple subcoalgebras $A = \bigoplus_j A_j$ gives decompositions of representations and more generally of G-sheaves on G-schemes with trivial G-operation into isotypic components corresponding to the A_j (see Remark 1(3)). After base extension one has decompositions of G_L-sheaves, we compare it with the decompositions coming from the decomposition of A_L into simple subcoalgebras.

Proposition 9. *Let X be a G-scheme with trivial operation, $\mathscr{F}$ a G-sheaf on X and let*

$$\mathscr{F} = \bigoplus_j \mathscr{F}_j, \qquad \mathscr{F}_L = \bigoplus_i \mathscr{F}_{L,i}$$

be the decompositions into isotypic components as a G-sheaf resp. G_L-sheaf. Then:

(i) $\mathscr{F}_L = \bigoplus_i \mathscr{F}_{L,i}$ *is a refinement of* $\mathscr{F}_L = \bigoplus_j (\mathscr{F}_j)_L$.
(ii) *The operation of Γ on $\mathscr{F}_L$ (see Remark 3) permutes the isotypic components $\mathscr{F}_{L,i}$ of $\mathscr{F}_L$. It is $\gamma^{-1}\mathscr{F}_{L,i} = \mathscr{F}_{L,\gamma(i)}$, if $V_{\gamma(i)} = V_i^\gamma$.*
(iii) $(\mathscr{F}_j)_L = \sum_{\gamma\in\Gamma} \gamma\mathscr{F}_{L,i}$, *if $\mathscr{F}_{L,i}$ is a summand of $(\mathscr{F}_j)_L$.*

Proof (Sketch of proof). Combine Remark 3, Proposition 8 and Lemma 2 with Corollary 1. □

Corollary 3. *Γ operates by $V_i \mapsto V_i^\gamma$ on the set $\{V_i \,|\, i \in I\}$ of isomorphism classes of irreducible representations of G_L. The subsets of $\{V_i \mid i \in I\}$, which occur by decomposing irreducible representations of G over K as representations over L, are exactly the Γ-orbits.*

For similar results in the representation theory of finite groups see e.g. [7, Vol. I, § 7B].

4 The Finite Subgroup Schemes of SL(2, *K*): Representations and Graphs

In this section K denotes a field of characteristic 0.

4.1 The Finite Subgroups of SL(2, *C*)

By the well known classification any finite subgroup $G \subset \mathrm{SL}(2, C)$, C an algebraically closed field of characteristic 0, is isomorphic to one of the following groups (presentations and character tables are listed in Sect. 6).

$\mathbb{Z}/n\mathbb{Z}$	(cyclic group of order n), $n \geq 1$
BD_n	(binary dihedral group of order $4n$), $n \geq 2$
BT	(binary tetrahedral group)
BO	(binary octahedral group)
BI	(binary icosahedral group).

4.2 Representation Graphs

In the following definition we will introduce the (extended) representation graph as an in general directed graph. A loop is defined to be an edge emanating from and terminating at the same vertex. In addition we will attach a natural number called multiplicity to any vertex, and for homomorphisms of graphs in addition we will require, that for any vertex of the target its multiplicity is the sum of the multiplicities of its preimages.

Definition 1. The extended representation graph $\mathrm{Graph}(G, V)$ associated to a finite subgroup scheme G of $\mathrm{GL}(n, K)$, V the given n-dimensional representation, is defined as the following directed graph:

- Vertices. A vertex of multiplicity n for each irreducible representation of G over K which decomposes over the algebraic closure of K into n irreducible representations.
- Edges. Vertices V_i and V_j are connected by $\dim_K \mathrm{Hom}_K^G(V_i, V \otimes_K V_j)$ directed edges from V_i to V_j. In particular any vertex V_i has $\dim_K \mathrm{Hom}_K^G(V_i, V \otimes_K V_i)$ directed loops.

Define the representation graph to be the graph, which arises by leaving out the trivial representation and all edges emanating from or terminating at the trivial representation.

We say that a graph is undirected, if between any two different vertices the numbers of directed edges of both directions coincide and for any vertex the number of directed loops is even.

Then one can form a graph having only undirected edges by defining *(number of undirected edges between V_i and V_j)* $:=$ *(number of directed edges from V_i to V_j)* $=$ *(number of directed edges from V_j to V_i)* for different vertices V_i, V_j and *(number of undirected loops of V_i)* $:= \frac{1}{2}$*(number of directed loops of V_i)* for any vertex V_i.

Remark 6. 1. For $G \subset \mathrm{SL}(2, K)$ the (extended) representation graph is undirected. There is the isomorphism $\mathrm{Hom}_K^G(V_i \otimes_K V, V_j) \cong \mathrm{Hom}_K^G(V_i, V \otimes_K V_j)$, which follows from the isomorphism $\mathrm{Hom}_K^G(V_i \otimes_K V, V_j) \cong \mathrm{Hom}_K^G(V_i, V^\vee \otimes_K V_j)$ and the fact that the 2-dimensional representation V given by inclusion $G \to \mathrm{SL}(2, K)$ is self-dual. Further, that the number of directed loops of any vertex is even, follows from the fact that over the algebraic closure C one has $\dim_C \mathrm{Hom}_C^G(U_i, V_C \otimes_C U_i) = 0$ for irreducible U_i over C.
2. There is a definition of (extended) representation graph with another description of the edges: vertices V_i and V_j are connected by a_{ij} edges from V_i to V_j, where $V \otimes_K V_j = a_{ij} V_i \oplus$ *other summands*. The two definitions coincide over algebraically closed fields, always one has $a_{ij} \leq \dim_K \mathrm{Hom}_K^G(V_i, V \otimes_K V_j)$, inequality comes from the presence of nontrivial automorphisms.

Definition 2. For a finite subgroup scheme $G \subset \mathrm{SL}(2, K)$, V the given 2-dimen-sional representation, define a $\mathbb{Z}$-bilinear form $\langle \cdot, \cdot \rangle$ on the representation ring of G by

$$\langle V_i, V_j \rangle := \dim_K \mathrm{Hom}_K^G(V_i, V \otimes_K V_j) - 2 \dim_K \mathrm{Hom}_K^G(V_i, V_j)$$

Remark 7. The form $\langle \cdot, \cdot \rangle$ determines and is determined by the extended representation graph (the second equation follows from the fact, that $\dim_K \mathrm{Hom}_K^G(V_i, V_i) =$ *multiplicity of* V_i):

$$\langle V_i, V_j \rangle = \langle V_j, V_i \rangle = \text{number of undirected edges between } V_i \text{ and } V_j, \quad \text{if } V_i \not\cong V_j$$
$$\tfrac{1}{2} \langle V_i, V_i \rangle = \text{number of undirected loops of } V_i - \text{multiplicity of } V_i$$

4.3 Representation Graphs and Field Extensions

Let $K \to L$ be a Galois extension, $\Gamma = \mathrm{Aut}_K(L)$ and let G be a finite subgroup scheme of $\mathrm{SL}(2, K)$.

An irreducible representation W of G over K decomposes as a representation of G_L over L into isotypic components $W = \bigoplus_i U_i$ which are Γ-conjugate by Proposition 9. Every U_i decomposes into irreducible components $U_i = V_i^{\oplus m}$ (the same m for all i because of Γ-conjugacy). In the following we will write $m(W, L/K)$ for this number. It is related to the Schur index in the representation theory of finite groups (see e.g. [7, Vol. II, § 74]).

Proposition 10. *For finite subgroup schemes G of $\mathrm{SL}(2, K)$ it is $m(W_j, L/K) = 1$ for every irreducible representation W_j of G. It follows that W_j decomposes over L into a direct sum $(W_j)_L \cong \bigoplus_i V_i$ of γ-conjugate irreducible representations V_i of G_L nonisomorphic to each other.*

Proof. We may assume L algebraically closed. Further we may assume that G is not cyclic. The natural 2-dimensional representation W given by inclusion $G \subset \mathrm{SL}(2, K)$ does satisfy $m(W, L/K) = 1$, because it is irreducible over L. Following the discussion below without using this proposition one obtains the graphs in Sect. 4.4 without multiplicities of vertices and edges but one knows which vertices over the algebraic closure may form a vertex over K and which vertices are connected. Argue that if an irreducible representation W_i satisfies $m(W_i, L/K) = 1$, then any irreducible W_j connected to W_i in the representation graph has to satisfy this property as well. □

There is a morphism of graphs $\mathrm{Graph}(G_L, W_L) \to \mathrm{Graph}(G, W)$ (resp. of the nonextended graphs, the following applies to them as well). For W_j an irreducible representation of G the base extension $(W_j)_L$ is a sum $(W_j)_L = \bigoplus_i V_i$ of irreducible representations of G_L nonisomorphic to each other by Proposition 10. The morphism $\mathrm{Graph}(G_L, W_L) \to \mathrm{Graph}(G, W)$ maps components of $(W_j)_L$ to W_j, thereby their multiplicities are added. Further, for irreducible representations $W_j, W_{j'}$ of G there is a bijection between the set of edges between W_j and $W_{j'}$ and the union of the sets of edges between the irreducible components of $(W_j)_L$ and $(W_{j'})_L$, again using Proposition 10 $(W_j)_L$ and $(W_{j'})_L$ are sums $(W_j)_L = \bigoplus_i V_i$, $(W_{j'})_L = \bigoplus_{i'} V_{i'}$ of irreducible representations of G_L nonisomorphic to each other and one has

$$\begin{aligned}\dim_K \operatorname{Hom}_K^G(W_j \otimes_K W, W_{j'}) &= \dim_L(\operatorname{Hom}_K^G(W_j \otimes_K W, W_{j'}) \otimes_K L) \\ &= \dim_L \operatorname{Hom}_L^G((W_j)_L \otimes_L W_L, (W_{j'})_L) \\ &= \dim_L \operatorname{Hom}_L^G(\textstyle\bigoplus_i V_i \otimes_L W_L, \bigoplus_{i'} V_{i'}) \\ &= \textstyle\sum_{i,i'} \dim_L \operatorname{Hom}_L^G(V_i \otimes_L W_L, V_{i'})\end{aligned}$$

The Galois group Γ operates on $\mathrm{Graph}(G_L, W_L)$ by graph automorphisms. Irreducible representations are mapped to conjugate representations and equivariant homomorphisms to the conjugate homomorphisms. The vertices of $\mathrm{Graph}(G, W)$ correspond to Γ-orbits of vertices of $\mathrm{Graph}(G_L, W_L)$ by Corollary 3.

Proposition 11. *The (extended) representation graph of G arises by identifying the elements of Γ-orbits of vertices of the (extended) representation graph of G_L, adding multiplicities. The edges between vertices W_j and $W_{j'}$ are in bijection with the edges between the isomorphism classes of irreducible components of $(W_j)_L$ and $(W_{j'})_L$.*

4.4 The Representation Graphs of the Finite Subgroup Schemes of SL(2, K)

As extended representation graph of a finite subgroup scheme of $\mathrm{SL}(2, K)$ with respect to the natural 2-dimensional representation the following graphs can occur. We list the extended representation graphs $\mathrm{Graph}(G, V)$ of the finite subgroups of $\mathrm{SL}(2, C)$ for C algebraically closed, their groups of automorphisms leaving the trivial representation fixed and the possible extended representation graphs for finite subgroup schemes over non algebraically closed K, which after suitable base extension become the graph $\mathrm{Graph}(G, V)$. We use the symbol $\circ$ for the trivial representation.

Cyclic Groups

(A_{2n}), $n \geq 1$

$\mathbb{Z}/2\mathbb{Z}$

$(A_{2n})'$

(A_{2n+1}), $n \geq 1$

$\mathbb{Z}/2\mathbb{Z}$

$(A_{2n+1})'$

(A_1)

$\{id\}$

Binary dihedral groups.

$(D_n), n \geq 5$

$\mathbb{Z}/2\mathbb{Z}$

$(D_n)'$

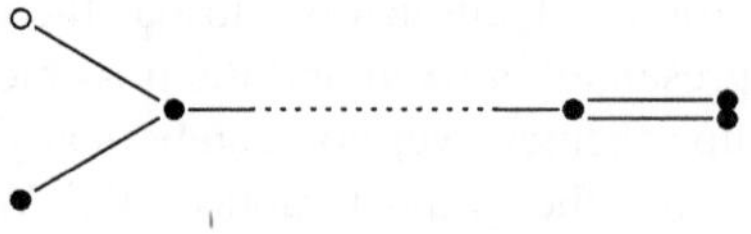

(D_4)

S_3

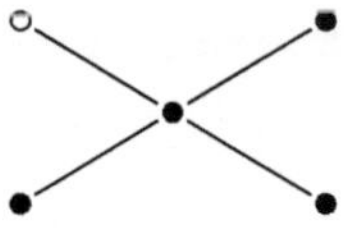

$(D_4)'$

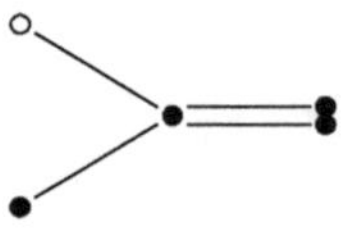

$(D_4)''$

Binary tetrahedral group.

(E_6)

$\mathbb{Z}/2\mathbb{Z}$

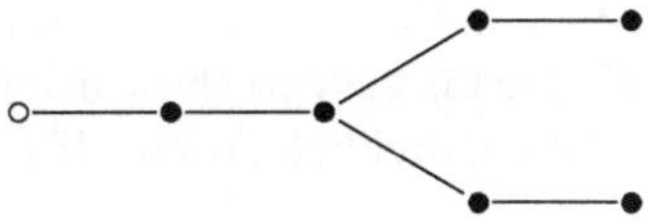

$(E_6)'$

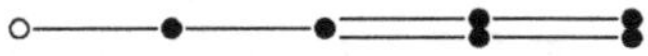

Binary octahedral group.

(E_7)

$\{id\}$

Binary icosahedral group.

(E_8)

$\{id\}$

Remark 8. Taking $\frac{2}{-\langle V,V\rangle}V$ for the isomorphism classes of irreducible representations V as simple roots one can form the Dynkin diagram with respect to the form $-\langle\cdot,\cdot\rangle$ (see e.g. [5, Groupes et algèbres de Lie]). Between (extended) representation graphs and (extended) Dynkin diagrams there is the correspondence

(A_n)	$(A_2)'$	$(A_{2n+1})'$	$(A_{2n+2})'$	(D_n)	$(D_n)'$	$(D_4)''$	(E_6)	$(E_6)'$	(E_7)	(E_8)
(A_n)	$(C_1) = (A_1)$	(C_{n+1})	(C_{n+1})	(D_n)	(B_{n-1})	(G_2)	(E_6)	(F_4)	(E_7)	(E_8)

A long time ago, the occurrence of the remaining Dynkin diagrams of types (B_n), (C_n), (F_4), (G_2) as resolution graphs had been observed in [13] with a slightly different assignment of the non extended diagrams to the resolutions of these singularities, see also [19].

4.5 *Finite Subgroups of* SL(2, *K*)

Given a field K of characteristic 0, it is a natural question, which of the finite subgroups $G \subset \mathrm{SL}(2, C)$, C the algebraic closure of K, are realisable over the subfield K as subgroups (not just as subgroup schemes), that is, there is an injective representation of the group G in $\mathrm{SL}(2, K)$.

For a finite subgroup G of $\mathrm{SL}(2, C)$ to occur as a subgroup of $\mathrm{SL}(2, K)$ it is necessary and sufficient that the given 2-dimensional representation in $\mathrm{SL}(2, C)$ is realisable over K. This is easy to show using the classification and the irreducible representations of the individual groups. If a representation of a group G over C is realisable over K, necessarily its character has values in K. For the finite subgroups of $\mathrm{SL}(2, C)$ and the natural representation given by inclusion this means:

$\mathbb{Z}/n\mathbb{Z}$: $\xi + \xi^{-1} \in K$, $\xi \in C$ a primitive n-th root of unity.
BD_n: $\xi + \xi^{-1} \in K$, $\xi \in C$ a primitive $2n$-th root of unity.
BT: no condition. BO: $\sqrt{2} \in K$. BI: $\sqrt{5} \in K$.

To formulate sufficient conditions, we introduce the following notation:

Definition 3 ([18, Part I, Chapter III, § 1]). For a field K the Hilbert symbol $((\cdot,\cdot))_K$ is the map $K^* \times K^* \to \{-1, 1\}$ defined by $((a,b))_K = 1$, if the equation $z^2 - ax^2 - by^2 = 0$ has a solution $(x, y, z) \in K^3 \setminus \{(0,0,0)\}$, and $((a,b))_K = -1$ otherwise.

Remark 9. It is $((-1,b))_K = 1$ if and only if $x^2 - by^2 = -1$ has a solution $(x, y) \in K^2$.

Theorem 2. *Let G be a finite subgroup of* $\mathrm{SL}(2, C)$ *such that the values of the character of the natural representation given by inclusion are contained in K. Then:*

(i) *If $G \cong \mathbb{Z}/n\mathbb{Z}$, then G is realisable over K.*
(ii) *If $G \cong BD_n$, let $\xi \in C$ be a primitive $2n$-th root of unity and $c := \frac{1}{2}(\xi + \xi^{-1})$. Then G is isomorphic to a subgroup of* $\mathrm{SL}(2, K)$ *if and only if $((-1, c^2-1))_K = 1$.*
(iii) *If $G \cong BT, BO$ or BI, then G is isomorphic to a subgroup of* $\mathrm{SL}(2, K)$ *if and only if $((-1,-1))_K = 1$.*

Proof. (i) For $n \geq 3$ let ξ be a primitive n-th root of unity and $c := \frac{1}{2}(\xi+\xi^{-1})$. By assumption $c \in K$. Then $\mathbb{Z}/n\mathbb{Z}$ is realisable over K, there is the representation

$$\mathbb{Z}/n\mathbb{Z} \to \mathrm{SL}(2, K), \qquad \bar{1} \mapsto \begin{pmatrix} 0 & -1 \\ 1 & 2c \end{pmatrix}$$

(ii) Let $G = BD_n = \langle \sigma, \tau \mid \tau^2 = \sigma^n = (\tau\sigma)^2 \rangle$ (then the element $\tau^2 = \sigma^n = (\tau\sigma)^2$ has order 2) and let ξ be a primitive $2n$-th root of unity. Then G is realisable as a subgroup of $\mathrm{SL}(2, K)$ if and only if the representation given by

$$\sigma \mapsto \begin{pmatrix} \xi & 0 \\ 0 & \xi^{-1} \end{pmatrix}, \qquad \tau \mapsto \begin{pmatrix} 0 & -1 \\ 1 & 0 \end{pmatrix} \tag{1}$$

is realisable over K.

The representation (1) is realisable over K if and only if there is a 2×2-*matrix* M_τ *over K having the properties*

$$\begin{aligned} &\det(M_\tau) = 1, \ \ \operatorname{ord}(M_\tau) = 4, \ \ (M_\tau M_\sigma)^2 = -\mathbb{1}, \\ &\textit{where } \ M_\sigma = \left(\begin{smallmatrix} 0 & -1 \\ 1 & 2c \end{smallmatrix}\right), \ \ c = \tfrac{1}{2}(\xi + \xi^{-1}). \end{aligned} \tag{2}$$

If the representation (1) is realisable over K, then with respect to a suitable basis it maps $\sigma \mapsto M_\sigma$ and the image of τ is a matrix satisfying the properties (2).

On the other hand, if M_τ is a matrix having these properties, then $\sigma \mapsto M_\sigma$, $\tau \mapsto M_\tau$ is a representation of G in $\mathrm{SL}(2, K)$, which is easily seen to be isomorphic to the representation (1).

There is a 2×2-*matrix* M_τ *over K having the properties (2) if and only if the equation*

$$x^2 + y^2 - 2cxy + 1 = 0 \tag{3}$$

has a solution $(x, y) \in K^2$.

A matrix $M_\tau = \left(\begin{smallmatrix} \alpha & \beta \\ \gamma & \delta \end{smallmatrix}\right)$ satisfies the conditions (2) if and only if $(\alpha, \beta, \gamma, \delta) \in K^4$ is a solution of $\alpha\delta - \beta\gamma - 1 = 0$, $\alpha + \delta = 0$, $\beta + 2c\delta - \gamma = 0$. Such an element of K^4 exists if and only if there exists a solution $(x, y) \in K^2$ of Eq. (3).

Equation (3) has a solution $(x, y) \in K^2$ *if and only if* $((-1, c^2 - 1))_K = 1$.

We write the equation $x^2 + y^2 - 2cxy + 1 = 0$ as $(x, y)\left(\begin{smallmatrix} 1 & -c \\ -c & 1 \end{smallmatrix}\right)\left(\begin{smallmatrix} x \\ y \end{smallmatrix}\right) = -1$. After diagonalisation $(x, y)\left(\begin{smallmatrix} 1 & 0 \\ 0 & 1-c^2 \end{smallmatrix}\right)\left(\begin{smallmatrix} x \\ y \end{smallmatrix}\right) = -1$ or $x^2 + (1 - c^2)y^2 + 1 = 0$. This equation has a solution $(x, y) \in K^2$ if and only if $((-1, c^2 - 1))_K = 1$.

(iii) Let $G = BT, BO$ or BI, that is $G = \langle a, b \mid a^3 = b^k = (ab)^2 \rangle$ for $k \in \{3, 4, 5\}$. Let ξ be a primitive $2k$-th root of unity and $c = \frac{1}{2}(\xi + \xi^{-1})$. As in (ii), using the subgroup $\langle b \rangle$ instead of $\langle \sigma \rangle$, we obtain:
G is isomorphic to a subgroup of $\mathrm{SL}(2, K)$ *if and only if there is a solution* $(x, y) \in K^2$ *of the equation*

$$x^2 + y^2 - 2cxy - x + 2cy + 1 = 0 \tag{4}$$

Next we show:
Equation (4) has a solution $(x, y) \in K^2$ *if and only if* $((-1, (2c)^2 - 3))_{K=1}$.

Equation (4) has a solution if and only if $(x, y, z)\begin{pmatrix} 1 & -c & -1/2 \\ -c & 1 & c \\ -1/2 & c & 1 \end{pmatrix}\left(\begin{smallmatrix} x \\ y \\ z \end{smallmatrix}\right) = 0$ has a solution $(x, y, z) \in K^3$ with $z \neq 0$. The existence of a solution with $z \neq 0$ is equivalent to the existence of a solution $(x, y, z) \in K^3 \setminus \{(0, 0, 0)\}$ (if $(x, y, 0)$ is a solution, then $(x, y, x - 2cy)$ as well). After diagonalisation:

$(x,y,z)\begin{pmatrix} 1 & 0 & 0 \\ 0 & 1 & 0 \\ 0 & 0 & 3-(2c)^2 \end{pmatrix}\begin{pmatrix} x \\ y \\ z \end{pmatrix} = 0$. The existence of a solution $(x,y,z) \in K^3 \setminus \{(0,0,0)\}$ for this equation is equivalent to $((-1,(2c)^2-3))_K = 1$.

For the individual groups we obtain:

BT: $c = \frac{1}{2}$, $((-1,-2))_K = 1$.
BO: $c = \frac{1}{\sqrt{2}}$, $((-1,-1))_K = 1$.
BI: $c = \frac{1}{4}(1 \pm \sqrt{5})$, $((-1, \frac{1}{2}(-3 \pm \sqrt{5})))_K = 1$.

Each of these conditions is equivalent to $((-1,-1))_K = 1$. For BI: $\frac{1}{2}(3 \pm \sqrt{5}) = (\frac{1}{2}(1 \pm \sqrt{5}))^2$. For BT one has maps between solutions (x,y) for $x^2+y^2 = -1$ corresponding to $((-1,-1))_K$ and (x',y') for $x'^2 + 2y'^2 = -1$ corresponding to $((-1,-2))_K$ given by $x = \frac{x'+1}{2y'} \leftrightarrow x' = \frac{x+y}{x-y}$, $y = \frac{x'-1}{2y'} \leftrightarrow y' = \frac{1}{x-y}$ for $x \neq y$ resp. $y' \neq 0$ and by $(x,x) \mapsto (0,x)$, $(x',0) \leftarrow (x',0)$. □

5 McKay Correspondence for $G \subset \mathrm{SL}(2,K)$

Let G be a finite subgroup scheme of $\mathrm{SL}(2,K)$, K a field of characteristic 0, and C the algebraic closure of K. There is the geometric quotient $\pi: \mathbb{A}^2_K \to \mathbb{A}^2_K/G$ and the natural morphism $\tau: \text{G-Hilb}_K \mathbb{A}^2_K \to \mathbb{A}^2_K/G$, which is the minimal resolution of this quotient singularity.

5.1 The Exceptional Divisor and the Intersection Graph

We define the exceptional divisor E by $E := \tau^{-1}(\overline{O})$ where $\overline{O} = \pi(O)$, O the origin of $\mathbb{A}^2_K$. In general E is not reduced, denote by E_{red} the underlying reduced subscheme.

Definition 4. The intersection graph of E_{red} is defined as the following undirected graph:

- Vertices. A vertex of multiplicity n for each irreducible component $(E_{\text{red}})_i$ of E_{red} which decomposes over the algebraic closure of K into n irreducible components.
- Edges. Different $(E_{\text{red}})_i$ and $(E_{\text{red}})_j$ are connected by $(E_{\text{red}})_i \cdot (E_{\text{red}})_j$ undirected edges. $(E_{\text{red}})_i$ has $\frac{1}{2}(E_{\text{red}})_i \cdot (E_{\text{red}})_i + \textit{multiplicity of}\,(E_{\text{red}})_i$ loops.

If K is algebraically closed, then the $(E_{\text{red}})_i$ are isomorphic to $\mathbb{P}^1_K$ and the self-intersection of each $(E_{\text{red}})_i$ is -2, because the resolution is crepant.

Let $K \to L$ be a Galois extension, $\Gamma = \mathrm{Aut}_K(L)$. Γ operates on the intersection graph of $(E_{\text{red}})_L$ by graph automorphisms. The irreducible components $(E_{\text{red}})_i$

of E_{red} correspond to Γ-orbits of irreducible components $(E_{\text{red}})_{L,k}$ of $(E_{\text{red}})_L$ by Proposition 6. For the intersection form one has

$$(E_{\text{red}})_i \cdot (E_{\text{red}})_j = ((E_{\text{red}})_i)_L \cdot ((E_{\text{red}})_j)_L = \sum_{kl} (E_{\text{red}})_{L,k} \cdot (E_{\text{red}})_{L,l}$$

where indices k and l run through the irreducible components of $((E_{\text{red}})_i)_L$ and $((E_{\text{red}})_j)_L$ respectively. Thus for the intersection graph there is a proposition similar to Proposition 11 for representation graphs.

Proposition 12. *The intersection graph of E_{red} arises by identifying the elements of Γ-orbits of vertices of the intersection graph of $(E_{\text{red}})_L$, adding multiplicities. The edges between vertices $(E_{\text{red}})_i$ and $(E_{\text{red}})_j$ are in bijection with the edges between the irreducible components of $((E_{\text{red}})_i)_L$ and $((E_{\text{red}})_j)_L$.*

5.2 Irreducible Components of E and Irreducible Representations of G

The basic statement of McKay correspondence is a bijection between the set of irreducible components of the exceptional divisor E and the set of isomorphism classes of nontrivial irreducible representations of the group scheme G.

Theorem 3. *There are bijections for intermediate fields $K \subseteq L \subseteq C$ between the set $\text{Irr}(E_L)$ of irreducible components of E_L and the set $\text{Irr}(G_L)$ of isomorphism classes of nontrivial irreducible representations of G_L having the property that for $K \subseteq L \subseteq L' \subseteq C$, if the bijection $\text{Irr}(E_L) \to \text{Irr}(G_L)$ for L maps $E_i \mapsto V_i$, then the bijection $\text{Irr}(E_{L'}) \to \text{Irr}(G_{L'})$ for L' maps irreducible components of $(E_i)_{L'}$ to irreducible components of $(V_i)_{L'}$.*

Proof. As described earlier, the Galois group $\Gamma = \text{Aut}_L(C)$ of the Galois extension $L \to C$, operates on the sets $\text{Irr}(G_C)$ and $\text{Irr}(E_C)$. In both cases elements of $\text{Irr}(G_L)$ and $\text{Irr}(E_L)$ correspond to Γ-orbits of elements of $\text{Irr}(G_C)$ and $\text{Irr}(E_C)$ by Corollary 3 and Proposition 6 respectively. This way a given bijection between the sets $\text{Irr}(G_C)$ and $\text{Irr}(E_C)$ defines a bijection between $\text{Irr}(G_L)$ and $\text{Irr}(E_L)$ on condition that the bijection is equivariant with respect to the operations of Γ. Checking this for the bijection of McKay correspondence over the algebraically closed field C constructed via stratification or via tautological sheaves will give bijections over intermediate fields L having the property of the theorem. This will be done in the process of proving Theorems 5 or 6. □

Moreover, in the situation of the theorem the Galois group $\Gamma = \text{Aut}_L(C)$ operates on the representation graph of G_C and on the intersection graph of $(E_{\text{red}})_C$. Then in both cases the graphs over L arise by identifying the elements of Γ-orbits of vertices of the graphs over C by Propositions 11 and 12. Therefore an isomorphism of the graphs over C, the bijection between the sets of vertices being Γ-equivariant, defines an isomorphism of the graphs over L.

For the algebraically closed field C this is the classical McKay correspondence for subgroups of $\mathrm{SL}(2, C)$ [8,11,14]. The statement, that there is a bijection of edges between given vertices $(E_{\mathrm{red}})_{L,i} \leftrightarrow V_i$ and $(E_{\mathrm{red}})_{L,j} \leftrightarrow V_j$, can be formulated equivalently in terms of the intersection form as $(E_{\mathrm{red}})_{L,i} . (E_{\mathrm{red}})_{L,j} = \langle V_i, V_j \rangle$.

Theorem 4. *The bijections $E_i \leftrightarrow V_i$ of Theorem 3 between irreducible components of E_L and isomorphism classes of nontrivial irreducible representations of G_L can be constructed such that $(E_{\mathrm{red}})_i \cdot (E_{\mathrm{red}})_j = \langle V_i, V_j \rangle$ or equivalently that these bijections define isomorphisms of graphs between the intersection graph of $(E_{\mathrm{red}})_L$ and the representation graph of G_L.*

We will consider two ways to construct bijections between nontrivial irreducible representations and irreducible components with the properties of Theorems 3 and 4: A stratification of G-Hilb$_K \mathbb{A}^2_K$ [10, 11] and the tautological sheaves on G-Hilb$_K \mathbb{A}^2_K$ [8, 12].

5.3 *Stratification of* G-Hilb$_K \mathbb{A}^2_K$

Let $S := K[x_1, x_2]$, let $O \in \mathbb{A}^2_K$ be the origin, $\mathfrak{m} \subset S$ the corresponding maximal ideal, $\overline{O} := \pi(O) \in \mathbb{A}^2_K/G$ with corresponding maximal ideal $\mathfrak{n} \subset S^G$, let $\overline{S} := S/\mathfrak{n}S$ with maximal ideal $\overline{\mathfrak{m}}$. An L-valued point of the fiber $E = \tau^{-1}(\overline{O})$ corresponds to a G-cluster defined by an ideal $I \subset S_L$ such that $\mathfrak{n}_L \subseteq I$ or equivalently an ideal $\overline{I} \subset \overline{S}_L = S_L/\mathfrak{n}_L S_L$. For such an ideal I define the representation $V(I)$ over L by

$$V(I) := \overline{I}/\overline{\mathfrak{m}}_L \overline{I}$$

Lemma 3. *For $\gamma \in \mathrm{Aut}_K(L)$: $V(\gamma^{-1} I) \cong V(I)^\gamma$.*

Proof. As an A_L-comodule $\overline{I} = \overline{I}_0 \oplus \overline{\mathfrak{m}}_L \overline{I}$, where $\overline{I}_0 \cong \overline{I}/\overline{\mathfrak{m}}_L \overline{I}$. Then $\gamma^{-1}\overline{I} = \gamma^{-1}\overline{I}_0 \oplus \overline{\mathfrak{m}}_L(\gamma^{-1}\overline{I})$ and $V(\gamma^{-1}I) = \gamma^{-1}\overline{I}/\overline{\mathfrak{m}}_L(\gamma^{-1}\overline{I}) \cong \gamma^{-1}\overline{I}_0 \cong \overline{I}_0^\gamma \cong V(I)^\gamma$ by Remark 5 applied to $\overline{I}_0 \subseteq \overline{S}_L$. □

Theorem 5. *There is a bijection $E_j \leftrightarrow V_j$ between the set* Irr(E) *of irreducible components of E and the set* Irr(G) *of isomorphism classes of nontrivial irreducible representations of G such that for any closed point $y \in E$: If $I \subset S_{\kappa(y)}$ is an ideal defining a $\kappa(y)$-valued point of the scheme $\{y\} \subset E$, then*

$$\mathrm{Hom}^G_{\kappa(y)}(V(I), (V_j)_{\kappa(y)}) \neq 0 \iff y \in E_j$$

and $V(I)$ is either irreducible or consists of two irreducible representations not isomorphic to each other. Applied to the situation after base extension $K \to L$, L

an algebraic extension of K, one obtains bijections $\mathrm{Irr}(E_L) \leftrightarrow \mathrm{Irr}(G_L)$ *having the properties of Theorems 3 and 4.*

Proof. In the case of algebraically closed K the theorem follows from [11].

In the general case denote by U_i the isomorphism classes of nontrivial irreducible representations of G_C over the algebraic closure C. Over C the theorem is valid, let $E_{C,i}$ be the component corresponding to U_i.

We show that this bijection is equivariant with respect to the operations of $\Gamma = \mathrm{Aut}_K(C)$. Let $x \in E_{C,i}$ be a closed point such that $x \notin E_{C,i'}$ for $i' \neq i$. Then for the corresponding C-valued point $\alpha\colon \mathrm{Spec} C \to E_{C,i}$ given by an ideal $I \subset S_C$ one has $V(I) \cong U_i$. By Corollary 2 the C-valued point corresponding to γx is α^γ given by the ideal $\gamma^{-1} I \subset S_C$. By Lemma 3 $V(\gamma^{-1} I) \cong U_{\gamma(i)}$, where $U_{\gamma(i)} = U_i^\gamma$. Therefore $\gamma x \in E_{\gamma(i)}$ and $\gamma E_i = E_{\gamma(i)}$.

For an irrreducible representation V_j of G define E_j to be the component of E, which decomposes over C into the irreducible components $E_{C,i}$ satisfying $U_i \subseteq (V_j)_C$. This method, applied to the situation after base extension $K \to L$, leads to bijections having the properties of Theorems 3 and 4.

We show that this bijection is given by the condition in the theorem. Let y be a closed point of E and α a $\kappa(y)$-valued point of the scheme $\{y\} \subset E$ given by an ideal $I \subset S_{\kappa(y)}$. $K \to \kappa(y)$ is an algebraic extension, embed $\kappa(y)$ into C. After base extension $\kappa(y) \to C$ one has the C-valued point $\alpha_C\colon \mathrm{Spec} C \to \{y\}_C$ given by $I_C \subset S_C$. Then $V(I)_C \cong V(I_C)$ and I_C corresponds to a closed point $z \in \{y\}_C \subset E_C$. Therefore

$$\begin{aligned} y \in E_j &\iff z \in E_{C,i} \text{ for some } i \text{ satisfying } U_i \subseteq (V_j)_C \\ &\iff \mathrm{Hom}_C^G(V(I_C), U_i) \neq 0 \text{ for some } i \text{ satisfying } U_i \subseteq (V_j)_C \\ &\iff \mathrm{Hom}_{\kappa(y)}^G(V(I), (V_j)_{\kappa(y)}) \neq 0 \end{aligned}$$

□

5.4 Tautological Sheaves

Let $0 \to \mathscr{I} \to \mathscr{O}_{\mathbb{A}^2_Y} \to \mathscr{O}_Z \to 0$ be the universal quotient of $Y := \text{G-Hilb}_K \mathbb{A}^2_K$. The projection $p\colon Z \to Y$ is a finite flat morphism, $p_* \mathscr{O}_Z$ is a locally free G-sheaf on Y with fibers $p_* \mathscr{O}_Z \otimes_{\mathscr{O}_Y} \kappa(y)$ isomorphic to the regular representation over $\kappa(y)$.

Let $V_0, \ldots, V_s$ the isomorphism classes of irreducible representations of G, V_0 the trivial representation. The G-sheaf $\mathscr{G} := p_* \mathscr{O}_Z$ on Y decomposes into isotypic components (see Remark 1(3) and Sect. 3.6)

$$\mathscr{G} \cong \bigoplus_{j=0}^{s} \mathscr{G}_j$$

where $\mathscr{G}_j$ is the component for V_j.

Definition 5. For any isomorphism class V_j of irreducible representations of G over K define the sheaf $\mathscr{F}_j$ on $Y = \text{G-Hilb}_K \mathbb{A}^2_K$ by

$$\mathscr{F}_j := \mathscr{H}\!om^G_{\mathscr{O}_Y}(V_j \otimes_K \mathscr{O}_Y, \mathscr{G}_j) = \mathscr{H}\!om^G_{\mathscr{O}_Y}(V_j \otimes_K \mathscr{O}_Y, \mathscr{G})$$

For a field extension $K \to L$ denote by $\mathscr{F}_{L,i}$ the sheaf $\mathscr{H}\!om^{G_L}_{\mathscr{O}_{Y_L}}(U_i \otimes_L \mathscr{O}_{Y_L}, \mathscr{G}_L)$ on Y_L, U_i an irreducible representation of G_L over L.

Remark 10. 1. For $K = \mathbb{C}$ the sheaves $\mathscr{F}_j$ were studied in [8, 12], they may be defined as well as $\mathscr{F}_j = \tau^* \mathscr{H}\!om^G_{\mathbb{A}^2_K/G}(V_j \otimes_K \mathscr{O}_{\mathbb{A}^2_K/G}, \pi_* \mathscr{O}_{\mathbb{A}^2_K})/(\mathscr{O}_Y\textit{-torsion})$ or $(p_* q^*(\mathscr{O}_{\mathbb{A}^2_K} \otimes_K V_j^\vee))^G$ using the canonical morphisms in the diagram

$$\begin{array}{ccccc} & & Z & & \\ & \swarrow^{p} & & \searrow^{q} & \\ Y & & & & \mathbb{A}^2_K \\ & \searrow_{\tau} & & \swarrow_{\pi} & \\ & & \mathbb{A}^2_K/G & & \end{array}$$

2. $\mathscr{F}_j$ is a locally free sheaf of rank $\dim_K V_j$.
3. For each j there is the natural isomorphism of G-sheaves $\mathscr{F}_j \otimes_{\text{End}^G_K(V_j)} V_j \xrightarrow{\sim} \mathscr{G}_j$.

Let $K \to L$ be a Galois extension and $U_0, \ldots, U_r$ be the isomorphism classes of irreducible representations of G_L over L. Then a decomposition $(V_j)_L = \bigoplus_{i \in I_j} U_i$ over L of an irreducible representation V_j of G over K gives a decomposition of the corresponding tautological sheaf

$$\begin{aligned}(\mathscr{F}_j)_L &= \mathscr{H}\!om^G_{\mathscr{O}_Y}(V_j \otimes_K \mathscr{O}_Y, \mathscr{G})_L \cong \mathscr{H}\!om^{G_L}_{\mathscr{O}_{Y_L}}((V_j \otimes_K \mathscr{O}_Y)_L, \mathscr{G}_L) \\ &\cong \mathscr{H}\!om^{G_L}_{\mathscr{O}_{Y_L}}(\textstyle\bigoplus_{i \in I_j} U_i \otimes_L \mathscr{O}_{Y_L}, \mathscr{G}_L) \cong \bigoplus_{i \in I_j} \mathscr{H}\!om^{G_L}_{\mathscr{O}_{Y_L}}(U_i \otimes_L \mathscr{O}_{Y_L}, \mathscr{G}_L) = \bigoplus_{i \in I_j} \mathscr{F}_{L,i}\end{aligned}$$

We have used the fact that the U_i occur with multiplicity 1 as it is the case for finite subgroup schemes of SL(2, K), see Proposition 10.

The tautological sheaves $\mathscr{F}_j$ can be used to establish a bijection between the set of irreducible components of E_{red} and the set of isomorphism classes of nontrivial irreducible representations of G by considering intersections $\mathscr{L}_j \cdot (E_{\text{red}})_{j'}$, i.e. the degrees of restrictions of the line bundles $\mathscr{L}_j := \bigwedge^{\text{rk}\mathscr{F}_j} \mathscr{F}_j$ to the curves $(E_{\text{red}})_{j'}$.

Theorem 6. *There is a bijection $E_j \leftrightarrow V_j$ between the set* Irr(E) *of irreducible components of E and the set* Irr(G) *of isomorphism classes of nontrivial irreducible representations of G such that*

$$\mathscr{L}_j \cdot (E_{\text{red}})_{j'} = \dim_K \text{Hom}^G_K(V_j, V_{j'})$$

where $\mathscr{L}_j = \bigwedge^{\text{rk}\mathscr{F}_j} \mathscr{F}_j$.

Applied to the situation after base extension $K \to L$, L an algebraic extension field of K, one obtains bijections $\mathrm{Irr}(E_L) \leftrightarrow \mathrm{Irr}(G_L)$ *having the properties of Theorems 3 and 4.*

Proof. In the case of algebraically closed K the theorem follows from [8].

In the general case denote by $U_0, \ldots, U_r$ the isomorphism classes of irreducible representations of G_C over the algebraic closure C, U_0 the trivial one. Over C the theorem is valid, let $E_{C,i}$ be the component corresponding to U_i, what means that $\mathscr{L}_{C,i}.(E_{\mathrm{red}})_{C,i'} = \delta_{ii'}$, where $\mathscr{L}_{C,i} = \bigwedge^{\mathrm{rk}\mathscr{F}_{C,i}} \mathscr{F}_{C,i}$.

To show that the bijection over C is equivariant with respect to the operations of $\Gamma = \mathrm{Aut}_K(C)$, one has to show that $\gamma_* \mathscr{L}_{C,i} \cong \mathscr{L}_{C,\gamma(i)}$, where $U_{\gamma(i)} = U_i^\gamma$. Then $\mathscr{L}_{C,i} \cdot E_{C,i'} = \gamma_* \mathscr{L}_{C,i} \cdot \gamma E_{C,i'} = \mathscr{L}_{C,\gamma(i)} \cdot \gamma E_{C,i'}$ and therefore $\gamma E_{C,i'} = E_{C,\gamma(i')}$. It is $\gamma_* \mathscr{L}_{C,i} \cong \mathscr{L}_{C,\gamma(i)}$, because using Lemma 1 and Remark 3

$$\gamma_* \mathscr{F}_{C,i} \cong \mathscr{H}\!om^{G_C}_{\mathscr{O}_{Y_C}}(\gamma_*(U_i \otimes_C \mathscr{O}_{Y_C}), \gamma_* \mathscr{G}_C) \cong \mathscr{H}\!om^{G_C}_{\mathscr{O}_{Y_C}}(U_i^\gamma \otimes_C \mathscr{O}_{Y_C}, \mathscr{G}_C) = \mathscr{F}_{C,\gamma(i)}$$

Since the bijection over C is equivariant with respect to the Γ-operations on $\mathrm{Irr}(G_C)$ and $\mathrm{Irr}(E_C)$, one can define a bijection $\mathrm{Irr}(G) \leftrightarrow \mathrm{Irr}(E)$: For $V_j \in \mathrm{Irr}(G)$ let E_j be the element of $\mathrm{Irr}(E)$ such that $(V_j)_C = \bigoplus_{i \in I_j} U_i$ and $(E_j)_C = \bigcup_{i \in I_j} E_{C,i}$ for the same subset $I_j \subseteq \{1, \ldots, r\}$. This method applied to the situation after base extension $K \to L$ leads to bijections having the properties of Theorems 3 and 4.

We show that this bijection is given by the construction of the theorem. It is $(\mathscr{F}_j)_C = \bigoplus_{i \in I_j} \mathscr{F}_{C,i}$ and therefore

$$\begin{aligned} \mathscr{L}_j \cdot (E_{\mathrm{red}})_{j'} &= (\mathscr{L}_j)_C \cdot ((E_{\mathrm{red}})_{j'})_C = \big(\textstyle\bigotimes_{i \in I_j} \mathscr{L}_{C,i}\big).\big(\textstyle\sum_{i' \in I_{j'}} (E_{\mathrm{red}})_{C,i'}\big) \\ &= \textstyle\sum_{i,i'} \mathscr{L}_{C,i} \cdot (E_{\mathrm{red}})_{C,i'} = \textstyle\sum_{i,i'} \dim_C \mathrm{Hom}_C^{G_C}(U_i, U_{i'}) \\ &= \dim_C \mathrm{Hom}_C^{G_C}((V_j)_C, (V_{j'})_C) = \dim_K \mathrm{Hom}_K^G(V_j, V_{j'}) \end{aligned}$$

□

5.5 Examples

Finite subgroups of SL(2, K). In the case of subgroups $G \subset \mathrm{SL}(2, K)$ the representation graph can be read off from the table of characters of the group G over an algebraically closed field, since in this case representations are conjugate if and only if the values of their characters are. We have the following graphs for the finite subgroups of $\mathrm{SL}(2, K)$:

- Cyclic group $\mathbb{Z}/n\mathbb{Z}$, $n \geq 1$. It is $\xi + \xi^{-1} \in K$, ξ a primitive n-th root of unity. Diagram (A_{n-1}) if $\xi \in K$, otherwise $(A_{n-1})'$.

- Binary dihedral group BD_n, $n \geq 2$. It is $c = \frac{1}{2}(\xi + \xi^{-1}) \in K$, ξ a primitive $2n$-th root of unity, and $((-1, c^2 - 1))_K = 1$. Diagram (D_{n+2}) if n even or $\sqrt{-1} \in K$, otherwise $(D_{n+2})'$.
- Binary tetrahedral group BT. It is $((-1,-1))_K = 1$. Diagram (E_6) if K contains a primitive 3rd root of unity, otherwise $(E_6)'$.
- Binary octahedral group BO. It is $((-1,-1))_K = 1$ and $\sqrt{2} \in K$. Diagram (E_7).
- Binary icosahedral group BI. It is $((-1,-1))_K = 1$ and $\sqrt{5} \in K$. Diagram (E_8).

Examples for the graphs $(A_n)'$, $(D_{2m+1})'$, $(E_6)'$:

$(A_n)'$ $\mathbb{Z}/(n+1)\mathbb{Z}$ over $\mathbb{Q}(\xi + \xi^{-1})$, ξ a primitive $(n+1)$-th root of unity.
$(D_{2m+1})'$ BD_{2m-1} over $\mathbb{Q}(\xi)$, ξ a primitve $2(2m-1)$-th root of unity.
$(E_6)'$ BT over $\mathbb{Q}(\sqrt{-1})$.

Abelian subgroup schemes. In the case of abelian subgroup schemes of $\mathrm{SL}(2, K)$ the graphs (A_n) and $(A_n)'$ occur.

- The cyclic group $G = \mathbb{Z}/n\mathbb{Z}$ is realisable as the subgroup of $\mathrm{SL}(2, K)$ generated by $g := \begin{pmatrix} 0 & -1 \\ 1 & \xi+\xi^{-1} \end{pmatrix}$, if the field K contains $\xi + \xi^{-1}$ for ξ a primitive n-th root of unity. If K does not contain ξ, then there are 1-dimensional representations over the algebraic closure that are not realisable over K, one has diagram $(A_{n-1})'$.
- For the subgroup scheme $G = \mu_n \subset \mathrm{SL}(2, K)$ the Hopf algebra $K[y]/\langle y^n \rangle$ decomposes into a direct sum of simple subcoalgebras $\langle y^j \rangle_K$ corresponding to 1-dimensional representations of G. Thus one has diagram (A_{n-1}).

The graph $(D_{2m})'$. Let $n \geq 2$, ε a primitive $4n$-th root of unity and $\xi = \varepsilon^2$. Put $K = \mathbb{Q}(\varepsilon + \varepsilon^{-1})$, $C = \mathbb{Q}(\varepsilon)$ and $\Gamma = \mathrm{Aut}_K(C) = \{id, \gamma\}$. One has the injective representation of $BD_n = \langle \sigma, \tau \mid \tau^2 = \sigma^n = (\tau\sigma)^2 \rangle$ in $\mathrm{SL}(2, C)$ given by

$$\sigma \mapsto \begin{pmatrix} \xi & 0 \\ 0 & \xi^{-1} \end{pmatrix}, \qquad \tau \mapsto \begin{pmatrix} 0 & -\varepsilon \\ \varepsilon^{-1} & 0 \end{pmatrix}$$

We will identify BD_n with its image in $\mathrm{SL}(2, C)$ and regard it as a subgroup scheme of $\mathrm{SL}(2, C)$.

Γ operates on $\mathrm{SL}(2, C)$, the K-automorphism $\gamma \in \Gamma$, $\gamma\colon \varepsilon \mapsto \varepsilon^{-1}$ of order 2 operates nontrivially on the closed points of BD_n by $\sigma \mapsto \sigma^{-1}$, $\tau \mapsto \tau\sigma$. The subgroup scheme $BD_n \subset \mathrm{SL}(2, C)$ is defined over K, let $G \subset \mathrm{SL}(2, K)$ such that $G_C = BD_n$. The closed points of G correspond to Γ-orbits of closed points of BD_n, they have the form $\{id\}$, $\{-id\}$, $\{\sigma^k, \sigma^{-k}\}$, $\{\tau\sigma^k, \tau\sigma^{-k+1}\}$.

The automorphism γ operates on the characters of BD_n trivially except that for even n it permutes two of the irreducible 1-dimensional representations. One has the graph $(D_{n+2})'$ for n even and the graph (D_{n+2}) for n odd.

6 Finite Subgroups of SL$(2, C)$: Presentations and Character Tables

6.1 Cyclic Groups

The irreducible representations are $\chi_j : \mathbb{Z}/n\mathbb{Z} \to C^*, \bar{i} \mapsto \xi^{ji}$ for $j \in \{0, \ldots, n-1\}$, where ξ is a primitive n-th root of unity.

Binary dihedral groups. $BD_n = \langle \sigma, \tau \mid \tau^2 = \sigma^n = (\tau\sigma)^2 \rangle$, $-id := (\tau\sigma)^2$.

BD_n, n odd

	id	$-id$	σ^k	τ	$\tau\sigma$
1	1	1	1	1	1
1$'$	1	1	1	-1	-1
1$''$	1	-1	$(-1)^k$	i	$-i$
1$'''$	1	-1	$(-1)^k$	$-i$	i
2^j	2	$(-1)^j 2$	$\xi^{kj} + \xi^{-kj}$	0	0

BD_n, n even

	id	$-id$	σ^k	τ	$\tau\sigma$
1	1	1	1	1	1
1$'$	1	1	1	-1	-1
1$''$	1	1	$(-1)^k$	1	-1
1$'''$	1	1	$(-1)^k$	-1	1
2^j	2	$(-1)^j 2$	$\xi^{kj} + \xi^{-kj}$	0	0

ξ a primitive $2n$-th root of unity and $j = 1, \ldots, n-1$

Binary tetrahedral group. $BT = \langle a, b \mid a^3 = b^3 = (ab)^2 \rangle$, $-id := (ab)^2$.

	id	$-id$	a	$-a$	b	$-b$	ab	
1	1	1	1	1	1	1	1	1
1$'$	1	1	ω	ω	ω^2	ω^2	1	$\mathbb{Z}/3\mathbb{Z}$
1$''$	1	1	ω^2	ω^2	ω	ω	1	$\mathbb{Z}/3\mathbb{Z}$
3	3	3	0	0	0	0	-1	A_4
2	2	-2	1	-1	1	-1	0	BT
2$'$	2	-2	ω	$-\omega$	ω^2	$-\omega^2$	0	BT
2$''$	2	-2	ω^2	$-\omega^2$	ω	$-\omega$	0	BT
	1	1	4	4	4	4	6	

ω a primitive 3rd root of unity

Binary octahedral group. $BO = \langle a, b \mid a^3 = b^4 = (ab)^2 \rangle$, $-id := (ab)^2$.

	id	$-id$	ab	a	$-a$	b	$-b$	b^2	
1	1	1	1	1	1	1	1	1	1
1$'$	1	1	-1	1	1	-1	-1	1	$\mathbb{Z}/2\mathbb{Z}$
2$'''$	2	2	0	-1	-1	0	0	2	S_3
3	3	3	1	0	0	-1	-1	-1	S_4
3$'$	3	3	-1	0	0	1	1	-1	S_4
2	2	-2	0	1	-1	$\sqrt{2}$	$-\sqrt{2}$	0	BO
2$'$	2	-2	0	1	-1	$-\sqrt{2}$	$\sqrt{2}$	0	BO
4	4	-4	0	-1	1	0	0	0	BO
	1	1	12	8	8	6	6	6	

Binary icosahedral group. $BI = \langle a, b \mid a^3 = b^5 = (ab)^2 \rangle$, $-id := (ab)^2$.

	id	$-id$	a	$-a$	b	$-b$	b^2	$-b^2$	ab	
1	1	1	1	1	1	1	1	1	1	1
3	3	3	0	0	μ^+	μ^+	μ^-	μ^-	-1	A_5
3′	3	3	0	0	μ^-	μ^-	μ^+	μ^+	-1	A_5
4′	4	4	1	1	-1	-1	-1	-1	0	A_5
5	5	5	-1	-1	0	0	0	0	0	A_5
2	2	-2	1	-1	μ^+	$-\mu^+$	$-\mu^-$	μ^-	0	BI
2′	2	-2	1	-1	μ^-	$-\mu^-$	$-\mu^+$	μ^+	0	BI
4	4	-4	-1	1	1	-1	-1	1	0	BI
6	6	-6	0	0	-1	1	1	-1	0	BI
	1	1	20	20	12	12	12	12	30	

$\mu^+ := \frac{1}{2}(1+\sqrt{5})$, $\mu^- := \frac{1}{2}(1-\sqrt{5})$

Acknowledgements The suggestion to investigate McKay correspondence over non algebraically closed fields is due to Victor Batyrev.

References

1. M. Blume, Construction of G-Hilbert schemes. Math. Nachr. **284**, 953–959 (2011). arXiv:math/0607577
2. N. Bourbaki, *Elements of mathematics, Algebra II*, ch. 4–7 (Springer, Berlin, 2003)
3. N. Bourbaki, *Éléments de mathématique, Algèbra*, ch. 8 (Hermann, Paris, 1973)
4. N. Bourbaki, *Elements of mathematics, Commutative Algebra*, ch 1–7 (Springer, 1989)
5. N. Bourbaki, *Éléments de mathématique, Groupes et algèbres de Lie*, ch. 4–6 (Hermann, Paris, 1968)
6. T. Bridgeland, A. King, M. Reid, The McKay correspondence as an equivalence of derived categories. J. Am. Math. Soc. **14**, 535–554 (2001). arXiv:math/9908027
7. C. Curtis, I. Reiner, *Methods of Representation Theory*, vol. I, II (Wiley-Interscience, New York, 1981, 1987)
8. G. Gonzalez-Sprinberg, J.-L. Verdier, Construction géométrique de la correspondance de McKay. Ann. Sci. École Norm. Sup. **16**, 409–449 (1983)
9. A. Grothendieck, *Techniques de construction et théorèmes d'existence en géométrie algébrique IV: Les schémas de Hilbert*, Séminaire Bourbaki 1960/61, No. 221
10. Y. Ito, I. Nakamura, McKay correspondence and Hilbert schemes. Proc. Jpn. Acad. Ser. A **72**(7), 135–138 (1996)
11. Y. Ito, I. Nakamura, Hilbert schemes and simple singularities, in *New trends in algebraic geometry* (Cambridge University Press, Cambridge, 1999), pp. 151–233
12. M. Kapranov, E. Vasserot, Kleinian singularities, derived categories and Hall algebras. Math. Ann. **316**, 565–576 (2000). arXiv:math/9812016
13. J. Lipman, Rational singularities with applications to algebraic surfaces and unique factorization. Publ. Math. IHES **36**, 195–279 (1969)
14. J. McKay, Graphs, singularities and finite groups, in *The Santa Cruz Conference on Finite Groups*. Proc. Symp. Pure Math. **37**, 183–186 (1980)
15. D. Mumford, *Geometric Invariant Theory* (Springer, Berlin/New York, 1965)

16. M. Reid, McKay correspondence, in *Proceedings Of Algebraic Geometry Symposium*, Kinosaki, Nov 1996, pp. 14–41. arXiv:alg-geom/9702016
17. M. Reid, *La correspondance de McKay*, Séminaire Bourbaki, vol. 1999/2000, Exposés 865–879, Société Mathématique de France, Astérisque **276**, 53–72 (2002). arXiv:math/9911165
18. J.P. Serre, *A Course in Arithmetic*. Graduate Texts in Mathematics, vol. 7 (Springer, New York, 1973)
19. P. Slodowy, *Simple Singularities and Simple Algebraic Groups*. Lecture Notes in Mathematics, vol. 815 (Springer, Berlin/New York, 1980)
20. M.E. Sweedler, *Hopf Algebras* (Benjamin, New York, 1969)

Gonality of Algebraic Curves and Graphs

Lucia Caporaso

Abstract We define d-gonal weighted graphs using "harmonic indexed" morphisms, and prove that a combinatorial locus of $\overline{M_g}$ contains a d-gonal curve if the corresponding graph is d-gonal and of Hurwitz type. Conversely the dual graph of a d-gonal stable curve is equivalent to a d-gonal graph of Hurwitz type. The hyperelliptic case is studied in detail. For $r \geq 1$, we show that the dual graph of a (d,r)-gonal stable is the underlying graph of a tropical curve admitting a degree-d divisor of rank at least r.

1 Introduction and Preliminaries

1.1 Introduction

In this paper we study the interplay between the theory of linear series on algebraic curves, and the theory of linear series on graphs.

A smooth curve C is d-gonal if it admits a linear series of degree d and rank 1; more generally, C is (d,r)-gonal if it admits a linear series of degree d and rank r. A stable, or singular, curve is defined to be (d,r)-gonal, if it is the specialization of a family of smooth (d,r)-gonal curves. This rather unwieldy definition is due to the fact that the divisor theory of singular curves is quite complex; for example, every reducible curve admits infinitely many divisors of degree d and rank r, for every d and $r \geq 0$. Moreover characterizing (d,r)-gonal curves is a well known difficult problem.

L. Caporaso (✉)
Dipartimento di Matematica e Fisica, Università Roma Tre, Largo San Leonardo Murialdo 1, 00146 Roma, Italy
e-mail: caporaso@mat.uniroma3.it

A. Frühbis-Krüger et al. (eds.), *Algebraic and Complex Geometry*, Springer Proceedings in Mathematics & Statistics 71, DOI 10.1007/978-3-319-05404-9_4,

On the other hand, the moduli space of Deligne-Mumford stable curves, $\overline{M_g}$, has a natural stratification into "combinatorial" loci, parametrizing curves having a certain weighted graph as dual graph; see [13]. It is thus natural to ask whether the existence in a combinatorial locus of a (d,r)-gonal curve can be detected uniquely from the corresponding graph and its divisor theory.

In fact, in recent times a theory for divisors on graphs has been set-up and developed in a purely combinatorial way, revealing some remarkable analogies with the algebro-geometric case; see [4, 6, 7] for example. One of the goals of this paper is to contribute to this development; we give a new definition for morphisms between graphs, which we call *indexed morphisms*, and then introduce *harmonic indexed morphisms*. Our definition is inspired by the theory of admissible coverings developed by J. Harris and D. Mumford in [16], and generalizes the combinatorial definition of *harmonic morphisms* given by M. Baker, S. Norine and H. Urakawa in [7] and [20] for weightless graphs; this is why we use the word "harmonic". Harmonic indexed morphisms have a well defined degree, and satisfy the Riemann-Hurwitz formula with an effective ramification divisors.

We say that a graph is d-gonal if it admits a non-degenerate harmonic indexed morphism, ϕ, of degree d to a tree; furthermore we say that it is of Hurwitz type if the Hurwitz existence problem naturally associated to ϕ has a positive solution; see Definition 6 for details. In particular, if $d \leq 3$ every d-gonal graph is of Hurwitz type. Then we prove the following:

Theorem 1. *If (G,w) is a d-gonal weighted stable graph of Hurwitz type, there exists a (stable) d-gonal curve whose dual graph is (G,w). Conversely, the dual graph of a stable d-gonal curve is equivalent to a d-gonal graph of Hurwitz type.*

This Theorem follows immediately from the more general Theorem 2, whose proof combines the theory of admissible coverings with properties of harmonic indexed morphisms.

Next, for all $r \geq 1$ we prove Theorem 3, which, in particular, states that *the dual graph of a (d,r)-gonal curve admits a refinement admitting a divisor of rank r and degree d.*

The proof of this theorem uses different methods than the previous one: the theory of stable curves, and a generalization, from [1], of Baker's specialization lemma [5, Lemma 2.8].

Testing whether a graph admits a divisor of given degree and rank involves only a finite number of steps, and can be done by a computer; hence Theorem 3 yields a handy necessary condition for a curve to be (d,r)-gonal.

This theorem has also consequences on tropical curves. In fact the moduli space of tropical curves of genus g, M_g^{trop}, has a stratification indexed by stable weighted graphs exactly as $\overline{M_g}$. Using our results we obtain that if a combinatorial stratum of $\overline{M_g}$ contains a (d,r)-gonal curve, so does the corresponding stratum of M_g^{trop}; see Sect. 3.1 for more details. The connections between the divisor theories of algebraic and tropical curves have been object of much interest in recent years; in fact some closely related issues are currently being investigated, under a completely different perspective, in a joint project of O. Amini, M. Baker, E. Brugallé and J. Rabinoff. We refer also to [8, 10–12] and [17] for some recent work on the relation between algebraic and tropical geometry.

The paper is organized in four sections; the first recalls definitions and results from algebraic geometry and graph theory needed in the sequel, mostly from [3, 6, 16] and [1]. In Sect. 2 we study the case $r = 1$ and prove Theorem 2 (and Theorem 1). The next section studies the case $r \geq 1$ and extends the analysis to tropical curves; the main result here is Theorem 3. In Sect. 4 we concentrate on the hyperelliptic case, and develop the basic theory by extending some of the results of [7]. It turns out that for this case the analogies between the algebraic and the combinatorial setting are stronger; see Theorem 4.

I wish to thank M. Baker, E. Brugallé, M. Chan, R Guralnick, and F. Viviani for enlightening discussions related to the topics in this paper. I am grateful to S. Payne for pointing out an error in the first version of Theorem 2.

1.2 Graphs and Dual Graphs of Curves

Details about the forthcoming topics may be found in [3] and [11].

Unless we specify otherwise, by the word "curve" we mean reduced, projective algebraic variety of dimension one over the field of complex numbers; we always assume that our curves have at most nodes as singularities. The genus of a curve is the arithmetic genus.

The graphs we consider, usually denoted by a "G" with some decorations, are connected graphs (no metric) admitting loops and multiple edges, unless differently stated. For the reader's convenience we recall some basic terminology from graph theory. Our conventions are chosen to fit both the combinatorial and algebro-geometric set up. For a graph G we denote by $V(G)$ the set of its vertices, by $E(G)$ the set of its edges and by $H(G)$ the set of its half-edges. The set of half-edges comes with a fixed-point-free involution whose orbits, written $\{h, \overline{h}\}$, bijectively correspond to $E(G)$, and with a surjective *endpoint* map $\epsilon : H(G) \to V(G)$. For $e \in E(G)$ corresponding to the half-edges $h, \overline{h}$ we often write $e = [h, \overline{h}]$.

A *loop-edge* is an edge $e = [h, \overline{h}]$ such that $\epsilon(h) = \epsilon(\overline{h})$.

A *leaf* is a pair, (v, e), of a vertex and an edge, where e is not a loop-edge and is the unique edge adjacent to v. We say that e is a *leaf-edge* and v is a *leaf-vertex*.

A *bridge* is an edge e such that $G \smallsetminus e$ is disconnected.

Let $v \in V(G)$; we denote by $E_v(G) \subset E(G)$, respectively by $H_v(G) \subset H(G)$, the set of edges, resp. of half-edges, adjacent to v.

In some cases we will need to consider graphs endowed with legs, then we will explicitly speak about *graphs with legs*. A leg of a graph G is a one-dimensional open simplex having exactly one endpoint $v \in V(G)$. We denote by $L(G)$ the set of legs of G, and by $L_v(G)$ the set of legs having v as endpoint.

The *valency*, $\mathrm{val}(v)$, of a vertex $v \in V(G)$ is defined as follows

$$\mathrm{val}(v) := |H_v(G)| + |L_v(G)|. \tag{1}$$

Let now X be a curve (having at most nodes as singularities), and let G_X be its so-called dual graph. So, the vertices of G_X correspond to the irreducible components of X, and we write $X = \cup_{v \in V(G_X)} C_v$ with C_v irreducible curve. The edges of G_X correspond to the nodes of X, and we denote the set of nodes of X by $X_{\mathrm{sing}} = \{N_e,\ e \in E(G_X)\}$. The endpoints of the edge e correspond to the components of X glued at the node N_e. Finally, the set of half-edges $H(G_X)$ is identified with the set of points of the normalization of X lying over the nodes, so that a pair $\{h, \overline{h}\} \subset H(G_X)$ corresponding to the edge $e \in E(G_X)$ is identified with a pair of points $p_h, p_{\overline{h}}$ on the normalization of X in such a way that, denoting by $v, \overline{v}$ the endpoints of e, with h adjacent to v and $\overline{h}$ adjacent to $\overline{v}$, we have that p_h lies on the normalization of C_v and $p_{\overline{h}}$ on the normalization of $C_{\overline{v}}$. This yields a handy description of X:

$$X = \frac{\sqcup_{v \in V(G_X)} C_v^{\nu}}{\{p_h = p_{\overline{h}},\ \forall h \in H(G_X)\}} \tag{2}$$

where C_v^{ν} denotes the normalization of C_v.

Next, let $(X; x_1, \ldots, x_b)$ be a *pointed* curve, i.e. X is a curve and $x_1, \ldots, x_b$ are nonsingular points of X. To $(X; x_1, \ldots, x_b)$ we associate a graph with legs, written

$$G_{(X; x_1, \ldots, x_b)},$$

by adding to the dual graph G_X described above one leg ℓ_i for each marked point x_i, so that the endpoint of ℓ_i is the vertex v such that $x_i \in C_v$.

A *weighted graph* is a pair (G, w) where G is a graph (possibly with legs) and w a weight function $w : V(G) \to \mathbb{Z}_{\geq 0}$. The genus of a weighted graph is

$$g_{(G,w)} := b_1(G) + \sum_{v \in V(G)} w(v).$$

A *tree* is a connected graph of genus zero (hence weights equal zero).

A weighted graph (G, w) with legs is *stable* (respectively *semistable*), if for every vertex v we have

$$w(v) + \mathrm{val}(v) \geq 3 \quad (\text{resp.} \geq 2).$$

Definition 1. Let (G, w) be a weighted graph of genus at least 2. Its *stabilization* is the stable graph obtained by removing from (G, w) all leaves (v, e) such that $w(v) = 0$ and all 2-valent vertices of weight zero (see below). We say that two graphs are *(stably) equivalent* if they have the same stabilization.

The stabilization does not change the genus.

As in the previous definition, we shall often speak about graphs obtained by "removing" a 2-valent vertex, v, from a given graph, G. By this we mean that after removing v, the topological space of the so-obtained graph is the same as that of G, but the sets of vertices and edges are different. The operation opposite to removing a 2-valent vertex is that of "inserting" a vertex (necessarily 2-valent) in the interior of an edge.

A *refinement* of a weighted graph (G, w) is a weighted graph obtained by inserting some weight zero vertices in the interior of some edges of G.

Let now X be a curve as before. The *(weighted) dual graph* of X is the weighted graph (G_X, w_X), with G_X as defined above, and for $v \in V(G_X)$ the value $w_X(v)$ is equal to the genus of the normalization of C_v.

It is easy to see that the genus of X is equal to the genus of (G_X, w_X).

The (weighted) dual graph of a pointed curve $(X; x_1, \ldots, x_b)$ is the graph with legs $(G_{(X;x_1,\ldots,x_b)}, w_X)$.

Remark 1. A pointed curve $(X; x_1, \ldots, x_b)$ is *stable*, or *semistable*, if and only if so is $(G_{(X;x_1,\ldots,x_b)}, w_X)$.

A curve X is *rational* (i.e. it has genus zero) if and only if (G_X, w_X) is a tree.

Remark 2. Let X be a curve of genus ≥ 2 and (G_X, w_X) its dual graph. There exists a unique stable curve X^s of genus g with a surjective map $\sigma : X \to X^s$, such that σ is birational away from some smooth rational components that get contracted to a point. X^s is called the *stabilization* of X. The dual graph of X^s is the stabilization of (G_X, w_X); see Definition 1.

For a stable graph (G, w) of genus g, we denote by $M^{\text{alg}}(G, w) \subset \overline{M_g}$ the locus of curves whose dual graph is (G, w), and we refer to it as a *combinatorial locus* of $\overline{M_g}$ (the superscript "alg" stands for algebraic, versus tropical, see Sect. 3.1). Of course, we have

$$\overline{M_g} = \bigsqcup_{(G,w) \text{ stable, genus } g} M^{\text{alg}}(G, w). \tag{3}$$

1.3 Admissible Coverings

Details about this subsection may be found in [15,16] and [3]. Let $\overline{M_g}$ be the moduli space of stable curves of genus $g \geq 2$ and $M_g \subset \overline{M_g}$ its open subset parametrizing smooth curves. We denote by $\overline{M^r_{g,d}}$ the closure in $\overline{M_g}$ of the locus, $M^r_{g,d}$, of smooth curves admitting a divisor of degree d and rank r; in symbols:

$$M^r_{g,d} := \{[X] \in M_g : W^r_d(X) \neq \emptyset\} \tag{4}$$

where $W^r_d(X)$ is the set of linear equivalence classes of divisors D on X such that $h^0(X, D) \geq r + 1$.

The case of hyperelliptic curves, $r = 1$ and $d = 2$, has traditionally a simpler notation: one denotes by $H_g \subset M_g$ the locus of hyperelliptic curves and by $\overline{H_g}$ its closure in $\overline{M_g}$. So, $\overline{H_g} = \overline{M^1_{g,2}}$.

Definition 2. Let X be a connected curve of genus $g \geq 2$.

If X is stable, then X is *hyperelliptic* if $[X] \in \overline{H_g}$; more generally X is (d,r)-*gonal*, respectively d-*gonal*, if $[X] \in \overline{M^r_{g,d}}$, resp. if $[X] \in \overline{M^1_{d,g}}$.

If X is arbitrary, we say X is hyperelliptic, (d,r)-gonal, or d-gonal if so is its stabilization.

A connected curve of genus $g \leq 1$ is d-gonal for all $d \geq 2$.

We recall the definition of admissible covering, due to J. Harris and D. Mumford [16, Sect. 4], and introduce some useful generalizations.

Definition 3. Let Y be a connected nodal curve of genus zero, and $y_1, \ldots, y_b$ be nonsingular points of Y; let X be a connected nodal curve.

(A) A *covering* (of Y) is a regular map $\alpha : X \to Y$ such that the following conditions hold:

a. $\alpha^{-1}(Y_{\text{sing}}) = X_{\text{sing}}$.
b. α is unramified away from X_{sing} and away from $y_1, \ldots, y_b$.
c. α has simple ramification (i.e. a single point with ramification index equals 2) over $y_1, \ldots, y_b$.
d. For every $N \in X_{\text{sing}}$ the ramification indices of α at the two branches of N coincide.

(B) A covering is called *semi-admissible* (resp. *admissible*) if the pointed curve $(Y; y_1, \ldots, y_b)$ is semistable (resp. stable), i.e. for every irreducible component D of Y we have

$$|D \cap \overline{Y \smallsetminus D}| + |D \cap \{y_1, \ldots, y_h\}| \geq 2 \quad (\text{resp.} \geq 3). \tag{5}$$

We shall write $\alpha : X \to (Y; y_1, \ldots, y_b)$ for a covering as above, and sometimes just $\alpha : X \to Y$. In fact the definition of a covering (without its being semi-admissible) does not need the points $y_1, \ldots, y_b$, as conditions (Ab) and (Ac) may be replaced by imposing that α has ordinary ramification away from X_{sing}. The following are simple consequences of the definition.

Remark 3. Let $\alpha : X \to Y$ be a covering.

(A) There exists an integer d such that for every irreducible component $D \subset Y$ the degree of $\alpha_{|D} : \alpha^{-1}(D) \to D$ is d. We say that d is the degree of α.
(B) Every irreducible component of X is nonsingular.
(C) If α is admissible of degree 2, then X is semistable.

In [16] the authors construct the moduli space $\overline{H_{d,b}}$ for admissible coverings, as a projective irreducible variety compactifiying the Hurwitz scheme (parametrizing admissible coverings having smooth range and target), and show that it has a natural morphism

$$\overline{H_{d,b}} \longrightarrow \overline{M_g}; \qquad [\alpha : X \to Y] \mapsto [X^s] \tag{6}$$

where X^s is the stabilization of X and g is its genus, so that $b = 2d + 2g - 2$. For example, if $d = 2$ we have $\overline{H_{2,2g+2}} \longrightarrow \overline{M_g}$.

Moreover, the image of $\overline{H_{2,2g+2}}$ coincides with the locus of hyperelliptic stable curves, $\overline{H_g}$, and more generally the image of (6) is the closure in $\overline{M_g}$ of the locus of d-gonal curves, here denoted by $\overline{M^1_{g,d}}$.

The description of an explicit admissible covering is in Example 3.

1.4 Divisors on Graphs

For any graph G, or any weighted graph (G, w), its divisor group, $\operatorname{Div} G$, or $\operatorname{Div}(G, w)$, is defined as the free abelian group generated by the vertices of G. We use the following notation for a divisor D on (G, w)

$$D = \sum_{v \in V(G)} D(v)v \tag{7}$$

where $D(v) \in \mathbb{Z}$. For loopless and weightless graphs we use the divisor theory developed in [6]. If G is a weighted graph with loops, we extend this theory as in [1]. We begin with a definition.

Definition 4. Let (G, w) be a weighted graph.

We denote by G^0 the loopless graph obtained from G by inserting a vertex in the interior of every loop-edge, and by (G^0, w^0) the weighted graph such that w^0 extends w and is equal to zero on all vertices in $V(G^0) \smallsetminus V(G)$.

We denote by G^w the weightless, loopless graph obtained from G^0 by adding $w(v)$ loops based at v for every $v \in V(G)$ and then inserting a vertex in the interior of every loop-edge.

Notice that (G, w), (G^0, w^0) and G^w have the same genus, and that $(G^0)^{w^0} = G^w$.

For every $D \in \operatorname{Div}(G, w)$ its rank, $r_{(G,w)}(D)$, is set equal to $r_{G^w}(D)$. Linearly equivalent divisors have the same rank. A weighted graph (G, w) of genus g has a canonical divisor $K_{(G,w)} = \sum_{v \in V(G)}(2w(v) - 2 + \operatorname{val}(v))v$ of degree $2g - 2$ such that the following Riemann-Roch formula holds [1, Thm. 3.8]

$$r_{(G,w)}(D) - r_{(G,w)}(K_{(G,w)} - D) = \deg D - g + 1.$$

Remark 4. A consequence of the Riemann-Roch formula is the fact that if $g \leq 1$ then for any divisor D of degree $d \geq 0$ we have $r_{(G,w)}(D) = d - g$.

For a weighted graph (G, w) we denote by $\mathrm{Jac}^d(G, w)$ the set of linear equivalence classes of degree-d divisors, and set

$$W_d^r(G, w) := \{[D] \in \mathrm{Jac}^d(G, w) : r_{(G,w)}(D) \geq r\}.$$

Definition 5. We say that a graph (G, w) is *divisorially d-gonal* if it admits a divisor of degree d and rank at least 1, that is if $W_d^1(G, w) \neq \emptyset$.

A *hyperelliptic* graph is a divisorially 2-gonal graph.

Example 1. Consider the following graph G with $n \geq 2$.

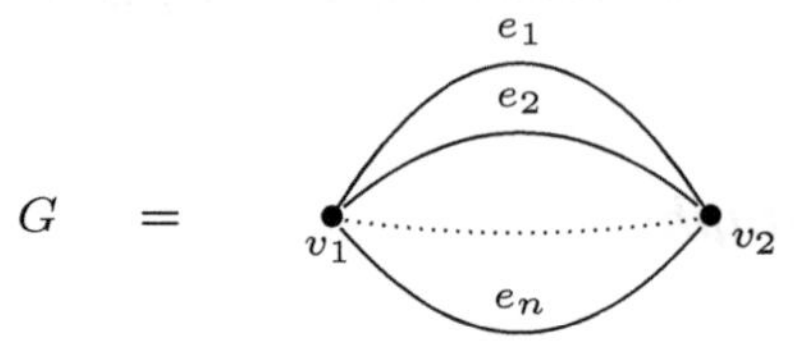

G is obviously hyperelliptic, as $r_G(v_1 + v_2) = 1$. Notice also that

$$r_G(2v_1) = \begin{cases} 1 & \text{if } n = 2 \\ 0 & \text{if } n \geq 3. \end{cases}$$

Now fix on G the weight function given by $w(v_1) = 0$ and $w(v_2) = 1$. Here is the picture of G^w (drawing weight-zero vertices by a "$\circ$")

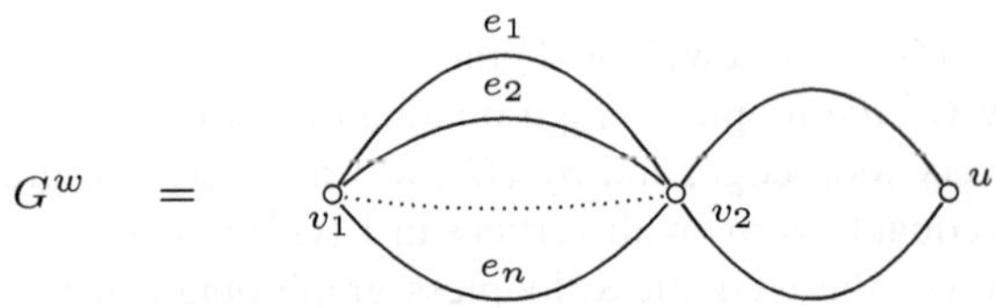

We have $r_{(G,w)}(v_1 + v_2) = r_{(G,w)}(u + v_1) = r_{(G,w)}(u + v_2) = 0$ for every $n \geq 2$. On the other hand

$$r_{(G,w)}(2v_1) = \begin{cases} 1 & \text{if } n = 2 \\ 0 & \text{if } n \geq 3 \end{cases}$$

and the same holds for $2v_2 \sim 2u$. Therefore (G, w) is hyperelliptic if and only if $n = 2$ (in fact $n \leq 2$). This example is generalized in Corollary 3

2 Admissible Coverings and Harmonic Morphisms

2.1 Harmonic Morphisms of Graphs

Let $\phi : G \to G'$ be a morphism; we denote by $\phi_V : V(G) \to V(G')$ the map induced by ϕ on the vertices. ϕ is a *homomorphism* if $\phi(E(G)) \subset E(G')$; in this case we denote by $\phi_E : E(G) \to E(G')$ and by $\phi_H : H(G) \to H(G')$ the induced maps on edges and half-edges. A morphism between weighted graphs (G, w) and (G', w') is defined as a morphism of the underlying graphs, so we write either $G \to G'$ or $(G, w) \to (G', w')$ depending on the situation.

In the next definition, extending the one in [7, Subsect. 2.1], we introduce some extra structure on morphisms between graphs.

Definition 6. Let (G, w) and (G', w') be loopless weighted graphs.

(A) An *indexed morphism* is a morphism $\phi : (G, w) \to (G', w')$ enriched by the assignment, for every $e \in E(G)$, of a non-negative integer, the *index* of ϕ at e, written $r_\phi(e)$, such that $r_\phi(e) = 0$ if and only if $\phi(e)$ is a point. An indexed morphism is *simple* if $r_\phi(e) \leq 1$ for every $e \in E(G)$. Let $e = [h, \overline{h}]$ with $h, \overline{h} \in H(G)$; we set $r_\phi(h) = r_\phi(\overline{h}) = r_\phi(e)$.

(B) An indexed morphism is *pseudo-harmonic* if for every $v \in V(G)$ there exists a number, $m_\phi(v)$, such that for every $e' \in E_{\phi_V(v)}(G')$ (and, redundantly for convenience, every $h' \in H_{\phi_V(v)}(G')$) we have

$$m_\phi(v) = \sum_{e \in E_v(G): \phi(e) = e'} r_\phi(e) = \sum_{h \in H_v(G): \phi(h) = h'} r_\phi(h). \tag{8}$$

(C) A pseudo-harmonic indexed morphism is *non-degenerate* if $m_\phi(v) \geq 1$ for every $v \in V(G)$.

(D) A pseudo-harmonic indexed morphism is *harmonic* if for every $v \in V(G)$ we have, writing $v' = \phi(v)$,

$$\sum_{e \in E_v(G)} (r_\phi(e) - 1) \leq 2\Big(m_\phi(v) - 1 + w(v) - m_\phi(v) w'(v')\Big). \tag{9}$$

In the sequel, all graph morphisms will be indexed morphisms, hence we shall usually omit the word "indexed".

For later use, let us observe that if $w' = \underline{0}$ (i.e. G' is weightless) condition (9) simplifies as follows

$$\sum_{e \in E_v(G)} (r_\phi(e) - 1) \leq 2(m_\phi(v) - 1 + w(v)). \tag{10}$$

Remark 5. Suppose that ϕ contracts a leaf-edge e whose leaf-vertex v has $w(v) = 0$. Then $r_\phi(e) = m_\phi(v) = 0$ and condition (9) is not satisfied on v. So, loosely speaking, a harmonic morphism contracts no weight-zero leaves.

Remark 6 (Relation with harmonic morphisms of [7]). For simple morphisms of weightless graphs our definition of harmonic morphism coincides with the one given in [7, Sec. 2.1] for morphisms which contract no leaves. Indeed, it is clear that any simple pseudo-harmonic morphism is harmonic in the sense of [7]. Conversely, a harmonic morphism in the sense of [7] satisfies (10) (with $w(v) = 0$) if and only if ϕ contracts no leaves; see the previous remark.

Lemma – Definition 1 *Let $\phi : (G, w) \to (G', w')$ be a pseudo-harmonic morphism. Then for every $e' \in E(G')$ and $v' \in V(G')$ we can define the degree of ϕ as follows*

$$\deg \phi = \sum_{e \in E(G):\phi(e)=e'} r_\phi(e) = \sum_{v \in \phi^{-1}(v')} m_\phi(v) \tag{11}$$

(i.e. the above summations are independent of the choice of e' and v').

Proof. Trivial extension of the proof of [7, Lm. 2.2 and Lm. 2.3].

Let $\phi : (G, w) \to (G', w')$ be a pseudo-harmonic morphism. As in [7, Subs. 2.3] we define a pull-back homomorphism $\phi^* : \mathrm{Div}(G', w') \to \mathrm{Div}(G, w)$ as follows: for every $v' \in V(G')$

$$\phi^* v' = \sum_{v \in \phi^{-1}(v')} m_\phi(v) v \tag{12}$$

and we extend this linearly to all of $\mathrm{Div}(G', w')$. By (11) we have

$$\deg D = \deg \phi \deg D'. \tag{13}$$

For a pseudo-harmonic morphism ϕ the *ramification divisor* R_ϕ is defined as follows.

$$R_\phi = \sum_{v \in V(G)} \Big(2\big(m_\phi(v) - 1 + w(v) - m_\phi(v) w'(v')\big) - \sum_{e \in E_v(G)} (r_\phi(e) - 1)\Big) v. \tag{14}$$

The next result, generalizing the analog in [7], implies that harmonic morphisms are characterized, among pseudo-harmonic morphisms, by a Riemann-Hurwitz formula with effective ramification divisor.

Proposition 1 (Riemann-Hurwitz). *Let $\phi : (G, w) \to (G' w')$ be a pseudo-harmonic morphism of weighted graphs of genus g and g' respectively. Then*

$$K_{(G,w)} = \phi^* K_{(G',w')} + R_\phi. \tag{15}$$

ϕ is harmonic if and only if $R_\phi \geq 0$ (equivalently $2g - 2 \geq \deg \phi (2g' - 2)$).

Proof. We write $K = K_{(G,w)}$ and $K' = K_{(G',w')}$. For every $v \in V(G)$ we have $K(v) = 2w(v) - 2 + \text{val}(v)$ (notation in (7)). Hence, writing $v' = \phi(v)$, by (12) we have

$$K(v) - \phi^* K'(v) = 2w(v) - 2 + \text{val}(v) - m_\phi(v)\Big(2w(v') - 2 + \text{val}(v')\Big)$$

$$= 2\Big(m_\phi(v) - 1 + w(v) - m_\phi(v)w(v')\Big) + \text{val}(v) - m_\phi(v)\,\text{val}(v').$$

On the other hand by (11)

$$\sum_{e \in E_v(G)} (r_\phi(e) - 1) = \sum_{e \in E_v(G)} r_\phi(e) - \text{val}(v) = m_\phi(v)\,\text{val}(v') - \text{val}(v).$$

The two above identities imply $K(v) - \phi^* K'(v) = R_\phi(v)$, so (15) is proved.

By definition, ϕ is harmonic if and only if its ramification R_ϕ divisor is effective. The equivalence in parenthesis follows from (13).

Remark 7. Other results proved in [7] for simple harmonic morphisms extend. In particular, if D' and E' are linearly equivalent divisors on (G', w'), their pull-backs $\phi^* D'$ and $\phi^* E'$ under a pseudo-harmonic morphisms ϕ are linearly equivalent.

2.2 *The Hurwitz Existence Problem*

Our goal is to use harmonic morphisms to characterize graphs that are dual graphs of d-gonal curves. This brings up the "Hurwitz existence problem", about the existence of branched coverings of $\mathbb{P}^1$ with prescribed ramification profiles; to state it precisely we need some terminology.

Let $d \geq 1$ be an integer and let $\underline{P} = \{P_1, \ldots, P_b\}$ be a set of partitions of d, so that we write $P_i = \{r_i^1, \ldots, r_i^{n_i}\}$ with $r_i^j \in \mathbb{Z}_{\geq 1}$ and $\sum_{j=1}^{n_i} r_i^j = d$.

We say that $\underline{P}$ is a *Hurwitz partition set*, or that $\underline{P}$ is of Hurwitz type, if the following condition holds. There exist b permutations $\sigma_1, \ldots, \sigma_b \in S_d$ (S_d the symmetric group) whose product is equal to the identity, such that σ_i is the product of n_i disjoint cycles of lengths given by P_i, and such that the subgroup $< \sigma_1, \ldots, \sigma_b >$ is transitive.

Notice that if $\underline{P}$ is of Hurwitz type and we add to it the *trivial* partition $\{1, 1, \ldots, 1\}$, the resulting partition set is again of Hurwitz type.

Remark 8. By the Riemann existence theorem, $\underline{P}$ is a Hurwitz partition set if and only if there exists a degree-d connected covering $\alpha : C \to \mathbb{P}^1$ with $q_1, \ldots, q_b \in \mathbb{P}^1$ such that α is unramified away from $q_1, \ldots, q_b$ and such that for all $i = 1, \ldots, b$ we have $\alpha^*(q_i) = \sum_{j=1}^{n_i} r_i^j p_i^j$. The genus g of C is determined by the Riemann-Hurwitz formula:

$$2g - 2 = -2d + \sum_{i=1}^{b}\sum_{j=1}^{n_i}(r_i^j - 1), \tag{16}$$

so that we shall also say that $\underline{P}$ is a *Hurwitz partition set of genus g and degree d*.

Remark 9. It is a fact that a partition set $\underline{P}$ satisfying (16) is not necessarily of Hurwitz type. Indeed, the so-called Hurwitz existence problem can be stated as follows: characterize Hurwitz partition sets among all $\underline{P}$ satisfying (16). This problem turns out to be very difficult and is open in general. Easy cases in which every $\underline{P}$ satisfying (16) is of Hurwitz type are $P_i = (2, 1, \ldots, 1)$ for every i, or $d \leq 3$, or $b \leq 2$.

On the other hand if $d = 4$ the partition set $\underline{P} = \{(3,1); (2,2); (2,2)\}$ is not of Hurwitz type, but the Riemann-Hurwitz formula (16) holds with $g = 0$; see [19] for this and other results on the Hurwitz existence problem.

Let now $\phi : (G, w) \to T$ be a non-degenerate pseudo-harmonic morphism, where T is a tree; let $v \in V(G)$. For any half-edge $h' \in H(T)$ in the image of some half-edge adjacent to v we define, using (8), a partition of $m_\phi(v)$:

$$P_{h'}(\phi, v) := \{r_\phi(h),\ \forall h \in H_v(G) : \phi(h) = h'\}.$$

Now we associate to v and ϕ the following partition set:

$$\underline{P}(\phi, v) = \{P_{h'}(\phi, v),\ \forall h' \in \phi_H(H_v(G))\}. \tag{17}$$

In the next definition we use the terminology of Remark 8.

Definition 7. **(A)** Let (G, w) be a loopless weighted graph. We say that (G, w) is *d-gonal* if it admits a non-degenerate, degree-d harmonic morphism $\phi : (G, w) \to T$ where T is a tree.

If such a ϕ has the property that for every $v \in V(G)$ the partition set $\underline{P}(\phi, v)$ is contained in a Hurwitz partition set of genus $w(v)$, we say that ϕ is a morphism of *Hurwitz type*, and that (G, w) is a d-gonal graph of *Hurwitz type*.

(B) Let (G, w) be any graph. We say that it is d-gonal, or of Hurwitz type, if so is (G^0, w^0), with (G^0, w^0) as in Definition 4.

Example 2. A harmonic morphism with indices at most equal to 2 is of Hurwitz type. Hence if $d \leq 3$ a d-gonal graph is always of Hurwitz type.

The following is one of the principal results of this paper, of which Theorem 1 is a special case. Recall the terminology introduced in Definition 1.

Theorem 2. *Let (G, w) be a d-gonal graph of Hurwitz type; then there exists a d-gonal curve whose dual graph is (G, w).*

Conversely, let X be a d-gonal curve; then its dual graph is equivalent to a d-gonal graph of Hurwitz type.

The proof of the first part of the theorem will be given in Sect. 2.4. The converse is easier, and will be proved earlier, in Corollary 1.

2.3 The Dual Graph-Map of a Covering

To prove Theorem 2 we shall associate to any covering $\alpha : X \to Y$ an indexed morphism of graphs, called the *dual graph-map* of α, and denoted by

$$\phi_\alpha : (G_X, w_X) \longrightarrow G_Y.$$

As all components of Y have genus zero, we omit the weight function for Y. We sometimes write just $G_X \to G_Y$ for simplicity.

We use the notation of Sect. 1.2; denote by

$$Y = \cup_{u \in V(G_Y)} D_u$$

the irreducible component decomposition of Y. For any $v \in V(G_X)$ we have that $\alpha(C_v)$ is an irreducible component of Y, hence there is a unique $u \in V(G_Y)$ such that $\alpha(C_v) = D_u$; this defines a map $\phi_{\alpha,V} : V(G_X) \to V(G_Y)$ mapping v to u.

Next, $E(G_X)$ and $E(G_Y)$ are identified with the set of nodes of X and Y. To define $\phi_{\alpha,E} : E(G_X) \to E(G_Y)$ let $e \in E(G_X)$; then e corresponds to the node N_e of X. The image $\alpha(N_e)$ is a node of Y, corresponding to a unique edge of G_Y, which we set to be the image of e under $\phi_{\alpha,E}$.

It is trivial to check that the pair $(\phi_{\alpha,V}, \phi_{\alpha,E})$ defines a morphism of graphs, $\phi_\alpha : G_X \to G_Y$.

Let us now define the indices of ϕ_α. For any $e \in E(G)$ let N_e be the corresponding node of X. By Definition 3, the restriction of α to each of the two branches of N_e has the form $u = x^r$ and $v = y^r$ where x and y are local coordinate at the branches of N_e, and u, v are local coordinates at the branches of $\alpha(N_e)$ (which is a node of Y). We set $r_{\phi_\alpha}(e) = r$.

If we need to keep track of the branch points of $\alpha : X \to (Y; y_1, \ldots, y_b)$, we endow the dual graph of Y with b legs, in the obvious way, and write $\phi_\alpha : G_X \to G_{(Y;y_1,\ldots,y_b)}$.

Example 3 (Dual graph-map for the admissible covering of an irreducible hyperelliptic curve). Let $X \in \overline{H_g}$ be an irreducible singular hyperelliptic curve. Such curves are completely characterized; we here choose X irreducible with g nodes, so that its normalization is $\mathbb{P}^1$. Let us describe an admissible covering $\alpha : Z \to Y$ which maps to X under the map (6). As we noticed in Remark 3, Z cannot be equal to X. In fact, Z is the "blow-up" of X at its g nodes, so that $Z = \cup_{i=0}^{g} C_i$ is the union of $g+1$ copies of $\mathbb{P}^1$, with one copy, C_0, corresponding to the normalization of X, and the remaining copies corresponding to the "exceptional" components. Hence $|C_i \cap C_0| = 2$ and $|C_i \cap C_j| = 0$ for all $i, j \neq 0$. Now, since X is hyperelliptic, its normalization C_0 has a two-to-one map to $\mathbb{P}^1$, written $\alpha_0 : C_0 \to D_0 \cong \mathbb{P}^1$, such that $\alpha_0(p_i) = \alpha_0(q_i) = t_i \in D_0$ for every pair $p_i, q_i \in C_0$ of points lying over the i-th node of X. Let $y_0, y_1 \in D_0$ be the two branch points of α_0.

We assume that in X the component C_0 is glued to C_i along the pair p_i, q_i. For $i \geq 1$ we pick a two-to-one map $\alpha_i : C_i \to D_i \cong \mathbb{P}^1$ such that the two points of

C_i glued to X have the same image, s_i, under α_i. Let $y_{2i}, y_{2i+1} \in D_i$ be the two branch points of α_i.

We define Y as the following nodal curve $Y := \sqcup_{i=0}^{g} D_i/_{\{t_i = s_i,\ \forall i=1,\ldots,g\}}$. Now, $(Y; y_{2i}, y_{2i+1},\ \forall i = 0, \ldots g)$ is stable, and it is clear that the α_i glue to an admissible covering $\alpha : Z \to Y$. The dual graphs and graph-map are in the following picture, where $g = 3$.

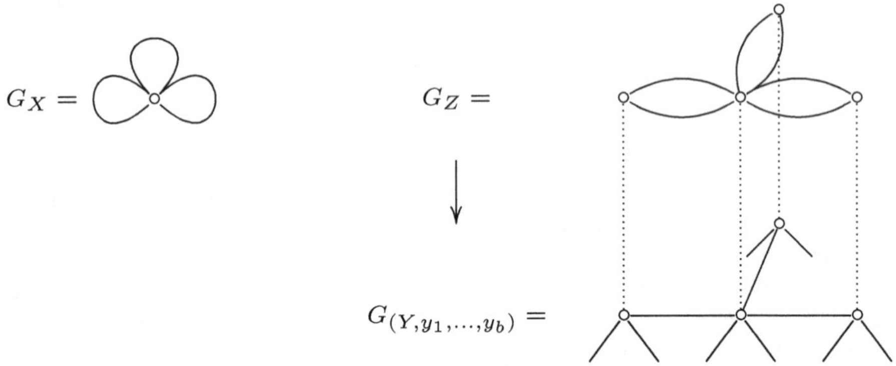

Lemma 1. *Let $\alpha : X \to Y$ be a covering and $\phi_\alpha : (G_X, w_X) \to G_Y$ the dual graph-map defined above. Then ϕ_α is a harmonic homomorphism of Hurwitz type.*

If $\deg \alpha = 2$ *and X has no separating nodes, then ϕ_α is simple.*

Proof. It is clear that G_Y has no loops. By Remark 3 (B), every component C_v of X is nonsingular, hence G_X has no loops.

Since α is a covering, we have that $\phi_{\alpha,V}$ and $\phi_{\alpha,E}$ are surjective, and ϕ_α does not contract any edge of G_X; hence ϕ_α is a homomorphism. We shall abuse notation by writing ϕ_α for $\phi_{\alpha,V}$, $\phi_{\alpha,H}$ and $\phi_{\alpha,E}$.

Let now $v \in V(G_X)$ and $h' \in H_{\phi(v)}(G_Y)$, so that h' corresponds to a point in the image of C_v via α, i.e. to a point in $D_{\phi(v)} \subset Y$. Consider the restriction of α to C_v:

$$\alpha_{|C_v} : C_v \longrightarrow D_{\phi(v)}.$$

This is a finite morphism, and it is clear that for every $h' \in H_{\phi(v)}(G_Y)$

$$\sum_{h \in H_v(G_X) : \phi(h) = h'} r_{\phi_\alpha}(h) = \deg \alpha_{|C_v}.$$

The right hand side above does not depend on h', hence we may set

$$m_{\phi_\alpha}(v) := \deg \alpha_{|C_v}. \tag{18}$$

Therefore ϕ_α is pseudo-harmonic. To prove that ϕ_α is harmonic we must prove that for every $v \in V(G_X)$ we have

$$\sum_{e \in E_v(G_X)} (r_{\phi_\alpha}(e) - 1) \leq 2(m_{\phi_\alpha}(v) - 1 + w_X(v)). \tag{19}$$

Let $R \in \mathrm{Div}(C_v)$ be the ramification divisor of the map $\alpha_{|C_v}$ above. Then, by the Riemann-Hurwitz formula applied to $\alpha_{|C_v}$ we have,

$$\deg R = 2(m_{\phi_\alpha}(v) - 1 + w_X(v)).$$

On the other hand the map $\alpha_{|C_v}$ has ramification index $r_{\phi_\alpha}(h)$ at all $p_h \in H_v(G_X)$, hence we must have

$$R - \sum_{h \in H_v(G_X)} (r_{\phi_\alpha}(h) - 1)p_h \geq 0$$

from which (19) follows. The fact that ϕ_α is of Hurwitz type follows immediately from Remark 8.

Assume $\deg \alpha = 2$ and X free from separating nodes. We must prove the indices of ϕ are all equal to one, i.e. that α_{C_v} does not ramify at the points p_h, for every $h \in H(G_X)$. By contradiction, suppose $\alpha_{|C_v}$ is ramified at p_h; hence, as $\deg \alpha = 2$, it is totally ramified at p_h, so that $\alpha^{-1}(\alpha(p_h)) \cap C_v = p_h$. Since α is an admissible covering, we have exactly the same situation at the other branch of N_e, i.e. at $p_{\overline{h}}$. Therefore

$$\alpha^{-1}(\alpha(N_e)) = \{N_e\}.$$

Now $\alpha(N_e)$ is a node of Y, and hence it is a separating node. So, the above identity implies that N_e is a separating node of X; a contradiction.

Corollary 1. *The second part of Theorem 2 holds.*

Proof. Let X be a d-gonal curve; we must prove that the dual graph of X is equivalent to a d-gonal graph of Hurwitz type. By hypothesis there exists an admissible covering $\hat{X} \to Y$ of degree d such that the stabilization of $\hat{X}$ is the same as the stabilization of X; see the end of Sect. 1.3. Therefore the dual graph of $\hat{X}$ is equivalent to the dual graph of X. By Lemma 1 the dual graph of $\hat{X}$ is of Hurwitz type, hence we are done.

The proof of the first part of Theorem 2 will be based on the next Proposition, which is a converse to Lemma 1.

Proposition 2. *Let (G, w) be a weighted graph of genus ≥ 2 and let T be a tree. Let $\phi : (G, w) \to T$ be a harmonic homomorphism of Hurwitz type. Then there exists a covering $\alpha : X \to Y$ whose dual graph map is ϕ.*

Proof. As ϕ is harmonic, for every $v \in V(G)$ condition (10) holds.

We will abuse notation and write ϕ also for the maps $V(G) \to V(T)$, $H(G) \to H(T)$ and $E(G) \to E(T)$ induced by ϕ. We begin by constructing two curves X and Y whose dual graphs are (G,w) and T.

For every $u \in V(T)$ we pick a pointed curve (D_u, Q_u) with $D_u \cong \mathbb{P}^1$, and such that the (distinct) points in Q_u are indexed by the half-edges adjacent to u:

$$Q_u = \{q_h, \ \forall h \in H_u(T)\}.$$

We have an obvious identification $\cup_{u \in V(T)} Q_u = H(T)$. To glue the curves D_u to a connected nodal curve Y we proceed as in Sect. 2.3, getting

$$Y = \frac{\sqcup_{u \in V(T)} D_u}{\{q_h = q_{\overline{h}}, \ \forall h \in H(T)\}}.$$

By construction, T is the dual graph of Y.

Now to construct X we begin by finding its irreducible components C_v with their gluing point sets P_v. Pick $v \in V(G)$ and $u = \phi(v) \in V(T)$. By hypothesis, $m_\phi(v) \geq 1$; we claim that there exists a morphism from a smooth curve C_v of genus $w(v)$ to D_u

$$\alpha_v : C_v \longrightarrow D_u \tag{20}$$

of degree equal to $m_\phi(v)$ such that for every $h' \in H_u(T)$ the pull-back of the divisor $q_{h'}$ has the form

$$\alpha_v^* q_{h'} = \sum_{\phi_H(h)=h'} r_\phi(h) p_h$$

for some points $\{p_h, \ h \in H(G)\} \subset C_v$; we set $P_v = \{p_h, \ h \in H(G)\}$.

Indeed, the degree of the ramification divisor of a degree-m morphism from a curve of genus $w(v)$ to $\mathbb{P}^1$ of is equal to $2(m - 1 + w(v))$. Therefore assumption (10) guarantees that the ramification conditions we are imposing are compatible; now as ϕ is of Hurwitz type, the Riemann Existence theorem yields that such an α_v exists; see Remark 8. Observe that α_v may have other ramification, in which case we can easily impose that any extra ramification and branch point lie $C_v \smallsetminus P_v$, respectively in $D_u \smallsetminus Q_u$, and that they are all simple.

Now that we have the pointed curves (C_v, P_v) for every $v \in V(G)$ such that C_v is a smooth curve of genus $w(v)$ we can define X:

$$X := \frac{\sqcup_{v \in V(G)} C_v}{\{p_h = p_{\overline{h}}, \ \forall h \in H(G)\}},$$

so, (G,w) is the dual graph of X.

Let us prove that the morphisms $\{\alpha_v, \ \forall v \in V(G)\}$ glue to a morphism $\alpha : X \to Y$. It suffices to check that for every pair $(p_h, p_{\overline{h}})$ we have $\alpha_v(p_h) = \alpha_{\overline{v}}(p_{\overline{h}})$, where $p_h \in C_v$ and $p_{\overline{h}} \in C_{\overline{v}}$. We have $\alpha_v(p_h) = q_{\phi(h)}$ and $\alpha_{\overline{v}}(p_{\overline{h}}) = q_{\phi(\overline{h})}$. Now, looking at the involution of $H(T)$ (see Sect. 1.2), we have $\phi(\overline{h}) = \overline{(\phi(h))}$, and hence $\alpha : X \to Y$ is well defined.

We now show that α is a covering. It is obvious that $\alpha^{-1}(Y_{\text{sing}}) = X_{\text{sing}}$. Next, for every node N_e of X, the ramification indices at the two branches, $p_h, p_{\overline{h}}$ where $[h, \overline{h}] = e$, are equal, as they are equal to $r_\phi(h)$ and $r_\phi(\overline{h})$. As we have imposed that α_v has only ordinary ramification points away from the nodes of X, condition (0c) of Definition 3 is satisfied. Therefore the map $\alpha : X \to Y$ is a covering; obviously α has ϕ as dual graph-map.

To deduce Theorem 2 from the previous Proposition we will need to construct a suitable homomorphism from a given morphism of Hurwitz type, which is done in the next Lemma.

Lemma 2. *Let $\phi : (G, w) \to T$ be a degree-d morphism of Hurwitz type. Then there exists a degree-d homomorphism $\hat{\phi} : (\hat{G}, \hat{w}) \to \hat{T}$ of Hurwitz type fitting in a commutative diagram*

$$\begin{array}{ccc} \widehat{G} & \xrightarrow{\ \widehat{\phi}\ } & \widehat{T} \\ \downarrow & & \downarrow \\ G & \xrightarrow{\ \phi\ } & T \end{array} \qquad (21)$$

whose vertical arrows are edge contractions, and such that $(\hat{G}, \hat{w})$ is equivalent to (G, w).

Proof. The picture after the proof illustrates the forthcoming construction. Since (G^0, w^0) is equivalent to (G, w) we can assume G loopless. Consider the set of "vertical" edges of ϕ:

$$E_\phi^{\text{ver}}(G) := \{e \in E(G) : \phi(e) \in V(G')\}$$

and set $E_\phi^{\text{hor}}(G) := E(G) \smallsetminus E_\phi^{\text{ver}}(G)$. Of course, if $E_\phi^{\text{ver}}(G) = \emptyset$ there is nothing to prove. So, let $e \in E_\phi^{\text{ver}}(G)$ and v_1, v_2 be its endpoints. We set $u = \phi(v_1) = \phi(v_2) = \phi(e)$ and write

$$\phi_V^{-1}(u) = \{v_1, v_2, \ldots, v_n\} \qquad (22)$$

with $n \geq 2$ and the v_i distinct. Set $m_i := m_\phi(v_i)$ for $i = 1, \ldots, n$.

We begin by constructing $\hat{G}$. First, we insert a weight zero vertex $\hat{v}_e$ in the interior of e, and denote by $\hat{e}_1, \hat{e}_2$ the two edges adjacent to it. Next, we attach $m_1 - 1$ leaves at v_1, $m_2 - 1$ leaves at v_2, and m_i leaves at v_i for all $i \geq 3$; all these leaf-vertices are given weight zero. We denote the j-th leaf-edge attached to v_i by $l^{(i)}_{e,j^{(i)}}$ and its leaf-vertex by $w^{(i)}_{e,j^{(i)}}$, with $j^{(i)} = 1, \ldots, m_i - 1$ if $i = 1, 2$ and $j^{(i)} = 1, \ldots, m_i$ if $i \geq 3$.

We repeat this construction for every $e \in E^{\text{ver}}_{\phi}(G)$, and we denote the so obtained graph by $\hat{G}$. We have identifications

$$E(\hat{G}) = E^{\text{hor}}_{\phi}(G) \sqcup \{\hat{e}_1, \hat{e}_2, \ \forall e \in E^{\text{ver}}_{\phi}(G)\} \sqcup \{l^{(i)}_{e,j^{(i)}} \ \ \forall e \in E^{\text{ver}}_{\phi}(G), \forall i, \forall j^{(i)}\}$$

and

$$V(\hat{G}) = V(G) \sqcup \{\hat{v}_e, \ \forall e \in E^{\text{ver}}_{\phi}(G)\} \sqcup \{w^{(i)}_{e,j^{(i)}} \ \ \forall e \in E^{\text{ver}}_{\phi}(G), \forall i, \forall j^{(i)}\}.$$

There is a contraction $\hat{G} \to G$ given by contracting, for every $e \in E^{\text{ver}}_{\phi}(G)$, the edge $\hat{e}_1$ and all leaf edges $l^{(i)}_{e,j^{(i)}}$. It is clear that G and $\hat{G}$ are equivalent.

Let us now construct $\hat{T}$; for every $e \in E^{\text{ver}}(G)$ we add to T a leaf based at $u = \phi(e)$; we denote by $\hat{l}_e$, and $\hat{w}_e$ the edge and vertex of this leaf. We let $\hat{T}$ be the tree obtained after repeating this process for every $e \in E^{\text{ver}}_{\phi}(G)$. There is a contraction $\hat{T} \to T$ given by contracting all leaf edges $\hat{l}_e$.

Let $G' := G - E^{\text{ver}}_{\phi}(G)$, so that G' is also a subgraph of $\hat{G}$. Denote by $\phi' : G' \to T$ the restriction of ϕ to G'; observe that ϕ' is a harmonic homomorphism. To construct $\hat{\phi} : \hat{G} \to \hat{T}$ we extend ϕ' as follows. For every $e \in E^{\text{ver}}_{\phi}(G)$ we set, with the above notations,

$$\hat{\phi}(\hat{e}_1) = \hat{\phi}(\hat{e}_2) = \hat{\phi}(l^{(i)}_{e,j^{(i)}}) = \hat{l}_e$$

and

$$\hat{\phi}(\hat{v}_e) = \hat{\phi}(w^{(i)}_{e,j^{(i)}}) = \hat{w}_e$$

for every i and $j^{(i)}$. Finally, we define the indices of $\hat{\phi}$

$$r_{\hat{\phi}}(\hat{e}) = \begin{cases} r_{\phi}(\hat{e}) & \text{if } \hat{e} \in E^{\text{hor}}_{\phi}(G) \\ 1 & \text{otherwise.} \end{cases}$$

It is clear that $\hat{\phi}$ is a homomorphism and that diagram (21) is commutative.

Let us check that $\hat{\phi}$ is pseudo-harmonic. Pick $e \in E^{\text{ver}}(G)$. Consider a leaf vertex $w^{(i)}_{e,j^{(i)}}$ of $\hat{G}$. Then it is clear that condition (8) holds with $m_{\hat{\phi}}(w^{(i)}_{e,j^{(i)}}) = 1$. Next, consider a vertex $\hat{v}_e$. It is again clear that condition (8) holds with $m_{\hat{\phi}}(\hat{v}_e) = 2$. Finally, consider the vertices $v_1, \ldots, v_n$ introduced in (22). Recall that $\hat{\phi}(v_i) = \phi(v_i) = u$ and condition (8) holds for any edge in $E(T) \subset E(\hat{T})$ adjacent to u with

$m_{\hat{\phi}}(v_i) = m_i$. We need to check that the same holds for the leaf-edges $\hat{l}_e \in E(\hat{T})$. For v_1 and any leaf $\hat{l}_e$ adjacent to $\hat{\phi}(v_1)$ we have

$$\sum_{\hat{e}\in E_{v_1}(\hat{G}):\hat{\phi}(\hat{e})=\hat{l}_e} r_{\hat{\phi}}(\hat{e}) = \sum_{j^{(1)}=1}^{m_1-1} r_{\hat{\phi}}(l^{(1)}_{e,j^{(1)}}) + r_{\hat{\phi}}(\hat{e}_1) = m_1 - 1 + 1 = m_1,$$

(as $r_{\hat{\phi}}(l^{(1)}_{e,j^{(1)}}) = r_{\hat{\phi}}(\hat{e}_1) = 1$) Similarly for v_2. Next, for v_i with $i = 3, \ldots, n$ we have

$$\sum_{\hat{e}\in E_{v_i}(\hat{G}):\hat{\phi}(e)=\hat{l}_e} r_{\hat{\phi}}(\hat{e}) = \sum_{j^{(i)}=1}^{m_i} r_{\hat{\phi}}(l^{(i)}_{e,j^{(i)}}) = m_i.$$

Since ϕ' is pseudo-harmonic there is nothing else to check; hence $\hat{\phi}$ is pseudo-harmonic. Now, to prove that $\hat{\phi}$ is harmonic we must check that condition (10) holds; since ϕ' is harmonic, this follows immediately from the fact that the index of $\hat{\phi}$ at each of the new edges is 1.

Finally, to prove that $\hat{\phi}$ is of Hurwitz type, pick a vertex of $\hat{G}$; if this vertex is of type $\hat{v}_e$ or $w^{(i)}_{e,j^{(i)}}$ then the associated partition set contains only the trivial partition, and hence it is obviously contained in some partition set of Hurwitz type. The remaining case is that of a vertex v of G. Then either $\underline{P}(\phi, v) = \underline{P}(\hat{\phi}, v)$ (if v is not adjacent to $e \in E^{\text{ver}}$), or $\underline{P}(\hat{\phi}, v)$ is obtained by adding the trivial partition to $\underline{P}(\phi, v)$; in both cases, since by hypothesis $\underline{P}(\phi, v)$ is contained in a partition set of Hurwitz type, so is $\underline{P}(\hat{\phi}, v)$.

The following picture illustrates $\hat{\phi}$ for a 3-gonal morphism ϕ. All indices of ϕ are set equal to 1, with the exception of the vertical edge e for which $r_\phi(e) = 0$.

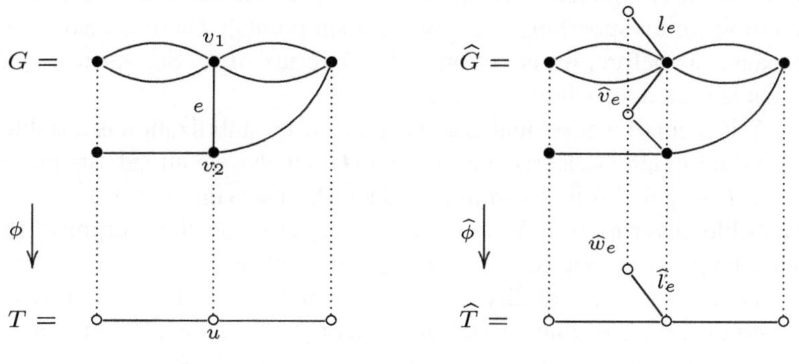

2.4 *Proof of Theorem 2*

By Corollary 1 we need only prove the first part of the Theorem. We first assume that G is free from loops.

By hypothesis we have a non-degenerate, degree-d, harmonic morphism $\phi : G \to T$ of Hurwitz type, where T is a tree. We let $\hat{\phi} : \hat{G} \to \hat{T}$ be a degree-d, harmonic homomorphism associated to ϕ by Lemma 2. Now $\hat{\phi} : \hat{G} \to \hat{T}$ satisfies all the assumptions of Proposition 2, hence there exists a covering $\alpha : \hat{X} \to \hat{Y}$ whose dual graph-map is $\hat{\phi} : \hat{G} \to \hat{T}$. We denote by $y_1, \dots, y_b \in Y$ the smooth branch points of α.

Suppose now that (G, w) is stable; we claim that α is admissible, i.e. that $(Y; y_1, \dots, y_b)$ is stable. We write $Y = \cup_{\hat{u} \in V(\hat{T})} D_{\hat{u}}$ as usual. For every branch point y_i we attach a leg to $\hat{T}$, having endpoint $\hat{u} \in V(\hat{T})$ such that $y_i \in D_{\hat{u}}$. We must prove that the graph $\hat{T}$ with these b legs has no vertex of valency less than 3. Pick a vertex of $\hat{T}$. There are two cases, either it is a vertex $u \in V(T)$ or it is a leaf vertex $\hat{w}_e$.

In the first case the preimage of u via $\hat{\phi}$ is made of vertices of the original graph G. So, pick $v \in V(G)$ with $\phi(v) = u$. The map $\alpha_v : C_v \to D_u$ has degree $m_\phi(v)$. If $m_\phi(v) = 1$, then, of course, $C_v \cong \mathbb{P}^1$ and we have $\text{val}(u) \geq \text{val}(v)$, and $\text{val}(v) \geq 3$ as G is stable; hence $\text{val}(u) \geq 3$ as wanted. Notice that this is the only place where we use that (G, w) is stable, the rest of the proof works for any d-gonal graph. If $m_\phi(v) \geq 2$ then the map α_v has at least two branch points, each of which corresponds to a leg adjacent to u. If α_v has more than two branch points, then u has more than two legs adjacent to it, hence we are done; if α_v has exactly two branch points, then, by Riemann-Hurwitz, $C_v \cong \mathbb{P}^1$ and hence $C_v \subsetneq X$ as X has genus ≥ 2. Therefore $C_v \cap \overline{X \smallsetminus C_v} \neq \emptyset$, and hence there is at least one edge of T adjacent to u, hence $\text{val}(u) \geq 3$.

Now consider a vertex of type $\hat{w}_e$. By construction, its preimage contains the vertex $\hat{v}_e$, for which $m_{\hat{\phi}}(\hat{v}_e) = 2$; hence the corresponding component of $\hat{X}$ maps two-to-one to the component corresponding to $\hat{w}_e$, and hence there are at least 2 legs attached to $\hat{w}_e$ (corresponding to the two branch points). There is also at least one edge because, as before, $\hat{w}_e$ is not an isolated vertex of $\hat{T}$. So, $\text{val}(\hat{w}_e) \geq 3$. This proves that α is an admissible covering.

Now, $\hat{X}$ is a curve whose dual graph is $(\hat{G}, \hat{w})$. Its stabilization is a stable curve, X, whose dual graph is clearly the original (G, w). As we already mentioned, the fact that X is d-gonal follows from [16, Sect 4], observing that X is the image of the admissible covering $\alpha : \hat{X} \to (Y; y_1, \dots, y_b)$ under the morphism (6). This concludes the proof in case (G, w) is stable and loopless.

Now let us drop the stability assumption on (G, w). If α is admissible, the previous argument yields that the stabilization of $\hat{X}$ is d-gonal. But the stabilization of $\hat{X}$ is the same as the stabilization of X, hence we are done.

Suppose α is not admissible; then there are two cases. First case: $\hat{T}$ has a vertex u of valency 1. By the previous part of the proof this can happen only if every vertex $v \in \phi_V^{-1}(u)$ has valency 1 and α induces an isomorphism $C_v \cong \mathbb{P}^1$; such components of $\hat{X}$ are called rational tails. We now remove the component D_u from $\hat{Y}$, and all the rational tails mapping to D_u from $\hat{X}$. Observe that this operation does not change the stabilization of $\hat{X}$. This corresponds to removing one leaf from $\hat{T}$ and all its preimages (all leaves) under ϕ. We repeat this process until there are no 1-valent vertices left.

Second case, $\hat{T}$ has a vertex u of valency 2. Again by the previous part this happens only if every $v \in \phi_{\overline{V}}^{-1}(u)$ has valency 2 and α induces an isomorphism $C_v \cong \mathbb{P}^1$. We collapse the component D_u of $\hat{Y}$ and all the exceptional components of $\hat{X}$ mapping to D_u. Again, this operation does not change the stabilization of $\hat{X}$. We repeat this process until there are no 2-valent vertices left.

In this way we arrive at two curves X' and $(Y'; y_1, \ldots, y_b)$, the latter being stable, endowed with a covering $\alpha' : X' \to Y'$ induced by α, by construction; indeed the process did not touch the branch points $y_1, \ldots, y_b$, which are now the smooth branch points of α'. The covering α' is admissible, hence the stabilization of X' is d-gonal (as before). Since the stabilization of X' is equal to the stabilization of X we are done. The loopless case is now proved.

We now suppose that G has some loop; let (G^0, w^0) be its loopless model. By Definition 7, (G^0, w^0) is d-gonal. The previous part yields that there exists a curve X^0 whose dual graph is (G^0, w^0) and whose stabilization is d-gonal. Since the stabilization of X is equal to the stabilization of X^0 we are done. Theorem 2 is proved. ■

Remark 10 (Hyperelliptic and 2-gonal graphs). It is easy to construct hyperelliptic (i.e. divisorially 2-gonal) graphs that are not 2-gonal; for example the weightless graph G in Example 1 for $n \geq 3$.

On the other hand every 2-gonal stable graph is hyperelliptic, by Theorem 2 and Proposition 4; see also Theorem 4. More generally, using Remark 7 one can prove directly that if a graph admits a pseudo-harmonic morphism of degree 2 to a tree, then it is hyperelliptic. We omit the details.

Example 4 (A 3-gonal graph which is not divisorially 3-gonal). In the following picture we have a pseudo-harmonic morphism ϕ of degree 3 from a weightless graph G of genus 5. There is one edge, joining v_2 and v_3, where the index is 2, and all other edges have index 1. The graph G is easily seen to be 3-gonal, but not divisorially 3-gonal, i.e. $W_3^1(G) = \emptyset$. We omit the details.

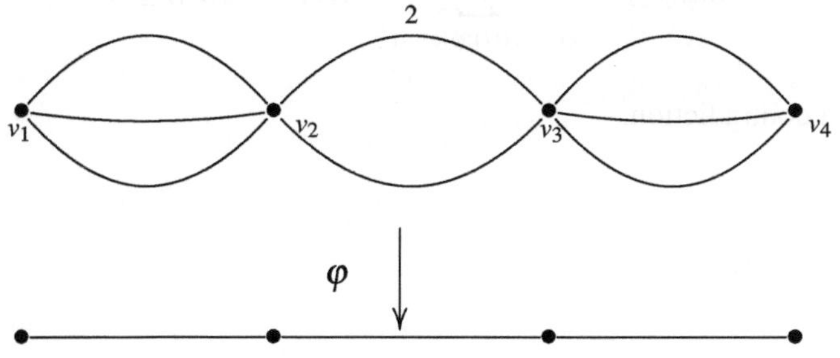

Example 5 (A divisorially 3-gonal graph which is not 3-gonal). In the graph G below, weightless of genus 5, we have

$$3v_1 \sim 3v_2 \sim -v_2 + 2v_0 + 2v_3 \sim 3v_3 \sim v_0 + v_2 + v_3 \sim 3v_4$$

so the graph is divisorially 3-gonal.

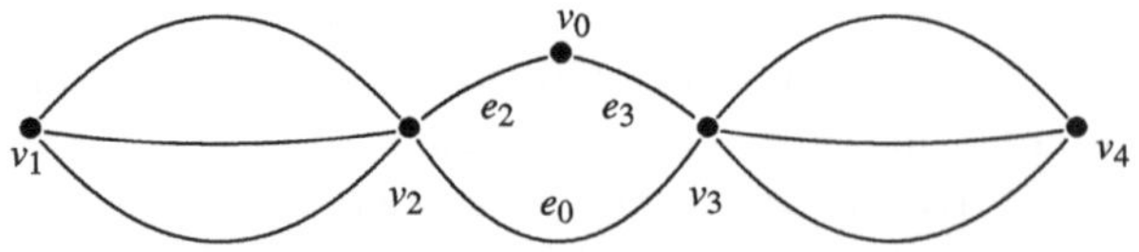

Let us show that G does not admit a non-degenerate pseudo-harmonic morphism of degree 3 to a tree. By contradiction, let $\phi : G \to T$ be such a morphism. Then the edges adjacent to v_1 cannot get contracted (if one of them is contracted, all of them will be contracted, for T has no loops; but if all of them get contracted then $m_\phi(v_1) = 0$, which is not possible). Therefore the three edges adjacent to v_1 are all mapped to the unique edge, e'_1, joining $\phi(v_1)$ with $\phi(v_2)$. Similarly, the edges adjacent to v_4 are all mapped to the unique edge e'_2 joining $\phi(v_4)$ with $\phi(v_3)$. Therefore, as ϕ as degree 3, all edges between v_1 and v_2, and all edges between v_3 and v_4 have index 1, hence $m_\phi(v_1) = m_\phi(v_2) = m_\phi(v_3) = m_\phi(v_4) = 3$.

Now, if $\phi(v_2) = \phi(v_3)$ then one easily checks that e_0 is contracted and e_2, e_3 are mapped to the same edge e'_3 of T, which is different from e'_1 and e'_2. Therefore we have $1 \leq r_\phi(e_i) \leq 2$ for $i = 1, 2$. But then by (8) we have

$$m_\phi(v_2) = \sum_{e \in E_{v_2}(G):\phi(e)=e'_3} r_\phi(e) = r_\phi(e_2) \leq 2$$

and this is a contradiction.

It remains to consider the case $\phi(v_2) \neq \phi(v_3)$, let $e'_0 = \phi(e_0)$. Then v_0 is either mapped to $\phi(v_2)$ by contracting e_2, or to $\phi(v_3)$ by contracting e_3 (for otherwise T would not be a tree). With no loss of generality, set $\phi(v_2) = \phi(v_0)$ so that $r_\phi(e_2) = 0$. Now, since $\phi(e_3) = \phi(e_0) = e'_0$ we have $r_\phi(e_0) \leq 2$. Hence

$$m_\phi(v_2) = \sum_{e \in E_{v_2}(G):\phi(e)=e'_0} r_\phi(e) = r_\phi(e_0) \leq 2$$

and this is a contradiction.

3 Higher Gonality and Applications to Tropical Curves

3.1 Basics on Tropical Curves

A (weighted) tropical curve is a weighted metric graph $\Gamma = (G, w, \ell)$ where (G, w) is a weighted graph and $\ell : E(G) \to \mathbb{R}_{>0}$. The divisor group $\text{Div}(\Gamma)$ is, as usual, the free abelian group generated by the points of Γ (viewed as a metric space). The weightless case has been carefully studied in [14], for example; the general case has been recently treated in [1], to which we refer for the definition of the rank $r_\Gamma(D)$ of any $D \in \text{Div}(\Gamma)$ and its basic properties. Here we just need the following facts. Given $\Gamma = (G, w, \ell)$ we introduce the tropical curve $\Gamma^w = (G^w, \underline{0}, \ell^w)$ such that G^w is as in Definition 4, the weight function is zero (hence denoted by $\underline{0}$), and ℓ^w is the extension of ℓ such that $\ell^w(e) = 1$ for every $e \in E(G^w) \smallsetminus E(G^0)$. We have a natural commutative diagram

$$\begin{array}{ccc} \text{Div}(G, w) & \hookrightarrow & \text{Div}(G^w) \\ \big\downarrow & & \big\downarrow \\ \text{Div}(\Gamma) & \hookrightarrow & \text{Div}(\Gamma^w) \end{array} \tag{23}$$

the above injections will be viewed as inclusions in the sequel. Then, for any $D \in \text{Div}(\Gamma)$ we have, by [1, Sect. 5]

$$r_\Gamma(D) = r_{\Gamma^w}(D). \tag{24}$$

So, the horizontal arrows of the above diagram preserve the rank. If the length functions on Γ and Γ^w are identically equal to 1, then, by [18, Thm 1.3], also the vertical arrows of the diagram preserve the rank.

For a tropical curve Γ we denote by $W^r_d(\Gamma)$ the set of equivalence classes of divisors of degree d and rank at least r; we say that Γ is (d,r)-gonal if $W^r_d(\Gamma) \neq \emptyset$.

The moduli space of equivalence classes of tropical curves of genus g is denoted by M_g^{trop}, and the locus in it of curves whose underlying weighted graph is (G, w) is denoted by $M^{\text{trop}}(G, w)$. This gives a partition

$$M_g^{\text{trop}} = \sqcup M^{\text{trop}}(G, w)$$

indexed by all stable graphs (G, w) of genus g.

3.2 From Algebraic Gonality to Combinatorial and Tropical Gonality

Theorem 3. *Let $X \in \overline{M^r_{g,d}}$ and let (G, w) be the dual graph of X. Then*

(A) *There exists a refinement $(\hat{G}, \hat{w})$ of (G, w), such that $W^r_d(\hat{G}, \hat{w}) \neq \emptyset$;*
(B) *There exists a tropical curve $\Gamma \in M^{\rm trop}(G, w)$ such that $W^r_d(\Gamma) \neq \emptyset$.*

Proof. By hypothesis there exists a family of curves, $f : \mathscr{X} \to B$, with B smooth, connected, of dimension one, such that there is a point $b_0 \in B$ over which the fiber of f is isomorphic to X, and the fiber over any other point of B is a smooth curve whose W^r_d is not empty. In the sequel we will work up to replacing B by an open neighborhood of b_0, or by an étale covering. Therefore we will also assume that f has a section.

For every $b \in B^* = B \smallsetminus \{b_0\}$ we have $W^r_d(X_b) \neq 0$ (X_b is the fiber of f over b). Write $f^* : \mathscr{X}^* \to B^*$ for the smooth family obtained by restricting f to $\mathscr{X} \smallsetminus X_0$. Recall that as b varies in B^* the $W^r_d(X_b)$ form a family ([2, Sect. 2] or [3, Ch. 21]), i.e. there exists a morphism of schemes

$$W^r_{d,f^*} \to B^* \tag{25}$$

whose fiber over b is $W^r_d(X_b)$.

Up to replacing B by a finite covering possibly ramified only over b_0, we may assume that the base change of the morphism (25) has a section. The base change of f to this covering may be singular (or even non normal) over b_0, but will still have smooth fiber away from b_0. Let $h : \mathscr{Z} \to B$ be the desingularization of the normalization of this base change of f. Then the fiber of h over b_0 is a semistable curve Z_0 whose stabilization is X; all remaining fibers are isomorphic to the original fibers of f. By construction, the morphism

$$W^r_{d,h^*} \to B^* \tag{26}$$

has a section, σ. By our initial assumption $h : \mathscr{Z} \to B$ is endowed with a section, hence, by [9, Prop. 8.4], σ corresponds to a line bundle $\mathscr{L}^* \in \operatorname{Pic} \mathscr{Z}^*$. Since $\mathscr{Z}$ is nonsingular $\mathscr{L}^*$ extends to some line bundle $\mathscr{L}$ on $\mathscr{Z}$, and we have, for every $b \in B$:

$$r(Z_b, \mathscr{L}_{|Z_b}) \geq r.$$

Let $(\hat{G}, \hat{w})$ be the dual graph of Z_0. We can apply the weighted specialization Lemma [1, Thm 4.9] to $\mathscr{Z} \to B$ with respect to the line bundle $\mathscr{L}$. This gives, viewing the multidegree $\underline{\deg}\, \mathscr{L}_{|Z_0}$ as a divisor on $\hat{G}$,

$$r_{(\hat{G},\hat{w})}(\underline{\deg}\, \mathscr{L}_{|Z_0}) \geq r(Z_b, \mathscr{L}_{|Z_b}) \geq r$$

and therefore $W^r_d(\hat{G}, \hat{w}) \neq \emptyset$.

Now, by construction $(\hat{G}, \hat{w})$ a refinement of (G, w) (the dual graph of X). Hence the first part is proved.

For the next part, consider the tropical curve $\hat{\Gamma} = (\hat{G}, \hat{w}, \hat{\ell})$ with $\hat{\ell}(e) = 1$ for every $e \in E(\hat{G})$. Let $D \in W^r_d(\hat{G}, \hat{w})$. Then D is also a divisor on $\hat{\Gamma}$ (cf. Diagram (23)). We claim that $r_{\hat{\Gamma}}(D) \geq r$.

We have, by definition,

$$r \leq r_{(\hat{G},\hat{w})}(D) = r_{\hat{G}^{\hat{w}}}(D).$$

Let $\hat{\Gamma}^{\hat{w}} = (\hat{G}^{\hat{w}}, \underline{0}, \hat{\ell}^{\hat{w}})$ be the tropical curve such that $\hat{\ell}^{\hat{w}}(e) = 1$ for every $e \in E(\Gamma^{\hat{w}})$; so D is also a divisor on $\hat{\Gamma}^{\hat{w}}$. By [18, Thm 1.3], we have

$$r_{\hat{G}^{\hat{w}}}(D) = r_{\hat{\Gamma}^{\hat{w}}}(D).$$

Now, as we noticed in (24) we have

$$r_{\hat{\Gamma}^{\hat{w}}}(D) = r_{\hat{\Gamma}}(D).$$

The claim is proved; therefore $W^r_d(\hat{\Gamma}) \neq \emptyset$.

The supporting graph $(\hat{G}, \hat{w})$ of $\hat{\Gamma}$ is not necessarily stable; its stabilization, obtained by removing every 2-valent vertex of weight zero, is the original (G, w), so that $\hat{\Gamma}$ is tropically equivalent to a curve $\Gamma \in M^{\text{trop}}_g(G, w)$. Since the underlying metric spaces of Γ and $\hat{\Gamma}$ coincide, we have

$$W^r_d(\Gamma) = W^r_d(\hat{\Gamma}) \neq \emptyset.$$

The statement is proved.

Corollary 2. *Every d-gonal stable weighted graph admits a divisorially d-gonal refinement.*

Proof. Let (G, w) be a d-gonal stable graph. By Theorem 2 there exists $X \in \overline{M^1_{g,d}}$ whose dual graph is (G, w). By Theorem 3 we are done.

The proof of Theorem 3 gives a more precise result, to state which we need some further terminology.

Let X be any curve. A *one-parameter smoothing* of X is a morphism $f : \mathscr{X} \to (B, b_0)$, where B is smooth connected with $\dim B = 1$, b_0 is a point of B such that $f^{-1}(b_0) = X$, and all other fibers of f are smooth curves. By definition, $\mathscr{X}$ is a surface having only singularities of type A_n at the nodes of X. To f we associate the following length function ℓ_f on G_X:

$$\ell_f : E(G_X) \longrightarrow \mathbb{R}_{>0}; \qquad e \mapsto n(e)$$

where $n(e)$ is the integer defined by the fact that $\mathscr{X}$ has a singularity of type $A_{n(e)-1}$ at the node of X corresponding to e. In particular, if $\mathscr{X}$ is nonsingular, then ℓ_f is constant equal to one. This defines the following tropical curve associated to f:

$$\Gamma_f = (G_X, w_X, \ell_f).$$

Similarly, we define a refinement of the dual graph of X by inserting $n(e)-1$ vertices of weight zero in e, for every $e \in E(G_X)$; we denote this refinement by (G_f, w_f). Now, if $\mathscr{Y} \to \mathscr{X}$ is the minimal resolution of singularities and $h : \mathscr{Y} \to B$ the composition with f, then (G_f, w_f) is the dual graph of the fiber of h over b_0; we denote by X_f this fiber.

For example, the surface $\mathscr{X}$ is nonsingular if and only if $X = X_f$, if and only if $(G_X, w_X) = (G_f, w_f)$

The following is a consequence the proof of Theorem 3, where X_f corresponds to the curve Z_0, while $(G_f, w_f) = (\hat{G}, \hat{w})$, and $\Gamma_f = \Gamma$.

Proposition 3. *Let $f : \mathscr{X} \to (B, b_0)$ be a one-parameter smoothing of the curve X. If the general fiber of f is (d, r)-gonal (i.e. if $W^r_d(f^{-1}(b)) \neq \emptyset$ for every $b \neq b_0$) then the following facts hold.*

1. $W^r_d(G_f, w_f) \neq \emptyset$.
2. $W^r_d(\Gamma_f) \neq \emptyset$.
3. $W^r_d(X_f) \neq \emptyset$.

Remark 11. The tropical curve Γ_f may be interpreted as a Berkovich skeleton of the generic fiber $\mathscr{X}_K$ of $\mathscr{X} \to B$, where K is the function field of B (note that Γ_f depends on $\mathscr{X}$). Then the theorem says that the Berkovich skeleton of a (d, r)-gonal smooth algebraic curve over K is a (d, r)-gonal tropical curve.

4 The Hyperelliptic Case

4.1 Hyperelliptic Weighted Graphs

Recall that a graph is hyperelliptic if it has a divisor of degree 2 and rank 1. Hyperelliptic graphs free from loops and weights have been thoroughly studied in [7]. In this subsection we extend some of their results to weighted graphs admitting loops.

Recall the notation of Definition 4. We will use the following terminology. A 2-valent vertex of is said to be *special* if its removal creates a loop. For example, given (G, w), every vertex in $V(G^w) \smallsetminus V(G)$ is special.

Lemma 3. *Let (G, w) be a weighted graph of genus g. Then (G, w) is hyperelliptic if and only if so is G^w if and only if so is (G^0, w^0).*

Proof. By Remark 4 we can assume $g \geq 2$. By definition, if G is hyperelliptic so is G^w. Conversely, assume G^w hyperelliptic and let $D \in \mathrm{Div}(G^w)$ be an effective divisor of degree 2 and rank 1. If $\mathrm{Supp}\, D \subset V(G)$ we are done, as $r_{(G,w)}(D) = r_{G^w}(D)$. Otherwise, suppose $D = u + u'$ with $u \in V(G^w) \smallsetminus V(G)$. So, u is a special vertex whose removal creates a loop based at a vertex v of G. As $r_{G^w}(u + u') = 1$, it is clear that $u' \neq v$ (e.g. by [1, Lm. 2.5(4)]), and a trivial direct checking yields that $u' = u$. Moreover, we have $2u \sim 2v$ and hence $r_{G^w}(2v) = 1$, by [1, Lm. 2.5(3)].

As $(G^0)^{w^0} = G^w$, the second double implication follows the first.

Let e be a non-loop edge of a weighted graph (G, w) and let $v_1, v_2 \in V(G)$ be its endpoints. Recall that the *(weighted) contraction* of e is defined as the graph (G_e, w_e) such that e is contracted to a vertex $\overline{v}$ of G_e, and $w_e(\overline{v}) = w(v_1) + w(v_2)$, whereas w_e is equal to w on every remaining vertex of G_e.

We denote by $(\overline{G}, \overline{w})$ the 2-edge-connected weighted graph obtained by contracting every bridge of G as described above.

By [7, Cor 5.11] a weightless, loopless graph is hyperelliptic if and only if so is $\overline{G}$. The following Lemma extends this fact to the weighted case.

Lemma 4. *Let (G, w) be a loopless weighted graph of genus at least 2. Then (G, w) is hyperelliptic if and only if so is $(\overline{G}, \overline{w})$.*

Proof. By Lemma 3, (G, w) is hyperelliptic if and only if so is G^w. Similarly, $(\overline{G}, \overline{w})$ is hyperelliptic if and only if so is $\overline{G}^{\overline{w}}$. Now, $\overline{G}^{\overline{w}}$ is obtained from G^w by contracting all of its bridges (indeed, the bridges of G and G^w are in natural bijection). Therefore, as we said above, G^w is hyperelliptic if and only if so is $\overline{G}^{\overline{w}}$. So we are done.

Recall, from [7], that a loopless, 2-edge-connected, weightless graph G is hyperelliptic if and only if it has an involution ι such that G/ι is a tree. If G has genus at least 2, this involution is unique and will be called the *hyperelliptic* involution. Furthermore, the quotient map $G \to G/\iota$ is a non-degenerate harmonic morphism, unless $|V(G)| = 2$; see [7, Thm 5.12 and Cor 5.15]. We are going to generalize this to the weighted case.

Remark 12. Let G be a loopless, 2-edge-connected hyperelliptic graph of genus ≥ 2 and ι its hyperelliptic involution. Let $v \in V(G)$ be a special vertex whose removal creates a loop based at the vertex u. Then $\iota(v) = v$, $\iota(u) = u$ and ι swaps the two edges adjacent to v.

Indeed, G/ι is a tree, hence the two edges adjacent to v are mapped to the same edge by $G \to G/\iota$. As v has valency 2 and u has valency at least 3 (G has genus at least 2), ι cannot swap v and u. Hence $\iota(v) = v$ and $\iota(u) = u$.

Lemma 5. *Let (G, w) be a loopless, 2-edge-connected weighted graph of genus at least 2. Then (G, w) is hyperelliptic if and only if G has an involution ι, the* hyperelliptic involution, *fixing every vertex of positive weight and such that G/ι is a tree.*

ι is unique and, if $|V(G)| \geq 3$, then the quotient $G \to G/\iota$ is a non-degenerate harmonic morphism of degree 2.

Proof. Assume that G has an involution as in the statement; then we extend ι to an involution ι^w of G^w by requiring that ι^w fix all the (special) vertices in $V(G^w) \smallsetminus V(G)$ and swap the two edges adjacent to them. It is clear that G^w/ι^w is the tree obtained by adding $w(v)$ leaves to the vertex of G/ι corresponding to every vertex $v \in V(G)$. Hence G^w is hyperelliptic, and hence so is (G, w) by Lemma 3.

Conversely, suppose G^w hyperelliptic and let ι^w be its hyperelliptic involution. Let $v \in V(G) \subset V(G^w)$ have positive weight. Then there is a 2-cycle in G^w attached at v; let e^+ and e^- be its two edges, and u its special vertex. By Remark 12 we know that ι^w fixes v and u and swaps e^+ and e^-. Notice that the image in G^w/ι^w of every such 2-cycle is a leaf.

We obtain that the restriction of ι^w to G is an involution of G, written ι, fixing all vertices of positive weight. Finally, the quotient G/ι is the tree obtained from G^w/ι^w by removing all the above leaves, so we are done.

As G is 2-edge-connected, by Remark 6 we can apply some results from [7]. In particular, the uniqueness of ι follows from Corollary 5.14. Next, if $|V(G)| \geq 3$ then $G \to G/\iota$ is harmonic and non-degenerate by Theorem 5.14 and Lemma 5.6.

Corollary 3. *Let (G, w) be a loopless, 2-edge-connected graph of genus at least 2, having exactly two vertices, v_1 and v_2. Then (G, w) is hyperelliptic if and only if either $|E(G)| = 2$, or $|E(G)| \geq 3$ and $w(v_1) = w(v_2) = 0$.*

Proof. Assume (G, w) hyperelliptic. Let $|E(G)| \geq 3$; by contradiction, suppose $w(v_1) \geq 1$. By Lemma 5 the hyperelliptic involution fixes v_1, and hence it fixes also v_2; therefore G/ι has two vertices. Since there are at least three edges between v_1 and v_2, such edges fall into at least two orbits under ι, and each such orbit is an edge of the quotient G/ι, which therefore cannot be a tree. This is a contradiction. The other implication is trivial; see Example 1.

4.2 Relating Hyperelliptic Curves and Graphs

Proposition 4. *Let X be a hyperelliptic stable curve. Then its dual graph (G_X, w_X) is hyperelliptic.*

Proof. We write $(G, w) = (G_X, w_X)$ for simplicity. By Theorem 3, there exists a hyperelliptic refinement, $(\hat{G}, \hat{w})$, of (G, w). Then the weightless graph $\hat{G}^{\hat{w}}$ is hyperelliptic. By Lemma 3 it is enough to prove that the weightless graph G^w is hyperelliptic. Now, one easily checks that G^w is obtained from $\hat{G}^{\hat{w}}$ by removing every non-special 2-valent vertex of weight zero, and possibly some special vertex of weight zero. On the other hand, by Lemma 3, the removal of any special vertex of weight zero does not alter being hyperelliptic. Therefore G^w is hyperelliptic if so is the graph obtained by removing every 2-valent vertex of weight zero from $\hat{G}^{\hat{w}}$. This follows from the following Lemma 6.

Lemma 6. *Let* $(\hat{G}, \hat{w})$ *be hyperelliptic of genus at least 2 and let* (G, w) *be the graph obtained from* $\hat{G}$ *by removing every 2-valent vertex of weight zero. Then G is hyperelliptic.*

Proof. By Lemma 4, contracting bridges does not alter being hyperelliptic, hence we may assume that $\hat{G}$ is 2-edge-connected. By Lemma 3 up to inserting some special vertices of weight zero we can also assume that $\hat{G}$ has no loops. Finally, we can assume that $\hat{G}$ has at least three vertices, for otherwise the result is trivial.

It suffices to prove that the loopless model (G^0, w^0) (see Definition 7) of (G, w) admits an involution ι fixing every vertex of positive weight and such that G^0/ι is a tree, by Lemma 5. As $(\hat{G}, \hat{w})$ is hyperelliptic, it admits such an involution, denoted by $\hat{\iota}$. Recall that the quotient map $\hat{G} \to \hat{G}/\hat{\iota}$ is a non-degenerate harmonic morphism.

Observe that G^0 is obtained from $\hat{G}$ by removing all the non-special 2-valent vertices of weight zero. Let $\hat{v} \in V(\hat{G})$ be such a vertex and write $\hat{e}_1, \hat{e}_2$ for the edges of $\hat{G}$ adjacent to $\hat{v}$. To prove our result it suffices to show that if one removes from a hyperelliptic graph either a non-special 2-valent vertex of weight zero fixed by the hyperelliptic involution, or a pair of non-special 2-valent vertices swapped by the hyperelliptic involution, then the resulting graph is hyperelliptic.

First, let $\hat{\iota}(\hat{v}) = \hat{v}$ and let (G', w') be the graph obtained by removing $\hat{v}$. We have $\hat{\iota}(\hat{e}_1) = \hat{e}_2$ (as $\hat{G} \to \hat{G}/\hat{\iota}$ is non-degenerate), and $\hat{v}$ is mapped to a leaf of $\hat{G}/\hat{\iota}$. Now, $V(G') = V(\hat{G}) \smallsetminus \{\hat{v}\}$, and $E(G') = \{e\} \cup E(\hat{G}) \smallsetminus \{\hat{e}_1, \hat{e}_2\}$ where e is the edge created by removing $\hat{v}$. We define the involution ι' of G' by restricting $\hat{\iota}$ on $V(G')$ and on $E(\hat{G}) \smallsetminus \{\hat{e}_1, \hat{e}_2\}$, and by setting $\iota'(e) = e$. Since ι' swaps the two endpoints of e (because so does $\hat{\iota}$), we have that e is contracted to a point by the quotient $G' \to G'/\iota'$. Therefore G'/ι' is the tree obtained from $\hat{G}/\hat{\iota}$ by removing the leaf corresponding to $\hat{v}$. It is clear that ι' fixes all vertices of positive weight, hence (G', w') is hyperelliptic.

Next, let $\hat{\iota}(\hat{v}) = \hat{v}' \neq \hat{v}$; with $\hat{v}$ and $\hat{v}'$ non-special and 2-valent, then the vertex of $\hat{G}/\hat{\iota}$ corresponding to $\{\hat{v}, \hat{v}'\}$ is 2-valent as well. Moreover, $\hat{v}$ and $\hat{v}'$ have weight zero, by Lemma 5. Let us show that the graph (G'', w'') obtained by removing $\hat{v}$ and $\hat{v}'$ is hyperelliptic. Now $\hat{\iota}$ maps $\hat{e}_1, \hat{e}_2$ to the two edges adjacent to $\hat{v}'$. We denote by e and e' the new edges of G''. We define ι'' on $V(G'') = V(\hat{G}) \smallsetminus \{\hat{v}, \hat{v}'\}$ by restricting $\hat{\iota}$; next, we define ι'' on $E(G'')$ so that $\iota''(e) = e'$ and ι'' coincides with $\hat{\iota}$ on the remaining edges. It is clear that ι'' is an involution fixing positive weight vertices and such that the quotient G''/ι'' is the tree obtained from $\hat{G}/\hat{\iota}$ by removing the 2-valent vertex corresponding to $\{\hat{v}, \hat{v}'\}$. We have thus proved that (G'', w'') is hyperelliptic. The proof is now complete.

Theorem 4. *Let (G, w) be a stable graph of genus $g \geq 2$. Then the following are equivalent.*

(A) $M^{\mathrm{alg}}(G, w)$ *contains a hyperelliptic curve.*
(B) (G, w) *is hyperelliptic and for every $v \in V(G)$ the number of bridges of G adjacent to v is at most $2w(v) + 2$.*
(C) *Assume $|V(G)| \neq 2$; the graph (G, w) is 2-gonal.*

Proof of the Lemma. (C) $\Rightarrow$ (A) by Theorem 2 and Example 2.

(A) $\Rightarrow$ (B). Let X be a hyperelliptic curve such that $(G_X, w_X) = (G, w)$. Then, by Proposition 4, (G, w) is hyperelliptic. Let $\alpha : \hat{X} \to Y$ be an admissible covering corresponding to X; by Remark 3 (C), $\hat{X}$ is semistable. Therefore the dual graph of $\hat{X}$, written $(\hat{G}, \hat{w})$, is a refinement of (G, w) (as X is the stabilization of $\hat{X}$).

Let $v \in V(G) \subset V(\hat{G})$ and $C_v \subset \hat{X}$ be the component corresponding to v, recall that C_v is nonsingular (by Remark 3) of genus $w(v)$. Now let $\hat{e} \in E(\hat{G})$ be a bridge of $\hat{G}$ adjacent to v. Then the corresponding node $N_{\hat{e}}$ of $\hat{X}$ is a separating node of $\hat{X}$, and hence $\alpha^{-1}(\alpha(N_{\hat{e}})) = N_{\hat{e}}$. This implies that the restriction of α to C_v ramifies at the point corresponding to $N_{\hat{e}}$. By the Riemann-Hurwitz formula, the number of ramification points of $\alpha_{|C_v}$ is at most $2w(v) + 2$, therefore the number of bridges of $\hat{G}$ adjacent to v is at most $2w(v) + 2$.

Now, by construction, we have a natural identification $E_v(\hat{G}) = E_v(G)$ which identifies bridges with bridges. Hence also the number of bridges of G adjacent to v is at most $2w(v) + 2$, and we are done.

(B) $\Rightarrow$ (C) *assuming* $|V(G)| \neq 2$. We can assume $|V(G)| \geq 3$ for the case $|V(G)| = 1$ is clear; see Example 3. Let us first assume that G has no loops. By Lemma 4, the 2-edge-connected graph $(\overline{G}, \overline{w})$ is hyperelliptic.

Suppose $|V(\overline{G})| > 2$. By Lemma 5, $\overline{G}$ has an involution $\bar{\iota}$ such that

$$\overline{\phi} : \overline{G} \longrightarrow \overline{T} := \overline{G}/\bar{\iota}$$

is a non-degenerate harmonic morphism of degree 2, with $\overline{T}$ a tree. Let us show that $\overline{\phi}$ corresponds to a non-degenerate pseudo-harmonic morphism of degree 2, $\phi : G \to T$, with T a tree, such that $r_\phi(e) = 2$ for every bridge e. Suppose that G has a unique bridge e, which is contracted to the vertex $\overline{v}$ of $\overline{G}$; let $\overline{u} = \overline{\phi}(\overline{v}) \in V(\overline{T})$. Let T be the tree obtained from $\overline{T}$ by replacing the vertex $\overline{u}$ by a bridge e' and its two endpoints in such a way that there exists a morphism $\phi : G \to T$ mapping e to e' fitting in a commutative diagram

$$\begin{array}{ccc} G & \longrightarrow & \overline{G} \\ \phi \downarrow & & \downarrow \overline{\phi} \\ T & \longrightarrow & \overline{T} \end{array} \qquad (27)$$

where the horizontal arrows are the maps contracting e and e' (it is trivial to check that such a ϕ exists). To make ϕ into an indexed morphism of degree 2 we set

$r_\phi(e) = 2$ and we set all other indices to be equal to 1. Since $\overline{\phi}$ was harmonic and non-degenerate, we have that ϕ is pseudo-harmonic and non-degenerate.

If G has any number of bridges, we iterate this construction one bridge at the time. This clearly yields a pseudo-harmonic, degree 2, non-degenerate morphism $\phi : G \to T$ where T is a tree.

We claim that condition (10) holds. Indeed, we have $r_\phi(e) = 2$ if and only if e is a bridge. Therefore (10) needs only be verified at the vertices of G that are adjacent to some bridge; notice that for any such vertex v we have $m_\phi(v) = 2$. Writing $\mathrm{brdg}(v)$ for the number of bridges adjacent to v, we have, as by hypothesis, $\mathrm{brdg}(v) \le 2w(v) + 2$,

$$\sum_{e \in E_v(G) \cap E_\phi^{\mathrm{hor}}(G)} (r_\phi(e) - 1) \le \mathrm{brdg}(v) \le 2w(v) + 2 = 2(w(v) + m_\phi(v) - 1).$$

This proves that (10) holds, that is, (G, w) is a 2-gonal graph. So we are done.

Suppose $|V(\overline{G})| = 2$, hence the bridges of G are leaf-edges. By Corollary 3, if $|E(\overline{G})| \ge 3$, then all the weights are zero, hence, as G is stable, $G = \overline{G}$, which is excluded. If $|E(\overline{G})| = 2$, then the vertices must be fixed by the hyperelliptic involution (for otherwise they would have weight zero by Lemma 5, contradicting that the genus be at least 2). But then $\overline{G}$ has clearly an involution $\overline{\iota}$ swapping its two edges and fixing the two vertices, whose quotient is a non-degenerate harmonic morphism of degree 2 to a tree, as in the previous part of the proof, which therefore applies also in the present case.

Suppose $|V(\overline{G})| = 1$. Then G is a tree, hence the identity map $G \to G$ with all indices equal to 2 is a pseudo-harmonic morphism, ϕ, of degree 2. Arguing as in the previous part we get ϕ is harmonic; so we are done.

Finally, suppose G admits some loops. Let (G^0, w^0) be the loopless model; then $|V(G^0)| \ge 3$. By the previous part we have that (G^0, w^0) is 2-gonal, hence so is (G, w).

(B) $\Rightarrow$ (A) *assuming* $|V(G)| = 2$. If G has loops, then $|V(G^0)| \ge 3$ and we can use the previous implications (B) $\Rightarrow$ (C) $\Rightarrow$ (A). So we assume G loopless. By [16], hyperelliptic curves with two components are easy to describe. Let $X = C_1 \cup C_2$ with C_i smooth, hyperelliptic of genus $w(v_i)$ and such that $X \in M^{\mathrm{alg}}(G, w)$. If $|E(G)| = 1$ for X to be hyperelliptic it suffices to glue $p_1 \in C_1$ to $p_2 \in C_2$ with p_i Weierstrass point of C_i for $i = 1, 2$.

If $|E(G)| = 2$ for X to be hyperelliptic it suffices to glue $p_1, q_1 \in C_1$ to $p_2, q_2 \in C_2$ with $h^0(C_i, p_i + q_i) \ge 2$ for $i = 1, 2$.

If $|E(G)| \ge 3$, by Corollary 3 all weights are zero. For X to be hyperelliptic it suffices to pick two copies of the same rational curve with $|E(G)|$ marked points, and glue the two copies at the corresponding marked points. The theorem is proved.

References

1. O. Amini, L. Caporaso, Riemann-Roch theory for weighted graphs and tropical curves. Adv. Math. **240**, 1–23 (2013)
2. E. Arbarello, M. Cornalba, Su una congettura di Petri. Comment. Math. Helv. **56**, 1–38 (1981)
3. E. Arbarello, M. Cornalba, P.A. Griffiths, *Geometry of Algebraic Curves. Vol. II*. With a contribution by J.D. Harris. Grundlehren der Mathematischen Wissenschaften, vol. 268 (Springer, Berlin/Heidelberg/New York, 2011)
4. R. Bacher, P. de la Harpe, T. Nagnibeda, The lattice of integral flows and the lattice of integral cuts on a finite graph. Bull. Soc. Math. Fr. **125**, 167–198 (1997)
5. M. Baker, Specialization of linear systems from curves to graphs. Algebra Number Theory **2**(6), 613–653 (2008)
6. M. Baker, S. Norine, Riemann-Roch and Abel-Jacobi theory on a finite graph. Adv. Math. **215**(2), 766–788 (2007)
7. M. Baker, S. Norine, Harmonic morphisms and hyperelliptic graphs. IMRN **2009**(15), 2914–2955 (2009)
8. M. Baker, S. Payne, J. Rabinoff, Nonarchimedean geometry, tropicalization, and metrics on curves (preprint). Available at arXiv:1104.0320
9. S. Bosch, W. Lütkebohmert, M. Raynaud, *Néron Models*. Ergebnisse der Mathematik, vol. 21 (Springer, Berlin/New York, 1990)
10. S. Brannetti, M. Melo, F. Viviani, On the tropical Torelli map. Adv. Math. **226**, 2546–2586 (2011)
11. L. Caporaso, in *Algebraic and tropical curves: comparing their moduli spaces*, ed. by G. Farkas, I. Morrison. Handbook of Moduli, vol. I. Advanced Lectures in Mathematics, vol. XXIV (2012), pp. 119–1160
12. M. Chan, *Tropical Hyperelliptic Curves*. J. Algebraic Combin. **37**(2), 331–359 (2013)
13. P. Deligne, D. Mumford, The irreducibility of the space of curves of given genus. Inst. Hautes Études Sci. Publ. Math. **36**, 75–120 (1969)
14. A. Gathmann, M. Kerber, A Riemann-Roch theorem in tropical geometry. Math. Z. **259**(1), 217–230 (2008)
15. J. Harris, I. Morrison, *Moduli of Curves*. Graduate Texts in Mathematics, vol. 187 (Springer, New York, 1998)
16. J. Harris, D. Mumford, On the Kodaira dimension of the moduli space of curves. Invent. Math. **67**(1), 23–88 (1982)
17. C.L. Lim, S. Payne, N. Potashnik, A note on Brill-Noether theory and rank determining sets for metric graphs. Int. Math. Res. Not. IMRN **2012**(23), 5484–5504 (2012)
18. Y. Luo, Rank-determining sets of metric graphs. J. Comb. Theory Ser. A **118**(6), 1775–1793 (2011)
19. E. Pervova, C. Petronio, On the existence of branched coverings between surfaces with prescribed branch data. I. Algebr. Geom. Topol. **6** 1957–1985 (2006)
20. H. Urakawa, A discrete analogue of the harmonic morphism and Green kernel comparison theorems. Glasg. Math. J. **42**(3), 319–334 (2000)

Caustics of Plane Curves, Their Birationality and Matrix Projections

Fabrizio Catanese

Dedicated to Klaus Hulek on the occasion of his 60th birthday.

Abstract After recalling the notion of caustics of plane curves and their basic equations, we first show the birationality of the caustic map for a general source point S in the plane. Then we prove more generally a theorem for curves D in the projective space of 3×3 symmetric matrices B. For a general 3×1 vector S the projection to the plane given by $B \to BS$ is birational on D, unless D is not a line and D is contained in a plane of the form $Delta_v := \{B | Bv = 0\}$.

1 Introduction and Setup

Given a plane curve C and a point S, a source of light (which could also lie at infinity, as the sun), the light rays L_P originating in S, and hitting the curve C in a point P, are reflected by the curve, and the caustic $\mathscr{C}$ of C is the envelope of the family of reflected rays Λ_P.

Our first Theorem 3 says that the correspondence between the curve C and the caustic curve $\mathscr{C}$ is birational, i.e., it is generically one to one, if the light source point S is chosen to be a general point.

AMS Classification: 14C25, 14F17, 14G35, 32M15

The present work took place in the realm of the DFG Forschergruppe 790 "Classification of algebraic surfaces and compact complex manifolds".

F. Catanese (✉)
Lehrstuhl Mathematik VIII, Mathematisches Institut der, Universität Bayreuth, NW II, Universitätsstr. 30, 95447 Bayreuth, Germany
e-mail: fabrizio.catanese@uni-bayreuth.de

A. Frühbis-Krüger et al. (eds.), *Algebraic and Complex Geometry*, Springer Proceedings in Mathematics & Statistics 71, DOI 10.1007/978-3-319-05404-9_5,

We learnt about this problem in [4], to which we refer for an account of the history of the theory of caustics and for references to the earlier works of von Tschirnhausen, Quetelet, Dandelin, Chasles, and more modern ones (as [2, 3]).

Our methods are from algebraic geometry, so we got interested in a generalization of this result, in which the special form of a certain curve D plays no role: we achieve this goal in Theorem 4.

Let us now describe the mathematical set up for the description of caustics.

Let $\mathbb{P}^2 = \mathbb{P}^2_{\mathbb{C}}$ and let $C \subset \mathbb{P}^2$ be a plane irreducible algebraic curve, whose normalization shall be denoted by C'.

Choose an orthogonality structure in the plane, i.e. two points, called classically the cyclic points, and let $\mathbb{P}^1_\infty$ be the line ('at infinity') joining them.

The two cyclic points determine a unique involution ι on $\mathbb{P}^1_\infty$ for which the cyclic points are fixed, hence an involution, called orthogonality, on the pencils of lines passing through a given point of the affine plane $\mathbb{P}^2 \setminus \mathbb{P}^1_\infty$.

Without loss of generality, we choose appropriate projective coordinates such that

$$\iota : (x, y, 0) \mapsto (-y, x, 0), Fix(\iota) = \{(1, \pm\sqrt{-1}, 0)\}.$$

Let $S \in \mathbb{P}^2$ be a light source point, and to each point $P \in \mathbb{P}^2 \setminus \{S\}$ associate the line $L_P := \overline{PS}$. In the case where $P \in C$, we define Λ_P, the reflected light ray, as the element of the pencil of lines through P determined by the condition that the cross ratio

$$CR(N_P, T_P, L_P, \Lambda_P) = -1,$$

ensuring the existence of a symmetry with centre P leaving the tangent line T_P to C at P and the normal line $N_P := \iota(T_P)$ fixed, and exchanging the incoming light ray L_P with the reflected light ray Λ_P.

We thus obtain a rational map of the algebraic curve C to the dual projective plane:

$$\Lambda : C \dashrightarrow (\mathbb{P}^2)^\vee.$$

Definition 1. The Caustic $\mathscr{C}$ of C is defined as the envelope of the family of lines $\{\Lambda_P\}$: in other words, setting $\Gamma := \Lambda(C), \mathscr{C} = \Gamma^\vee$.

Remark 1. since the biduality map $\Gamma \dashrightarrow \Gamma^\vee$ is birational (cf. [7], pages 151–152), the map $C \dashrightarrow \mathscr{C}$ is birational iff $\Lambda : C \dashrightarrow \Gamma$ is birational. Moreover, by the biduality theorem, the class of the caustic $\mathscr{C}$ is the degree of Γ, and the degree of $\mathscr{C}$ is the class of Γ.

We shall quickly see in the next section the basic calculations which give the class of $\mathscr{C}$, i.e. the degree of Γ, in the case where C and S are general (more precise Plücker type formulae which show how the singularities of the curve C

and the special position of S make these numbers decrease are to be found in [4] and [5]).

In Sect. 3 we show our fist result, that Λ is birational onto its image for general choice of the source point S, if C is not a line (in this case Γ is a line, and the caustic is a point). The next section recalls a well known lemma about lines contained in the determinantal variety Δ which is the secant variety of the Veronese surface V.

This lemma plays a crucial role in the proof of our main result, which says the following (see Theorem 4 for more details):

Theorem 1. *Let $D \subset \mathbb{P} := \mathbb{P}(Sym^2(\mathbb{C}^3))$ be a curve.*

Then, for general $S \in \mathbb{P}^2$, the projection $\pi_S : \mathbb{P} = \mathbb{P}(Sym^2(\mathbb{C}^3)) \dashrightarrow \mathbb{P}^2$ given by $\pi_S(B) := BS$ has the property that its restriction to D, $\pi_S|_D$, is birational onto its image, unless (and this is indeed an exception) D is a curve contained in a plane $\Delta(S') = \{B|\ BS' = 0\}$ (contained in the determinantal hypersurface $\Delta = \{B|\det(B) = 0\}$) and D is not a line.

This result suggests the investigation of a more general situation concerning the birationality of linear projections given by matrix multiplications.

Problem 1. Given a linear space $\mathbb{P}$ of matrices B, and a linear space $\mathbb{P}'$ of matrices S, consider the matrix multiplication $\pi_S(B) = BS$. For which algebraic subvarieties $D \subset \mathbb{P}$ is the restriction of the projection $\pi_S|_D$ birational onto its image for a general choice of $S \in \mathbb{P}'$?

2 Equations in Coordinates

Let $f(x_0, x_1, x_2) = 0$ be the equation of C in the appropriate system of homogeneous coordinates, let $d := deg(f)$, and let $F := (f_0(x), f_1(x))$ be the first part of the gradient of f. For a point $x = (x_0, x_1, x_2)$ we define

$$(F, x) := f_0(x)x_0 + f_1(x)x_1, \quad \{F \wedge x\} := f_0(x)x_1 - f_1(x)x_0.$$

Then the tangent line T_P at a point P with coordinates x is the transpose of the row vector $(f_0(x), f_1(x), f_2(x))$.

The normal line N_P is orthogonal to the tangent line, hence it has the form $N_P = {}^t(-f_1(x), f_0(x), f_3(x))$, and the condition that $P \in N_P$ forces the unknown rational function $f_3(x)$ to fulfill $-f_1(x)x_0 + f_0(x)x_1 + f_3(x)x_2 \equiv 0$, thus

${}^tN_P$ is the row vector

$${}^tN_P = (-x_2 f_1(x), x_2 f_0(x), -\{F \wedge x\}).$$

We find now the line L_P as the line in the pencil spanned by T_P and N_P passing through S: as such the line L_P is a column vector which is a linear combination

$\lambda T_P + \mu N_P$; the condition that $S \in L_P$ then determines $\lambda = -{}^t N_P \cdot S, \mu = {}^t T_P \cdot S$, where S is the transpose of the vector (s_0, s_1, s_2).

Hence we get

$$L_P(S) = A(P)S, \quad A(P) := -T_P\, {}^t N_P + N_P\, {}^t T_P,$$

in particular the matrix $A(P)$ is skew symmetric.

To obtain the reflected ray $\Lambda(P)$ it is sufficient, by definition, to change the sign of λ, and we get therefore:

$$\Lambda_P(S) = B(P)S, \quad B(P) := T_P\, {}^t N_P + N_P\, {}^t T_P.$$

Remark 2. (1) The matrices $A(P)$ and $B(P)$ are functions which are defined for all general points P of the plane.

(2) The matrix $B(P)$ is symmetric and has rank at most two, since its image is generated by N_P and T_P; moreover we have

$$B(P)P = 0, A(P)P = 0, \quad \forall P \in C.$$

(3) Assume that C is not a line passing through a cyclic point: then the matrix $B(P)$ has precisely rank two on the non empty open set where $f_1^2 + f_0^2 \neq 0$ and $x_2 \neq 0$; the former condition clearly holds for a general point $P \in C$, otherwise the dual curve of C would be contained in a line $y_0 = \pm\sqrt{-1}y_1$.
(4) The entries of the matrix $B(x)$ are given by polynomials of degree $2d - 1$.

By the preceding remark follows easily the classical theorem asserting that

Theorem 2. *The class of the caustic, i.e., the degree of Γ, equals $d(2d-1)$, for a general curve C and a general choice of S.*

In fact C has degree d, and $B(x)S$ is given by polynomials of degree $2d - 1$ in x, which have no base points on a general curve C.

3 Birationality of the Caustic Map

Theorem 3. *If C is not a line, then the caustic map $C \dashrightarrow \mathscr{C}$ is birational, for general choice of S.*

Proof. As already remarked, the caustic map is birational iff the map $\Lambda : C \dashrightarrow \Gamma$ is birational. Observe that Λ defines a morhism $C' \to \Gamma$ which we also denote by Λ.

The matrix B, whose entries are polynomials of degree $2d - 1$, yields a map

$$B : C' \to D \subset \mathbb{P}^5 = \mathbb{P}(Sym^2(\mathbb{C}^3)).$$

Lemma 1. *$B : C' \to D := \Phi(C)$ is birational.*

Proof. It suffices to recall Remark 2: for a general point $P \in C$, $B(P)$ has rank exactly two, and $B(P)P = 0$. Hence $P = \ker(B(P))$, and the matrix $B(P)$ determines the point $P \in \mathbb{P}^2$. ∎

We have now a projection $\mathbb{P}(Sym^2(\mathbb{C}^3)) \dashrightarrow \mathbb{P}^2$ given by

$$\pi_S(B) := BS.$$

Consider the linear subspace

$$W := \{B | B_{0,0} + B_{1,1} = 0\}.$$

We observe preliminarily that the curve D is contained in the linear subspace W since, setting for convenience $f_i := f_i(x)$, the matrix $B(x)$ has the following entries:

$$B_{0,0} = -2x_2 f_0 f_1, \quad B_{1,1} = 2x_2 f_0 f_1.$$

Then our main result follows from the next assertion, that, for a general choice of $S \in \mathbb{P}^2$, the projection π_S yields a birational map of D onto $\Gamma := \pi_S(D)$.

In order to prove this, we set up the following notation:

$$\Delta_S := \{B | BS = 0\}, \quad \Delta := \{B | \det(B) = 0\} = \cup_S \Delta_S.$$

Observe that Δ is the secant variety of the Veronese surface

$$V := \{B | \ rank(B) = 1\}.$$

Observe that the curve D is contained in the linear subspace W, is contained in Δ but not contained in the Veronese surface V.

We are working inside the subspace W, and we observe first of all that the centre of the projection π_S restricted to W is the linear space

$$W_S := \Delta_S \cap W.$$

Observe moreover that $\Delta \cap W = \cup_S W_S$.

Now, the projection π_S is not birational on D if and only if, for a general $B \in D$, there exists another $B' \in D$, $B \neq B'$, such that the chord (i.e., secant line) $B * B'$ intersects W_S in a point B'' (observe that the general point $B \in D$ does not lie in the line W_S).

There are two possible cases:

Case I: B'' is independent of the point $B \in D$.

Case II: B'' moves as a rational function of the point $B \in D$, hence the points B'' sweep the line W_S.

Lemma 2. *The assumption that case I holds for each $S \in \mathbb{P}^2$ leads to a contradiction.*

Proof of the Lemma. Under our assumption, for each S there is a point $B''(S)$ such that infinitely many chords of D meet W_S in $B''(S)$.

Let us see what happens if we specialize S to be a general point $P \in C$.

The first alternative is

(I-1) $B''(P) = B(P)$: in this case, for each point $B^1 \in D$ there is $B^2 \in D$ such that $B(P), B^1, B^2$ are collinear. Since this happens for each choice of $B(P), B^1$, every secant is a trisecant, hence, by the well known trisecant lemma (cf. [1], page 110), D is a plane curve of order at least three.

Take now a general $S \in \mathbb{P}^2$: since $B''(S)$ is on a secant to D, $B''(S)$ belongs to the secant variety Σ of D (here a plane Π), but we claim that it is not in D. In fact, if there were a point $P \in C'$ such that $B''(S) = B(P)$, then $B(P)S = 0$ contradicting that S is a general point. Hence we obtain that the plane Π intersects Δ in a bigger locus than D: since Δ is a cubic hypersurface, it follows that $\Pi \subset \Delta$.

By Proposition 1 it follows that either there is a point S' such that $S' \in \ker(B), \forall B \in \Pi$, or there is a line $L \in \mathbb{P}^2$ such that $\ker(B) \in L, \forall B \in \Pi$: both cases imply that the curve C must be contained in a line, a contradiction.

The second alternative is

(I-2) $B'' := B''(P) \neq B(P)$. Then there is a point $B' \in D$ (possibly infinitely near) such that B' is a linear combination of B'' and $B := B(P)$.

However, since $BP = 0, B''P = 0$, and $B \neq B''$, then also for their linear combination B' we have $B'P = 0$. The consequence is, since $B'P = B'P' = 0$, that B' has rank one. Therefore, if B' is not infinitely near, B' cannot be a general point of D, hence B' is independent of P: but then $C \subset \ker(B')$, and since we assume that C is not a line, we obtain $B' = 0$, a contradiction.

If P' is infinitely near to the point $P \in C$, i.e., P, P' span the tangent line to C at P, and B, B' span the tangent line to D at $B = B(P)$, we work over the ring of tangent vectors $\mathbb{C}[\epsilon]/(\epsilon^2)$, and we observe that

$$(B + \epsilon B')(P + \epsilon P') = 0 \Rightarrow BP' = 0.$$

For $P \in C$ general this is a contradiction, since $BP' = 0, BP = 0$ imply that $B = B(P)$ has rank one. ∎

Lemma 3. *The assumption that case II holds for general $S \in \mathbb{P}^2$ leads to a contradiction.*

Proof of the Lemma. As we already observed, for general S, B'' moves as a rational function of the point $B \in D$, hence the points B'' sweep the line W_S. Therefore the line W_S is contained in the secant variety Σ of the curve D. As this happens for general S, and $\Delta \cap W = \cup_S W_S$, it follows that the threefold $\Delta \cap W$ is contained in the secant variety Σ.

Since Σ is irreducible, and has dimension at most three, it follows that we have equality

$$\Delta \cap W = \Sigma.$$

We conclude that, for P_1, P_2 general points of C, the line joining $B(P_1)$ and $B(P_2)$ is contained in Δ.

By Proposition 1, and since $ker(B(P_1)) = P_1, ker(B(P_2)) = P_2$, we have that the matrices in the pencil $\lambda_1 B(P_1) + \lambda_2 B(P_2)$ send the span of P_1, P_2 to its orthogonal subspace.

This condition is equivalent to

$$^t P_1(B(P_2))P_1 = 0 \ \forall P_1, P_2 \in C$$

($^t P_2(B(P_1))P_2 = 0$ follows in fact since P_1, P_2 are general).

Fix now a general point P_2: then we have a quadratic equation for C, hence C is contained in a conic.

A little bit more of attention: the matrix $B(P_2)$ has rank two, hence the quadratic equation defines a reducible conic, and, C being irreducible, C is a line, a contradiction. ■

4 Linear Subspaces Contained in the Determinantal Cubic $\Delta := \{B \mid \det(B) = 0\}$

Proposition 1. *Let $\lambda B_0 + \mu B_1$ be a line contained in the determinantal hypersurface Δ of the projective space of symmetric 3×3 matrices.*

Then the line is contained in a maximal projective subspace contained in Δ, which is either of the type

$$\Delta_S := \{B | BS = 0\},$$

for some $S \in \mathbb{P}^2$, or of the type

$$\Delta(L) := \{B | BL \subset L^{\perp}\} = \{B | B_{|L} \equiv 0\},$$

for some line $L \subset \mathbb{P}^2$.

Proof. A pencil of reducible conics either has at most one (non infinitely near) base point $S \in \mathbb{P}^2$, or it has a line L as fixed component.

In the first case the pencil is $\subset \Delta_S$, in the second case it is contained in the subspace $\Delta(L)$ consisting of the conics of the form $L + L'$, where L' is an arbitrary line in the plane. ■

Remark 3. Even if the result above follows right away from the classification of pencils of conics, it is useful to recall the arguments which will be used in the sequel.

For instance, we observe that the hyperplane sections of the Veronese surface V are smooth conics, hence no line is contained in V.

5 Birationality of Certain Matrix Projections of Curves

In this final section we want to show the validity of a much more general statement:

Theorem 4. *Let $D \subset \mathbb{P} := \mathbb{P}(Sym^2(\mathbb{C}^3))$ be a curve and $B : C' \to D \subset \mathbb{P}$ be its normalization.*

Then, for general $S \in \mathbb{P}^2$, the projection $\pi_S : \mathbb{P} = \mathbb{P}(Sym^2(\mathbb{C}^3)) \dashrightarrow \mathbb{P}^2$ given by $\pi_S(B) := BS$ has the property that its restriction to D, $\pi_S|_D$ is birational onto its image, unless D is a curve contained in a plane $\Delta(S')$ and is not a line.

In the latter case, each projection $\pi_S|_D$ has as image the line $(S')^\perp$ and is not birational.

Proof. Let $\mathcal{G} := Gr(1, \mathbb{P})$ be the Grassmann variety of lines $\Lambda \subset \mathbb{P}$: $\mathcal{G}$ has dimension 8.

Define, for $S \in \mathbb{P}^2$, $\mathcal{G}_S := \{\Lambda \in \mathcal{G} | \Lambda \cap \Delta_S \neq \emptyset\}$. Indeed, these 6-dimensional submanifolds of $\mathcal{G}$ are the fibres of the second projection of the incidence correspondence

$$I \subset \mathcal{G} \times \mathbb{P}^2, \; I := \{(\Lambda, S) | \Lambda \cap \Delta_S \neq \emptyset\}.$$

In turn I is the projection of the correspondence

$$J \subset \mathcal{G} \times \Delta \times \mathbb{P}^2, \; J := \{(\Lambda, B, S) | B \in \Lambda, \; BS = 0\}.$$

Recall further that $\Delta \setminus V$ has a fibre bundle structure

$$\mathcal{K} : \Delta \setminus V \to \mathbb{P}^2$$

such that $\mathcal{K}(B) := ker(B)$, and with fibre over S equal to $\Delta_S \setminus V$.

Remark 4. 1. Observe that for matrices $B \in V$ we can write them in the form $B = x\,{}^t x$, for a suitable vector x, and in this case $\ker(B) = x^\perp$, $Im(B) = \langle\langle x \rangle\rangle$.
2. In any case, since the matrices B are symmetric, we have always

$$Im(B) = \ker(B)^\perp.$$

Consider now the fibres of $I \to \mathcal{G}$: for a general line Λ, its fibre $\mathcal{S}(\Lambda)$ is

1. If $\Lambda \cap \Delta \neq \Lambda$, $\Lambda \cap \Delta \subset \Delta \setminus V$, then $\mathcal{S}(\Lambda)$ consists of at most three points;

2. If $\Lambda \cap \Delta \neq \Lambda$, $(\Lambda \cap V) \neq \emptyset$, then $\mathscr{S}(\Lambda)$ consists of a line $x^\perp$ and at most one further point;
3. If $\Lambda \subset \Delta$ is of the form $\Lambda \subset \Delta_S$, then $\mathscr{S}(\Lambda)$ consists of one or two lines containing S;
4. If $\Lambda \subset \Delta$ is of the form $\Lambda \subset \Delta(L)$, $\mathscr{S}(\Lambda)$ consists of the line L.
   Since, if $\Lambda \subset \Delta(L)$, the conics in Λ consist of L plus a line L' moving in the pencil of lines through a given point P.

We let

$$U \subset \mathscr{G} \times \mathbb{P}, U := \{(\Lambda, B) | B \in \Lambda\}$$

be the universal tautological $\mathbb{P}^1$-bundle, and we denote by $p : U \to \mathbb{P}$ the second projection.

Recall now that the secant variety Σ of D is defined as follows: we have a rational map $\psi : C' \times C' \dashrightarrow \mathscr{G}$ associating to the pair (s, t) the line $B(s) * B(t)$ joining the two image points $B(s), B(t)$.

Then one denotes by U' the pull back of the universal bundle, and defines Σ as the closure of the image $p(U')$.

The condition that for each $S \in \mathbb{P}^2$ the projection π_S is not birational on D means that, if Y is the closure of the image of ψ, then $Y \cap \mathscr{G}_S$ has positive dimension.

This implies that the correspondence

$$I_D := \{(\Lambda_y, S) | y \in Y, \ \Lambda_y \cap \Delta_S \neq \emptyset\} \subset Y \times \mathbb{P}^2$$

has dimension at least three and surjects onto $\mathbb{P}^2$.

Projecting I_D on the irreducible surface Y, we obtain that all the fibres have positive dimension, and we infer that each secant line Λ_y has a fibre $\mathscr{S}(\Lambda_y)$ of positive dimension.

There are two alternatives:

(i) A general secant Λ_y is not contained in Δ, but intersects the Veronese surface V.
(ii) Each secant line $\Lambda_y \subset \Delta$.

Step (I): the theorem holds true if $D \subset V$.

Proof of step I.
In this case any element of D is of the form $B(t) = x(t)^t x(t)$, and

$$\pi_S(B(t)) = x(t)[{}^t x(t) S] = (x(t), S) x(t) = x(t).$$

Hence, for each S, the projection π_S is the inverse of the isomorphism

$$\phi : x \in \mathbb{P}^2 \to V, \phi(x) = x\,{}^t x.$$

■

We may therefore assume in the sequel that D is not contained in V.

Step (II): the theorem holds in case (i).

Proof of step II.

Observe preliminarly that, in case (i), $D \not\subset \Delta$; else we could take two smooth points $B_1, B_1 \in D \cap (\Delta \setminus V)$, and the secant line $B_1 * B_2$ could not fulfill (i).

Choose then a point $B_0 \in D$, $B_0 \in \mathbb{P} \setminus \Delta$, hence w.l.o.g. we may assume that B_0 is the identity matrix I.

Since any other point $B(t) \in D$ is on the line joining B_0 with a point $x(t)^t x(t) \in V$, we may write locally around a point of C'

$$B(t) = I + \xi(t)^t \xi(t),$$

where $\xi(t)$ is a vector valued holomorphic function.

Now, for each s, t, the secant line $B(t) * B(s)$ meets the Veronese surface V.

Since $B(t)$ cannot have rank equal to 1, there exists λ such that

$$\lambda B(t) + B(s) = \lambda(I + \xi(t)^t \xi(t)) + (I + \xi(s)^t \xi(s))$$

has rank equal to 1, i.e.,

$$K_\lambda := \ker(\lambda B(t) + B(s)) = \{v | [\lambda(I + \xi(t)^t \xi(t)) + (I + \xi(s)^t \xi(s))]v = 0\} =$$

$$\{v | (\lambda + 1)v + \lambda \xi(t)(\xi(t), v) + \xi(s)(\xi(s), v) = 0\}$$

has dimension 2.

Let us now make the assumption:

(**) two general points $\xi(t), \xi(s)$ are linearly independent.

The above formula shows however that, under assumption (**), it must be that v is a linear combination of $\xi(t), \xi(s)$. This is clear if $\lambda + 1 \neq 0$, otherwise v is orthogonal to the span of $\xi(t), \xi(s)$, contradicting that the kernel has dimension 2.

Hence $K_\lambda = \langle\langle \xi(t), \xi(s) \rangle\rangle$ and the condition that $\xi(t) \in K_\lambda$ yields

$$(\lambda + 1)\xi(t) + \lambda \xi(t)(\xi(t), \xi(t)) + \xi(s)(\xi(s), \xi(t)) = 0$$

and implies

$$(***) \ \forall s, t \ (\xi(s), \xi(t)) = 0.$$

(***) says that $K_\lambda = \langle\langle \xi(t), \xi(s) \rangle\rangle$ is an isotropic subspace, which can have at most dimension 1.

Hence assumption (**) is contradicted, and we conclude that it must be:

$$(****) \ \xi(t) = f(t)u,$$

where u is an isotropic vector and $f(t)$ is a scalar function.

Even if this situation can indeed occur, we are done since in this case the matrix in V is unique, $u^t u$, each secant Λ_y contains $u^t u$, hence $\mathscr{S}(\Lambda_y) = u^\perp \cup T_y$ where T_y is a finite set. Therefore, for general S, the fibre $\{y | \Lambda_y \cap \Delta_S \neq \emptyset\}$ is a finite set. ■

Step III: the theorem holds true in case (ii).

Proof of Step III.

Consider the general secant line Λ_y. We have two treat two distinct cases.

Case (3): $\Lambda_y \subset \Delta$ is of the form $\Lambda_y \subset \Delta_S$ (then $\mathscr{S}(\Lambda_y)$ consists of one or two lines containing S).

Case (4): $\Lambda_y \subset \Delta$ is of the form $\Lambda_y \subset \Delta(L)$ (then $\mathscr{S}(\Lambda_y)$ consists of the line L).

In case (3), this means that two general matrices $B(s), B(t)$ have a common kernel $S(s,t)$. Since the general matrix $B(t)$ is in $\Delta \setminus V$, its rank equals 2 and $S(s,t) = S(t)\ \forall s$.

Hence the curve D is contained in a plane Δ_S. In this case however $Im B(t) \subset S^\perp$ and every projection $\pi_{S'}(B(t)) = B(t)S'$ lands in the line $S^\perp$, so that the projection cannot be birational, unless our curve D is a line.

In case (4) for two general matrices $B(s), B(t)$ there exists a line $L = L(s,t)$ such that $B(s), B(t) \in \Delta(L)$.

Since two such general matrices have rank equal to 2, and $B(t)L \subset L^\perp$, $B(s)L \subset L^\perp$, if $v(t) \in \ker B(t)$ it follows that $v(t) \in L$ (since $\ker B(t) \cap L \neq \emptyset$). Therefore, if $B(t) \neq B(s)$, then $L(t,s) = \langle\langle v(t), v(s)\rangle\rangle$.

However, the above conditions $B(t)L \subset L^\perp, B(s)L \subset L^\perp$ are then equivalent to

$$(B(t)v(s), v(s)) =^t v(s)B(t)v(s) = 0,\ \forall t, s.$$

Fixing t this is a quadratic equation in $v(s)$, but, since the curve D is irreducible, and $B(t)$ has rank equal to 2, we see that the vectors $v(s)$ belong to a line. Therefore the line $L = L(s,t)$ is independent of s, t and the conclusion is that the curve D is contained in the plane $\Delta(L)$.

In suitable coordinates for $\mathbb{P}^2$, we may assume that $L = \langle\langle e_2, e_3\rangle\rangle$ and $L^\perp = \langle\langle e_1\rangle\rangle$.

Choosing then $S = e_1$, we obtain an isomorphic projection, since for a matrix

$$B = \begin{pmatrix} a & b & c \\ b & 0 & 0 \\ c & 0 & 0 \end{pmatrix}$$

we have

$$B(e_1) = \begin{pmatrix} a \\ b \\ c \end{pmatrix}.$$

■

Remark 5. The referee suggested some arguments to simplify the proofs.

For Theorem 3, this is the proposal:

(a) Firstly, in the case of the caustic, the curve D parametrizes the reducible conics of the form $T_P + N_P$, where T_P is the tangent to the curve C at P, and N_P is the normal.

If S is a general point in $\mathbb{P}^2$, then the degree of D equals the number of such conics passing through S, hence, if ν is the degree of the curve $\mathscr{N}$ of normal lines, μ is the degree of $C' \to \mathscr{N}$, then

$$deg(D) = deg(C^\vee) + \mu\nu.$$

The above formula shows that $deg(D) \geq 4$ if C is not a line.

In fact, then $deg(C^\vee) \geq 2$, while in general $\nu \geq 1$ (the normal N_P contains P). But, if $\nu = 1$, then the dual curve of $\mathscr{N}$, the evolute, is a point, so C is a circle, but in this case $\mu = 2$.

(b) Therefore, if one shows that D is contained in a plane π, then the plane π is contained in the cubic hypersurface Δ, hence we can apply Proposition 1.

(c) In turn, to show that D is a plane curve, it is necessary and sufficient to show that two general tangent lines to D meet, which follows if one proves that:

(d) For each secant line there is a cone over D and with vertex a point B'', such that the secant line passes through B''

(since then the two tangent lines are coplanar).

In case (I), (d) follows since then, for each general S, there is a point $B''(S)$ such that a curve of secants passes through $B''(S)$, and we get a cone over D with vertex $B''(S)$. Varying S, the point $B''(S)$ must vary, since $B''(S)S = 0$; hence the cone varies, and we get that for each secant (d) holds true.

In case (II), as we have shown, the secant variety of D equals $W \cap \Delta$, which is the secant variety of the rational normal quartic $W \cap V$: but the singular locus of the secant variety of $W \cap V$ equals $W \cap V$ and contains D, hence $W \cap V = D$, a contradiction.

The argument suggested for Theorem 4 requires some delicate verification, so we do not sketch it here.

Acknowledgements I would like to thank Alfrederic Josse and Francoise Pène for stimulating email correspondence and for spotting a mistake in my first naive attempt to prove birationality of the caustic map for general source S, thus pushing me to find the proof of Theorem 3, which I announced to them in an e-mail on December 20, 2012.

At the moment of writing up the references for the present article, I became aware, by searching on the arXiv, that they have written an independent and different proof of birationality of the caustic map for general source, in [6].

Thanks to the referee for helpful comments, and for sketching alternative arguments, which are reproduced in the remark above.

References

1. E. Arbarello, M. Cornalba, P.A. Griffiths, J. Harris, *Geometry of Algebraic Curves. Volume I*. Grundlehren der mathematischen Wissenschaften, vol. 267 (Springer, New York, 1985), XVI, 386 p.
2. J.W. Bruce, P.J. Giblin, C.G. Gibson, On caustics of plane curves. Am. Math. Mon. **88**, 651–667 (1981)
3. J.W. Bruce, P.J. Giblin, C.G. Gibson, Source genericity of caustics by reflexion in the plane. Q. J. Math. Oxf. II. Ser. **33**, 169–190 (1982)
4. A. Josse, F. Pène, On the degree of caustics by reflection. Commun. Alg. (to appear). arXiv: 1201.0621
5. A. Josse, F. Pène, Francoise Degree and class of caustics by reflection for a generic source. C. R. Math. Acad. Sci. **351**, 295–297 (2013)
6. A. Josse, F. Pène, Degree and class of caustics by reflection for a generic source. arXiv: 1301.1846
7. R.J. Walker, *Algebraic Curves*. Princeton Mathematical Series, vol. 13 (Princeton University Press/Oxford University Press X, Princeton/London, 1950), 201 p. Reprinted also by Dover and Springer

Limits of Pluri–Tangent Planes to Quartic Surfaces

Ciro Ciliberto and Thomas Dedieu

Abstract We describe, for various degenerations $S \to \Delta$ of quartic $K3$ surfaces over the complex unit disk (e.g., to the union of four general planes, and to a general Kummer surface), the limits as $t \in \Delta^*$ tends to 0 of the Severi varieties $V_\delta(S_t)$, parametrizing irreducible δ-nodal plane sections of S_t. We give applications of this to (i) the counting of plane nodal curves through base points in special position, (ii) the irreducibility of Severi varieties of a general quartic surface, and (iii) the monodromy of the universal family of rational curves on quartic $K3$ surfaces.

1 Introduction

Our objective in this paper is to study the following:

Question 1. Let $f : S \to \Delta$ be a projective family of surfaces of degree d in $\mathbf{P}^3$, with S a smooth threefold, and Δ the complex unit disc (usually called a *degeneration* of the general $S_t := f^{-1}(t)$, for $t \neq 0$, which is a smooth surface, to the *central fibre* S_0, which is in general supposed to be singular). What are the limits of tangent, bitangent, and tritangent planes to S_t, for $t \neq 0$, as t tends to 0?

Similar questions make sense also for degenerations of plane curves, and we refer to [24, pp. 134–135] for a glimpse on this subject. For surfaces, our contribution is based on foundational investigations by Caporaso and Harris [9, 10],

C. Ciliberto (✉)
Dipartimento di Matematica, Università degli Studi di Roma Tor Vergata, Via della Ricerca Scientifica, 00133 Roma, Italy
e-mail: cilibert@mat.uniroma2.it

T. Dedieu
Institut de Mathématiques de Toulouse (CNRS UMR 5219), Université Paul Sabatier, 31062 Toulouse Cedex 9, France
e-mail: thomas.dedieu@m4x.org

A. Frühbis-Krüger et al. (eds.), *Algebraic and Complex Geometry*, Springer Proceedings in Mathematics & Statistics 71, DOI 10.1007/978-3-319-05404-9_6,

and independently by Ran [30–32], which were both aimed at the study of the so-called *Severi varieties*, i.e. the families of irreducible plane nodal curves of a given degree. We have the same kind of motivation for our study; the link with Question 1 resides in the fact that nodal plane sections of a surface S_t in $\mathbf{P}^3$ are cut out by those planes that are tangent to S_t.

Ultimately, our interest resides in the study of Severi varieties of nodal curves on $K3$ surfaces. The first interesting instance of this is the one of plane sections of smooth quartics in $\mathbf{P}^3$, the latter being primitive $K3$ surfaces of genus 3. For this reason, we concentrate here on the case $d = 4$. We consider a couple of interesting degenerations of such surfaces to quite singular degree 4 surfaces, and we answer Question 1 in these cases.

The present paper is of an explorative nature, and hopefully shows, in a way we believe to be useful and instructive, how to apply some general techniques for answering some specific questions. On the way, a few related problems will be raised, which we feel can be attacked with the same techniques. Some of them we solve (see below), and the other ones we plan to make the object of future research.

Coming to the technical core of the paper, we start from the following key observation due to Caporaso and Harris, and Ran (see Sect. 3.4 for a complete statement). Assume the central fibre S_0 is the transverse union of two smooth surfaces, intersecting along a smooth curve R. Then the limiting plane of a family of tangent planes to the general fibre S_t, for $t \neq 0$, is: (i) either a plane that is tangent to S_0 at a smooth point, or (ii) a tangent plane to R. Furthermore, the limit has to be counted with multiplicity 2 in case (ii).

Obviously, this is not enough to deal directly with all possible degenerations of surfaces. Typically, one overcomes this by applying a series of base changes and blow–ups to $S \to \Delta$, thus producing a semistable model $\tilde{S} \to \Delta$ of the initial family, such that it is possible to provide a complete answer to Question 1 for $S \to \Delta$ by applying a suitable extended version of the above observation to $\tilde{S} \to \Delta$. We say that $S \to \Delta$ is *well behaved* when it is possible to do so, and $\tilde{S} \to \Delta$ is then said to be a *good model* of $S \to \Delta$.

We give in Sect. 3.4 a rather restrictive criterion to ensure that a given semistable model is a good model, which nevertheless provides the inspiration for constructing a good model for a given family. We conjecture that there are suitable assumptions, under which a family is well behaved. We do not seek such a general statement here, but rather prove various incarnations of this principle, thus providing a complete answer to Question 1 for the degenerations we consider. Specifically, we obtain:

Theorem 1. *Let* $f : S \to \Delta$ *be a family of general quartic surfaces in* $\mathbf{P}^3$ *degenerating to a* tetrahedron S_0, *i.e. the union of four independent planes. The singularities of* S *consist in four ordinary double points on each* edge *of* S_0. *The limits in* $|\mathcal{O}_{S_0}(1)|$ *of* δ*-tangent planes to* S_t, *for* $t \neq 0$, *are:*

($\delta = 1$) *the* 24 *webs of planes passing through a singular point of* S, *plus the* 4 *webs of planes passing through a* vertex *of* S_0, *the latter counted with multiplicity* 3;

($\delta = 2$) *the* 240 *pencils of planes passing through two double points of the total space S that do not belong to an edge of S_0, plus the* 48 *pencils of planes passing through a vertex of S_0 and a double point of S that do not belong to a common edge of S_0 (with multiplicity* 3*), plus the* 6 *pencils of planes containing an edge of S_0 (with multiplicity* 16*);*

($\delta = 3$) *the* 1,024 *planes containing three double points of S but no edge of S_0, plus the* 192 *planes containing a vertex of S_0 and two double points of S, but no edge of S_0 (with multiplicity* 3*), plus the* 24 *planes containing an edge of S_0 and a double point of S not on this edge (with multiplicity* 16*), plus the* 4 faces *of S_0 (with multiplicity* 304*).*

Theorem 2. *Let $f : S \to \Delta$ be a family of general quartic surfaces degenerating to a general* Kummer surface *S_0. The limits in $|\mathcal{O}_{S_0}(1)|$ of δ-tangent planes to S_t, for $t \neq 0$, are:*

($\delta = 1$) *the dual surface $\check{S}_0$ to the Kummer (which is itself a Kummer surface), plus the* 16 *webs of planes containing a node of S_0 (with multiplicity* 2*);*

($\delta = 2$) *the* 120 *pencils of planes containing two nodes of S_0, each counted with multiplicity* 4*;*

($\delta = 3$) *the* 16 *planes tangent to S_0 along a* contact conic *(with multiplicity* 80*), plus the* 240 *planes containing exactly three nodes of S_0 (with multiplicity* 8*).*

We could also answer Question 1 for degenerations to a general union of two smooth quadrics, as well as to a general union of a smooth cubic and a plane; once the much more involved degeneration to a tetrahedron is understood, this is an exercise. We do not dwell on this here, and we encourage the interested reader to treat these cases on his own, and to look for the relations between these various degenerations. However, a mention to the degeneration to a *double quadric* is needed, and we treat this in Sect. 6.

Apparent in the statements of Theorems 1 and 2 is the strong enumerative flavour of Question 1, and actually we need information of this kind (see Proposition 4) to prove that the two families under consideration are well behaved. Still, we hope to find a direct proof in the future.

As a matter of fact, Caporaso and Harris' main goal in [9, 10] is the computation of the degrees of Severi varieties of irreducible nodal plane curves of a given degree, which they achieve by providing a recursive formula. Applying the same strategy, we are able to derive the following statement (see Sect. 9):

Theorem 3. *Let a, b, c be three independent lines in the projective plane, and consider a degree* 12 *divisor Z cut out on $a + b + c$ by a general quartic curve. The sub–linear system $\mathcal{V}$ of $|\mathcal{O}_{\mathbf{P}^2}(4)|$ parametrizing curves containing Z has dimension* 3*.*

For $1 \leqslant \delta \leqslant 3$, we let $\mathcal{V}_\delta$ be the Zariski closure in $\mathcal{V}$ of the locally closed subset parametrizing irreducible δ-nodal curves. Then $\mathcal{V}_\delta$ has codimension δ in $\mathcal{V}$, and degree 21 *for $\delta = 1$, degree* 132 *for $\delta = 2$, degree* 304 *for $\delta = 3$.*

Remarkably, one first proves a weaker version of this (in Sect. 9), which is required for the proof of Theorem 1, given in Sect. 5. Then, Theorem 3 is a corollary of Theorem 1.

It has to be noted that Theorems 1 and 2 display a rather coarse picture of the situation. Indeed, in describing the good models of the degenerations, we interpret all limits of nodal curves as elements of the limit $\mathfrak{O}(1)$ of $|\mathcal{O}_{S_t}(1)|$, for $t \neq 0$, inside the relative Hilbert scheme of curves in S. We call $\mathfrak{O}(1)$ the *limit linear system* of $|\mathcal{O}_{S_t}(1)|$, for $t \neq 0$ (see Sect. 3.2), which in general is no longer a $\mathbf{P}^3$, but rather a degeneration of it. While in $|\mathcal{O}_{S_0}(1)|$, which is also a limit of $|\mathcal{O}_{S_t}(1)|$, for $t \neq 0$, there are in general elements which do not correspond to curves (think of the plane section of the tetrahedron with one of its faces), all elements in $\mathfrak{O}(1)$ do correspond to curves, and this is the right ambient to locate the limits of nodal curves. So, for instance, each face appearing with multiplicity 304 in Theorem 1 is much better understood once interpreted as the contribution given by the 304 curves in $\mathscr{V}_3$ appearing in Theorem 3.

It should also be stressed that the analysis of a semistable model of $S \to \Delta$ encodes information about several flat limits of the S_t's in $\mathbf{P}^3$, as $t \in \Delta^*$ tends to 0 (each flat limit corresponds to an irreducible component of the limit linear system $\mathfrak{O}(1)$), and an answer to Question 1 for such a semistable model would provide answers for all these flat limits at the same time. Thus, in studying Question 1 for degenerations of quartic surfaces to a tetrahedron, we study simultaneously degenerations to certain rational quartic surfaces, e.g., to certain *monoid* quartic surfaces that are projective models of the faces of the tetrahedron, and to sums of a *self–dual* cubic surface plus a suitable plane. For degenerations to a Kummer, we see simultaneously degenerations to double quadratic cones, to sums of a smooth quadric and a double plane (the latter corresponding to the projection of the Kummer from one of its nodes), etc.

Though we apply the general theory (introduced in Sect. 3) to the specific case of degenerations of singular plane sections of general quartics, it is clear that, with some more work, the same ideas can be applied to attack similar problems for different situations, e.g., degenerations of singular plane sections of general surfaces of degree $d > 4$, or even singular higher degree sections of (general or not) surfaces of higher degree. For example, we obtain Theorem 3 thinking of the curves in $\mathscr{V}$ as cut out by quartic surfaces on a plane embedded in $\mathbf{P}^3$, and letting this plane degenerate. By the way, this is the first of a series of results regarding no longer triangles, but general configurations of lines, which can be proved, we think, by using the ideas in this paper. On the other hand, for general primitive $K3$ surfaces of any genus $g \geqslant 2$, there is a whole series of known enumerative results [3, 6, 29, 35], yet leaving some open space for further questions, which also can be attacked in the same way.

Another application of our analysis of Question 1 is to the irreducibility of families of singular curves on a given surface. This was indeed Ran's main motivation in [30–32], since he applied these ideas to give an alternative proof to Harris' one [22, 24] of the irreducibility of Severi varieties of plane curves. The analogous question for the family of irreducible δ–nodal curves in $|\mathcal{O}_S(n)|$, for S a general primitive $K3$ surface of genus $g \geqslant 3$ is widely open.

In [11] one proves that for any non negative $\delta \leqslant g$, with $3 \leqslant g \leqslant 11$ and $g \neq 10$, the *universal Severi variety* $\mathcal{V}_g^{n,\delta}$, parametrizing δ–nodal members of $|\mathcal{O}_S(n)|$, with S varying in the moduli space $\mathcal{B}_g$ of primitive $K3$ surfaces of genus g in $\mathbf{P}^g$, is irreducible for $n = 1$. One may conjecture that all universal Severi varieties $\mathcal{V}_g^{n,\delta}$ are irreducible (see [13]), and we believe it is possible to obtain further results in this direction using the general techniques presented in this paper. For instance, the irreducibility of $\mathcal{V}_3^{1,\delta}$, $0 < \delta \leqslant 3$, which is well known and easy to prove (see Proposition 32), could also be deduced with the degeneration arguments developed here.

Note the obvious surjective morphism $p : \mathcal{V}_g^{n,\delta} \to \mathcal{B}_g$. For $S \in \mathcal{B}_g$ general, one can consider $V_g^{n,\delta}(S)$ the *Severi variety* of δ–nodal curves in $|\mathcal{O}_S(n)|$ (i.e. the fibre of p over $S \in \mathcal{B}_g$), which has dimension $g - \delta$ (see [11, 15]). Note that the irreducibility of $\mathcal{V}_g^{n,\delta}$ does not imply the one of the Severi varieties $V^{n,\delta}(S)$ for a general $S \in \mathcal{B}_g$; by the way, this is certainly not true for $\delta = g$, since $V^{n,g}(S)$ has dimension 0 and degree bigger than 1, see [3, 35]. Of course, $V^{1,1}(S)$ is isomorphic to the *dual variety* $\check{S} \subset \check{\mathbf{P}}^g$, hence it is irreducible. Generally speaking, the smaller δ is with respect to g, the easier it is to prove the irreducibility of $V^{n,\delta}(S)$: partial results along this line can be found in [26] and [27, Appendix A]. To the other extreme, the curve $V^{1,g-1}(S)$ is not known to be irreducible for $S \in \mathcal{B}_g$ general. In the simplest case $g = 3$, this amounts to proving the irreducibility of $V^{1,2}(S)$ for a general quartic S in $\mathbf{P}^3$, which is the nodal locus of $\check{S}$. This has been commonly accepted as a known fact, but we have not been able to find any proof of this in the current literature. We give one with our methods (see Theorem 4).

Finally, in Sect. 10.2, we give some information about the monodromy group of the finite covering $\mathcal{V}_3^{1,3} \to \mathcal{B}_3$, by showing that it contains some *geometrically interesting* subgroups. Note that a remarkable open question is whether the monodromy group of $\mathcal{V}_g^{1,g} \to \mathcal{B}_g$ is the full symmetric group for all $g \geqslant 2$.

The paper is organized as follows. In Sect. 3, we set up the machinery: we give general definitions, introduce limit linear systems, state our refined versions of Caporaso and Harris' and Ran's results, introduce limit Severi varieties. In Sect. 4, we state some known results for proper reference, mostly about the degrees of the singular loci of the dual to a projective variety. In Sects. 5 and 8, we give a complete description of limit Severi varieties relative to general degenerations of quartic surfaces to tetrahedra and Kummer surfaces respectively; Theorems 1 and 2 are proved in Sects. 5.8 and 8.4 respectively. In Sect. 6 we briefly treat other degenerations of quartics. Section 7 contains some classical material concerning Kummer quartic surfaces, as well as a few results on the monodromy action on their nodes (probably known to the experts but for which we could not find any proper reference): they are required for our proof of Theorem 4 and of the results in Sect. 10.2. Section 9 contains the proof of a preliminary version of Theorem 3; it is useful for Sect. 5, and required for Sect. 10. Section 10 contains Theorem 4 and the aforementioned results on the monodromy.

2 Conventions

We will work over the field $\mathbf{C}$ of complex numbers. We denote the linear equivalence on a variety X by $\sim_X$, or simply by $\sim$ when no confusion is likely. Let G be a group; we write $H \leq G$ when H is a subgroup of G.

We use the classical notation for projective spaces: if V is a vector space, then $\mathbf{P}V$ is the space of lines in V, and if $\mathscr{E}$ is a locally free sheaf on some variety X, we let $\mathbf{P}(\mathscr{E})$ be $\mathbf{Proj}\,(\mathrm{Sym}\,\mathscr{E}^\vee)$. We denote by $\check{\mathbf{P}}^n$ the projective space dual to $\mathbf{P}^n$, and if X is a closed subvariety of $\mathbf{P}^n$, we let $\check{X}$ be its dual variety, i.e. the Zariski closure in $\check{\mathbf{P}}^n$ of the set of those hyperplanes in $\mathbf{P}^n$ that are tangent to the smooth locus of X.

By a *node*, we always mean an ordinary double point. Let $\delta \geqslant 0$ be an integer. A *nodal* (resp. *δ-nodal*) variety is a variety having nodes as its only possible singularities (resp. precisely δ nodes and otherwise smooth). Given a smooth surface S together with an effective line bundle L on it, we define the *Severi variety* $V_\delta(S, L)$ as the Zariski closure in the linear system $|L|$ of the locally closed subscheme parametrizing irreducible δ-nodal curves.

We usually let H be the line divisor class on $\mathbf{P}^2$; when $\mathbf{F}_n = \mathbf{P}(\mathscr{O}_{\mathbf{P}^1} \oplus \mathscr{O}_{\mathbf{P}^1}(n))$ is a *Hirzebruch surface*, we let F be the divisor class of its ruling over $\mathbf{P}^1$, we let E be an irreducible effective divisor with self-intersection $-n$ (which is unique if $n > 0$), and we let H be the divisor class of $F + nE$.

When convenient (and if there is no danger of confusion), we will adopt the following abuse of notation: let $\varepsilon : Y \to X$ be a birational morphism, and C (resp. D) a divisor (resp. a divisor class) on X; we use the same symbol C (resp. D) to denote the proper transform $(\varepsilon_*)^{-1}(C)$ (resp. the pull-back $\varepsilon^*(D)$) on Y.

For example, let L be a line in $\mathbf{P}^2$, and H the divisor class of L. We consider the blow-up $\varepsilon_1 : X_1 \to \mathbf{P}^2$ at a point on L, and call E_1 the exceptional divisor. The divisor class H on X_1 is $\varepsilon_1^*(H)$, and L on X_1 is linearly equivalent to $H - E_1$. Let then $\varepsilon_2 : X_2 \to X_1$ be the blow-up of X_1 at the point $L \cap E_1$, and E_2 be the exceptional divisor. The divisor E_1 (resp. L) on X_2 is linearly equivalent to $\varepsilon_2^*(E_1) - E_2$ (resp. to $H - 2E_1 - E_2$).

In figures depicting series of blow–ups, we indicate with a big black dot those points that have been blown up.

3 Limit Linear Systems and Limit Severi Varieties

In this section we explain the general theory upon which this paper relies. We build on foundational work by Caporaso and Harris [9, 10] and Ran [30–32], as reinvestigated by Galati ([17, 18]), see also the detailed discussion in [Galati-Knutsen].

3.1 Setting

In this paper we will consider flat, proper families of surfaces $f : S \to \Delta$, where $\Delta \subset \mathbf{C}$ is a disc centered at the origin. We will denote by S_t the (schematic) fibre of f over $t \in \Delta$. We will usually consider the case in which the *total space* S is a smooth threefold, f is smooth over $\Delta^* = \Delta - \{0\}$, and S_t is irreducible for $t \in \Delta^*$. The *central fibre* S_0 may be singular, but we will usually consider the case in which S_0 is reduced and with *local normal crossing* singularities. In this case the family is called *semistable*.

Another family of surfaces $f' : S' \to \Delta$ as above is said to be a *model of* $f : S \to \Delta$ if there is a commutative diagram

$$\begin{array}{ccccccc} S' & \longleftarrow & \bar{S}' & \overset{p}{\dashrightarrow} & \bar{S} & \longrightarrow & S \\ {\scriptstyle f'}\downarrow & \square & \downarrow & & \downarrow & \square & \downarrow{\scriptstyle f} \\ \Delta & \underset{t^{d'} \leftarrow t}{\longleftarrow} & \Delta & = & \Delta & \underset{t \mapsto t^{d}}{\longrightarrow} & \Delta \end{array}$$

where the two squares marked with a '$\square$' are Cartesian, and p is a birational map, which is an isomorphism over Δ^*. The family $f' : S' \to \Delta$, if semistable, is a *semistable model of* $f : S \to \Delta$ if in addition $d' = 1$ and p is a morphism. The *semistable reduction theorem* of [28] asserts that $f : S \to \Delta$ always has a semistable model.

Example 1 (Families of surfaces in $\mathbf{P}^3$*).* Consider a *linear pencil* of degree k surfaces in $\mathbf{P}^3$, generated by a *general* surface S_∞ and a *special* one S_0. This pencil gives rise to a flat, proper family $\varphi : \mathscr{S} \to \mathbf{P}^1$, with $\mathscr{S}$ a hypersurface of type $(k, 1)$ in $\mathbf{P}^3 \times \mathbf{P}^1$, isomorphic to the blow–up of $\mathbf{P}^3$ along the *base locus* $S_0 \cap S_\infty$ of the pencil, and S_0, S_∞ as fibres over $0, \infty \in \mathbf{P}^1$, respectively.

We will usually consider the case in which S_0 is reduced, its various components may have isolated singularities, but meet transversely along smooth curves contained in their respective smooth loci. Thus S_0 has local normal crossing singularities, except for finitely many isolated *extra singularities* belonging to one, and only one, component of S_0.

We shall study the family $f : S \to \Delta$ obtained by restricting $\mathscr{S}$ to a disk $\Delta \subset \mathbf{P}^1$ centered at 0, such that $\mathscr{S}_t$ is smooth for all $t \in \Delta^*$, and we will consider a semistable model of $f : S \to \Delta$. To do so, we resolve the singularities of S which occur in the central fibre of f, at the points mapped by $\mathscr{S}_0 \to S_0 \subset \mathbf{P}^3$ to the intersection points of S_∞ with the double curves of S_0 (they are the singular points of the curve $S_0 \cap S_\infty$). These are *ordinary double points* of S, i.e. singularities analytically equivalent to the one at the origin of the hypersurface $xy = zt$ in $\mathbf{A}^4$. Such a singularity is resolved by a single blow–up, which produces an exceptional divisor $F \cong \mathbf{P}^1 \times \mathbf{P}^1$, and then it is possible to contract F in the direction of either one of its rulings without introducing any singularity: the result is called a *small* *resolution* of the ordinary double point. If S_0 has no extra singularities, the small resolution process provides a semistable model. Otherwise we will have to deal with the extra singularities, which are in any case smooth points of the total space. We will do this when needed.

Let $\tilde{f} : \tilde{S} \to \Delta$ be the semistable model thus obtained. One has $\tilde{S}_t \cong S_t$ for $t \in \Delta^*$. If S_0 has irreducible components $Q_1, \dots, Q_r$, then $\tilde{S}_0$ consists of irreducible components $\tilde{Q}_1, \dots, \tilde{Q}_r$ which are suitable blow–ups of $Q_1, \dots, Q_r$, respectively. If q is the number of ordinary double points of the original total space S, we will denote by $E_1, \dots, E_q$ the exceptional curves on $\tilde{Q}_1, \dots, \tilde{Q}_r$ arising from the small resolution process.

Going back to the general case, we will say that $f : S \to \Delta$ is *quasi–semistable* if S_0 is reduced, with *local normal crossing* singularities, except for finitely many isolated *extra singularities* belonging to one, and only one, component of S_0, as in Example 1.

Assume then that S_0 has irreducible components $Q_1, \dots, Q_r$, intersecting transversally along the double curves $R_1, \dots, R_p$, which are Cartier divisors on the corresponding components.

Lemma 1 (Triple Point Formula, [8, 16]). *Assume $f : S \to \Delta$ is quasi–semistable. Let Q, Q' be irreducible components of S_0 intersecting along the* double curve *R. Then*

$$\deg(N_{R|Q}) + \deg(N_{R|Q'}) + \mathrm{Card}\left\{\begin{array}{c}\textit{triple points of } S_0\\ \textit{along } R_s\end{array}\right\} = 0,$$

where a triple point *is the intersection $R \cap Q''$ with a component Q'' of S_0 different from Q, Q'.*

Remark 1 (See [8, 16]). There is a version of the Triple Point Formula for the case in which the central fibre is not reduced, but its support has local normal crossings. Then, if the multiplicities of Q, Q' are m, m' respectively, one has

$$m' \deg(N_{R|Q}) + m \deg(N_{R|Q'}) + \mathrm{Card}\left\{\begin{array}{c}\text{triple points of } S_0\\ \text{along } R_s\end{array}\right\} = 0,$$

where each triple point $R \cap Q''$ has to be counted with the multiplicity m'' of Q'' in S_0.

3.2 *Limit Linear Systems*

Let us consider a quasi–semistable family $f : S \to \Delta$ as in Sect. 3.1. Suppose there is a fixed component free line bundle $\mathscr{L}$ on the total space S, restricting to a line bundle $\mathscr{L}_t$ on each fibre S_t, $t \in \Delta$. We assume $\mathscr{L}$ to be ample, with $\mathrm{h}^0(S_t, \mathscr{L}_t)$

constant for $t \in \Delta$. If W is an effective divisor supported on the central fibre S_0, we may consider the line bundle $\mathscr{L}(-W)$, which is said to be obtained from $\mathscr{L}$ by *twisting* by W. For $t \in \Delta^*$, its restriction to S_t is the same as $\mathscr{L}_t$, but in general this is not the case for S_0; any such a line bundle $\mathscr{L}(-W)|_{S_0}$ is called a *limit line bundle* of $\mathscr{L}_t$ for $t \in \Delta^*$.

Remark 2. Since $\mathrm{Pic}(\Delta)$ is trivial, the divisor $S_0 \subset S$ is linearly equivalent to 0. So if W is a divisor supported on S_0, one has $\mathscr{L}(-W) \cong \mathscr{L}(mS_0 - W)$ for all integers m. In particular if $W + W' = S_0$ then $\mathscr{L}(-W) \cong \mathscr{L}(W')$.

Consider the subscheme $\mathrm{Hilb}(\mathscr{L})$ of the *relative Hilbert scheme* of curves of S over Δ, which is the Zariski closure of the set of all curves $C \in |\mathscr{L}_t|$, for $t \in \Delta^*$. We assume that $\mathrm{Hilb}(\mathscr{L})$ is a component of the relative Hilbert scheme, a condition satisfied if $\mathrm{Pic}(S_t)$ has no torsion, which will always be the case in our applications. One has a natural projection morphism $\varphi : \mathrm{Hilb}(\mathscr{L}) \to \Delta$, which is a projective bundle over Δ^*; actually $\mathrm{Hilb}(\mathscr{L})$ is isomorphic to $\mathfrak{P} := \mathbf{P}(f_*(\mathscr{L}))$ over Δ^*. We call the fibre of φ over 0 the *limit linear system* of $|\mathscr{L}_t|$ as $t \in \Delta^*$ tends to 0, and we denote it by $\mathfrak{L}$.

Remark 3. In general, *the limit linear system is not a linear system*. One would be tempted to say that $\mathfrak{L}$ is nothing but $|\mathscr{L}_0|$; this is the case if S_0 is irreducible, but it is in general no longer true when S_0 is reducible. In the latter case, there may be non–zero sections of $\mathscr{L}_0$ whose zero–locus contains some irreducible component of S_0, and accordingly points of $|\mathscr{L}_0|$ which do not correspond to points in the Hilbert scheme of curves (see, e.g., Example 2 below).

In any event, $\mathrm{Hilb}(\mathscr{L})$ is a birational modification of $\mathfrak{P}$, and $\mathfrak{L}$ is a suitable degeneration of the projective space $|\mathscr{L}_t|$, $t \in \Delta^*$. One has:

Lemma 2. *Let $\mathfrak{P}' \to \Delta$ be a flat and proper morphism, isomorphic to $\mathbf{P}(f_*(\mathscr{L}))$ over Δ^*, and such that $\mathfrak{P}'$ is a Zariski closed subset of the relative Hilbert scheme of curves of S over Δ. Then $\mathfrak{P}' = \mathrm{Hilb}(\mathscr{L})$.*

Proof. The two Zariski closed subsets $\mathfrak{P}'$ and $\mathrm{Hilb}(\mathscr{L})$ are irreducible, and coincide over Δ^*. □

In passing from $\mathbf{P}(f_*(\mathscr{L}))$ to $\mathrm{Hilb}(\mathscr{L})$, one has to perform a series of blow–ups along smooth centres contained in the central fibre, which correspond to spaces of non–trivial sections of some (twisted) line bundles which vanish on divisors contained in the central fibre. The exceptional divisors one gets in this way give rise to components of $\mathfrak{L}$, and may be identified with birational modifications of sublinear systems of twisted linear systems restricted to S_0, as follows from Lemma 3 below. We will see examples of this later (the first one in Example 2).

Lemma 3. (i) *Let X be a connected variety, $\mathscr{L}$ a line bundle on X, and σ a non zero global section of $\mathscr{L}$ defining a subscheme Z of X. Then the projectivized tangent space to $\mathbf{P}H^0(X, \mathscr{L})$ at $\langle\sigma\rangle$ canonically identifies with the* restricted linear system

$$\mathbf{P}\operatorname{Im}\big(\mathrm{H}^0(X,\mathscr{L}) \to \mathrm{H}^0(Z,\mathscr{L}|_Z)\big),$$

also called the trace *of* $|\mathscr{L}|$ *on* Z *(which in general is not the complete linear system* $|\mathscr{L}\otimes\mathscr{O}_Z|$*).*

(ii) *More generally, let* $\mathfrak{l}$ *be a linear subspace of* $\mathbf{P}\mathrm{H}^0(X,\mathscr{L})$ *with fixed locus scheme* F *defined by the system of equations* $\{\sigma = 0\}_{\langle\sigma\rangle\in\mathfrak{l}}$. *Then the projectivized normal bundle of* $\mathfrak{l}$ *in* $\mathbf{P}\mathrm{H}^0(X,\mathscr{L})$ *canonically identifies with*

$$\mathfrak{l}\times\mathbf{P}\operatorname{Im}\big(\mathrm{H}^0(X,\mathscr{L}) \to \mathrm{H}^0(F,\mathscr{L}|_F)\big).$$

Proof. Assertion (i) comes from the identification of the tangent space of $\mathbf{P}\mathrm{H}^0(X,\mathscr{L})$ at $\langle\sigma\rangle$ with the cokernel of the injection $\mathrm{H}^0(X,\mathscr{O}_X) \to \mathrm{H}^0(X,\mathscr{L})$, given by the multiplication by σ. As for (ii), note that the normal bundle of $\mathfrak{l}$ in $\mathbf{P}\mathrm{H}^0(X,\mathscr{L})$ splits as a direct sum of copies of $\mathscr{O}_{\mathfrak{l}}(1)$, hence the associated projective bundle is trivial. Then the proof is similar to that of (i). □

Example 2 (See [20]). Consider a family of degree k surfaces $f : S \to \Delta$ arising, as in Example 1, from a pencil generated by a general surface S_∞ and by $S_0 = F \cup P$, where P is a plane and F a general surface of degree $k-1$. One has a semistable model $\tilde{f} : \tilde{S} \to \Delta$ of this family, as described in Example 1, with $\tilde{S}_0 = F \cup \tilde{P}$, where $\tilde{P} \to P$ is the blow–up of P at the $k(k-1)$ intersection points of S_∞ with the smooth degree $k-1$ plane curve $R := F \cap P$ (with exceptional divisors E_i, for $1 \leqslant i \leqslant k(k-1)$).

We let $\mathscr{L} := \mathscr{O}_{\tilde{S}}(1)$ be the pull–back by $\tilde{S} \to S$ of $\mathscr{O}_S(1)$, obtained by pulling back $\mathscr{O}_{\mathbf{P}^3}(1)$ via the map $S \to \mathbf{P}^3$. The component $\mathrm{Hilb}(\mathscr{L})$ of the Hilbert scheme is gotten from the projective bundle $\mathbf{P}(f_*(\mathscr{O}_{\tilde{S}}(1)))$, by blowing up the point of the central fibre $|\mathscr{O}_{S_0}(1)|$ corresponding to the 1–dimensional space of non–zero sections vanishing on the plane P. The limit linear system $\mathfrak{L}$ is the union of $\mathfrak{L}_1$, the blown–up $|\mathscr{O}_{S_0}(1)|$, and of the exceptional divisor $\mathfrak{L}_2 \cong \mathbf{P}^3$, identified as the twisted linear system $|\mathscr{O}_{S_0}(1)\otimes\mathscr{O}_{S_0}(-P)|$. The corresponding twisted line bundle restricts to the trivial linear system on F, and to $|\mathscr{O}_{\tilde{P}}(k)\otimes\mathscr{O}_{\tilde{P}}(-\sum_{i=1}^{k(k-1)} E_i)|$ on $\tilde{P}$.

The components $\mathfrak{L}_1$ and $\mathfrak{L}_2$ of $\mathfrak{L}$ meet along the exceptional divisor $\mathfrak{E} \cong \mathbf{P}^2$ of the morphism $\mathfrak{L}_1 \to |\mathscr{O}_{S_0}(1)|$. Lemma 3 shows that the elements of $\mathfrak{E} \subset \mathfrak{L}_1$ identify as the points of $|\mathscr{O}_R(1)| \cong |\mathscr{O}_P(1)|$, whereas the plane $\mathfrak{E} \subset \mathfrak{L}_2$ is the set of elements $\Gamma \in |\mathscr{O}_{\tilde{P}}(k)\otimes\mathscr{O}_{\tilde{P}}(-\sum_{i=1}^{k(k-1)} E_i)|$ containing the proper transform $\hat{R} \cong R$ of R on $\tilde{P}$. The corresponding element of $|\mathscr{O}_R(1)|$ is cut out on $\hat{R}$ by the further component of Γ, which is the pull–back to $\tilde{P}$ of a line in P.

3.3 Severi Varieties and Their Limits

Let $f : S \to \Delta$ be a semistable family as in Sect. 3.1, and $\mathscr{L}$ be a line bundle on S as in Sect. 3.2. We fix a non–negative integer δ, and consider the locally closed subset $\mathring{V}_\delta(S,\mathcal{L})$ of Hilb($\mathcal{L}$) formed by all curves $D \in |\mathcal{L}_t|$, for $t \in \Delta^*$, such that D is irreducible, nodal, and has exactly δ nodes. We define $V_\delta(S,\mathcal{L})$ (resp. $V_\delta^{\mathrm{cr}}(S,\mathcal{L})$) as the Zariski closure of $\mathring{V}_\delta(S,\mathcal{L})$ in Hilb($\mathcal{L}$) (resp. in $\mathbf{P}(f_*(\mathcal{L}))$). This is the *relative Severi variety* (resp. the *crude relative Severi variety*). We may write $\mathring{V}_\delta$, V_δ, and V_δ^{cr}, rather than $\mathring{V}_\delta(S,\mathcal{L})$, $V_\delta(S,\mathcal{L})$, and $V_\delta^{\mathrm{cr}}(S,\mathcal{L})$, respectively.

We have a natural map $f_\delta : V_\delta \to \Delta$. If $t \in \Delta^*$, the fibre $V_{\delta,t}$ of f_δ over t is the *Severi variety* $V_\delta(S_t,\mathcal{L}_t)$ of δ–nodal curves in the linear system $|\mathcal{L}_t|$ on S_t, whose degree, independent on $t \in \Delta^*$, we denote by $d_\delta(\mathcal{L})$ (or simply by d_δ). We let $\mathfrak{V}_\delta(S,\mathcal{L})$ (or simply $\mathfrak{V}_\delta$) be the central fibre of $f_\delta : V_\delta \to \Delta$; it is the *limit Severi variety* of $V_\delta(S_t,\mathcal{L}_t)$ as $t \in \Delta^*$ tends to 0. This is a subscheme of the limit linear system $\mathfrak{L}$, which, as we said, has been studied by various authors. In particular, one can describe in a number of situations its various irreducible components, with their multiplicities (see Sect. 3.4 below). This is what we will do for several families of quartic surfaces in $\mathbf{P}^3$.

In a similar way, one defines the *crude limit Severi variety* $\mathfrak{V}_\delta^{\mathrm{cr}}(S,\mathcal{L})$ (or $\mathfrak{V}_\delta^{\mathrm{cr}}$), sitting in $|\mathcal{L}_0|$.

Remark 4. For $t \in \Delta^*$, the *expected dimension* of the Severi variety $V_\delta(S_t,\mathcal{L}_t)$ is $\dim(|L_t|) - \delta$. We will always assume that the dimension of (all components of) $V_\delta(S_t,\mathcal{L}_t)$ equals the expected one for all $t \in \Delta^*$. This is a strong assumption, which will be satisfied in all our applications.

Notation 1. Let $f : S \to \Delta$ be a family of degree k surfaces in $\mathbf{P}^3$ as in Example 1, and let $\tilde{f} : \tilde{S} \to \Delta$ be a semistable model of $f : S \to \Delta$. We consider the line bundle $\mathcal{O}_S(1)$, defined as the pull–back of $\mathcal{O}_{\mathbf{P}^3}(1)$ via the natural map $S \to \mathbf{P}^3$, and let $\mathcal{O}_{\tilde{S}}(1)$ be its pull–back on $\tilde{S}$. We denote by $\mathfrak{V}_{n,\delta}(\tilde{S})$ (resp. $\mathfrak{V}_{n,\delta}(S)$), or simply $\mathfrak{V}_{n,\delta}$, the limit Severi variety $\mathfrak{V}_\delta(\tilde{S},\mathcal{O}_{\tilde{S}}(n))$ (resp. $\mathfrak{V}_\delta(S,\mathcal{O}_S(n))$). Similar notation $\mathfrak{V}_{n,\delta}^{\mathrm{cr}}(\tilde{S})$ (resp. $\mathfrak{V}_{n,\delta}^{\mathrm{cr}}(S)$), or $\mathfrak{V}_{n,\delta}^{\mathrm{cr}}$, will be used for the crude limit.

3.4 Description of the Limit Severi Variety

Let again $f : S \to \Delta$ be a semistable family as in Sect. 3.1, and $\mathcal{L}$ a line bundle on S as in Sect. 3.2. The local machinery developed in ([17, 18, Galati-Knutsen]) enables us to identify the components of the limit Severi variety, with their multiplicities. As usual, we will suppose that S_0 has irreducible components $Q_1, \ldots, Q_r$, intersecting transversally along the double curves $R_1, \ldots, R_p$. We will also assume that there are q exceptional curves $E_1, \ldots, E_q$ on S_0, arising from a small resolution of an original family with singular total space, as discussed in Sect. 3.1.

Notation 2. Let $\underline{\mathbf{N}}$ be the set of sequences $\underline{\tau} = (\tau_m)_{m \geqslant 2}$ of non–negative integers with only finitely many non–vanishing terms. We define two maps $\nu,\ \mu : \underline{\mathbf{N}} \to \mathbf{N}$ as follows:

$$\nu(\underline{\tau}) = \sum_{m\geqslant 2} \tau_m \cdot (m-1), \quad \text{and} \quad \mu(\underline{\tau}) = \prod_{m\geqslant 2} m^{\tau_m}.$$

Given a p-tuple $\underline{\boldsymbol{\tau}} = (\underline{\tau}_1, \ldots, \underline{\tau}_p) \in \underline{\mathbf{N}}^p$, we set

$$\nu(\underline{\boldsymbol{\tau}}) = \nu(\underline{\tau}_1) + \cdots + \nu(\underline{\tau}_p), \quad \text{and} \quad \mu(\underline{\boldsymbol{\tau}}) = \mu(\underline{\tau}_1)\cdots\mu(\underline{\tau}_p),$$

thus defining two maps $\nu,\ \mu : \underline{\mathbf{N}}^p \to \mathbf{N}$. Given $\boldsymbol{\delta} = (\delta_1, \ldots, \delta_r) \in \mathbf{N}^r$, we set

$$|\boldsymbol{\delta}| := \delta_1 + \cdots + \delta_r.$$

Given a subset $I \subset \{1, \ldots, q\}$, $|I|$ will denote its cardinality.

Definition 1. Consider a divisor W on S, supported on the central fibre S_0, i.e. a linear combination of $Q_1, \ldots, Q_r$. Fix $\boldsymbol{\delta} \in \mathbf{N}^r$, $\underline{\boldsymbol{\tau}} \in \underline{\mathbf{N}}^p$, and $I \subseteq \{1, \ldots, r\}$. We let $\mathring{V}(W, \boldsymbol{\delta}, I, \underline{\boldsymbol{\tau}})$ be the Zariski locally closed subset in $|\mathscr{L}(-W) \otimes \mathscr{O}_{S_0}|$ parametrizing curves D such that:

(i) D neither contains any curve R_l, with $l \in \{1, \ldots, p\}$, nor passes through any triple point of S_0;
(ii) D contains the exceptional divisor E_i, with multiplicity 1, if and only if $i \in I$, and has a node on it;
(iii) $D - \sum_{i\in I} E_i$ has δ_s nodes on Q_s, for $s \in \{1, \ldots, r\}$, off the singular locus of S_0, and is otherwise smooth;
(iv) For every $l \in \{1, \ldots, p\}$ and $m \geqslant 2$, there are exactly $\tau_{l,m}$ points on R_l, off the intersections with $\sum_{i\in I} E_i$, at which D has an *m-tacnode* (see below for the definition), with reduced tangent cone equal to the tangent line of R_l there.

We let $V(W, \boldsymbol{\delta}, I, \underline{\boldsymbol{\tau}})$ be the Zariski closure of $\mathring{V}(W, \boldsymbol{\delta}, I, \underline{\boldsymbol{\tau}})$ in $|\mathscr{L}(-W) \otimes \mathscr{O}_{X_0}|$.

Recall that an *m-tacnode* is an A_{2m-1}-double point, i.e. a plane curve singularity locally analytically isomorphic to the hypersurface of $\mathbf{C}^2$ defined by the equation $y^2 = x^{2m}$ at the origin. Condition (iv) above requires that D is a divisor having $\tau_{l,m}$ m–th order tangency points with the curve R_l, at points of R_l which are not triple points of S_0.

Notation 3. In practice, we shall not use the notation $V(W, \boldsymbol{\delta}, I, \underline{\boldsymbol{\tau}})$, but rather a more expressive one like, e.g., $V(W, \delta_{Q_1} = 2, E_1, \tau_{R_1,2} = 1)$ for the variety parametrizing curves in $|\mathscr{L}(-W)\otimes\mathscr{O}_{S_0}|$, with two nodes on Q_1, one simple tacnode along R_1, and containing the exceptional curve E_1.

Proposition 1. *([17, 18, Galati-Knutsen]). Let $W, \boldsymbol{\delta}, I, \underline{\boldsymbol{\tau}}$ be as above, and set $|\boldsymbol{\delta}| + |I| + \nu(\underline{\boldsymbol{\tau}}) = \delta$. Let V be an irreducible component of $V(W, \boldsymbol{\delta}, I, \underline{\boldsymbol{\tau}})$. If*

(i) The linear system $|\mathscr{L}(-W) \otimes \mathscr{O}_{X_0}|$ has the same dimension as $|\mathscr{L}_t|$ for $t \in \Delta^*$, and
(ii) V has (the expected) codimension δ in $|\mathscr{L}(-W) \otimes \mathscr{O}_{X_0}|$,

then V is an irreducible component of multiplicity $\mu(V) := \mu(\underline{\tau})$ of the limit Severi variety $\mathfrak{V}_\delta(S, \mathscr{L})$.

Remark 5. Same assumptions as in Proposition 1. If there is at most one tacnode (i.e. all $\tau_{l,m}$ but possibly one vanish, and this is equal to 1), the relative Severi variety V_δ is smooth at the general point of V (see [17, 18, Galati-Knutsen]), and thus V belongs to only one irreducible component of V_δ. There are other cases in which such a smoothness property holds (see [9]).

If V_δ is smooth at the general point $D \in V$, the multiplicity of V in the limit Severi variety $\mathfrak{V}_\delta$ is the minimal integer m such that there are local analytic *m–multisections* of $V_\delta \to \Delta$, i.e. analytic smooth curves in V_δ, passing through D and intersecting the general fibre $V_{\delta,t}$, $t \in \Delta^*$, at m distinct points.

Proposition 1 still does not provide a complete picture of the limit Severi variety. For instance, curves passing through a triple point of S_0 could play a role in this limit. It would be desirable to know that one can always obtain a semistable model of the original family, where *every* irreducible component of the limit Severi variety is realized as a family of curves of the kind stated in Definition 1.

Definition 2. Let $f : S \to \Delta$ be a semistable family as in Sect. 3.1, $\mathscr{L}$ a line bundle on S as in Sect. 3.2, and δ a positive integer. The *regular part of the limit Severi variety* $\mathfrak{V}_\delta(S, \mathscr{L})$ is the cycle in the limit linear system $\mathfrak{L} \subset \mathrm{Hilb}(\mathscr{L})$

$$\mathfrak{V}_\delta^{\mathrm{reg}}(S, \mathscr{L}) := \sum_{W} \sum_{|\delta|+|I|+\nu(\underline{\tau})=\delta} \mu(\underline{\tau}) \cdot \left(\sum_{V \in \mathrm{Irr}^\delta(V(W,\delta,I,\underline{\tau}))} V \right) \tag{1}$$

(sometimes simply denoted by $\mathfrak{V}_\delta^{\mathrm{reg}}$), where:

(i) W varies among all effective divisors on S supported on the central fibre S_0, such that $\mathrm{h}^0(\mathscr{L}_0(-W)) = \mathrm{h}^0(\mathscr{L}_t)$ for $t \in \Delta^*$;
(ii) $\mathrm{Irr}^\delta(Z)$ denotes the set of all codimension δ irreducible components of a scheme Z.

Proposition 1 asserts that the cycle $Z(\mathfrak{V}_\delta) - \mathfrak{V}_\delta^{\mathrm{reg}}$ is effective, with support disjoint in codimension 1 from that of $\mathfrak{V}_\delta^{\mathrm{reg}}$ (here, $Z(\mathfrak{V}_\delta)$ is the cycle associated to $\mathfrak{V}_\delta$). We call the irreducible components of the support of $\mathfrak{V}_\delta^{\mathrm{reg}}$ the *regular components* of the limit Severi variety.

Let $\tilde{f} : \tilde{S} \to \Delta$ be a semistable model of $f : S \to \Delta$, and $\tilde{\mathscr{L}}$ the pull–back on $\tilde{S}$ of $\mathscr{L}$. There is a natural map $\mathrm{Hilb}(\tilde{\mathscr{L}}) \to \mathrm{Hilb}(\mathscr{L})$, which induces a morphism $\phi : \tilde{\mathfrak{L}} \to |\mathscr{L}_0|$.

Definition 3. The semistable model $\tilde{f} : \tilde{S} \to \Delta$ is a *δ–good model* of $f : S \to \Delta$ (or simply *good model*, if it is clear which δ we are referring at), if the following equality of cycles holds

$$\phi_*(\mathfrak{V}_\delta^{\mathrm{reg}}(\tilde{S}, \tilde{\mathscr{L}})) = \mathfrak{V}_\delta^{\mathrm{cr}}(S, \mathscr{L}).$$

Note that the cycle $\mathfrak{V}^{\mathrm{cr}}_{\delta}(S,\mathscr{L}) - \phi_*(\mathfrak{V}^{\mathrm{reg}}_{\delta}(\tilde{S},\tilde{\mathscr{L}}))$ is effective. The family $f : S \to \Delta$ is said to be *δ–well behaved* (or simply *well behaved*) if it has a δ-good model. A semistable model $\tilde{f} : \tilde{S} \to \Delta$ of $f : S \to \Delta$ as above is said to be *δ–absolutely good* if $\mathfrak{V}_{\delta}(\tilde{S},\tilde{\mathscr{L}}) = \mathfrak{V}^{\mathrm{reg}}_{\delta}(\tilde{S},\tilde{\mathscr{L}})$ as cycles in the relative Hilbert scheme. It is then a δ–good model both of itself, and of $f : S \to \Delta$.

Theorems 1 and 2 will be proved by showing that the corresponding families of quartic surfaces are well behaved.

Remark 6. Suppose that $f : S \to \Delta$ is δ–well behaved, with δ–good model $\tilde{f} : \tilde{S} \to \Delta$. It is possible that some components in $\mathfrak{V}^{\mathrm{reg}}_{\delta}(\tilde{S},\tilde{\mathscr{L}})$ are contracted by $\mathrm{Hilb}(\tilde{\mathscr{L}}) \to |\mathscr{L}_0|$ to varieties of smaller dimension, and therefore that their push–forwards are zero. Hence these components of $\mathfrak{V}_{\delta}(\tilde{S})$ are *not visible* in $\mathfrak{V}^{\mathrm{cr}}_{\delta}(S)$. They are however usually visible in the crude limit Severi variety of another model $f' : S' \to \Delta$, obtained from $\tilde{S}$ via an appropriate twist of $\mathscr{L}$. The central fibre S'_0 is then a flat limit of S_t, as $t \in \Delta^*$ tends to 0, different from S_0.

Conjecture 1. Let $f : S \to \Delta$ be a semistable family of surfaces, endowed with a line bundle $\mathscr{L}$ as above, and δ a positive integer. Then:

(Weak version) Under suitable assumptions (to be discovered), $f : S \to \Delta$ is δ–well behaved.

(Strong version) Under suitable assumptions (to be discovered), $f : S \to \Delta$ has a δ–absolutely good semistable model.

The local computations in [18] provide a criterion for absolute goodness:

Proposition 2. *Assume there is a semistable model $\tilde{f} : \tilde{S} \to \Delta$ of $f : S \to \Delta$, with a limit linear system $\tilde{\mathfrak{L}}$ free in codimension $\delta + 1$ of curves of the following types:*

(i) Curves containing double curves of $\tilde{S}_0$;
(ii) Curves passing through a triple point of $\tilde{S}_0$;
(iii) Non–reduced curves.

If in addition, for $W, \boldsymbol{\delta}, I, \underline{\boldsymbol{\tau}}$ as in Definition 1, every irreducible component of $V(W, \boldsymbol{\delta}, I, \underline{\boldsymbol{\tau}})$ has the expected codimension in $|\mathscr{L}_0(-W)|$, then $\tilde{f} : \tilde{S} \to \Delta$ is δ–absolutely good, which implies that $f : S \to \Delta$ is δ–well behaved.

Unfortunately, in the cases we shall consider conditions (i)–(iii) in Proposition 2 are violated (see Propositions 15 and 23), which indicates that further investigation is needed to prove the above conjectures. The components of the various $V(W, \boldsymbol{\delta}, I, \underline{\boldsymbol{\tau}})$ have nevertheless the expected codimension, and we are able to prove that our examples are well–behaved, using additional enumerative information.

Absolute goodness seems to be a property hard to prove, except when the dimension of the Severi varieties under consideration is 0, equal to the expected one (and even in this case, we will need extra enumerative information for the proof). We note in particular that the δ–absolute goodness of $\tilde{f} : \tilde{S} \to \Delta$ implies that it is a δ–good model of every model $f' : S' \to \Delta$, obtained from $\tilde{S}$ via a twist of $\tilde{\mathscr{L}}$ corresponding to an irreducible component of the limit linear system $\tilde{\mathfrak{L}}$.

3.5 An Enumerative Application

Among the applications of the theory described above, there are the ones to enumerative problems, in particular to the computation of the degree d_δ of Severi varieties $V_\delta(S_t, \mathscr{L}_t)$, for the general member S_t of a family $f : S \to \Delta$ as in Sect. 3.1, with $\mathscr{L}$ a line bundle on S as in Sect. 3.2.

Let $t \in \Delta^*$ be general, and let m_δ be the dimension of $V_\delta(S_t, \mathscr{L}_t)$, which we assume to be $m_\delta = \dim(|\mathscr{L}_t|) - \delta$. Then d_δ is the number of points in common of $V_\delta(S_t, \mathscr{L}_t)$ with m_δ *sufficiently general* hyperplanes of $|\mathscr{L}_t|$. Given $x \in S_t$,

$$H_x := \{[D] \in |\mathscr{L}_t| \text{ s.t. } x \in D\}$$

is a plane in $|\mathscr{L}_t|$. It is well known, and easy to check (we leave this to the reader), that if $x_1, \dots, x_{m_\delta}$ are *general points* of S_t, then $H_{x_1}, \dots, H_{x_{m_\delta}}$ are *sufficiently general* planes of $|\mathscr{L}_t|$ with respect to $V_\delta(S_t, \mathscr{L}_t)$. Thus d_δ *is the number of δ–nodal curves in $|\mathscr{L}_t|$ passing through m_δ general points of S_t.*

Definition 4. In the above setting, let V be an irreducible component of the limit Severi variety $\mathfrak{V}_\delta(S, \mathscr{L})$, endowed with its reduced structure. We let $Q_1, \dots, Q_r$ be the irreducible components of S_0, and $\mathbf{n} = (n_1, \dots, n_r) \in \mathbf{N}^r$ be such that $|\mathbf{n}| := n_1 + \dots + n_r = m_\delta$. Fix a collection Z of $n_1, \dots, n_r$ general points on $Q_1, \dots, Q_r$ respectively. The **n**–*degree* of V is the number $\deg_{\mathbf{n}}(V)$ of points in V corresponding to curves passing through the points in Z.

Note that in case $m_\delta = 0$, the above definition is somehow pointless: in this case, $\deg_{\mathbf{n}}(V)$ is simply the number of points in V. By contrast, when V has positive dimension, it is possible that $\deg_{\mathbf{n}}(V)$ be zero for various **n**'s. This is related to the phenomenon described in Remark 6 above. We will see examples of this below.

By flatness, the following result is clear:

Proposition 3. *Let $\tilde{f} : \tilde{S} \to \Delta$ be a semistable model, and name $P_1, \dots, P_{\tilde{r}}$ the irreducible components of $\tilde{S}_0$, in such a way that $P_1, \dots, P_r$ are the proper transforms of $Q_1, \dots, Q_r$ respectively.*

(i) For every $\tilde{\mathbf{n}} = (n_1, \dots, n_r, 0, \dots, 0) \in \mathbf{N}^{\tilde{r}}$ such that $|\tilde{\mathbf{n}}| = m_\delta$, one has

$$d_\delta \geqslant \sum_{V \in \mathrm{Irr}(\mathfrak{V}_\delta^{\mathrm{reg}}(\tilde{S}, \tilde{\mathscr{L}}))} \mu(V) \cdot \deg_{\tilde{\mathbf{n}}}(V) \tag{2}$$

(recall the definition of $\mu(V)$ in Proposition 1).

(ii) If equality holds in (2) *for every* $\tilde{\mathbf{n}}$ *as above, then* $\tilde{f} : \tilde{S} \to \Delta$ *is a* δ*–good model of* $f : S \to \Delta$ *endowed with* $\mathscr{L}$.

4 Auxiliary Results

In this section we collect a few results which we will use later.

First of all, for a general surface S of degree k in $\mathbf{P}^3$, we know from classical projective geometry the degrees $d_{\delta,k}$ of the Severi varieties $V_\delta(S, \mathscr{O}_S(1))$, for $1 \leqslant \delta \leqslant 3$. For $K3$ surfaces, this fits in a more general framework of known numbers (see [3, 6, 29, 35]). One has:

Proposition 4 ([33, 34]). *Let* S *be a general degree* k *hypersurface in* $\mathbf{P}^3$. *Then*

$$\begin{aligned} d_{1,k} &= k(k-1)^2, \\ d_{2,k} &= \frac{1}{2}k(k-1)(k-2)(k^3-k^2+k-12), \\ d_{3,k} &= \frac{1}{6}k(k-2)(k^7-4k^6+7k^5-45k^4+114k^3-111k^2+548k-960). \end{aligned}$$

For $k = 4$, *these numbers are* 36, 480, 3,200 *respectively.*

Note that $V_1(S, \mathscr{O}_S(1))$ identifies with the *dual surface* $\check{S} \subset \check{\mathbf{P}}^3$. The following is an extension of the computation of $d_{1,k}$ for surfaces with certain singularities. This is well–known and the details can be left to the reader.

Proposition 5. *Let* S *be a degree* k *hypersurface in* $\mathbf{P}^3$, *having* ν *and* κ *double points of type* A_1 *and* A_2 *respectively as its only singularities. Then*

$$\deg(\check{S}) = k(k-1)^2 - 2\nu - 3\kappa.$$

The following topological formula is well-known (see, e.g., [2, Lemme VI.4]).

Lemma 4. *Let* $p : S \to B$ *be a surjective morphism of a smooth projective surface onto a smooth curve. One has*

$$\chi_{\mathrm{top}}(S) = \chi_{\mathrm{top}}(F_{\mathrm{gen}})\chi_{\mathrm{top}}(B) + \sum_{b \in \mathrm{Disc}(p)} \big(\chi_{\mathrm{top}}(F_b) - \chi_{\mathrm{top}}(F_{\mathrm{gen}})\big),$$

where F_{gen} *and* F_b *respectively denote the fibres of* p *over the generic point of* B *and a closed point* $b \in B$, *and* $\mathrm{Disc}(p)$ *is the set of points above which* p *is not smooth.*

As a side remark, note that it is possible to give a proof of the Proposition 5 based on Lemma 4. This can be left to the reader.

Propositions 4 and 5 are sort of Plücker formulae for surfaces in $\mathbf{P}^3$. The next proposition provides analogous formulae for curves in a projective space of any dimension.

Proposition 6. *Let $C \subset \mathbf{P}^N$ be an irreducible, non–degenerate curve of degree d and of genus g, the normalization morphism of which is unramified. Let $\tau \leqslant N$ be a non-negative integer, and assume $2\tau < d$. Then the Zariski closure of the locally closed subset of $\check{\mathbf{P}}^N$ parametrizing τ–*tangent *hyperplanes to C (i.e. planes tangent to C at τ distinct points) has degree equal to the coefficient of $u^\tau v^{d-2\tau}$ in*

$$(1 + 4u + v)^g (1 + 2u + v)^{d-\tau-g}.$$

Proof. Let $\nu : \bar{C} \to C$ be the normalization of C, and let $\mathfrak{g}$ be the g^N_μ on $\bar{C}$ defined as the pull–back on $\bar{C}$ of the hyperplane linear series on C. Since ν is unramified, the degree of the subvariety of $\check{\mathbf{P}}^N$ parametrizing τ-tangent hyperplanes to C is equal to the number of divisors having τ double points in a general sublinear series g^τ_μ of $\mathfrak{g}$. This number is computed by a particular instance of de Jonquières' formula, see [1, p. 359]. □

The last result we shall need is:

Lemma 5. *Consider a smooth, irreducible curve R, contained in a smooth surface S in $\mathbf{P}^3$. Let $\check{R}_S$ be the irreducible curve in $\check{\mathbf{P}}^3$ parametrizing planes tangent to S along R. Then the dual varieties $\check{S}$ and $\check{R}$ both contain $\check{R}_S$, and do not intersect transversely at its general point.*

Proof. Clearly $\check{R}_S$ is contained in $\check{S} \cap \check{R}$. If either $\check{S}$ or $\check{R}$ are singular at the general point of $\check{R}_S$, there is nothing to prove. Assume that $\check{S}$ and $\check{R}$ are both smooth at the general point of $\check{R}_S$. We have to show that they are tangent there. Let $x \in R$ be general. Let H be the tangent plane to S at x. Then $H \in \check{R}_S$ is the general point. Now, the biduality theorem (see, e.g., [23, Example 16.20]) says that the tangent plane to $\check{S}$ and of $\check{R}$ at H both coincide with the set of planes in $\mathbf{P}^3$ containing x, hence the assertion. □

5 Degeneration to a Tetrahedron

We consider a family $f : \mathcal{S} \to \Delta$ of surfaces in $\mathbf{P}^3$, induced (as in Example 1 and in Sect. 3.2) by a pencil generated by a general quartic surface S_∞ and a *tetrahedron* S_0 (i.e. S_0 is the union of four independent planes, called the *faces* of the tetrahedron), together with the pull-back $\mathcal{O}_{\mathcal{S}}(1)$ of $\mathcal{O}_{\mathbf{P}^3}(1)$. We will prove that it is δ–well behaved for $1 \leqslant \delta \leqslant 3$ by constructing a suitable good model.

The plan is as follows. We construct the good model in Sect. 5.1, and complete its description in Sect. 5.2. We then construct the corresponding limit linear system: the core of this is Sects. 5.3–5.6, are devoted to the study of the geometry of the exceptional components of the limit linear system (alternatively, of the geometry of the corresponding flat limits of the smooth quartic surfaces $S_t, t \in \Delta^*$); eventually, we complete the description in Sect. 5.7. We then identify the limit Severi varieties in Sect. 5.8.

5.1 A Good Model

The outline of the construction is as follows:

(I) *We first make a small resolution of the singularities of S as in Example 1;*
(II) *Then we perform a degree 6 base change;*
(III) *Next we resolve the singularities of the total space arisen with the base change, thus obtaining a new semistable family $\pi : X \to \Delta$;*
(IV) *Finally we will flop certain double curves in the central fibre X_0, thus obtaining a new semistable family $\varpi : \bar{X} \to \Delta$.*

The central fibre of the intermediate family $\pi : X \to \Delta$ is pictured in Fig. 1 (p. 143; we provide a cylindrical projection of a real picture of X_0, the dual graph of which is topologically an $\mathbf{S}^2$ sphere), and the flops are described in Fig. 2 (p. 144). The reason why we need to make the degree 6 base change is, intuitively, the following: a degree 3 base change is needed to understand the contribution to the limit Severi variety of curves passing through a *vertex* (i.e. a triple point) of the tetrahedron, while an additional degree 2 base change enables one to understand the contributions due to the *edges* (i.e. the double lines) of the tetrahedron.

Steps (I) and (II)

The singularities of the initial total space S consist of four ordinary double points on each edge of S_0. We consider (cf. Example 1) the small resolution $\tilde{S} \to S$ obtained by arranging for every edge the four (-1)–curves two by two on the two adjacent faces. We call $\tilde{f} : \tilde{S} \to \Delta$ the new family.

Let $p_1, \ldots, p_4$ be the triple points of $\tilde{S}_0$. For each $i \in \{1, \ldots, 4\}$, we let P_i be the irreducible component of $\tilde{S}_0$ which is opposite to the vertex p_i: it is a plane blown-up at six points. For distinct $i, j \in \{1, \ldots, 4\}$, we let E_{ij}^+ and E_{ij}^- be the two (-1)-curves contained in P_i and meeting P_j. We call z_{ij}^+ and z_{ij}^- the two points cut out on P_i by E_{ji}^+ and E_{ji}^- respectively.

Let now $\bar{f} : \bar{S} \to \Delta$ be the family obtained from $\tilde{f} : \tilde{S} \to \Delta$ by the base change $t \in \Delta \mapsto t^6 \in \Delta$. The central fibre $\bar{S}_0$ is isomorphic to $\tilde{S}_0$, so we will keep the above notation for it.

Step (III)

As a first step in the desingularization of $\bar{S}$, we perform the following sequence of operations for all $i \in \{1, \ldots, 4\}$. The total space $\bar{S}$ around p_i is locally analytically isomorphic to the hypersurface of $\mathbf{C}^4$ defined by the equation $xyz = t^6$ at the origin. We blow-up $\bar{S}$ at p_i. The blown–up total space locally sits in $\mathbf{C}^4 \times \mathbf{P}^3$. Let $(\xi : \eta : \zeta : \vartheta)$ be the homogeneous coordinates in $\mathbf{P}^3$. Then the new total space is locally defined in $\mathbf{C}^4 \times \mathbf{P}^3$ by the equations

$$\xi^4\eta\zeta = \vartheta^6 x^3, \quad \xi\eta^4\zeta = \vartheta^6 y^3, \quad \xi\eta\zeta^4 = \vartheta^6 z^3, \quad \text{and} \quad \xi\eta\zeta = \vartheta^3 t^3. \tag{3}$$

The equation of the exceptional divisor (in the exceptional $\mathbf{P}^3$ of the blow–up of $\mathbf{C}^4$) is $\xi\eta\zeta = 0$, hence this is the union of three planes meeting transversely at a point p_i' in $\mathbf{P}^3$. For i, j distinct in $\{1, \ldots, 4\}$, we call A_j^i the exceptional planes meeting the proper transform of P_j (which, according to our conventions, we still denote by P_j, see Sect. 2).

The equation of the new family around the point p_i' given by $\bigcap_{j \neq i} A_j^i$ is $\xi\eta\zeta = t^3$ (which sits in the affine chart $\vartheta = 1$). Next we blow-up the points p_i', for $i \in \{1, \ldots, 4\}$. The new exceptional divisor T^i at each point p_i' is isomorphic to the cubic surface with equation $\xi\eta\zeta = t^3$ in the $\mathbf{P}^3$ with coordinates $(\xi : \eta : \zeta : t)$. Note that T^i has three A_2–double points, at the vertices of the triangle $t = 0,\ \xi\eta\zeta = 0$.

Next we have to get rid of the singularities of the total space along the double curves of the central fibre. First we take care of the curves $C_{hk} := P_h \cap P_k$, for h, k distinct in $\{1, \ldots, 4\}$. The model we constructed so far is defined along such a curve by an equation of the type $\xi\eta = \vartheta^6 z^3$, (as it follows, e.g., from the third equation in (3) by setting $\zeta = 1$). The curve C_{hk} is defined by $\xi = \eta = \vartheta = 0$. If $i \in \{1, \ldots, 4\} - \{h, k\}$, the intersection point $p_{hki} := C_{hk} \cap A_h^i \cap A_k^i$ is cut out on C_{hk} by the hyperplane with equation $z = 0$. Away from the p_{hki}'s, with $i \in \{1, \ldots, 4\} - \{h, k\}$, the points of C_{hk} are double points of type A_5 for the total space. We blow–up along this curve: this introduces new homogeneous coordinates $(\xi_1 : \eta_1 : \vartheta_1)$, with new equations for the blow–up

$$\xi_1^5\eta_1 = \vartheta_1^6\xi^4 z^3, \quad \xi_1\eta_1^5 = \vartheta_1^6\eta^4 z^3, \quad \text{and} \quad \xi_1\eta_1 = \vartheta_1^2\vartheta^4 z^3.$$

The exceptional divisor is defined by $\xi_1\eta_1 = 0$, and is the transverse union of two ruled surfaces: we call W_{hk}' the one that meets P_h, and W_{kh}' the other. The affine chart we are interested in is $\vartheta_1 = 1$, where the equation is $\xi_1\eta_1 = \vartheta^4 z^3$. We then blow–up along the curve $\xi_1 = \eta_1 = \vartheta = 0$, which gives in a similar way the new equation $\xi_2\eta_2 = \vartheta^2 z^3$ with the new coordinates $(\xi_2 : \eta_2 : \vartheta_2)$. The exceptional divisor consists of two ruled surfaces, and we call W_{hk}'' (resp. W_{kh}'') the one that meets W_{hk}' (resp. W_{kh}'). Finally, by blowing-up along the curve $\xi_2 = \eta_2 = \vartheta = 0$, we obtain a new equation $\xi_3\eta_3 = \vartheta_3^2 z^3$, with new coordinates $(\xi_3 : \eta_3 : \vartheta_3)$. The exceptional divisor is a ruled surface, with two A_2–double points at its intersection points with the curves $C_{hk}^i := A_h^i \cap A_k^i$, with $i \in \{1, \ldots, 4\} - \{h, k\}$. We call it either W_{hk} or W_{kh}, with no ambiguity.

The final step of our desingularization process consists in blowing–up along the 12 curves C_{hk}^i, with pairwise distinct $h, k, i \in \{1, \ldots, 4\}$. The total space is given along each of these curves by an equation of the type $\xi\eta = \vartheta^3 t^3$ in the variables

(ξ,η,θ,t), obtained from the last equation in (3) by setting $\zeta = 1$. The curve C^i_{hk} is defined by the local equations $\xi = \eta = t = 0$, which shows that they consist of A_2–double points for the total space. They also contain an A_2–double point of W_{hk} and T^i respectively. A computation similar to the above shows that the blow–up along these curves resolves all singularities in a single move. The exceptional divisor over C^i_{hk} is the union of two transverse ruled surfaces: we call V^i_{hk} the one that meets A^i_h, and V^i_{kh} the other.

At this point, we have a semistable family $\pi : X \to \Delta$, whose central fibre is depicted in Fig. 1: for each double curve we indicate its self–intersections in the two components of the central fibre it belongs to. This is obtained by applying the Triple Point Formula (see Lemma 1).

Step (IV)

For our purposes, we need to further blow-up the total space along the 12 curves $\Gamma^i_{hk} := V^i_{hk} \cap V^i_{kh}$. This has the drawback of introducing components with multiplicity two in the central fibre, namely the corresponding exceptional divisors. To circumvent this, we will flop these curves as follows.

Let $\hat{\pi} : \hat{X} \to \Delta$ be the family obtained by blowing-up X along the Γ^i_{hk}'s. We call W^i_{hk} (or, unambiguously, W^i_{kh}) the corresponding exceptional divisors: they appear with multiplicity two in the central fibre $\hat{X}_0$. By applying the Triple Point Formula as in Remark 1, one checks that the surfaces W^i_{kh} are all isomorphic to $\mathbf{P}^1 \times \mathbf{P}^1$. Moreover, it is possible to contract W^i_{hk} in the direction of the ruling cut out by V^i_{hk} and V^i_{kh}, as indicated on Fig. 2. We call $\hat{X} \to \bar{X}$ the contraction of the 12 divisors W^i_{hk} in this way, and $\varpi : \bar{X} \to \Delta$ the corresponding semistable family of surfaces.

Even though $\bar{X} \dashrightarrow X$ is only a birational map, we have a birational morphism $\bar{X} \to \bar{S}$ over Δ.

5.2 *Identification of the Components of the Central Fibre*

Summarizing, the irreducible components of the central fibre $\bar{X}_0$ are the following:

(i) The 4 surfaces P_i, with $1 \leqslant i \leqslant 4$.

Each P_i is a plane blown–up at $6 + 3$ points, and H (i.e. the pull-back of a general line in the plane, recall our conventions in Sect. 2) is the restriction class of $\mathcal{O}_{\bar{X}}(1)$ on P_i. For $j, k \in \{1,\dots,4\} - \{i\}$, we set

$$L_{ij} := P_i \cap W'_{ij} \quad \text{and} \quad G^k_i := P_i \cap A^k_i,$$

as indicated in Fig. 3. In addition to the three (-1)–curves G^k_i, we have on P_i the six exceptional curves E^+_{ij}, E^-_{ij}, for all $j \in \{1,\dots,4\} - \{i\}$, with E^+_{ij}, E^-_{ij} intersecting

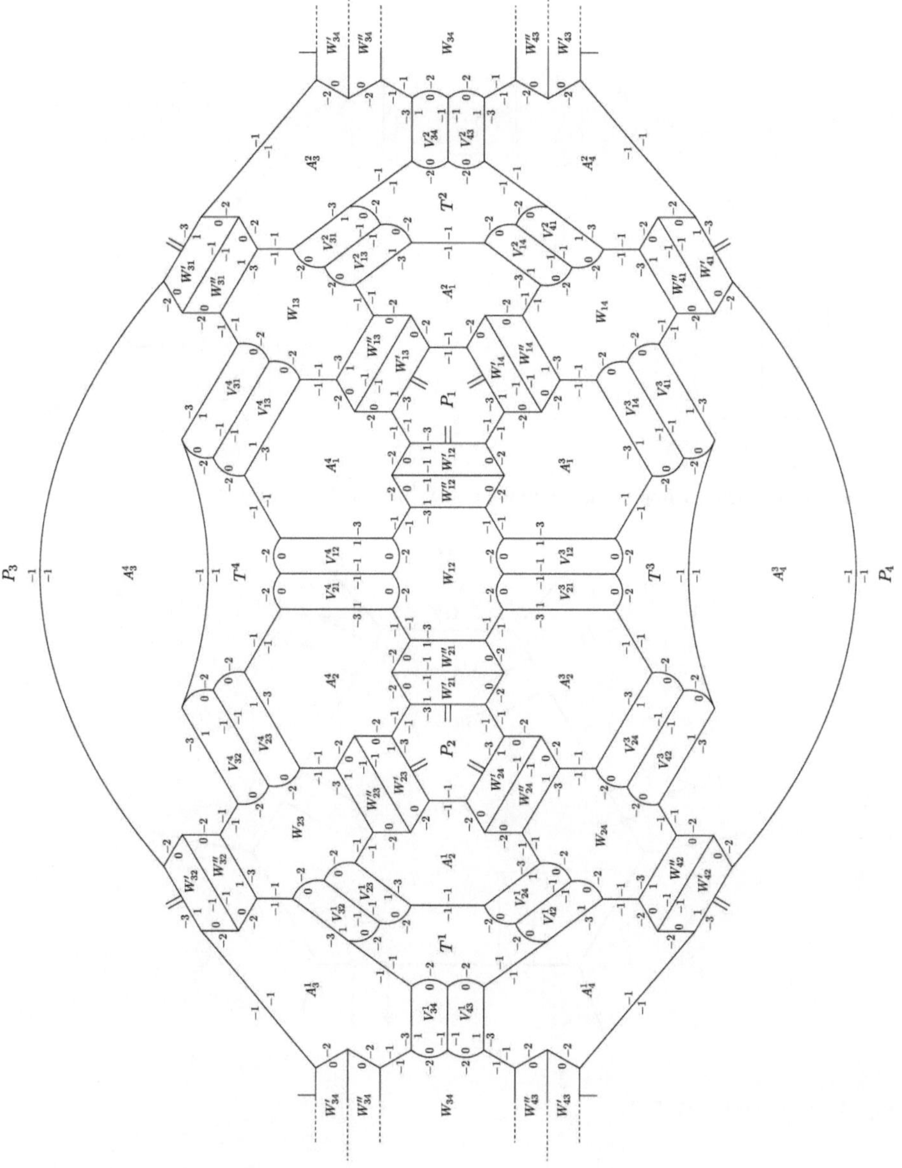

Fig. 1 Planisphere of the model X_0 of the degeneration into four planes

L_{ij} at one point. Moreover, for $j \in \{1, \dots, 4\} - \{i\}$, we have on L_{ij} the two points $z_{ji}^{\pm}$ defined as the strict transform of the intersection $E_{ji}^{\pm} \cap L_{ij}$ in $\tilde{S}$. We will denote by Z_i the 0–dimensional scheme of length 6 given by $\sum_{j \neq i}(z_{ji}^{+} + z_{ji}^{-})$. We let $\mathscr{I}_{Z_i} \subset \mathscr{O}_{P_i}$ be its defining sheaf of ideals.

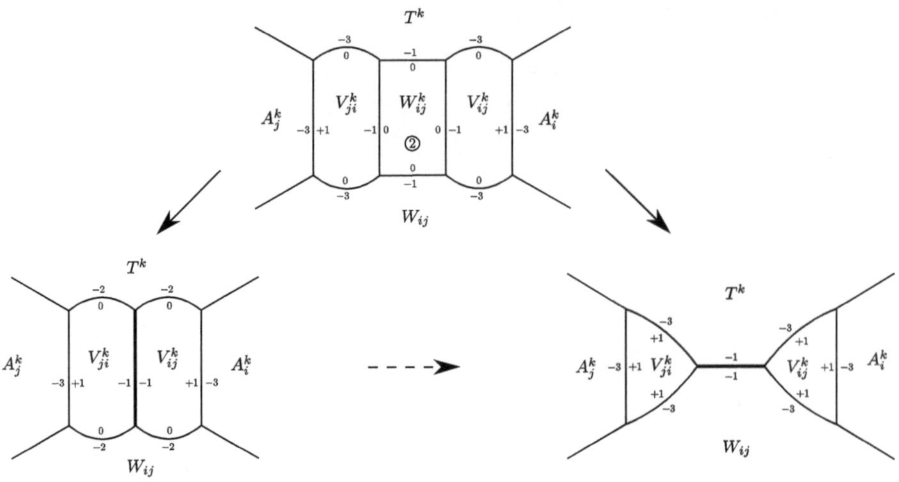

Fig. 2 One elementary flop of the birational transformation $X \dashrightarrow \bar{X}$

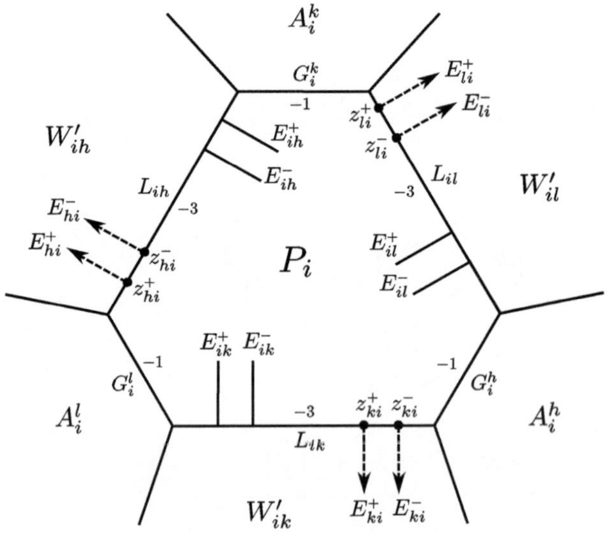

Fig. 3 Notations for $P_i \subset \bar{X}_0$

(ii) The 24 surfaces W'_{ij}, W''_{ij}, with $i, j \in \{1, \ldots, 4\}$ distinct.

Each of them is isomorphic to $\mathbf{F}_1$. We denote by $|F|$ the ruling. Note that the divisor class F corresponds to the restriction of $\mathcal{O}_{\bar{X}}(1)$.

(iii) The 6 surfaces W_{ij}, with $i, j \in \{1, \ldots, 4\}$ distinct.

For each $k \in \{1, \ldots, 4\} - \{i, j\}$, we set

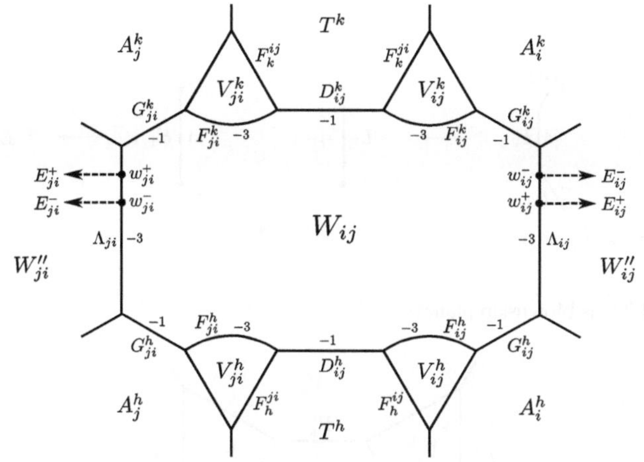

Fig. 4 Notations for $W_{ij} \subset \bar{X}_0$

$$\Lambda_{ij} := W''_{ij} \cap W_{ij}, \quad G^k_{ij} := W_{ij} \cap A^k_i, \quad F^k_{ij} = W_{ij} \cap V^k_{ij}, \quad D^k_{ij} = W_{ij} \cap T^k,$$

and define similarly Λ_{ji}, G^k_{ji}, F^k_{ji} (D^k_{ij} may be called D^k_{ji} without ambiguity). This is indicated in Fig. 4.

A good way of thinking to the surfaces W_{ij} is to consider them as (non–minimal) rational ruled surfaces, for which the two curves Λ_{ji} and Λ_{ij} are sections which do not meet, and the two rational chains

$$G^k_{ji} + F^k_{ji} + 2D^k_{ij} + F^k_{ij} + G^k_{ij}, \quad k \in \{1, \ldots, 4\} - \{i, j\},$$

are two disjoint reducible fibres of the ruling $|F|$. One has furthermore $\mathcal{O}_{W_{ij}}(F) = \mathcal{O}_{\bar{X}}(1) \otimes \mathcal{O}_{W_{ij}}$.

The surface W_{ij} has the length 12 anticanonical cycle

$$\Lambda_{ji} + G^k_{ji} + F^k_{ji} + D^k_{ij} + F^k_{ij} + G^k_{ij} + \Lambda_{ij} + G^h_{ij} + F^h_{ij} + D^h_{ij} + F^h_{ji} + G^h_{ji} \qquad (4)$$

cut out by $\bar{X}_0 - W_{ij}$, where we fixed k and h such that $\{i, j, k, h\} = \{1, \ldots, 4\}$. It therefore identifies with a plane blown–up as indicated in Fig. 5: consider a general triangle L_1, L_2, L_3 in $\mathbf{P}^2$, with vertices a_1, a_2, a_3, where a_1 is opposite to L_1, etc.; then blow–up the three vertices a_s, and call E_s the corresponding exceptional divisors; eventually blow–up the six points $L_r \cap E_s$, $r \neq s$, and call E_{rs} the corresponding exceptional divisors. The obtained surface has the anticanonical cycle

$$L_1 + E_{13} + E_1 + E_{23} + L_2 + E_{21} + E_1 + E_{31} + L_3 + E_{32} + E_2 + E_{12}, \quad (5)$$

which we identify term-by-term and in this order with the anticanonical cycle (4) of W_{ij}.

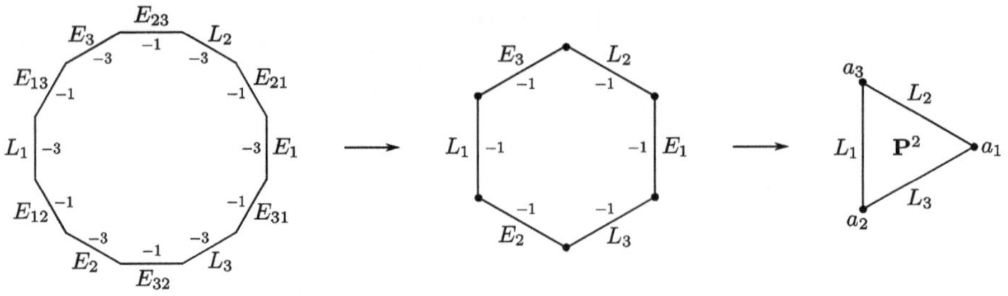

Fig. 5 W_{ij} and T^k as blown–up planes

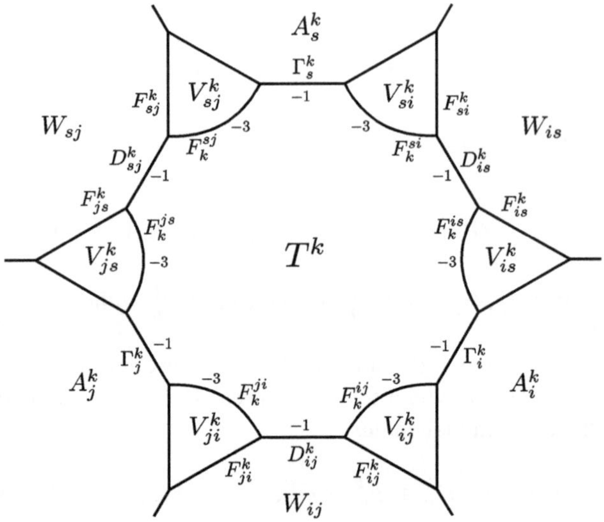

Fig. 6 Notations for $T^k \subset \bar{X}_0$

We let H be, as usual, (the transform of) a general line in the plane

$$H \sim_{W_{ij}} \Lambda_{ji} + \sum\nolimits_{k \notin \{i,j\}} (2G_{ji}^k + F_{ji}^k + D_{ji}^k). \tag{6}$$

The ruling $|F|$ is the strict transform of the pencil of lines through the point a_1, hence

$$|F| = \left|H - (\Lambda_{ij} + G_{ij}^k + G_{ij}^h)\right|, \quad \text{with} \quad \{1, \ldots, 4\} = \{i, j, k, h\}. \tag{7}$$

(iv) The 4 surfaces T^k, with $1 \leqslant k \leqslant 4$.

Here we set $\Gamma_i^k = T^k \cap A_i^k$ for $i \in \{1, \ldots, 4\} - \{k\}$, and $F_k^{ij} = T^k \cap V_{ij}^k$ for $i, j \in \{1, \ldots, 4\} - \{k\}$ distinct. Also recall that $D_{ij}^k = T^k \cap W_{ij}$ for $i, j \in \{1, \ldots, 4\} - \{k\}$ distinct. This is indicated in Fig. 6.

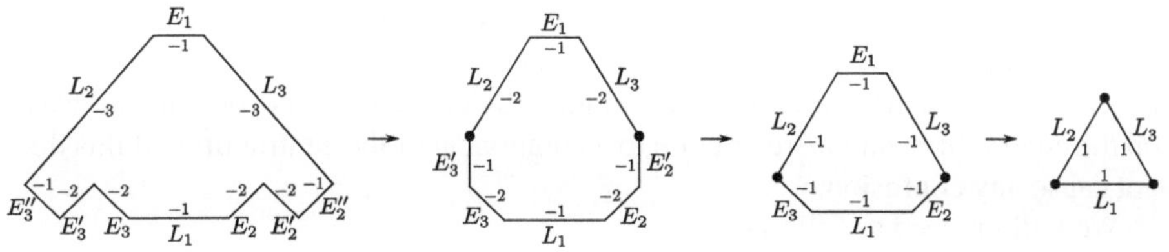

Fig. 7 A_j^k as a blown-up plane

Each T^k identifies with a plane blown–up as indicated in Fig. 5, as in the case of the W_{ij}'s: it has the length 12 anticanonical cycle

$$F_k^{js} + D_{sj}^k + F_k^{sj} + \Gamma_s^k + F_k^{si} + D_{is}^k + F_k^{is} + \Gamma_i^k + F_k^{ij} + D_{ij}^k + F_k^{ji} + \Gamma_j^k \tag{8}$$

(where we fixed indices s, i, j such that $\{s, i, j, k\} = \{1, \ldots, 4\}$) cut out by $\bar{X}_0 - T^k$ on T^k, which we identify term-by-term and in this order with the anticanonical cycle (5). This yields

$$H \sim_{T^k} F_k^{js} + (2D_{sj}^k + F_k^{sj} + \Gamma_s^k) + (2\Gamma_j^k + F_k^{ji} + D_{ij}^k). \tag{9}$$

We have on T^k the proper transform of a pencil of (bitangent) conics that meet the curves Γ_s^k and D_{ij}^k in one point respectively, and do not meet any other curve in the anticanonical cycle (8): we call this pencil $|\Phi_s^k|$, and we have

$$|\Phi_s^k| = \left|2H - (F_k^{sj} + D_{sj}^k + 2\Gamma_s^k) - (F_k^{ji} + \Gamma_j^k + 2D_{ij}^k)\right|.$$

The restriction of $\mathcal{O}_{\bar{X}}(1)$ on T^k is trivial.

(v) The 12 surfaces A_i^k, with $i, k \in \{1, \ldots, 4\}$ distinct.

Each of them identifies with a blown–up plane as indicated in Fig. 7. It is equipped with the ruling $|H - \Gamma_i^k|$, the members of which meet the curves G_i^k and Γ_i^k at one point respectively, and do not meet any other curve in the length 8 anticanonical cycle cut out by $\bar{X}_0 - A_i^k$ on A_i^k. The restriction of $\mathcal{O}_{\bar{X}}(1)$ on A_i^k is trivial.

(vi) The 24 surfaces V_{ij}^k with $i, j, k \in \{1, \ldots 4\}$ distinct.

These are all copies of $\mathbf{P}^2$, on which the restriction of $\mathcal{O}_{\bar{X}}(1)$ is trivial.

5.3 The Limit Linear System, I: Construction

According with the general principles stated in Sect. 3.2, we shall now describe the limit linear system of $|\mathcal{O}_{\bar{X}_t}(1)|$ as $t \in \Delta^*$ tends to 0. This will suffice for the proof, presented in Sect. 5.7, that $\varpi : \bar{X} \to \Delta$ is a δ–good model for $1 \leqslant \delta \leqslant 3$.

We start with $\mathfrak{P} := \mathbf{P}(\varpi_*(\mathcal{O}_{\bar{X}}(1)))$, which is a $\mathbf{P}^3$–bundle over Δ, whose fibre at $t \in \Delta$ is $|\mathcal{O}_{\bar{X}_t}(1)|$. We set $\mathcal{L} = \mathcal{O}_{\bar{X}}(1)$, and $|\mathcal{O}_{\bar{X}_t}(1)| = |\mathcal{L}_t|$; note that $|\mathcal{L}_0| \cong |\mathcal{O}_{S_0}(1)|$. We will often use the same notation to denote a divisor (or a divisor class) on the central fibre and its restriction to a component of the central fibre, if this does not cause any confusion.

We will proceed as follows:

(I) We first blow–up $\mathfrak{P}$ at the points π_i corresponding to the irreducible components P_i of S_0, for $i \in \{1, \dots, 4\}$ (the new central fibre then consists of $|\mathcal{O}_{S_0}(1)| \cong \mathbf{P}^3$ blown–up at four independent points, plus the four exceptional $\mathbf{P}^3$'s);

(II) Next, we blow–up the total space along the proper transforms ℓ_{ij} of the six lines of $|\mathcal{O}_{S_0}(1)|$ joining two distinct points π_i, π_j, with $i, j \in \{1, \dots, 4\}$, corresponding to pencils of planes with base locus an edge of S_0 (the new central fibre is the proper transform of the previous one, plus the six exceptional $\mathbf{P}(\mathcal{O}_{\mathbf{P}^1} \oplus \mathcal{O}_{\mathbf{P}^1}(1)^{\oplus 2})$'s);

(III) Finally, we further blow–up along the proper transforms of the planes Π_k corresponding to the webs of planes passing through the vertices p_k of S_0, for $k \in \{1, \dots, 4\}$ (this adds four more exceptional divisors to the central fibre, for a total of 15 irreducible components).

In other words, we successively blow–up $\mathfrak{P}$ along all the cells of the tetrahedron dual to S_0 in $\mathfrak{P}_0$, by increasing order of dimension.

Each of these blow–ups will be interpreted in terms of suitable twisted linear systems as indicated in Remark 3. It will then become apparent that every point in the central fibre of the obtained birational modification of $\mathfrak{P}$ corresponds to a curve in $\bar{X}_0$ (see Sect. 5.7), and hence that this modification is indeed the limit linear system $\mathfrak{L}$.

Step (I)

In $\mathrm{H}^0(\bar{X}_0, \mathcal{O}_{\bar{X}_0}(1))$ there is for each $i \in \{1, \dots, 4\}$ the 1–dimensional subspace of sections vanishing on P_i, which corresponds to the sections of $\mathrm{H}^0(S_0, \mathcal{O}_{S_0}(1))$ vanishing on the plane P_i. As indicated in Remark 3, in order to construct the limit linear system, we have to blow up the corresponding points $\pi_i \in |\mathcal{L}_0|$. Let $\mathfrak{P}' \to \mathfrak{P}$ be this blow–up, and call $\tilde{\mathfrak{L}}_i$, $1 \leqslant i \leqslant 4$, the exceptional divisors. Each $\tilde{\mathfrak{L}}_i$ is a $\mathbf{P}^3$, and can be interpreted as the trace of the linear system $|\mathcal{L}_0(-P_i)|$ on X_0 (see Lemma 3 and Example 2). However, any section of $\mathrm{H}^0(\bar{X}_0, \mathcal{L}_0(-P_i))$ still vanishes on components of $\bar{X}_0$ different from P_i. By subtracting all of them with the appropriate multiplicities (this computation is tedious but not difficult and can be left to the reader), one sees that $\tilde{\mathfrak{L}}_i$ can be identified as the linear system $\mathfrak{L}_i := |\mathcal{L}_0(-M_i)|$, where

$$M_i := 6P_i + \sum_{j\neq i}(5W'_{ij} + 4W''_{ij} + 3W_{ij} + 2W''_{ji} + W'_{ji})+$$

$$+\sum_{k\neq i}\left(2T^k + 4A^k_i + \sum_{j\notin\{i,k\}}\left(3V^k_{ij} + 2V^k_{ji} + A^k_j\right) + \sum_{\{j<\bar{j}\}\cap\{i,k\}=\emptyset}\left(V^k_{j\bar{j}} + V^k_{\bar{j}j}\right)\right). \tag{10}$$

With the notation introduced in Sect. 5.2, one has:

Lemma 6. *The restriction class of $\mathscr{L}_0(-M_i)$ to the irreducible components of $\bar{X}_0$ is as follows:*

(i) On P_i, we find $4H - \sum_{j\neq i}(E^+_{ij} + E^-_{ij})$;
(ii) On P_j, $j \neq i$, we find $E^+_{ji} + E^-_{ji}$;
(iii) For each $j \neq i$, we find $2F$ on each of the surfaces W'_{ij}, W''_{ij}, W_{ij}, W''_{ji}, W'_{ji}.
(iv) On the remaining components the restriction is trivial.

Proof. This is a tedious but standard computation. As a typical sample we prove (iii), and leave the remaining cases to the reader. Set $\{h,k\} = \{1,\dots,4\} - \{i,j\}$. Then, recalling (6) and (7), we see that the restriction of $\mathscr{L}_0(-M_i)$ to W_{ij} is the line bundle determined by the divisor class

$$F + \left(W''_{ji} - W''_{ij} + \sum_{k\notin\{i,j\}}(2A^k_j + V^k_{ji} + T^k - A^k_i)\right)\Big|_{W_{ij}}$$

$$\sim F + \Lambda_{ji} - \Lambda_{ij} + (2G^k_{ji} + F^k_{ji} + D^k_{ij} - G^k_{ij}) + (2G^h_{ji} + F^h_{ji} + D^h_{ij} - G^h_{ij})$$

$$= F + \left(\Lambda_{ji} + (2G^k_{ji} + F^k_{ji} + D^k_{ij}) + (2G^h_{ji} + F^h_{ji} + D^h_{ij})\right) - (\Lambda_{ij} + G^k_{ij} + G^h_{ij})$$

$$= 2F.$$

□

From this, we deduce that $\mathfrak{L}_i$ identifies with its restriction to P_i:

Proposition 7. *There is a natural isomorphism*

$$\mathfrak{L}_i \cong \left|\mathscr{O}_{P_i}\left(4H - \sum\nolimits_{j\neq i}(E^+_{ij} + E^-_{ij})\right) \otimes \mathscr{I}_{Z_i}\right|. \tag{11}$$

Proof. For each $j \neq i$, the restriction of $\mathfrak{L}_i$ to P_j has $E^+_{ji} + E^-_{ji}$ as its only member. This implies that its restriction to W'_{ji} has only one member as well, which is the sum of the two curves in $|F|$ intersecting E^+_{ji} and E^-_{ji} respectively. On W''_{ji}, we then only have the sum of the two curves in $|F|$ intersecting the two curves on W'_{ji} respectively, and so on W_{ij}, W''_{ij}, and W'_{ij}. Now the two curves on W'_{ij} impose the two base points z_{ji}^+ and z_{ji}^- to the restriction of $\mathfrak{L}_i$ to P_i. The right hand side in (11) being 3–dimensional, this ends the proof with (i) of Lemma 6. □

Step (II)

Next, we consider the blow–up $\mathfrak{P}'' \to \mathfrak{P}'$ along the proper transforms ℓ_{ij} of the six lines of $|\mathscr{L}_0|$ joining two distinct points π_i, π_j, with $i, j \in \{1, \ldots, 4\}$, corresponding to the pencils of planes in $|\mathscr{O}_{S_0}(1)|$ respectively containing the lines $P_i \cap P_j$. The exceptional divisors are isomorphic to $\mathbf{P}(\mathscr{O}_{\mathbf{P}^1} \oplus \mathscr{O}_{\mathbf{P}^1}(1)^{\oplus 2})$; we call them $\tilde{\mathfrak{L}}_{ij}$, $1 \leqslant i < j \leqslant 4$. Arguing as in Step (I) and leaving the details to the reader, we see that $\tilde{\mathfrak{L}}_{ij}$ is in a natural way a birational modification (see Sect. 5.5 below) of the complete linear system $\mathfrak{L}_{ij} := \left|\mathscr{L}_0(-M_{ij})\right|$, where

$$\begin{aligned} M_{ij} := {} & 3W_{ij} + (2W''_{ji} + W'_{ji}) + (2W''_{ij} + W'_{ij}) + \\ & + \sum_{k \notin \{i,j\}} \left(2T^k + \sum_{s \neq k} A^k_s + 2\left(V^k_{ji} + V^k_{ij}\right) + \sum_{\substack{s \in \{i,j\} \\ r \notin \{i,j,k\}}} \left(V^k_{sr} + V^k_{rs}\right) \right). \end{aligned} \tag{12}$$

We will denote by $k < h$ the two indices in $\{1, \ldots, 4\} - \{i, j\}$, and go on using the notations introduced in Sect. 5.2.

Lemma 7. *The restriction class of $\mathscr{L}_0(-M_{ij})$ to the irreducible components of $\bar{X}_0$ is as follows:*

(i) *On P_k (resp. P_h) we find $H - G^h_k$ (resp. $H - G^k_h$);*
(ii) *On each of the surfaces W'_{kh}, W''_{kh}, W_{kh}, W''_{hk}, and W'_{hk}, we find F;*
(iii) *On A^k_h (resp. A^h_k) we find $H - \Gamma^k_h$ (resp. $H - \Gamma^h_k$);*
(iv) *On T^k (resp. T^h), we find Φ^k_h (resp. Φ^h_k);*
(v) *On P_i (resp. P_j), we find $E^+_{ij} + E^-_{ij}$ (resp. $E^+_{ji} + E^-_{ji}$);*
(vi) *On W'_{ij}, W''_{ij}, W''_{ji}, W'_{ji}, we find $2F$;*
(vii) *On W_{ij}, with H as in (6), we find*

$$\begin{aligned} 4H - 2\left(\Lambda_{ij} + G^k_{ij} + G^h_{ij}\right) - \left(F^k_{ji} + G^k_{ji} + D^k_{ij}\right) \\ - \left(F^h_{ji} + G^h_{ji} + D^h_{ij}\right) - D^k_{ij} - D^h_{ij}; \end{aligned}$$

(viii) *On the remaining components the restriction is trivial.*

Proof. As for Lemma 6, this is a tedious but not difficult computation. Again we make a sample verification, proving (vii) above. The restriction class is

$$F + \left(W''_{ji} + \sum_{l=k,h} \left(2A^l_j + V^l_{ji} + T^l + V^l_{ij} + 2A^l_i\right) + W''_{ij}|_{W_{ij}} \right)\Bigg|_{W_{ij}}$$

$$\sim F + \Lambda_{ji} + \sum_{l=k,h} \left(2G_{ji}^{l} + F_{ji}^{l} + D_{ij}^{l} + F_{ij}^{l} + 2G_{ij}^{l}\right) + \Lambda_{ij}$$

which, by taking into account the identification of Fig. 5, i.e. with (6) and (7), is easily seen to be equivalent to the required class. □

Let $w_{ij}^{\pm} \in W_{ij}$ be the two points cut out on W_{ij} by the two connected chains of curves in $|F|_{W_{ij}'} \times |F|_{W_{ij}''}$ meeting $E_{ij}^{\pm}$ respectively. We let $w_{ji}^{\pm} \in W_{ij}$ be the two points defined in a similar fashion by starting with $E_{ji}^{\pm}$. Define the 0-cycle $Z_{ij} = w_{ij}^{+} + w_{ij}^{-} + w_{ji}^{+} + w_{ji}^{-}$ on W_{ij}, and let $\mathscr{I}_{Z_{ij}} \subset \mathscr{O}_{W_{ij}}$ be its defining sheaf of ideals.

Proposition 8. *There is a natural isomorphism between $\mathfrak{L}_{ij}$ and its restriction to W_{ij}, which is the 3–dimensional linear system*

$$\Big|\mathscr{O}_{W_{ij}}\Big(4H - 2\left(\Lambda_{ij} + G_{ij}^{k} + G_{ij}^{h}\right) - \left(F_{ji}^{k} + G_{ji}^{k} + D_{ij}^{k}\right) \\ - \left(F_{ji}^{h} + G_{ji}^{h} + D_{ij}^{h}\right) - D_{ij}^{k} - D_{ij}^{h}\Big) \otimes \mathscr{I}_{Z_{ij}}\Big|, \qquad (13)$$

where we set $\{1, \ldots, 4\} = \{i, j, h, k\}$, and H as in (6).

Proof. Consider a triangle L_1, L_2, L_3 in $\mathbf{P}^2$, with vertices a_1, a_2, a_3, where a_1 is opposite to L_1, etc. Consider the linear system $\mathscr{W}$ of quartics with a double point at a_1, two simple base points infinitely near to a_1 not on L_2 and L_3, two base points at a_2 and a_3 with two infinitely near base points along L_3 and L_2 respectively, two more base points along L_1. There is a birational transformation of W_{ij} to the plane (see Fig. 5) mapping (13) to a linear system of type $\mathscr{W}$. One sees that two independent conditions are needed to impose to the curves of $\mathscr{W}$ to contain the three lines L_1, L_2, L_3 and the residual system consists of the pencil of lines through a_1. This proves the dimensionality assertion (see Sect. 5.5 below for a more detailed discussion).

Consider then the restriction of $\mathfrak{L}_{ij}$ to the chain of surfaces

$$P_j + W_{ji}' + W_{ji}'' + W_{ij} + W_{ij}'' + W_{ij}' + P_i.$$

By taking into account (v)–(vii), of Lemma 7, we see that each divisor C of this system determines, and is determined, by its restriction C' on W_{ij}, since C consists of C' plus four rational tails matching it.

The remaining components of $\bar{X}_0$ on which $\mathfrak{L}_{ij}$ is non–trivial, all sit in the chain

$$T^k + A_h^k + P_h + W_{lk}' + W_{lk}'' + W_{hk} + W_{kh}'' + W_{kh}' + P_k + A_k^h + T^h. \qquad (14)$$

The restrictions of $\mathfrak{L}_{ij}$ to each irreducible component of this chain is a base point free pencil of rational curves, hence $\mathfrak{L}_{ij}$ restricts on (14) to the 1–dimensional system of connected chains of rational curves in these pencils: we call it $\mathfrak{N}^{kh}$. Given a curve in $\mathfrak{L}_{ij}$, it cuts T^k and T^h in one point each, and there is a unique chain of rational curves in $\mathfrak{N}^{kh}$ matching these two points. □

Step (III)

Finally, we consider the blow-up $\mathfrak{P}''' \to \mathfrak{P}''$ along the proper transforms of the three planes that are strict transforms of the webs of planes in $|\mathscr{O}_{S_0}(1)|$ containing a vertex p_k, with $1 \leqslant k \leqslant 4$. For each k, the exceptional divisor $\bar{\mathfrak{L}}^k$ is a birational modification (see Sect. 5.6 below) of the complete linear system $\mathfrak{L}^k := \left|\mathscr{L}_0(-M^k)\right|$, where

$$M^k := 2T^k + \sum_{s\neq k} A_s^k + \sum_{\{s<r\}\not\ni k} \left(V_{sr}^k + V_{rs}^k\right).$$

Lemma 8. *The restriction class of $\mathscr{L}_0(-M^k)$ to the irreducible components of $\bar{X}_0$ is as follows:*

(i) On P_i, $i \neq k$, we find $H - G_i^k$;
(ii) On A_i^k, $i \neq k$, we find $H - \Gamma_i^k$;
(iii) On P_k, as well as on the chains $W'_{ik} + W''_{ik} + W_{ik} + W''_{ki} + W'_{ki}$, $i \neq k$, we find the restriction class of $\mathscr{L}_0$;
(iv) On T^k, we find

$$3H - (F_k^{sj} + D_{sj}^k + 2\Gamma_s^k) - (F_k^{ji} + D_{ij}^k + 2\Gamma_j^k) - (F_k^{is} + D_{is}^k + 2\Gamma_i^k),$$

with $\{s,i,j,k\} = \{1,\dots,4\}$, and H as in (9);
(v) On the remaining components it is trivial.

Proof. We limit ourselves to a brief outline of how things work for T^k. The restriction class is

$$\left(\sum_{r\neq k} A_r^k + \sum_{\{r<r'\}\not\ni k} \left(V_{rr'}^k + 2W_{rr'} + V_{r'r}^k\right)\right)\Bigg|_{T^k}$$

which is seen to be equal to the required class with the identification of Figs. 5 and 6, i.e. with H as in (9). □

Proposition 9. *There is a natural isomorphism between $\mathfrak{L}^k$ and its restriction to T^k, which is the 3–dimensional linear system*

$$\left|3H - (F_k^{sj} + D_{sj}^k + 2\Gamma_s^k) - (F_k^{ji} + D_{ij}^k + 2\Gamma_j^k) - (F_k^{is} + D_{is}^k + 2\Gamma_i^k)\right|,$$

where we set $\{s,i,j,k\} = \{1,\dots,4\}$, and H as in (9).

Proof. This is similar (in fact, easier) to the proof of Proposition 8, so we will be sketchy here. The dimensionality assertion will be discussed in Sect. 5.6 below.

For each $i \neq k$, the restriction of $\mathfrak{L}^k$ to each irreducible component of the chain

$$A_i^k + P_i + W'_{ik} + W''_{ik} + W_{ik} + W''_{ki} + W'_{ki} \tag{15}$$

is a base point free pencil of rational curves, and $\mathfrak{L}^k$ restricts on (15) to the 1–dimensional system of connected chains of rational curves in these pencils, that we will call $\mathfrak{N}_i^k$.

Now the general member of $\mathfrak{L}^k$ consists of a curve in $\mathfrak{L}^k|_{T^k}$, which uniquely determines three chains of rational curves in $\mathfrak{N}_i^k$, $i \neq k$, which in turn determine a unique line in $|\mathcal{O}_{P^k}(H)|$. □

5.4 The Linear Systems $\mathfrak{L}_i$

Let a, b, c be three independent lines in $\mathbf{P}^2$, and consider a 0–dimensional scheme Z cut out on $a + b + c$ by a general quartic curve. Consider the linear system $\mathscr{P}$ of plane quartics containing Z. This is a linear system of dimension 3. Indeed containing the union of the three lines a, b, c is one condition for the curves in $\mathscr{P}$ and the residual system is the 2–dimensional complete linear system of all lines in the plane.

Proposition 7 shows that $\mathfrak{L}_i$ can be identified with a system of type $\mathscr{P}$. We denote by $\sigma_i : P_i \dashrightarrow \mathbf{P}^3$ (or simply by σ) the rational map determined by $\mathfrak{L}_i$ and by Y its image, which is the same as the image of the plane via the rational map determined by the linear system $\mathscr{P}$.

Proposition 10. *The map $\sigma : P_i \dashrightarrow Y$ is birational, and Y is a* monoid *quartic surface, with a triple point p with tangent cone consisting of a triple of independent planes through p, and with no other singularity.*

Proof. The triple point $p \in Y$ is the image of the curve $C = \sum_{i=j}^{3}(2D_i^j + L_{ij})$ (alternatively, of the sides of the triangle a, b, c). By subtracting C to $\mathfrak{L}_i$ one gets a homaloidal net, mapping to the net of lines in the plane. This proves the assertion. □

Remark 7. The image of $\bar{X}$ by the complete linear system $|\mathscr{L}(-M_i)|$ provides a model $f' : S' \to \Delta$ of the initial family $f : S \to \Delta$, such that the corresponding flat limit of $S'_t \cong S_{t^6}$ with $t \neq 0$, is $S'_0 = Y$ the quartic monoid image of the face P_i of the tetrahedron via σ. The map $\bar{X}_0 \to S'_0$ contracts all other irreducible components of $\bar{X}_0$ to the triple point of the monoid.

Remark 8. Theorem 3 says that the degree of the dual surface of the monoid Y is 21.

The strict transform of $\tilde{\mathfrak{L}}_i$ in $\mathfrak{P}_0'''$ (which we still denote by $\tilde{\mathfrak{L}}_i$, see Sect. 2) can be identified as a blow–up of $\mathfrak{L}_i \cong \mathscr{P}$: first blow–up the three points corresponding to the three non–reduced curves $2a+b+c$, $2b+a+c$, $2c+a+b$. Then blow–up the proper transforms of the three pencils of lines with centres at A, B, C plus the fixed part $a+b+c$. We will interpret this geometrically in Sect. 5.7, using Lemma 3.

5.5 The Linear Systems $\mathfrak{L}_{ij}$

Next, we need to study some of the geometric properties of the linear systems $\mathfrak{L}_{ij}$ as in Proposition 8. Consider the rational map $\varphi_{ij} : W_{ij} \dashrightarrow \mathbf{P}^3$ (or simply φ) determined by $\mathfrak{L}_{ij}$. Alternatively, one may consider the rational map, with the same image W (up to projective transformations), determined by the planar linear system $\mathscr{W}$ of quartics considered in the proof of Proposition 8.

Proposition 11. *The map φ is birational onto its image, which is a quartic surface $W \subset \mathbf{P}^3$, with a double line D, and two triple points on D.*

Proof. First we get rid of the four base points in Z_{ij} by blowing them up and taking the proper transform $\bar{\mathfrak{L}}_{ij}$ of the system. Let $u : \bar{W} \to W_{ij}$ be this blow–up, and let $I_{ij}^{\pm}$ (resp. $I_{ji}^{\pm}$) be the two (-1)–curves that meet Λ_{ij} (resp. Λ_{ji}).

The strict transform $\bar{\mathfrak{L}}_{ij} := u^*(\mathfrak{L}_{ij}) - \left(I_{ij}^{+} + I_{ij}^{-} + I_{ji}^{+} + I_{ij}^{-}\right)$, has self–intersection 4. Set, as usual, $\{1,\dots,4\} = \{i,j,h,k\}$ and consider the curves

$$
\begin{aligned}
C_{ji} &:= \Lambda_{ji} + (2G_{ji}^{k} + F_{ji}^{k}) + (2G_{ji}^{h} + F_{ji}^{h}) \\
\text{and} \quad C_{ij} &:= \Lambda_{ij} + (2G_{ij}^{k} + F_{ij}^{k}) + (2G_{ij}^{h} + F_{ij}^{h}).
\end{aligned} \tag{16}
$$

One has

$$
\bar{\mathfrak{L}}_{ij} \cdot C_{\mathfrak{s}} = 0, \quad p_a(C_{\mathfrak{s}}) = 0, \quad C_{\mathfrak{s}}^2 = -3, \quad \text{for} \quad \mathfrak{s} \in \{(ij),(ji)\}.
$$

By mapping $\bar{W}$ to W_{ij}, and this to the plane as in Fig. 5 with (4) and (5) identified, one sees that C_{ji} goes to the line L_1 and C_{ij} to the union of the two lines L_2, L_3. The considerations in the proof of Proposition 8 show that $\bar{\mathfrak{L}}_{ij}$ has no base points on $C_{ji} \cup C_{ij}$ (i.e., $\mathscr{W}$ has only the prescribed base points along the triangle $L_1+L_2+L_3$). On the other hand, the same considerations show that the base points of $\bar{\mathfrak{L}}_{ij}$ may only lie on $C_{ji} \cup C_{ij}$. This shows that $\bar{\mathfrak{L}}_{ij}$ is base points free, and the associated morphism $\bar{\varphi} : \bar{W} \to \mathbf{P}^3$ contracts C_{ji} and C_{ij} to points c_1 and c_2 respectively.

The points c_1 and c_2 are distinct, since subtracting the line L_1 from the planar linear system $\mathscr{W}$ does not force subtracting the whole triangle $L_1+L_2+L_3$ to the system. By subtracting C_{ji} from $\bar{\mathfrak{L}}_{ij}$, the residual linear system is a linear system of rational curves with self–intersection 1, mapping W_{ij} birationally to the plane. Indeed, this residual linear system corresponds to the residual linear system of L_1

with respect to $\mathscr{W}$, which is the linear system of plane cubics, with a double point at a_1, two simple base points infinitely near to a_1 not on L_2 and L_3, two base points at a_2 and a_3, and this is a homaloidal system. This shows that c_1 is a triple point of W and that $\bar{\varphi}$ is birational. The same for c_2. Finally $\bar{\varphi}$ maps (the proper transforms of) D_{ij}^k and D_{ij}^h both to the unique line D containing c_1 and c_2. □

Remark 9. The subpencil of $\mathfrak{L}_{ij}$ corresponding to planes in $\mathbf{P}^3$ that contain the line D corresponds to the subpencil of curves in $\mathscr{W}$ with the triangle $L_1 + L_2 + L_3$ as its fixed part, plus the pencil of lines through a_1. In this subpencil we have two special curves, namely $L_1 + 2L_2 + L_3$ and $L_1 + L_2 + 2L_3$. This shows that the tangent cone to W at the general point of D is fixed, formed by two planes.

Remark 10. The image of $\bar{X}$ via the complete linear system $|\mathscr{L}(-M_{ij})|$ provides a model $f' : S' \to \Delta$ of the initial family $f : S \to \Delta$, such that the corresponding flat limit of $S'_t \cong S_{t^6}$ with $t \neq 0$, is $S'_0 = W$ the image of W_{ij} via φ. The map $\bar{X}_0 \to S'_0$ contracts the chain (14) to the double line of W, and the two connected components of $\bar{X}_0 - W_{ij}$ minus the chain (14) (cf. Fig. 1) to the two triple points of W respectively.

Corollary 1. *The exceptional divisor $\tilde{\mathfrak{L}}_{ij}$ of $\mathfrak{P}'' \to \mathfrak{P}'$ is naturally isomorphic to the blow–up of the complete linear system $\mathfrak{L}_{ij} \cong |\mathscr{O}_W(1)|$ along its subpencil corresponding to planes in $\mathbf{P}^3$ containing the line D.*

Proof. This is a reformulation of the description of $\mathfrak{P}'' \to \mathfrak{P}'$ (cf. Step (II) in Sect. 5.3 above), taking into account Propositions 8 and 11. □

The divisor $\tilde{\mathfrak{L}}_{ij} \subset \mathfrak{P}''_0$ is a $\mathbf{P}(\mathscr{O}_{\mathbf{P}^1}(1)^{\oplus 2} \oplus \mathscr{O}_{\mathbf{P}^1})$, and its structure of $\mathbf{P}^2$–bundle over $\mathbf{P}^1$ is the minimal resolution of indeterminacies of the rational map $\mathfrak{L}_{ij} \dashrightarrow |\mathscr{O}_D(1)|$, which sends a general divisor $C \in \mathfrak{L}_{ij}$ to its intersection point with D. The next Proposition provides an identification of the general fibres of $\tilde{\mathfrak{L}}_{ij}$ over $|\mathscr{O}_D(1)| = \mathbf{P}^1$ as certain linear systems.

Proposition 12. *The projection of W from a general point of D is a double cover of the plane, branched over a sextic B which is the union*

$$B = B_0 + B_1 + B_2$$

of a quartic B_0 with a node p, and of its tangent cone $B_1 + B_2$ at p, such that the two branches of B_0 at p both have a flex there (see Fig. 8; the intersection $B_i \cap B_0$ is concentrated at the double point p, for $1 \leqslant i \leqslant 2$).

Proof. Let us consider a double cover of the plane as in the statement. It is singular. Following [7, §4], we may obtain a resolution of singularities as a double cover of a blown–up plane with non–singular branch curve. We will then observe that it identifies with $\bar{W}$ blown–up at two general *conjugate* points on D_{ij}^k and D_{ij}^h respectively (here *conjugate* means that the two points are mapped to the same

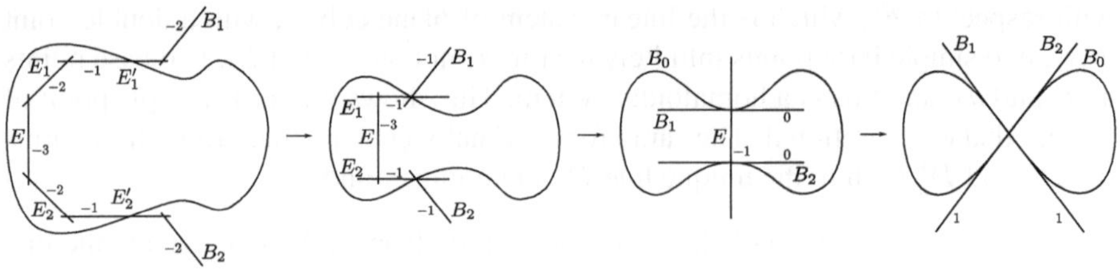

Fig. 8 Desingularization of the branch curve of the projection of W_{ij}

point x of D by $\bar{\varphi}$). We will denote by $\tilde{W}$ the surface $\bar{W}$ blown–up at two such points, and by I'_x, I''_x the two exceptional divisors.

First note that our double plane is rational, because it has a pencil of rational curves, namely the pull-back of the pencil of lines passing through p (eventually this will correspond to the pencil of conics cut out on W by the planes through D).

In order to resolve the singularities of the branch curve (see Fig. 8), we first blow–up p, pull–back the double cover and normalize it. Since p has multiplicity 4, which is even, the exceptional divisor E of the blow–up does not belong to the branch curve of the new double cover, which is the proper transform B (still denoted by B according to our general convention). Next we blow–up the two double points of B which lie on E, and repeat the process. Again, the two exceptional divisors E_1, E_2 do not belong to the branch curve. Finally we blow–up the two double points of B (which lie one on E_1 one on E_2, off E), and repeat the process. Once more, the two exceptional divisors E'_1, E'_2 do not belong to the branch curve which is the union of B_0, B_1 and B_2 (which denote here the proper transforms of the curves with the same names on the plane). This curve is smooth, so the corresponding double cover is smooth.

The final double cover has the following configuration of negative curves: B_1 (resp. B_2) is contained in the branch divisor, so over it we find a (-1)-curve; E'_1 (resp. E'_2) meets the branch divisor at two points, so its pull-back is a (-2)-curve; E_1 (resp. E_2) does not meet the branch divisor, so its pull-back is the sum of two disjoint (-2)-curves; similarly, the pull-back of E is the sum of two disjoint (-3)-curves. In addition, there are four lines through p tangent to B_0 and distinct from B_1 and B_2. After the resolution, they are curves with self-intersection 0 and meet the branch divisor at exactly one point with multiplicity 2. The pull-back of any such a curve is the transverse union of two (-1)-curves, each of which meets transversely one component of the pull-back of E.

This configuration is precisely the one we have on $\tilde{W}$, after the contraction of the four (-1)-curves G^k_{ji}, G^k_{ij}, G^h_{ji}, and G^h_{ij}. Moreover, the pull-back of the line class of $\mathbf{P}^2$ is the pull–back to $\tilde{W}$ of $\mathfrak{L}_{ij}(-(I'_x + I''_x))$. □

Corollary 2. *In the general fibre of the generic* $\mathbf{P}^2$ *bundle structure of* $\tilde{\mathfrak{L}}_{ij}$*, the Severi variety of* 1*–nodal (resp.* 2*–nodal) irreducible curves is an irreducible curve of degree* 10 *(resp. the union of* 16 *distinct points).*

Proof. This follows from the fact that the above mentioned Severi varieties are respectively the dual curve $\check{B}_0$ of a plane quartic as in Proposition 12, and the set of ordinary double points of $\check{B}_0$. One computes the degrees using Plücker formulae. □

5.6 The Linear Systems $\mathfrak{L}^k$

Here we study some geometric properties of the linear systems $\mathfrak{L}^k$ appearing in the third step of Sect. 5.3.

Consider a triangle L_1, L_2, L_3 in $\mathbf{P}^2$, with vertices a_1, a_2, a_3, where a_1 is opposite to L_1, etc. Consider the linear system $\mathscr{T}$ of cubics through a_1, a_2, a_3 and tangent there to L_3, L_1, L_2 respectively. By Proposition 9, there is a birational transformation of T^k to the plane (see Fig. 5) mapping $\mathfrak{L}^k$ to $\mathscr{T}$. We consider the rational map $\phi_k : T^k \dashrightarrow \mathbf{P}^3$ (or simply ϕ) determined by the linear system $\mathfrak{L}^k$, or, alternatively, the rational map, with the same image T (up to projective transformations), determined by the planar linear system $\mathscr{T}$. The usual notation is $\{1,\ldots,4\} = \{i,j,s,k\}$.

Proposition 13. *The map $\phi : T^k \to T \subset \mathbf{P}^3$ is a birational morphism, and T is a cubic surface with three double points of type A_2 as its only singularities. The minimal resolution of T is the blow–down of T^k contracting the (-1)–curves $D^k_{ij}, D^k_{is}, D^k_{js}$. This cubic contains exactly three lines, each of them containing two of the double points.*

Proof. The linear system $\mathscr{T}$ is a system of plane cubics with six simple base points, whose general member is clearly irreducible. This implies that $\phi : T^k \to T \subset \mathbf{P}^3$ is a birational morphism and T is a cubic surface. The linear system $\mathfrak{L}^k$ contracts the three chains of rational curves

$$C_1 = F^{js}_k + 2D^k_{sj} + F^{sj}_k, \quad C_2 = F^{si}_k + 2D^k_{is} + F^{is}_k, \quad C_3 = F^{ij}_k + 2D^k_{ij} + F^{ji}_k,$$

which map in the plane to the sides of the triangle L_1, L_2, L_3. By contracting the (-1)–curves D^k_{ij}, D^k_{is}, D^k_{js}, the three curves C_1, C_2, C_3 are mapped to three (-2)–cycles contracted by ϕ to double points of type A_2.

The rest follows from the classification of cubic hypersurfaces in $\mathbf{P}^3$ (see, e.g., [5]). The three lines on T are the images via ϕ of the three exceptional divisors $\Gamma^k_i, \Gamma^k_j, \Gamma^k_s$. □

Remark 11. We now see that the image of $\bar{X}$ by the complete linear system $|\mathscr{L}(-M^k)|$ provides a model $f' : S' \to \Delta$ of the initial family $f : S \to \Delta$, such that the corresponding flat limit of $S'_t \cong S_{t^6}$ with $t \neq 0$, is $S'_0 = T + P$, where T is the image of T^k via ϕ, and P is the plane in $\mathbf{P}^3$ through the three lines contained in T, image of P_k by the map associated to $\mathfrak{L}^k$. The three other faces of the initial tetrahedron S_0 are contracted to the three lines in T respectively.

Proposition 14. *The dual surface $\check{T} \subset \check{\mathbf{P}}^3$ to T is itself a cubic hypersurface with three double points of type A_2 as its only singularities. Indeed, the Gauss map γ_T fits into the commutative diagram*

$$\begin{array}{ccc} & T^k & \\ {\scriptstyle \phi}\swarrow & & \searrow{\scriptstyle \check{\phi}} \\ T & \underset{\gamma_T}{\dashrightarrow} & \check{T} \end{array}$$

where $\check{\phi}$ is the morphism associated to the linear system

$$\left|3H - (F_k^{sj} + \Gamma_s^k + 2D_{sj}^k) - (F_k^{ji} + \Gamma_j^k + 2D_{ij}^k) - (F_k^{is} + \Gamma_i^k + 2D_{is}^k)\right|,$$

which is mapped to the linear system $\mathscr{T}'$ of cubics through a_1, a_2, a_3 and tangent there to L_2, L_3, L_1 respectively, by the birational map $T^k \dashrightarrow \mathbf{P}^2$ identifying $\mathfrak{L}^k$ with $\mathscr{T}$.

Proof. The dual hypersurface $\check{T}$ has degree 3 by Proposition 5. Let p be a double point of T. The tangent cone to T at p is a rank 2 quadric, with vertex a line L_p. A local computation shows that the limits of all tangent planes to T at smooth points tending to p are planes through L_p. This means that γ_T is not well defined on the minimal resolution of T, which is the blow–down of T^k contracting the (-1)–curves $D_{ij}^k, D_{is}^k, D_{js}^k$, its indeterminacy points being exactly the three points images of these curves. The same local computation also shows that γ_T is well defined on T^k, hence γ_T fits in the diagram as stated.

In $\mathscr{T}$ there are the three curves $2L_1+L_3, 2L_2+L_1, 2L_3+L_2$, which implies that for any given line $\ell \subset T$ there is a plane Π_ℓ in $\mathbf{P}^3$ tangent to T at the general point of ℓ (actually one has $\Pi_\ell \cap T = 3\ell$). Then γ_T contracts each of the three lines contained in T to three different points, equivalently $\check{\phi}$ contracts to three different points the three curves $\Gamma_i^k, \Gamma_j^k, \Gamma_s^k$. Being $\check{T}$ a (weak) Del Pezzo surface, this implies that $\check{\phi}$ must contract the three chains of rational curves $F_k^{sj} + 2\Gamma_s^k + F_k^{si}$, $F_k^{is} + 2\Gamma_i^k + F_k^{ij}$, and $F_k^{ji} + 2\Gamma_j^k + F_j^{js}$, because they have 0 intersection with the anticanonical system, and the rest of the assertion follows. □

Recalling the description of $\mathfrak{P}''' \to \mathfrak{P}$, one can realize $\tilde{\mathfrak{L}}^k$ as a birational modification of $\mathfrak{L}^k \cong |\mathscr{O}_T(1)|$: first blow–up the point corresponding to the plane containing the three lines of T, then blow–up the strict transforms of the three lines in $|\mathscr{O}_T(1)|$ corresponding to the three pencils of planes respectively containing the three lines of T. Notice that $\tilde{\mathfrak{L}}^k$ has a structure of $\mathbf{P}^1$–bundle on the blow–up of $\mathbf{P}^2$ at three non–coplanar points, as required.

Alternatively, we have in $\mathscr{T}$ the four curves $C_0 = L_1 + L_2 + L_3, C_1 = 2L_1 + L_3, C_2 = 2L_2 + L_1, C_3 = 2L_3 + L_2$, corresponding to four independent points $c_0 \ldots, c_3$ of $\mathscr{T}$. Then $\tilde{\mathfrak{L}}^k$ is the blow–up of $\mathscr{T}$ at c_0, further blown–up along the proper transforms of the lines $\langle c_0, c_1\rangle$, $\langle c_0, c_2\rangle$, and $\langle c_0, c_3\rangle$. Via the map $\tilde{\mathfrak{L}}^k \to \mathscr{T}$, the projection of the $\mathbf{P}^1$–bundle structure corresponds to the projection of $\mathscr{T}$ from c_0 to the plane spanned by c_1, c_2, c_3.

This will be interpreted using Lemma 3 in Sect. 5.7 below.

5.7 The Limit Linear System, II: Description

We are now ready to prove:

Proposition 15. *The limit linear system of* $|\mathscr{L}_t| = |\mathscr{O}_{\bar{X}_t}(1)|$ *as* $t \in \Delta^*$ *tends to* 0 *is* $\mathfrak{P}_0'''$.

Proof. The identification of $\mathfrak{P}'''$ as $\mathrm{Hilb}(\mathscr{L})$ will follow from the fact that every point in $\mathfrak{P}_0'''$ corresponds to a curve in $\bar{X}_0$ (see Lemma 2). Having the results of Sects. 5.3–5.6 at hand, we are thus left with the task of describing how the various components of the limit linear system intersect each other. We carry this out by analyzing, with Lemma 3, the birational modifications operated on the components $\mathfrak{P}_0$, $\tilde{\mathfrak{L}}_i$, $\tilde{\mathfrak{L}}_{ij}$, and $\tilde{\mathfrak{L}}^k$, during the various steps of the construction of $\mathfrak{P}'''$ (see Sect. 5.3).

(I) In $\mathfrak{P}_0'$, the strict transform of $\mathfrak{P}_0$ (which we shall go on calling $\mathfrak{P}_0$, according to the conventions set in Sect. 2) is the blow up of $|\mathscr{L}_0| \cong |\mathscr{O}_{S_0}(1)|$ at the four points corresponding to the faces of S_0. For each $i \in \{1, \dots, 4\}$, the corresponding exceptional plane is the intersection $\mathfrak{P}_0 \cap \tilde{\mathfrak{L}}_i$, and it identifies with the subsystem of $\mathfrak{L}_i$ consisting of curves

$$L + \sum_{\substack{j \neq i \\ \{i,j,k,h\}=\{1,\dots,4\}}} (L_{ij} + G_i^h + G_i^k), \quad L \in |\mathscr{O}_{P_i}(H)|,$$

together with six rational tails respectively joining $E_{ji}^{\pm}$ to $z_{ji}^{\pm}$, $j \neq i$.

(II) For each $\{i \neq j\} \subset \{1, \dots, 4\}$, the intersection $\mathfrak{P}_0 \cap \tilde{\mathfrak{L}}_{ij} \subset \mathfrak{P}_0''$ identifies as the exceptional $\mathbf{P}^1 \times \mathbf{P}^1$ of both the blow–up of $\mathfrak{P}_0 \subset \mathfrak{P}_0'$ along the line ℓ_{ij}, and the blow–up $\tilde{\mathfrak{L}}_{ij} \to \bar{\mathfrak{L}}_{ij}$ described in Corollary 1. As a consequence, it parametrizes the curves

$$C + \Phi + D_{ij}^k + D_{ij}^h + C_{ji} + C_{ij} \tag{17}$$

($\{i, j, k, h\} = \{1, \dots, 4\}$, C_{ji} and C_{ij} as in (16)), where C is a chain in $\mathfrak{N}^{kh}$, and $\Phi \in |F|_{W_{ij}}$ is the proper transform by φ_{ij} of a conic through the two triple points of W (cf. Proposition 11), together with four rational tails respectively joining $E_{ij}^{\pm}$ and $E_{ji}^{\pm}$ to $w_{ij}^{\pm}$ and $w_{ji}^{\pm}$. The two components C and Φ are independent one from another, and respectively move in a 1–dimensional linear system.

The intersection $\tilde{\mathfrak{L}}_{ij} \cap \tilde{\mathfrak{L}}_i \subset \mathfrak{P}_0''$ is a $\mathbf{P}^2$. In $\tilde{\mathfrak{L}}_{ij}$, it identifies as the proper transform via $\tilde{\mathfrak{L}}_{ij} \to \bar{\mathfrak{L}}_{ij}$ of the linear system of curves

$$C_{ij} + C, \quad C \in |\mathscr{L}_0(-M_{ij}) \otimes \mathscr{O}_{W_{ij}}(-C_{ij})|, \quad \text{and } C_{ij} \text{ as in (16)}, \tag{18}$$

while in $\tilde{\mathfrak{L}}_i$ it is the exceptional divisor of the blow–up of $\tilde{\mathfrak{L}}_i \subset \mathfrak{P}'_0$ at the point corresponding to the curve

$$2(L_{ij}+G_i^h+G_i^k)+(L_{ih}+G_i^j+G_i^k)+(L_{ik}+G_i^j+G_i^h), \quad \{i,j,k,h\} = \{1,\dots,4\}.$$

It follows that it parametrizes sums of a curve as in (18), plus the special member of $\mathfrak{N}^{kh}$ consisting of double curves of $\bar{X}_0$ and joining the two points $D_{ij}^k \cap F_{ij}^k$ and $D_{ij}^h \cap F_{ij}^h$.

(III) For each $k \in \{1,\dots,4\}$, the intersection $\Pi_k = \tilde{\mathfrak{L}}^k \cap \mathfrak{P}_0$ is a $\mathbf{P}^2$ blown up at three non colinear points. Seen in $\mathfrak{P}_0$, it identifies as the blow–up of the web of planes in $|\mathscr{L}_0| \cong |\mathscr{O}_{S_0}(1)|$ passing through the vertex k of S_0, at the three points corresponding to the faces of S_0 containing this very vertex. In $\tilde{\mathfrak{L}}^k$ on the other hand, it is the strict transform of the exceptional $\mathbf{P}^2$ of the blow–up $\tilde{\mathscr{T}} \to \mathscr{T} \cong \mathfrak{L}^k$ at the point $[a+b+c]$. It therefore parametrizes the curves

$$L + \sum_{i \neq k}\Big(\Gamma_i^k + \sum_{j \notin \{i,k\}} (F_k^{ij} + D_{ij}^k)\Big), \tag{19}$$

where L is a line in P_k, together with three rational tails joining respectively $L \cap L_{ki}$ to Γ_i^k, $i \neq k$.

For $i \neq k$, $\tilde{\mathfrak{L}}^k \cap \tilde{\mathfrak{L}}_i$ is a $\mathbf{P}^1 \times \mathbf{P}^1$, identified as the exceptional divisor of both the blow–up $\tilde{\mathfrak{L}}^k \to \tilde{\mathscr{T}}$ along the strict transform of the line parametrizing planes in $\mathscr{T} \cong |\mathscr{O}_T(1)|$ containing the line $\phi^k(\Gamma_i^k)$, and the blow–up of $\tilde{\mathfrak{L}}_i \subset \mathfrak{P}''_0$ along the strict transform of the line parametrizing curves

$$L + G_i^k + \sum_{\substack{j \neq i \\ \{i,j,k,h\}=\{1,\dots,4\}}} (L_{ij} + G_i^h + G_i^k), \quad L \in |\mathscr{O}_{P_i}(H - G_i^k)|. \tag{20}$$

It therefore parametrizes sums of

$$\Phi + \Big(\Gamma_i^k + \sum_{j \notin \{i,k\}} (F_k^{ij} + 2D_{ij}^k + F_k^{ji}) + C\Big) \tag{21}$$

(where $\Phi \in \mathfrak{N}_i^k$, and the second summand is a member of $\mathfrak{L}^k|_{T^k}$), plus the fixed part $L_{ki} + E_{ki}^+ + E_{ki}^- + \sum_{j \notin \{i,k\}}(G_k^j + \Phi_j)$, where Φ_j is the special member of $\mathfrak{N}_j^k$ consisting of double curves of $\bar{X}_0$ and joining the two points $G_k^j \cap L_{k\bar{j}}$ on P_k and $F_k^{\bar{j}j} \cap \Gamma_{\bar{j}}^k$ on T^k, for each $j \notin \{i,k\}$, with $\bar{j}$ such that $\{i,k,j,\bar{j}\} = \{1,\dots,4\}$. The two curves Φ and C are independent one from another, and respectively move in a 1–dimensional linear system.

For each $j \notin \{k, i\}$, $\tilde{\mathfrak{L}}^k \cap \tilde{\mathfrak{L}}_{ij}$ is an $\mathbf{F}_1$, and identifies as the blow–up of the plane in $\mathfrak{L}^k$ corresponding to divisors in $|\mathcal{O}_T(1)|$ passing through the double point $\phi^k(\Gamma_i^k) \cap \phi^k(\Gamma_j^k)$, at the point $\left[\sum_{i \neq k} \phi^k(\Gamma_i^k)\right]$; it also identifies as the exceptional divisor of the blow–up of $\tilde{\mathfrak{L}}_{ij} \subset \mathfrak{P}_0''$ along the $\mathbf{P}^1$ corresponding to the curves as in (17), with Φ the only member of $|F|_{W_{ij}}$ containing D_{ij}^k. We only need to identify the curves parametrized by the exceptional curve of this $\mathbf{F}_1$; they are as in (19), with L corresponding to a line in the pencil $|\mathcal{O}_{P_k}(H - G_k^s)|$, $s \notin \{i, j, k\}$.

In conclusion, $\mathfrak{P}'''$ is an irreducible Zariski closed subset of the relative Hilbert scheme of $\bar{X}$ over Δ, and this proves the assertion. □

5.8 The Limit Severi Varieties

We shall now identify the regular parts of the limit Severi varieties $\mathfrak{V}_{1,\delta}(\bar{X}) = \mathfrak{V}_\delta(\bar{X}, \mathcal{L})$ for $1 \leqslant \delta \leqslant 3$ (see Definition 2). To formulate the subsequent statements, we use Notation 3 and the notion of **n**–degree introduced in Sect. 3.5.

We will be interested in those **n** that correspond to a choice of $3 - \delta$ general base points on the faces P_i of S_0, with $1 \leq i \leq 4$. These choices can be identified with 4–tuples $\mathbf{n} = (n_1, n_2, n_3, n_4) \in \mathbf{N}^4$ with $|\mathbf{n}| = 3 - \delta$ (by choosing n_i general points on P_i). The vector **n** is non–zero only if $1 \leqslant \delta \leqslant 2$. For $\delta = 1$ (resp. for $\delta = 2$), to give **n** is equivalent to give two indices $i, j \in \{1, \ldots, 4\}^2$ (resp. an $i \in \{1, \ldots, 4\}$): we let $\mathbf{n}_{i,j}$ (resp. $\mathbf{n}_i$) be the 4–tuple corresponding to the choice of general base points on P_i and P_j respectively if $i \neq j$, and of two general base points on P_i if $i = j$ (resp. a general base point on P_i).

Proposition 16 (Limits of 1-nodal curves). *The regular components of the limit Severi variety $\mathfrak{V}_{1,1}(\bar{X})$ are the following (they all appear with multiplicity 1):*

(i) The proper transforms of the 24 planes $V(E) \subset |\mathcal{O}_{S_0}(1)|$, where E is any one of the (-1)–curves $E_{ij}^{\pm}$, for $1 \leqslant i, j \leqslant 4$ and $i \neq j$. The $\mathbf{n}_{hk}$–degree is 1 if $h \neq k$; when $h = k$, it is 1 if $h \notin \{i, j\}$, and 0 otherwise;

(ii) The proper transforms of the four degree 3 surfaces $V(M^k, \delta_{T^k} = 1) \subset \mathfrak{L}^k$, $1 \leqslant k \leqslant 4$. The $\mathbf{n}_{ij}$–degree is 3 if $i \neq j$; when $i = j$, it is 3 if $k = i$, and 0 otherwise;

(iii) The proper transforms of the four degree 21 surfaces $V(M_i, \delta_{P_i} = 1) \subset \mathfrak{L}_i$, $1 \leqslant i \leqslant 4$. The $\mathbf{n}_{hk}$–degree is 21 if $h = k = i$, and 0 otherwise;

(iv) The proper transforms of the six surfaces in $V(M_{ij}, \delta_{W_{ij}} = 1) \subset \mathfrak{L}_{ij}$, $1 \leqslant i < j \leqslant 4$. They have $\mathbf{n}_{hk}$–degree 0 for every $h, k \in \{1, \ldots, 4\}^2$.

Proof. This follows from (1), and from Propositions 14 and 29. Proposition 29 tells us that $V(M_i, \delta_{P_i} = 1)$ has degree at least 21 in $\mathfrak{L}_i$ for $1 \leqslant i \leqslant 4$; the computations in Remark 12 (a) below yield that it cannot be strictly larger than 21 (see also the proof of Corollary 4), which proves Theorem 3 for $\delta = 1$. The $\mathbf{n}_{hk}$–degree computation is straightforward. □

Remark 12. **(a)** The degree of the dual of a smooth surface of degree 4 in $\mathbf{P}^3$ is 36. It is instructive to identify, in the above setting, the 36 limiting curves passing through two general points on the proper transform of S_0 in $\bar{X}$. This requires the $\mathbf{n}_{hk}$–degree information in Proposition 16. If we choose the two points on different planes, 24 of the 36 limiting curves through them come from (i), and 4 more, each with multiplicity 3, come from (ii). If the two points are chosen in the same plane, then we have 12 contributions from (i), only one contribution, with multiplicity 3, from (ii), and 21 more contributions form (iii). No contribution ever comes from (iv) if we choose points on the faces of the tetrahedron.

(b) We have here an illustration of Remark 6: the components $V(M_i, \delta_{P_i} = 1)$ are mapped to points in $|\mathcal{O}_{S_0}(1)|$, hence they do not appear in the crude limit $\mathfrak{V}^{\mathrm{cr}}_{1,1}(S)$ (see Corollary 3 below); they are however visible in the crude limit Severi variety of the degeneration to the quartic monoid corresponding to the face P_i. In a similar fashion, to see the component $V(M_{ij}, \delta_{W_{ij}} = 1)$ one should consider the flat limit of the S_t, $t \in \Delta^*$, given by the surface W described in Proposition 11.

Corollary 3 (Theorem 1 for $\delta = 1$). *Consider a family $f : S \to \Delta$ of general quartic surfaces in $\mathbf{P}^3$ degenerating to a tetrahedron S_0. The singularities of the total space S consist in 24 ordinary double points, four on each edge of S_0 (see Sect. 3.1). It is 1–well behaved, with good model $\varpi : \bar{X} \to \Delta$. The limit in $|\mathcal{O}_{S_0}(1)|$ of the dual surfaces $\check{S}_t$, $t \in \Delta^*$ (which is the crude limit Severi variety $\mathfrak{V}^{\mathrm{cr}}_{1,1}(S)$), consists in the union of the 24 webs of planes passing through a singular point of S, and of the 4 webs of planes passing through a vertex of S_0, each counted with multiplicity 3.*

Proof. The only components of $\mathfrak{V}^{\mathrm{reg}}_{1,1}(\bar{X})$ which are not contracted to lower dimensional varieties by the morphism $\mathfrak{P}''' \to \mathfrak{P}$ are the ones in (i) and in (ii) of Proposition 16. Their push–forward in $\mathfrak{P}_0 \cong |\mathcal{O}_{S_0}(1)|$ has total degree 36. The assertion follows. □

Corollary 4. *Consider a family $f' : S' \to \Delta$ of general quartic surfaces in $\mathbf{P}^3$, degenerating to a monoid quartic surface Y with tangent cone at its triple point p consisting of a triple of independent planes (see Remark 7). This family is 1–well behaved, with good model $\varpi : \bar{X} \to \Delta$. The crude limit Severi variety $\mathfrak{V}^{\mathrm{cr}}_{1,1}(S')$ consists in the surface $\check{Y}$ (which has degree 21), plus the plane $\check{p}$ counted with multiplicity 15.*

Proof. We have a morphism $\mathfrak{P}''' \to \mathbf{P}(\varpi_*(\mathcal{L}(-M_i))) \cong \mathbf{P}(f'_*(\mathcal{O}_{S'}(1)))$. The push–forward by this map of the regular components of $\mathfrak{V}_{1,1}(\bar{X})$ are $\check{Y}$ for $V(M_i, \delta_{P_i} = 1)$, $3 \cdot \check{p}$ for $V(M^i, \delta_{T^i} = 1)$, $\check{p}$ for each of the 12 $V(E)$ corresponding to a (-1)–curve $E^{\pm}_{hk}$ with $i \in \{h, k\}$, and 0 otherwise. The degree of $V(M_i, \delta_{P_i} = 1)$ in $\mathfrak{L}_i$ is at least 21 by Proposition 29, so the total degree of the push–forward in $|\mathcal{O}_{S'_0}(1)|$ of the regular components of $\mathfrak{V}_{1,1}(\bar{X})$ is at least 36. The assertion follows. □

Proposition 17 (Limits of 2-nodal curves). *The regular components of the limit Severi variety $\mathfrak{V}_{1,2}(\bar{X})$ are the following (they all appear with multiplicity* 1*):*

(i) $V(E, E')$ *for each set of two curves* $E, E' \in \{E_{ij}^{\pm}, 1 \leqslant i, j \leqslant 4, i \neq j\}$ *that do not meet the same edge of the tetrahedron* S_0*. The* $\mathbf{n}_h$*–degree is* 1 *if* $P_h \subset S_0$ *does not contain the two edges met by* E, E'*, and* 0 *otherwise;*
(ii) $V(M^k, \delta_{T^k} = 1, E)$ *for* $k = 1, \dots, 4$ *and* $E \in \{E_{ij}^{\pm}, 1 \leqslant i, j \leqslant 4, i \neq j, k \in \{i, j\}\}$*, which is a degree* 3 *curve in* $\mathfrak{L}^k$*. The* $\mathbf{n}_h$*–degree is* 3 *if* P_h *does not contain both the edge met by* E *and the vertex corresponding to* T^k*, it is* 0 *otherwise;*
(iii) $V(M_{ij}, \delta_{W_{ij}} = 2)$ *for* $1 \leqslant i < j \leqslant 4$*, which has* $\mathbf{n}_h$*-degree* 16 *for* $h \notin \{i, j\}$*, and* 0 *otherwise;*
(iv) $V(M_i, \delta_{P_i} = 2)$ *for* $1 \leqslant i \leqslant 4$*, which has* $\mathbf{n}_h$*-degree* 132 *for* $h = i$*, and* 0 *otherwise;*
(v) $V(M_{ij}, \delta_{W_{ij}} = 1, E)$ *for* $1 \leqslant i < j \leqslant 4$*, and* $E \in \left\{E_{\bar{i}\bar{j}}^{\pm}, \{\bar{i}, \bar{j}\} \cup \{i, j\} = \{1, \dots, 4\}\right\}$*, which is a curve of* $\mathbf{n}_h$*-degree* 0 *for* $1 \leqslant h \leqslant 4$*.*

Proof. It goes as the proof of Proposition 16. Again, Proposition 30 asserts that $V(M_i, \delta_{P_i} = 2)$ has degree at least 132 in $\mathfrak{L}_i$, but it follows from the computations in Remark 13 (a) below that it is exactly 132, which proves Theorem 3 for $\delta = 2$. □

Remark 13. **(a)** The degree of the Severi variety $V_2(\Sigma, \mathcal{O}_\Sigma(1))$ for a general quartic surface Σ is 480 (see Proposition 4). Hence if we fix a general point x on one of the components P_h of S_0 we should be able to see the 480 points of the limit Severi variety $\mathfrak{V}_{1,2}$ through x. The $\mathbf{n}_h$–degree information in Proposition 17 tells us this.

For each choice of two distinct edges of S_0 spanning a plane distinct from P_h, and of two (-1)-curves E and E' meeting these edges, we have a curve containing x in each of the items of type (i) . This amounts to a total of 192 such curves.

For each choice of a vertex and an edge of S_0, such that they span a plane distinct from P_h, there are 3 curves containing x in each of the four corresponding items (ii). This amounts to a total of 108 such curves.

For each choice of an edge of S_0 not contained in P_h, there are 16 curves containing x in the corresponding item (ii). This gives a contribution of 48 curves.

Finally, there are 132 plane quartics containing x in the item (ii) for $i = h$. Adding up, one finds the right number 480.

(b) Considerations similar to the ones in Remark 12 (b) could be made here, but we do not dwell on this.

Corollary 5 (Theorem 1 for $\delta = 2$). *Same setting as in Corollary 3. The family* $f : S \to \Delta$ *is* 2*–well behaved, with good model* $\varpi : \bar{X} \to \Delta$*. The crude limit Severi variety* $\mathfrak{V}_{1,2}^{\mathrm{cr}}(S)$ *consists of the image in* $|\mathcal{O}_{S_0}(1)|$ *of:*

(i) The 240 *components in* (i) *of Proposition 17, which map to as many lines in* $|\mathcal{O}_{S_0}(1)|$*;*

(ii) The 48 *components in* (ii) *of Proposition 17, each mapping* 3 : 1 *to as many lines in* $|\mathcal{O}_{S_0}(1)|$;

(iii) The 6 *components in* (ii) *of Proposition 17, respectively mapping* 16 : 1 *to the dual lines of the edges of* S_0.

Proof. The components in question are the only ones not contracted to points by the morphism $\mathfrak{P}_0''' \to |\mathcal{O}_{S_0}(1)|$, and their push–forward sum up to a degree 480 curve. □

Corollary 6. *Same setting as in Corollary 4; the family* $f' : S' \to \Delta$ *is 2–well behaved, with good model* $\varpi : \bar{X} \to \Delta$. *The crude limit Severi variety* $\mathfrak{V}^{\mathrm{cr}}_{1,2}(S')$ *consists of the ordinary double curve of the surface* $\check{Y}$, *which has degree* 132, *plus a sum (with multiplicities) of lines contained in the dual plane* $\check{p}$ *of the vertex of* Y.

Proof. It is similar to that of Corollary 4. The lines of $\mathfrak{V}^{\mathrm{cr}}_{1,2}(S')$ contained in $\check{p}$ are the push–forward by $\mathfrak{P}_0''' \to |\mathcal{O}_Y(1)|$ of the regular components of $\mathfrak{V}_{1,2}(\bar{X})$ listed in Remark 13 (a), with the exception of $V(M_i, \delta_{P_i} = 2)$. They sum up (with their respective multiplicities) to a degree 348 curve, while $V(M_i, \delta_{P_i} = 2)$ has degree at least 132 in $\mathfrak{L}_i$ by Proposition 30. □

Proposition 18 (Limits of 3-nodal curves). *The family* $\varpi : \bar{X} \to \Delta$ *is absolutely 3–good, and the limit Severi variety* $\mathfrak{V}_{1,3}(\bar{X})$ *is reduced, consisting of:*

(i) *The* 1,024 *points* $V(E, E', E'')$, *for* $E, E', E'' \in \{E^{\pm}_{ij}, 1 \leqslant i < j \leqslant 4\}$ *such that the span of the three corresponding double points of* S *is not contained in a face of* S_0;

(ii) *The* 192 *schemes* $V(M^k, \delta_{Tk} = 1, E, E')$, *for* $1 \leqslant k \leqslant 4$ *and* $E, E' \in \{E^{\pm}_{ij}, 1 \leqslant i < j \leqslant 4\}$, *such that the two double points of* S *corresponding to* E *and* E' *and the vertex with index* k *span a plane which is not a face of* S_0. *They each consist of* 3 *points;*

(iii) *The* 24 *schemes* $V(M_{ij}, \delta_{W_{ij}} = 2, E)$, *for* $1 \leqslant i < j \leqslant 4$, *and* $E \in \{E^{\pm}_{ij}, 1 \leqslant i < j \leqslant 4\}$, *such that the double point of* S *corresponding to* E *does not lie on the edge* $P_i \cap P_j$ *of* S_0, *and that these two together do not span a face of* S_0. *They each consist of* 16 *points;*

(iv) *The* 4 *schemes* $V(M_i, \delta_{P_i} = 3)$, *each consisting of* 304 *points.*

Proof. The list in the statement enumerates all regular components of the limit Severi variety $\mathfrak{V}_{1,3}(\bar{X})$ with their degrees (as before, Corollary 14 only gives 304 as a lower bound for the degree of (iv)). They therefore add up to a total of at least 3,200 points, which implies, by Proposition 18, that $\mathfrak{V}_{1,3}(\bar{X})$ has no component besides the regular ones, and that those in (iv) have degree exactly 304. Reducedness then follows from Remark 21, (b). □

In conclusion, all the above degenerations of quartic surfaces constructed from $\bar{X} \to \Delta$ with a twist of $\mathcal{L}$ are 3–well behaved, with $\bar{X}$ as a good model. In particular:

Corollary 7 (Theorem 1 for $\delta = 3$). *Same setting as in Corollary 3. The limits in $|\mathcal{O}_{S_0}(1)|$ of 3-tangent planes to S_t, for $t \in \Delta^*$, consist of:*

(i) *The* 1,024 *planes (each with multiplicity 1) containing three double points of S but no edge of S_0;*
(ii) *The* 192 *planes (each with multiplicity* 3*) containing a vertex of S_0 and two double points of S, but no edge of S_0;*
(iii) *The* 24 *planes (each with multiplicity* 16*) containing an edge of S_0 and a double point of S on the opposite edge;*
(iv) *The* 4 *faces of S_0 (each with multiplicity* 304*).*

6 Other Degenerations

The degeneration of a general quartic we considered in Sect. 5 is, in a sense, one of the most intricate. There are *milder* ones, e.g. to:

(i) A general union of a cubic and a plane;
(ii) A general union of two quadrics (this is an incarnation of a well known degeneration of $K3$ surfaces described in [12]).

Though we encourage the reader to study in detail the instructive cases of degenerations (i) and (ii), we will not dwell on this here, and only make the following observation about degeneration (ii). Let $f : S \to \Delta$ be such a degeneration, with central fibre $S_0 = Q_1 \cup Q_2$, where Q_1, Q_2 are two general quadrics meeting along a smooth quartic elliptic curve R. Then the limit linear system of $|\mathcal{O}_{S_t}(1)|$ as $t \in \Delta^*$ tends to 0 is just $|\mathcal{O}_{S_0}(1)|$, so that $f : S \to \Delta$ endowed with $\mathcal{O}_S(1)$ is absolutely good.

On the other hand, there are also degenerations to special singular irreducible surfaces, as the one we will consider in Sect. 7 below. In the subsequent sub–section, we will consider for further purposes another degeneration, the central fibre of which is still a (smooth) $K3$ surface.

6.1 Degeneration to a Double Quadric

Let $Q \subset \mathbf{P}^3$ be a smooth quadric and let B be a general curve of type $(4, 4)$ on Q. We consider the double cover $p : S_0 \to Q$ branched along B. This is a $K3$ surface and there is a smooth family $f : S \to \Delta$ with general fibre a general quartic surface and central fibre S_0. The pull–back to S_0 of plane sections of Q which are bitangent to B fill up a component $\mathfrak{V}$ of multiplicity 1 of the crude limit Severi variety $\mathfrak{V}_2^{\mathrm{cr}}$. Note that $\mathfrak{V}_2^{\mathrm{cr}}$ naturally sits in $|\mathcal{O}_{S_0}(1)| \cong \check{\mathbf{P}}^3$ in this case, hence one can unambiguously talk about its degree. Although it makes sense to conjecture that $\mathfrak{V}$ is irreducible, we will only prove the following weaker statement:

Proposition 19. *The curve* $\mathfrak{V}$ *contains an irreducible component of degree at least 36.*

We point out the following immediate consequence, which will be needed in Sect. 10.1 below:

Corollary 8. *If* X *is a general quartic surface in* $\mathbf{P}^3$*, then the Severi variety* $V_2(X, \mathscr{O}_X(1))$ *(which naturally sits in* $|\mathscr{O}_X(1)| \cong \check{\mathbf{P}}^3$*) has an irreducible component of degree at least 36.*

To prove Proposition 19 we make a further degeneration to the case in which B splits as $B = D + H$, where D is a general curve of type $(3,3)$ on Q, and H is a general curve of type $(1,1)$, i.e. a general plane section of $Q \subset \mathbf{P}^3$. Then the limit of $\mathfrak{V}$ contains the curve $\mathfrak{W} := \mathfrak{W}_{D,H}$ in $\check{\mathbf{P}}^3$ parametrizing those planes in $\mathbf{P}^3$ tangent to both H and D (i.e., $\mathfrak{W}$ is the intersection curve of the dual surfaces $\check{H}$ and $\check{D}$). Note that $\check{H}$ is the quadric cone circumscribed to the quadric $\check{Q}$ and with vertex the point $\check{P}$ orthogonal to the plane P cutting out H on Q, while $\check{D}$ is a surface scroll, the degree of which is 18 by Proposition 6, hence $\deg(\mathfrak{W}) = 36$. To prove Proposition 19, it suffices to prove that:

Lemma 9. *The curve* $\mathfrak{W}$ *is irreducible.*

To show this, we need a preliminary information. Let us consider the irreducible, locally closed subvariety $\mathscr{U} \subset |\mathscr{O}_Q(4)|$ of dimension 18, consisting of all curves $B = D + H$, where D is a smooth, irreducible curve of type $(3,3)$, and H is a plane section of Q which is not tangent to D. Consider $\mathscr{I} \subset \mathscr{U} \times \check{\mathbf{P}}^3$ the Zariski closure of the set of all pairs $(D + H, \Pi)$ such that the plane Π is tangent to both D and H, i.e. $\check{\Pi} \in \check{H} \cap \check{D}$. We have the projections $p_1 : \mathscr{I} \to \mathscr{U}$ and $p_2 : \mathscr{I} \to \check{\mathbf{P}}^3$. The curve $\mathfrak{W}$ is a general fibre of p_1.

Lemma 10. *The variety* $\mathscr{I}$ *contains a unique irreducible component* $\mathscr{J}$ *of dimension 19 which dominates* $\check{\mathbf{P}}^3$ *via the map* p_2*.*

Proof. Let Π be a general plane of $\mathbf{P}^3$. Consider the conic $\Gamma := \Pi \cap Q$, and fix distinct points $q_1, \ldots, q_6$ on Γ. There is a plane P tangent to Γ at q_1, and a cubic surface F passing through $q_3, \ldots, q_6$ and tangent to Γ at q_2; moreover P and F can be chosen general enough for $D + H$ to belong to $\mathscr{U}$, where $H = P \cap Q$ and $D = F \cap Q$. Then $(D + H, \Pi) \in \mathscr{I}$, which proves that p_2 is dominant.

Let $\mathscr{F}_\Pi$ be the fibre of p_2 over Π. The above argument shows that there is a dominant map $\mathscr{F}_\Pi \dashrightarrow \Gamma^2 \times \mathrm{Sym}^4(\Gamma)$ whose general fibre is an open subset of $\mathbf{P}^1 \times \mathbf{P}^9$: precisely, if $((q_1, q_2), q_3 + \ldots + q_6) \in \Gamma^2 \times \mathrm{Sym}^4(\Gamma)$ is a general point, the $\mathbf{P}^1$ is the linear system of plane sections of Q tangent to Γ at q_1, and the $\mathbf{P}^9$ is the linear subsystem of $|\mathscr{O}_Q(3)|$ consisting of curves passing through $q_3, \ldots, q_6$ and tangent to Γ at q_2. The existence and unicity of $\mathscr{J}$ follow. □

Now we consider the commutative diagram

$$\begin{array}{ccc} \mathcal{J}' & \xrightarrow{\nu} & \mathcal{J} \\ {\scriptstyle p'}\downarrow & \searrow & \downarrow{\scriptstyle p_1} \\ \mathcal{U}' & \xrightarrow[f]{} & \mathcal{U} \end{array} \tag{22}$$

where ν is the normalization of $\mathcal{J}$, and $f \circ p'$ is the Stein factorization of $p_1 \circ \nu : \mathcal{J}' \to \mathcal{U}$. The morphism $f : \mathcal{U}' \to \mathcal{U}$ is finite of degree h, equal to the number of irreducible components of the general fibre of p_1, which is $\mathfrak{W}$. The irreduciblity of $\mathcal{J}$ implies that the monodromy group of $f : \mathcal{U}' \to \mathcal{U}$ acts transitively on the set of components of $\mathfrak{W}$.

Proof (of Lemma 9). We need to prove that $h = 1$. To do this, fix a general $D \in |\mathcal{O}_Q(3)|$, and consider the curve $\mathfrak{W} = \mathfrak{W}_{D,H}$, with H general, which consists of h components. We can move H to be a section of Q by a general tangent plane Z. Then the quadric cone $\check{H}$ degenerates to the tangent plane $T_{\check{Q},z}$ to $\check{Q}$ at $z := \check{Z}$, counted with multiplicity 2.

We claim that, for $z \in \check{Q}$ general, the intersection of $T_{\check{Q},z}$ with $\check{D}$ is irreducible. Indeed, since $\check{D}$ is a scroll, a plane section of $\check{D}$ is reducible if and only if it contains a ruling, i.e. if and only if it is a tangent plane section of $\check{D}$. Since $\check{D} \neq \check{Q}$, the biduality theorem implies the claim.

The above assertion implies $h \leqslant 2$. If equality holds, the general curve $\mathfrak{W}$ consists of two curves which, by transitivity of the monodromy action of f, are both unisecant to the lines of the ruling of $\check{D}$.

To see that this is impossible, let us degenerate D as $D_1 + D_2$, where D_1 is a general curve of type $(2, 1)$ and D_2 is a general curve of type $(1, 2)$ on Q. Then $\check{D}$ accordingly degenerates and its limit contains as irreducible components $\check{D}_1$ and $\check{D}_2$, which are both scrolls of degree 4 (though we will not use it, we note that $D_1 \cdot D_2 = 5$ and the (crude) limit of $\check{D}$ in the above degeneration consists of the union of $\check{D}_1$, $\check{D}_2$, and of the five planes dual to the points of $D_1 \cap D_2$, each of the latter counted with multiplicity 2). We denote by $\mathfrak{D}$ either one of the curves D_1, D_2.

Let again H be a general plane section of Q. We claim that the intersection of $\check{\mathfrak{D}}$ with $\check{H}$ does not contain any unisecant curve to the lines of the ruling of $\check{\mathfrak{D}}$. This clearly implies that the general curve $\mathfrak{W}$ cannot split into two unisecant curves to the lines of the ruling of $\check{D}$, thus proving that $h = 1$.

To prove the claim, it suffices to do it for specific $\mathfrak{D}$, Q and H. For $\mathfrak{D}$ we take the rational normal cubic with affine parametric equations $x = t, y = t^2, z = t^3$, with $t \in \mathbf{C}$. For Q we take the quadric with affine equation $x^2 + y^2 - xz - y = 0$, and for H the intersection of Q with the plane $z = 0$. Let (p, q, r) be affine coordinates in the dual space, so that (p, q, r) corresponds to the plane with equation $px + qy + rz + 1 = 0$ (i.e., we take as plane at infinity in the dual space the orthogonal to the origin). Then the affine equation of $\check{\mathfrak{D}}$ is gotten by eliminating t in the system

$$pt + qt^2 + rt^3 + 1 = 0, \quad p + 2qt + 3rt^2 = 0, \tag{23}$$

which defines the ruling ρ_t of $\check{\mathfrak{D}}$ orthogonal to the tangent line to $\mathfrak{D}$ at the point with coordinates (t, t^2, t^3), $t \in \mathbf{C}$. The affine equation of $\check{H}$ is gotten by imposing that the system

$$px + qy + qz + 1 = 0, \quad x^2 + y^2 - xz - y = 0, \quad z = 0,$$

has one solution with multiplicity 2; the resulting equation is $p^2 - 4q - 4 = 0$. Adding this to (23) means intersecting $\check{H}$ with ρ_t; for $t \neq 0$, the resulting system can be written as

$$p^2t^2 + 8pt - 4(t^2 - 3) = 0, \quad q = \frac{p^2}{4} - 1, \quad r = \frac{4 - p^2}{6t} - \frac{p}{3t^2}.$$

For a general $t \in \mathbf{C}$, the first equation gives two values of p and the remaining equations the corresponding values of q and r, i.e., we get the coordinates (p, q, r) of the two intersection points of $\check{H}$ and ρ_t. Now we note that the discriminant of $p^2t^2 + 8pt - 4(t^2 - 3)$ as a polynomial in p is $16t^2(t^2 + 1)$, which has the two simple solutions $\pm i$. This implies that the projection on $\mathfrak{D} \cong \mathbf{P}^1$ of the curve cut out by $\check{H}$ on $\check{\mathfrak{D}}$ has two simple ramification points. In particular $\check{H} \cap \check{\mathfrak{D}}$ is locally irreducible at these points, and it cannot split as two unisecant curves to the lines of the ruling. This proves the claim and ends the proof of the Lemma. □

7 Kummer Quartic Surfaces in $\mathbf{P}^3$

This section is devoted to the description of some properties of quartic *Kummer surfaces* in $\mathbf{P}^3$. They are quartic surfaces with 16 ordinary double points $p_1, \ldots, p_{16}$ as their only singularities. Alternatively a Kummer surface X is the image of the Jacobian $J(C)$ of a smooth genus 2 curve C, via the degree 2 morphism $\vartheta : J(C) \to X \subset \mathbf{P}^3$ determined by the complete linear system $|2C|$, where $C \subset J(C)$ is the Abel–Jacobi embedding, so that $(J(C), C)$ is a principally polarised abelian surface (see, e.g., [4, Chap. 10]). Since ϑ is composed with the $\pm$ involution on $J(C)$, the 16 nodes of X are the images of the 16 points of order 2 of $J(C)$. By projecting from a node, Kummer surfaces can be realised as double covers of the plane, branched along the union of six distinct lines tangent to one single conic (see, e.g., [2, Chap. VIII, Exercises]). We refer to the classical book [25] for a thorough description of these surfaces (see also [14, Chap. 10]).

7.1 The 16_6 Configuration and Self–Duality

An important feature of Kummer surfaces is that they carry a so-called 16_6–*configuration* (see [19], as well as the above listed references). Let X be such a surface. There are exactly 16 distinct planes Π_i tangent to X along a *contact conic* Γ_i, for $1 \leqslant i \leqslant 16$. The contact conics are the images of the 16 symmetric theta divisors $C_1, \ldots, C_{16}$ on $J(C)$. Each of them contains exactly 6 nodes of X, coinciding with the branch points of the map $\vartheta|_{C_i} : C_i \cong C \to \Gamma_i \cong \mathbf{P}^1$ determined by the canonical g^1_2 on C.

Two conics Γ_i, Γ_j, $i \neq j$, intersect at exactly two points, which are double points of X: they are the nodes corresponding to the two order 2 points of $J(C)$ where C_i and C_j meet. Since the restriction map $\mathrm{Pic}^0(J(C)) \to \mathrm{Pic}^0(C)$ is an isomorphism, there is no pair of points of $J(C)$ contained in three different theta divisors. This implies that, given a pair of nodes of X, there are exactly two contact conics containing both of them. In other words, if we fix an $i \in \{1, \ldots, 16\}$, the map from $\{1, \ldots, 16\} - \{i\}$ to the set of pairs of distinct nodes of X on Γ_i, which maps j to $\Gamma_i \cap \Gamma_j$, is bijective. This yields that each node of X is contained in exactly 6 conics Γ_i. The configuration of 16 nodes and 16 conics with the above described incidence property is called a 16_6–configuration.

Let $\tilde{X}$ be the minimal smooth model of X, $E_1, \ldots, E_{16}$ the (-2)-curves over the nodes $p_1, \ldots, p_{16}$ of X respectively, and D_i the proper transform of the conic Γ_i, for $1 \leqslant i \leqslant 16$. Since $\tilde{X}$ is a $K3$ surface and the D_i's are rational curves, the latter are (-2)-curves. The 16_6–configuration can be described in terms of the existence of the two sets

$$\mathscr{E} = \{E_1, \ldots, E_{16}\} \quad \text{and} \quad \mathscr{D} = \{D_1, \ldots, D_{16}\}$$

of 16 pairwise disjoint (-2)–curves, enjoying the further property that each curve of a given set meets exactly six curves of the other set, transversely at a single point.

Proposition 20. *Let X be a Kummer surface. Then its dual $\check{X} \subset \check{\mathbf{P}}^3$ is also a Kummer surface.*

Proof. By Proposition 5, we have $\deg(\check{X}) = 4$. Because of the singularities on X, the Gauss map $\gamma_X : X \dashrightarrow \check{X}$ is not a morphism. However we get an elimination of indeterminacies

$$\begin{array}{ccc} & \tilde{X} & \\ {\scriptstyle f} \swarrow & & \searrow {\scriptstyle g} \\ X & \underset{\gamma_X}{\dashrightarrow} & \check{X} \end{array}$$

by considering the minimal smooth model $\tilde{X}$ of X. The morphism f is the contraction of the 16 curves in $\mathscr{E}$, and g maps each E_i to a conic which is the dual of the tangent cone to X at the node corresponding to E_i. On the other hand, since γ_X contracts each of the curves $\Gamma_1, \ldots, \Gamma_{16}$ to a point, then g contracts the curves in $\mathscr{D}$ to as many ordinary double points of $\check{X}$. The assertion follows. □

7.2 The Monodromy Action on the Nodes

Let $\mathscr{K}^\circ$ be the locally closed subset of $|\mathscr{O}_{\mathbf{P}^3}(4)|$ whose points correspond to Kummer surfaces and let $\pi : \mathscr{X} \to \mathscr{K}^\circ$ be the universal family: over $x \in \mathscr{K}^\circ$, we have the corresponding Kummer surface $X = \pi^{-1}(x)$. We have a subscheme $\mathscr{N} \subset \mathscr{X}$ such that $p := \pi|_{\mathscr{N}} : \mathscr{N} \to \mathscr{K}^\circ$ is a finite morphism of degree 16: the fibre $p^{-1}(x)$ over $x \in \mathscr{K}^\circ$ consists of the nodes of X. We denote by $G_{16,6} \subset \mathfrak{S}_{16}$ the monodromy group of $p : \mathscr{N} \to \mathscr{K}^\circ$

There is in addition another degree 16 finite covering $q : \mathscr{G} \to \mathscr{K}^\circ$: for $x \in \mathscr{K}^\circ$, the fibre $q^{-1}(x)$ consists of the set of the contact conics on X. Proposition 20 implies that the monodromy group of this covering is isomorphic to $G_{16,6}$. Then we can consider the commutative square

$$\begin{array}{ccc} \mathscr{I} & \xrightarrow{q'} & \mathscr{N} \\ {\scriptstyle p'}\downarrow & & \downarrow{\scriptstyle p} \\ \mathscr{G} & \xrightarrow[q]{} & \mathscr{K}^\circ \end{array} \tag{24}$$

where $\mathscr{I}$ is the incidence correspondence between nodes and conics. Note that p', q' are both finite of degree 6, with isomorphic monodromy groups (see again Proposition 20).

Here, we collect some results on the monodromy groups of the coverings appearing in (24). They are probably well known to the experts, but we could not find any reference for them.

Lemma 11. *The monodromy group of $q' : \mathscr{I} \to \mathscr{N}$ and of $p' : \mathscr{I} \to \mathscr{G}$ is the full symmetric group $\mathfrak{S}_6$.*

Proof. It suffices to prove only one of the two assertions, e.g. the one about p'. Let X be a general Kummer surface and let e be a node of X. As we noticed, by projecting from e, we realise X as a double cover of $\mathbf{P}^2$ branched along 6 lines tangent to a conic E, which is the image of the (-2)–curve over e. These 6 lines are the images of the six contact conics through e, i.e. the fibre over q'. Since X is general, these 6 tangent lines are general. The assertion follows. □

Corollary 9. *The group $G_{16,6}$ acts transitively, so $\mathscr{G}$ and $\mathscr{N}$ are irreducible.*

Proof. It suffices to prove that the monodromy of $p : \mathscr{N} \to \mathscr{K}^\circ$ is transitive. This follows from Lemma 11 and from the fact that any two nodes of a Kummer surface lie on some contact conic. □

It is also possible to deduce the transitivity of the monodromy action of p and q from the irreducibility of the Igusa quartic solid, which parametrizes quartic Kummer surfaces with one marked node (see, e.g., [14, Chap. 10]). The following is stronger:

Proposition 21. *The group $G_{16,6}$ acts 2–transitively.*

Proof. Again, it suffices to prove the assertion for $p : \mathscr{N} \to \mathscr{K}^\circ$. By Corollary 9, proving that the monodromy is 2–transitive is equivalent to showing that the stabilizer of a point in the general fibre of p acts transitively on the remaining points of the fibre. Let X be a general Kummer surface and $e \in X$ a node. Consider the projection from e, which realizes X as a double cover of $\mathbf{P}^2$ branched along 6 lines tangent to a conic E. The 15 nodes on X different from e correspond to the pairwise intersections of the 6 lines. Moving the tangent lines to E one leaves the node e fixed, while acting transitively on the others. □

Look now at the pull back $q^*(\mathscr{N})$. Of course $\mathscr{J}$ is a component of $q^*(\mathscr{N})$. We set $\mathscr{W} = q^*(\mathscr{N}) - \mathscr{J}$, and the morphism $p' : \mathscr{W} \to \mathscr{G}$ which is finite of degree 10. We let $H_{16,6} \subseteq \mathfrak{S}_{10}$ be the monodromy of this covering.

Lemma 12. *The group $H_{16,6}$ acts transitively, i.e. $\mathscr{W}$ is irreducible.*

Proof. Let $a, b \in X$ be two nodes not lying on the contact conic Γ. There is a contact conic Γ' that contains both a and b; it meets Γ transversely in two points, distinct from a and b, that we shall call c and d. Now a monodromy transformation that fixes Γ' and fixes c and d necessarily fixes Γ. It therefore suffices to find a monodromy transformation fixing Γ' which fixes c and d, and sends a to b. Such a transformation exists by Lemma 11. □

Proposition 22. *Let X be a general Kummer surface. Then:*

(i) *$G_{16,6}$ acts transitively the set of unordered triples of distinct nodes belonging to a contact conic;*
(ii) *The action of $G_{16,6}$ on the set of unordered triples of distinct nodes not belonging to a contact conic has at most two orbits.*

To prove this, we need to consider degenerations of Kummer surfaces when the principally polarised abelian surface $(J(C), C)$ becomes non–simple, e.g., when C degenerates to the union of two elliptic curves E_1, E_2 transversally meeting at a point. In this case the linear system $|2(E_1+E_2)|$ on the abelian surface $A = E_1 \times E_2$, is still base point free, but it determines a degree 4 morphism $\vartheta : A \to \mathbf{Q} \cong \mathbf{P}^1 \times \mathbf{P}^1 \subset \mathbf{P}^3$ (where $\mathbf{Q} \subset \mathbf{P}^3$ is a smooth quadric), factoring through the *product Kummer surface* $\mathsf{X} = A/\pm$, and a double cover $\mathsf{X} \to \mathbf{Q}$ branched along a curve of bidegree $(4,4)$ which is a union of 8 lines; the lines in question are $L_{1a} = \mathbf{P}^1 \times \{a\}$ (resp. $L_{2b} = \{b\} \times \mathbf{P}^1$) where a (resp. b) ranges among the four branch points of the morphism $E_1 \to (E_1/\pm) \cong \mathbf{P}^1$ (resp. $E_2 \to (E_2/\pm) \cong \mathbf{P}^1$). We call the former *horizontal lines*, and the latter *vertical lines*. Each of them has four marked points: on a line L_{1a} (resp. L_{2b}), these are the four points $L_{2b} \cap L_{1a}$ where b (resp. a) varies as above. One thus gets 16 points, which are the limits on X of the 16 nodes of a

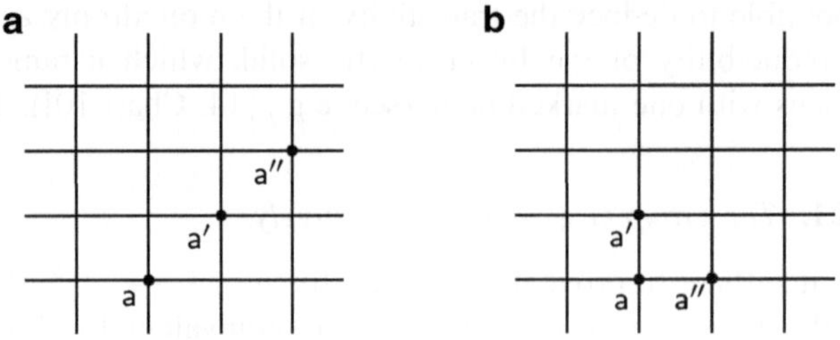

Fig. 9 Limits in a product Kummer surface of three double points not on a double conic

general Kummer surface X. The limits on X of the 16 contact conics on a general Kummer surface X are the 16 curves $L_{1a} + L_{2b}$. On such a curve, the limits of the six double points on a contact conic on a general Kummer surface are the six marked points on L_{1a} and L_{2b} that are distinct from $L_{1a} \cap L_{2b}$.

Proof (of Proposition 22). Part (i) follows from Lemma 11. As for part (ii), consider three distinct nodes a, a' and a'' (resp. b, b' and b'') of X that do not lie on a common conic of the 16_6 configuration on X. We look at their limits a, a′ and a″ (resp. b, b′ and b″) on the product Kummer surface X; they are in one of the two configurations (a) and (b) described in Fig. 9.

The result follows from the fact that the monodromy of the family of product Kummer surfaces acts as the full symmetric group $\mathfrak{S}_4$ on the two sets of vertical and horizontal lines respectively. Hence the triples in configuration (a) (resp. in configuration (b)) are certainly in one and the same orbit. □

8 Degeneration to a Kummer Surface

We consider a family $f : S \to \Delta$ of surfaces in $\mathbf{P}^3$ induced (as explained in Sect. 3.1) by a pencil generated by a general quartic surface S_∞ and a general Kummer surface S_0. We will describe a related δ-good model $\varpi : \bar{X} \to \Delta$ for $1 \leq \delta \leq 3$.

8.1 The Good Model

Our construction is as follows:

(I) We first perform a degree 2 base change on $f : S \to \Delta$;
(II) Then we resolve the singularities of the new family;
(III) We blow–up the proper transforms of the 16 contact conics on S_0.

The base change is useful to analyze the contribution of curves passing through a node of S_0.

Steps (I) and (II)

The total space S is smooth, analytically–locally given by the equation

$$x^2 + y^2 + z^2 = t$$

around each of the double points of S_0. We perform a degree 2 base change on f, and call $\bar{f} : \bar{S} \to \Delta$ the resulting family. The total space $\bar{S}$ has 16 ordinary double points at the preimages of the nodes of S_0.

We let $\varepsilon_1 : X \to \bar{S}$ be the resolution of these 16 points, gotten by a simple blow–up at each point. We have the new family $\pi : X \to \Delta$, with $\pi = \bar{f} \circ \varepsilon_1$. The new central fibre X_0 consists of the minimal smooth model $\tilde{S}_0$ of S_0, plus the exceptional divisors $Q_1, \ldots, Q_{16}$. These are all isomorphic to a smooth quadric $\mathbf{Q} \cong \mathbf{P}^1 \times \mathbf{P}^1 \subset \mathbf{P}^3$. We let $E_1, \ldots, E_{16}$ be the exceptional divisors of $\tilde{S}_0 \to S_0$. Each Q_i meets $\tilde{S}_0$ transversely along the curve E_i, and two distinct Q_i, Q_j do not meet.

Step (III)

As in Sect. 7.1, we let $D_1, \ldots, D_{16}$ be the proper transforms of the 16 contact conics $\Gamma_1, \ldots, \Gamma_{16}$ on S_0: they are pairwise disjoint (−2)-curves in X_0. We consider the blow-up $\varepsilon_2 : \bar{X} \to X$ of X along them. The surface $\tilde{S}_0$ is isomorphic to its strict transform on $\bar{X}_0$. Let $W_1, \ldots, W_{16}$ be the exceptional divisors of ε_2. Each W_i meets $\tilde{S}_0$ transversely along the (strict transform of the) curve D_i. Note that, by the Triple Point Formula 1, one has $\deg(N_{D_i|W_i}) = -\deg(N_{D_i|\tilde{S}_0}) - 6 = -4$, so that W_i is an $\mathbf{F}_4$–Hirzebruch surface, and D_i is the negative section on it.

We call $\tilde{Q}_1, \ldots, \tilde{Q}_{16}$ the strict transforms of $Q_1, \ldots, Q_{16}$ respectively. They respectively meet $\tilde{S}_0$ transversely along (the strict transforms of) $E_1, \ldots, E_{16}$. For $1 \leqslant i \leqslant 16$, there are exactly six curves among the D_j's that meet E_i: we call them $D_1^i, \ldots, D_6^i$. The surface $\tilde{Q}_i$ is the blow-up of Q_i at the six intersection points of E_i with $D_1^i, \ldots, D_6^i$: we call $'G_1^i, \ldots, 'G_6^i$ respectively the six corresponding (−1)-curves on $\tilde{Q}_i$. Accordingly, $\tilde{Q}_i$ meets transversely six W_j's, that we denote by $W_i^1, \ldots, W_i^6$, along $'G_1^i, \ldots, 'G_6^i$ respectively. The surface $\tilde{Q}_i$ is disjoint from the remaining W_j's.

For $1 \leqslant j \leqslant 16$, we denote by $E_j^1, \ldots, E_j^6$ the six E_i's that meet D_j. There are correspondingly six $\tilde{Q}_i$'s that meet W_j: we denote them by $\tilde{Q}_j^1, \ldots, \tilde{Q}_j^6$, and let $G_j^1, \ldots, G_j^6$ be their respective intersection curves with W_j. Note the equality of sets

$$\{'G_{s}^{i},\ 1 \leqslant i \leqslant 16,\ 1 \leqslant s \leqslant 6\} = \left\{G_{j}^{s},\ 1 \leqslant j \leqslant 16,\ 1 \leqslant s \leqslant 6\right\}.$$

We shall furthermore use the following notation (cf. Sect. 2). For $1 \leqslant j \leqslant 16$, we let F_{W_j} (or simply F) be the divisor class of the ruling on W_j, and H_{W_j} (or simply H) the divisor class $D_j + 4F_{W_j}$. Note that $G_j^s \sim_{W_j} F$ and $'G_i^s \sim_{W_i^s} F$, for $1 \leqslant i, j \leqslant 16$ and $1 \leqslant s \leqslant 6$. We write H_0 for the pull-back to $\tilde{S}_0$ of the plane section class of $S_0 \subset \mathbf{P}^3$. For $1 \leqslant i \leqslant 16$, we let L_i' and L_i'' be the two rulings of Q_i, and H_{Q_i} (or simply H) be the divisor class $L_i' + L_i''$; we use the same symbols for their respective pull-backs in $\tilde{Q}_i$. When designing one of these surfaces by $\tilde{Q}_j^s$, we use the obvious notation $L_j^{s\prime}$ and $L_j^{s\prime\prime}$.

8.2 The Limit Linear System

We shall now describe the limit linear system of $|\mathcal{O}_{\bar{X}_t}(1)|$ as $t \in \Delta^*$ tends to 0, and from this we will see that $\bar{X}$ is a good model of S over Δ. We start with $\mathfrak{P} = \mathbf{P}(\varpi_*(\mathcal{O}_{\bar{X}}(1)))$, which is a $\mathbf{P}^3$–bundle over Δ, whose fibre at $t \in \Delta$ is $|\mathcal{O}_{\bar{X}_t}(1)|$; we set $\mathcal{L} = \mathcal{O}_{\bar{X}}(1)$, and $|\mathcal{O}_{\bar{X}_t}(1)| = |\mathcal{L}_t|$ for $t \in \Delta$. Note that $|\mathcal{L}_0| \cong |\mathcal{O}_{S_0}(1)|$.

We will proceed as follows:

(I) *We first blow–up $\mathfrak{P}$ at the points of $\mathfrak{P}_0 \cong |\mathcal{L}_0|$ corresponding to planes in $\mathbf{P}^3$ containing at least three distinct nodes of S_0 (i.e. either planes containing exactly three nodes, or planes in the 16_6 configuration)*;

(II) *Then we blow–up the resulting variety along the proper transforms of the lines of $|\mathcal{L}_0|$ corresponding to pencils of planes in $\mathbf{P}^3$ containing two distinct nodes of S_0*;

(III) *Finally we blow–up along the proper transforms of the planes of $|\mathcal{L}_0|$ corresponding to webs of planes in $\mathbf{P}^3$ containing a node of S_0.*

The description of these steps parallels the one in Sect. 5.3, so we will be sketchy here.

Step (Ia)

The $\binom{16}{3} - 16\binom{6}{3} = 240$ planes in $\mathbf{P}^3$ containing exactly three distinct nodes of S_0 correspond to the 0-dimensional subsystems

$$\left|H_0 - E_{s'} - E_{s''} - E_{s'''}\right|_{\tilde{S}_0} \tag{25}$$

of $|H_0| \cong |\mathcal{L}_0|$, where $\{s', s'', s'''\}$ ranges through all subsets of cardinality 3 of $\{1, \ldots, 16\}$ such that the nodes $p_{s'}, p_{s''}, p_{s'''}$ corresponding to the (–2)–curves $E_{s'}, E_{s''}, E_{s'''}$ do not lie in a plane of the 16_6 configuration of S_0. We denote by

$C_{s's''s'''}$ the unique curve in the system (25) and we set $H_{s's''s'''} = C_{s's''s'''} + E_{s'} + E_{s''} + E_{s'''}$, which lies in $|H_0|$.

The exceptional component $\tilde{\mathfrak{L}}_{s's''s'''}$ of the blow-up of $\mathfrak{P}$ at the point corresponding to $H_{s's''s'''}$ can be identified with the 3–dimensional complete linear system

$$\mathfrak{L}_{s's''s'''} := \left|\mathscr{L}_0(-\tilde{Q}_{s'} - \tilde{Q}_{s''} - \tilde{Q}_{s'''})\right|,$$

which is isomorphic to the projectivization of the kernel of the surjective map

$$\mathfrak{f}: \left(\bigoplus_{s\in\{s',s'',s'''\}} \mathrm{H}^0\big(\tilde{Q}_s, \mathscr{O}_{\tilde{Q}_s}(H)\big)\right) \oplus \mathrm{H}^0\big(\tilde{S}_0, \mathscr{O}_{\tilde{S}_0}(C_{s's''s'''})\big)$$
$$\longrightarrow \bigoplus_{s\in\{s',s'',s'''\}} \mathrm{H}^0\big(\mathscr{O}_{E_s}(-E_s)\big) \cong \mathrm{H}^0\big(\mathbf{P}^1, \mathscr{O}_{\mathbf{P}^1}(2)\big)^{\oplus 3}$$

mapping $(\varsigma', \varsigma'', \varsigma''', \varsigma)$ to $(\varsigma' - \varsigma, \varsigma'' - \varsigma, \varsigma''' - \varsigma)$.

The typical element of $\mathfrak{L}_{s's''s'''}$ consists of

(i) The curve $C_{s's''s'''}$ on $\tilde{S}_0$, plus
(ii) One curve in $|\mathscr{O}_{\tilde{Q}_s}(H)|$ for each $s \in \{s', s'', s'''\}$, matching $C_{s's''s'''}$ along E_s, plus
(iii) Two rulings in each W_j (i.e. a member of $|\mathscr{O}_{W_j}(2F)| = |\mathscr{L}_0 \otimes \mathscr{O}_{W_j}|$), $1 \leqslant j \leqslant 16$, matching along the divisor D_j, while
(iv) The restriction to $\tilde{Q}_s$ is trivial for every $s \in \{1, \ldots, 16\} - \{s', s'', s'''\}$.

The strict transform of $\mathfrak{P}_0$ is isomorphic to the blow–up of $|H_0|$ at the point corresponding to $H_{s's''s'''}$. By Lemma 3, the exceptional divisor $\mathscr{H}_{s's''s'''} \cong \mathbf{P}^2$ of this blow-up identifies with the pull-back linear series on $H_{s's''s'''}$ of the 2-dimensional linear system of lines in the plane spanned by $p_{s'}, p_{s''}, p_{s'''}$ (note that in this linear series there are three linear subseries corresponding to sections vanishing on the curves $E_{s'}, E_{s''}, E_{s'''}$ which are components of $H_{s',s'',s'''}$).

The divisor $\mathscr{H}_{s's''s'''}$ is cut out on the strict transform of $|H_0|$ by $\tilde{\mathfrak{L}}_{s's''s'''}$, along the plane $\Pi \subset \mathfrak{L}_{s's''s'''}$ given by the equation $\varsigma = 0$ in the above notation. The identification of $\mathscr{H}_{s's''s'''}$ with Π is not immediate. It would become more apparent by blowing up the curves $C_{s's''s'''}$ in the central fibre; we will not do this here, because we do not need it, and we leave it to the reader (see Step (Ib) for a similar argument). However, we note that $\ker(\mathfrak{f}) \cap \{\varsigma = 0\}$ coincides with the $\mathbf{C}^3$ spanned by three non–zero sections $(\varsigma_{s'}, 0, 0, 0), (0, \varsigma_{s''}, 0, 0), (0, 0, \varsigma_{s'''}, 0)$, where ς_s vanishes exactly on E_s for each $s \in \{s', s'', s'''\}$. These three sections correspond to three points $\pi_{s'}, \pi_{s''}, \pi_{s'''}$ in Π. In the identification of Π with $\mathscr{H}_{s's''s'''}$ the points $\pi_{s'}, \pi_{s''}, \pi_{s'''}$ are mapped to the respective pull–backs on $H_{s's''s'''}$ of the three lines $\ell_{s''s'''} = \langle p_{s''}, p_{s'''}\rangle, \ell_{s's'''} = \langle p_{s'}, p_{s'''}\rangle, \ell_{s's''} = \langle p_{s'}, p_{s''}\rangle$.

Step (Ib)

The 16 planes of the 16_6 configuration correspond to the 0-dimensional subsystems

$$\left|H_0 - E_j^1 - \cdots - E_j^6\right|_{\tilde{S}_0} \subset \left|H_0\right| \cong \left|\mathscr{L}_0\right| \qquad (1 \leqslant j \leqslant 16),$$

consisting of the only curve $2D_j$. The blow-up of $\mathfrak{P}$ at these points introduces 16 new components $\tilde{\mathfrak{L}}^j$, $1 \leqslant j \leqslant 16$, in the central fibre, respectively isomorphic to the linear systems

$$\mathfrak{L}^j := \left|\mathscr{L}_0(-2W_j - \tilde{Q}_j^1 - \cdots - \tilde{Q}_j^6)\right|.$$

The corresponding line bundles restrict to the trivial bundle on all components of $\bar{X}_0$ but W_j and $\tilde{Q}_j^s$, for $1 \leqslant s \leqslant 6$, where the restriction is to $\mathscr{O}_{W_j}(2H)$ and to $\mathscr{O}_{\tilde{Q}_j^s}(H - 2G_j^s)$, respectively.

For each $s \in \{1, \ldots, 16\}$, the complete linear system $|H - 2G_j^s|_{\tilde{Q}_j^s}$ is 0-dimensional, its only divisor is the strict transform in $\tilde{Q}_j^s$ of the unique curve in $|H|_{Q_j^s}$ that is singular at the point $D_j \cap Q_j^s$. This is the union of the proper transforms of the two curves in $|L_j^{s\prime}|_{Q_j^s}$ and $|L_j^{s\prime\prime}|_{Q_j^s}$ through $D_j \cap Q_j^s$, and it cuts out a 0-cycle Z_j^s of degree 2 on G_j^s. We conclude that

$$\mathfrak{L}^j \cong \left|\mathscr{O}_{W_j}(2H) \otimes \mathscr{I}_{Z_j}\right|, \quad \text{for} \quad 1 \leqslant j \leqslant 16, \tag{26}$$

where $\mathscr{I}_{Z_j} \subset \mathscr{O}_{W_j}$ is the defining sheaf of ideals of the 0-cycle $Z_j := Z_j^1 + \cdots + Z_j^6$ supported on the six fibres $G_j^1, \ldots, G_j^6$ of the ruling of W_j. We shall later study the rational map determined by this linear system on W_j (see Proposition 24).

For each j, the glueing of $\tilde{\mathfrak{L}}^j$ with the strict transform of $|H_0|$ is as follows: the exceptional plane $\mathscr{H}^j$ on the strict transform of $|H_0|$ identifies with $|\mathscr{O}_{D_j}(H_0)| \cong |\mathscr{O}_{\mathbf{P}^1}(2)|$ by Lemma 3, and the latter naturally identifies as the 2-dimensional linear subsystem of $\left|\mathscr{O}_{W_j}(2H) \otimes \mathscr{I}_{Z_j}\right|$ consisting of divisors of the form

$$2D_j + G_j^1 + \cdots + G_j^6 + \Phi, \quad \Phi \in |\mathscr{O}_{W_j}(2F)|.$$

Step (II)

Let $\mathfrak{P}'$ be the blow-up of $\mathfrak{P}$ at the $240 + 16$ distinct points described in the preceding step. The next operation is the blow-up $\mathfrak{P}'' \to \mathfrak{P}'$ along the $\binom{16}{2}$ pairwise disjoint respective strict transforms of the pencils

$$\left|H_0 - E_{s'} - E_{s''}\right|_{\tilde{S}_0}, \quad 1 \leqslant s' < s'' \leqslant 16. \tag{27}$$

To describe the exceptional divisor $\tilde{\mathfrak{L}}_{s's''}$ of $\mathfrak{P}'' \to \mathfrak{P}'$ on the proper transform of (27), consider the 3–dimensional linear system $\mathfrak{L}_{s's''} := \left|\mathscr{L}_0(-\tilde{Q}_{s'} - \tilde{Q}_{s''})\right|$, isomorphic to the projectivization of the kernel of the surjective map

$$\left(\bigoplus_{s\in\{s',s''\}} \mathrm{H}^0\big(\tilde{Q}_s, \mathscr{O}_{\tilde{Q}_s}(H)\big)\right) \oplus \mathrm{H}^0\big(\tilde{S}_0, \mathscr{O}_{\tilde{S}_0}(H_0 - E_{s'} - E_{s''})\big) \to \bigoplus_{s\in\{s',s''\}} \mathrm{H}^0\big(E_s, \mathscr{O}_{E_s}(-E_s)\big) \quad (28)$$

mapping $(\varsigma', \varsigma'', \varsigma)$ to $(\varsigma' - \varsigma, \varsigma'' - \varsigma)$. Then $\tilde{\mathfrak{L}}_{s's''}$ identifies as the blow–up of $\mathfrak{L}_{s's''}$ along the line defined by $\varsigma = 0$ in the above notation; in particular it is isomorphic to $\mathbf{P}\big(\mathscr{O}_{\mathbf{P}^1} \oplus \mathscr{O}_{\mathbf{P}^1}(1)^{\oplus 2}\big)$, with $\mathbf{P}^2$–bundle structure

$$\rho_{s's''} : \tilde{\mathfrak{L}}_{s's''} \to \left|H_0 - E_{s'} - E_{s''}\right|_{\tilde{S}_0}$$

induced by the projection of the left-hand side of (28) on its last summand, as follows from Lemma 3. The typical element of $\tilde{\mathfrak{L}}_{s's''}$ consists of

(i) A member C of $|H_0 - E_{s'} - E_{s''}|_{\tilde{S}_0}$, plus
(ii) Two curves in $|H|_{\tilde{Q}'_s}$ and $|H|_{\tilde{Q}'_s}$ respectively, matching C along $E_{s'}$ and $E_{s''}$, together with
(iii) Rational tails on the W_j's (two on those W_j meeting neither $\tilde{Q}_{s'}$ nor $\tilde{Q}_{s''}$, one on those W_j meeting exactly one component among $\tilde{Q}_{s'}$ and $\tilde{Q}_{s''}$, and none on the two W_j's meeting both $\tilde{Q}_{s'}$ and $\tilde{Q}_{s''}$) matching C along D_j.

The image by $\rho_{s's''}$ of such a curve is the point corresponding to its component (i).

Remark 14. The image of $\bar{X}$ via the complete linear system $|\mathscr{L}(-\tilde{Q}_{s'} - \tilde{Q}_{s''})|$ provides a model $f' : S' \to \Delta$ of the initial family $f : S \to \Delta$, with central fibre the transverse union of two double planes $\Pi_{s'}$ and $\Pi_{s''}$. For $s \in \{s', s''\}$, the plane Π_s is the projection of $\tilde{Q}_s$ from the point $p_{\bar{s}}$ corresponding to the direction of the line $\ell_{s,s'}$ in $|\mathscr{O}_{\tilde{Q}_s}(H)|^\vee \cong |\mathscr{L}_0(-\tilde{Q}s)|^\vee$, where $\{s, \bar{s}\} = \{s', s''\}$; there is a *marked conic* on Π_s, corresponding to the branch locus of this projection. The restriction to E_s of the morphism $\tilde{Q}_s \to \Pi_s$ is a degree 2 covering $E_s \to \Pi_{s'} \cap \Pi_{s''} =: L_{s's''}$. The two marked conics on $\Pi_{s'}$ and $\Pi_{s''}$ intersect at two points on the line $L_{s's''}$, which are the two branch points of both the double coverings $E_{s'} \to L_{s's''}$ and $E_{s'} \to L_{s's''}$. These points correspond to the two points cut out on $E_{s'}$ (resp. $E_{s''}$) by the two curves D_j that correspond to the two double conics of S_0 passing through $p_{s'}$ and $p_{s''}$. There are in addition six distinguished points on $L_{s's''}$, corresponding to the six pairs of points cut out on $E_{s'}$ (resp. $E_{s''}$) by the six curves $C_{s's''s'''}$ on $\tilde{S}_0$.

Step (III)

The last operation is the blow-up $\mathfrak{P}''' \to \mathfrak{P}''$ along the 16 disjoint surfaces that are the strict transforms of the 2-dimensional linear systems

$$\left|H_0 - E_s\right|_{\tilde{S}_0}, \quad 1 \leqslant s \leqslant 16.$$

We want to understand the exceptional divisor $\tilde{\mathfrak{L}}_s$. Consider the linear system $\mathfrak{L}_s := \left|\mathscr{L}_0(-\tilde{Q}_s)\right|$, which identifies with the projectivization of the kernel of the surjective map

$$\mathfrak{f}_s : \mathrm{H}^0(\tilde{Q}_s, \mathscr{O}_{\tilde{Q}_s}(H)) \oplus \mathrm{H}^0(\tilde{S}_0, \mathscr{O}_{\tilde{S}_0}(H_0 - E_s)) \to \mathrm{H}^0(E_s, \mathscr{O}_{E_s}(-E_s))$$
$$(\varsigma', \varsigma) \mapsto (\varsigma' - \varsigma)$$

(itself isomorphic to $\mathrm{H}^0(\tilde{Q}_s, \mathscr{O}_{\tilde{Q}_s}(H))$, by the way). Blow–up $\mathfrak{L}_s$ at the point ξ corresponding to $\varsigma = 0$; one thus gets a $\mathbf{P}^1$–bundle over the plane $|H_0 - E_s|_{\tilde{S}_0}$. Then $\tilde{\mathfrak{L}}_s$ is obtained by further blowing–up along the proper transforms of the lines joining ξ with the $6+\left[\binom{15}{2} - 6\binom{5}{2}\right] = 51$ points of $|H_0 - E_s|$ we blew–up in Step (I). The typical member of $\tilde{\mathfrak{L}}_s$ consists of two members of $|H_0 - E_s|_{\tilde{S}_0}$ and $|H|_{\tilde{Q}_s}$ respectively, matching along E_s, together with rational tails on the surfaces W_j.

Remark 15. The image of $\bar{X}$ by the complete linear system $|\mathscr{L}(-\tilde{Q}_s)|$ provides a model $f' : S' \to \Delta$ of the initial family $f : S \to \Delta$, with central fibre the transverse union of a smooth quadric Q, and a double plane Π branched along six lines tangent to the conic $\Gamma := \Pi \cap Q$ (i.e. the projection of S_0 from the node p_s). There are 15 *marked points* on Π, namely the intersection points of the six branch lines of the double covering $S_0 \to \Pi$.

Conclusion

We shall now describe the curves parametrized by the intersections of the various components of $\mathfrak{P}_0'''$, thus proving:

Proposition 23. *The central fibre $\mathfrak{P}_0'''$ is the limit linear system of $|\mathscr{L}_t| = |\mathscr{O}_{\bar{X}_t}|$ as $t \in \Delta^*$ tends to* 0.

Proof. We analyze step by step the effect on the central fibre of the birational modifications operated on $\mathfrak{P}$ in the above construction, each time using Lemma 3 without further notification.

(I) At this step, recall (cf. Sect. 2) that $\mathfrak{P}_0 \subset \mathfrak{P}'$ denotes the proper transform of $\mathfrak{P}_0 \subset \mathfrak{P}$ in the blow–up $\mathfrak{P}' \to \mathfrak{P}$. For each $\{s', s'', s'''\} \subset \{1, \ldots, 16\}$ such that $\langle p', p'', p''' \rangle$ is a plane that does not belong to the 16_6 configuration, the intersection

$\tilde{\mathfrak{L}}_{s's''s'''} \cap \mathfrak{P}_0 \subset \mathfrak{P}'$ is the exceptional $\mathbf{P}^2$ of the blow–up of $|\mathscr{L}_0| \cong |\mathscr{O}_{S_0}(1)|$ at the point corresponding to $H_{s's''s'''}$. Its points, but those lying on one of the three lines joining two points among $\pi_{s'}, \pi_{s''}, \pi_{s'''}$ which also have been blown–up (the notation is that of Step (Ia)), correspond to the trace of the pull–back of $|\mathscr{O}_{S_0}(1)|$ on $C_{s's''s'''} + E_{s'} + E_{s''} + E_{s'''}$.

For each $j \in \{1, \dots, 16\}$, the intersection $\tilde{\mathfrak{L}}^j \cap \mathfrak{P}_0 \subset \mathfrak{P}'$ is a plane, the points of which correspond to curves $2D_j + G_j^1 + \dots + G_j^6 + \Phi$ of $\bar{X}_0$, $\Phi \in |\mathscr{O}_{W_j}(2F)|$, except for those points on the six lines corresponding to the cases when Φ contains one of the six curves $G_j^1, \dots, G_j^6$.

(II) Let $\{s' \neq s''\} \subset \{1, \dots, 16\}$. The intersection $\tilde{\mathfrak{L}}_{s's''} \cap \mathfrak{P}_0 \subset \mathfrak{P}''$ is a $\mathbf{P}^1 \times \mathbf{P}^1$; the first factor is isomorphic to the proper transform of the line $|H_0 - E_{s'} - E_{s''}|_{\tilde{S}_0}$ in $\mathfrak{P}_0$, while the second is isomorphic to the line $\{\varsigma = 0\} \subset \mathfrak{L}_{s's''}$ in the notation of Step (II) above. Then the points in $\tilde{\mathfrak{L}}_{s's''} \cap \mathfrak{P}_0 \subset \mathfrak{P}''$ correspond to curves $C + E_{s'} + E_{s''}$ in $\bar{X}_0$, with $C \in |H_0 - E_{s'} - E_{s''}|_{\tilde{S}_0}$, exception made for the points with second coordinate $[\varsigma_{s'} : 0 : 0]$ or $[0 : \varsigma_{s''} : 0]$ in $\mathfrak{L}_{s's''}$, where $\varsigma_s \in \mathrm{H}^0(\mathscr{O}_{\tilde{Q}_s}(H))$ vanishes on E_s for each $s \in \{s', s''\}$.

Let $s''' \notin \{s', s''\}$ be such that $\langle p', p'', p''' \rangle$ is a plane outside the 16_6 configuration. The intersection $\tilde{\mathfrak{L}}_{s's''} \cap \tilde{\mathfrak{L}}_{s's''s'''} \subset \mathfrak{P}''$ is the $\mathbf{P}^2$ preimage of the point corresponding to $C_{s's''s'''}$ in $|H_0 - E_{s'} - E_{s''}|_{\tilde{S}_0}$ via $\rho_{s's''}$, and parametrizes curves $C_{s's''s'''} + E_{s'''} + C' + C''$ + rational tails, with $C' \in |H|_{\tilde{Q}_{s'}}$ and $C'' \in |H|_{\tilde{Q}_{s''}}$ matching $C_{s's''s'''}$ along $E_{s'}$ and $E_{s''}$ respectively.

On the other hand, for $s''' \notin \{s', s''\}$ such that $\langle p', p'', p''' \rangle$ belongs to the 16_6 configuration, let $j \in \{1, \dots, 16\}$ be such that $2D_j$ is cut out on S_0 by $\langle p', p'', p''' \rangle$, and set $\tilde{Q}_{s'} = \tilde{Q}_j^1$ and $\tilde{Q}_{s''} = \tilde{Q}_j^2$; then $\tilde{\mathfrak{L}}_{s's''} \cap \tilde{\mathfrak{L}}_{s's''s'''} \subset \mathfrak{P}''$ is the preimage by $\rho_{s's''}$ of the point corresponding to D_j in $|H_0 - E_{s'} - E_{s''}|_{\tilde{S}_0}$, and parametrizes the curves

$$2D_j + (G_j^1 + C') + (G_j^2 + C'') + \sum_{s=3}^{6} (G_j^s + E_j^s),$$

where $C' \in |H - G_j^1|_{\tilde{Q}_{s'}}$ is the proper transform by $\tilde{Q}_{s'} \to Q_{s'}$ of a member of $|H|_{Q_{s'}}$ tangent to $E_{s'}$ at $D_j \cap E_{s'}$, and similarly for C''.

(III) Let $s \in \{1, \dots, 16\}$. The intersection $\tilde{\mathfrak{L}}_s \cap \mathfrak{P}_0 \subset \mathfrak{P}'''$ is isomorphic to the plane $|H_0 - E_s|_{\tilde{S}_0}$ blown–up at the 51 points corresponding to the intersection of at least two lines among the 15 $|H_0 - E_s - E_{s'}|$, $s' \neq s$. Each point of the non–exceptional locus of this surface corresponds to a curve $C + E_s \subset \bar{X}_0$, with $C \in |H_0 - E_s|_{\tilde{S}_0}$.

Let $s' \in \{1, \dots, 16\} - \{s\}$. The intersection $\tilde{\mathfrak{L}}_s \cap \tilde{\mathfrak{L}}_{ss'} \subset \mathfrak{P}'''$ is an $\mathbf{F}_1$, isomorphic to the blow–up at ξ of the plane in $\mathfrak{L}_s$ projectivization of the kernel of the restriction of $\mathfrak{f}_s$ to $\mathrm{H}^0(\mathscr{O}_{\tilde{Q}_s}(H)) \oplus \mathrm{H}^0(\mathscr{O}_{\tilde{S}_0}(H_0 - E_s - E_{s'}))$. It has the structure of a $\mathbf{P}^1$–bundle over $|H_0 - E_s - E_{s'}|$, and its points correspond to curves $C + E_{s'} + C_s$ +rational tails, with $C_s \in |H|_{\tilde{Q}_s}$ matching with $C \in |H_0 - E_s - E_{s'}|$ along E_s; note that the points on the exceptional section correspond to the curves $C + E_{s'} + E_s$ + rational tails.

Let $s'' \in \{1, \dots, 16\} - \{s, s'\}$, and assume the plane $\langle p', p'', p''' \rangle$ is outside the 16_6 configuration. Then $\tilde{\mathfrak{L}}_s \cap \tilde{\mathfrak{L}}_{ss's''} \subset \mathfrak{P}'''$ is a $\mathbf{P}^1 \times \mathbf{P}^1$, the two factors of which are respectively isomorphic to the projectivization of the kernel of the restriction of $\mathfrak{f}_s$ to $\mathrm{H}^0(\mathcal{O}_{\tilde{Q}_s}(H)) \oplus \mathrm{H}^0(\mathcal{O}_{\tilde{S}_0}(H_0 - E_s - E_{s'} - E_{s''}))$, and to the line $\langle \pi_{s'}, \pi_{s''} \rangle$ in $\mathfrak{L}_{ss's''}$ (with the notations of Step (Ib)). It therefore parametrizes the curves

$$C_{ss's''} + E_{s'} + E_{s''} + C + \text{rational tails},$$

where $C \in |H|_{\tilde{Q}_s}$ matches $C_{ss's''}$ along E_s.

Let $j \in \{1, \dots, 16\}$ be such that W_j intersects $\tilde{Q}_s$, and set $\tilde{Q}_j^1 = \tilde{Q}_s$. Then $\tilde{\mathfrak{L}}_s \cap \tilde{\mathfrak{L}}^j \subset \mathfrak{P}'''$ is a $\mathbf{P}^1 \times \mathbf{P}^1$, the two factors of which are respectively isomorphic to the pencil of pull–backs to $\tilde{Q}_s$ of members of $|H|_{Q_s}$ tangent to E_s at the point $D_j \cap E_s$, and to the subpencil $2D_j + 2G_j^1 + G_j^2 + \cdots + G_j^6 + |F|_{W_j}$ of $\mathfrak{L}^j$. It parametrizes curves

$$2D_j + (G_j^1 + C) + \sum\nolimits_{s=2}^{6} (G_j^s + E_j^s),$$

where $C \in |H - G_j^1|_{\tilde{Q}_s}$ is the proper transform of a curve on Q_s tangent to E_s at $D_j \cap E_s$.

It follows from the above analysis that the points of $\mathfrak{P}_0'''$ all correspond in a canonical way to curves on $\bar{X}_0$, which implies our assertion by Lemma 2. □

8.3 The Linear System $\mathfrak{L}^j$

In this section, we study the rational map φ_j (or simply φ) determined by the linear system $\mathfrak{L}^j = \left|\mathcal{O}_{W_j}(2H) \otimes \mathcal{I}_{Z_j}\right|$ on W_j, for $1 \leqslant j \leqslant 16$.

Let $u_j : \bar{W}_j \to W_j$ be the blow–up at the 12 points in the support of Z_j. For $1 \leqslant s \leqslant 6$, we denote by $\hat{G}_j^s$ the strict transform of the ruling G_j^s, and by $I_j^{s'}, I_j^{s''}$ the two exceptional curves of u_j meeting $\hat{G}_j^s$. Then the pull–back via u_j induces a natural isomorphism

$$\left|\mathcal{O}_{W_j}(2H) \otimes \mathcal{I}_{Z_j}\right| \cong \left|\mathcal{O}_{\bar{W}_j}(2H - \sum_{s=1}^{6} (I_j^{s'} + I_j^{s''}))\right|;$$

we denote by $\bar{\mathfrak{L}}^j$ the right hand side linear system.

Proposition 24. *The linear system* $\bar{\mathfrak{L}}^j$ *determines a* $2:1$ *morphism*

$$\bar{\varphi} : \bar{W}_j \to \Sigma \subset \mathbf{P}^3,$$

where Σ is a quadric cone. The divisor $\tilde{D}_j := D_j + \hat{F}_j^1 + \cdots + \hat{F}_j^6$ is contracted by $\bar{\varphi}$ to the vertex of Σ. The branch curve B of $\bar{\varphi}$ is irreducible, cut out on Σ by a quartic surface; it is rational, with an ordinary six–fold point at the vertex of Σ.

Before the proof, let us point out the following corollary, which we will later need.

Corollary 10. *The Severi variety of irreducible δ–nodal curves in $|\mathcal{O}_{W_j}(2H) \otimes \mathcal{I}_{Z_j}|$ is isomorphic to the subvariety of $\check{\mathbf{P}}^3$ parametrizing δ–tangent planes to B, for $\delta = 1, \ldots, 3$. They have degree 14, 60, and 80, respectively.*

For the proof of Proposition 24 we need two preliminary lemmas.

Lemma 13. *The linear system $|\mathcal{O}_{W_j}(2H) \otimes \mathcal{I}_{Z_j}| \subset |2H_{W_j}|$ has dimension 3.*

Proof. The 0-cycle Z_j is cut out on $G_j^1 + \cdots + G_j^6$ by a general curve in $|2H|$. Let then

$$\sigma \in \bigoplus_{s=1}^{6} \mathrm{H}^0(G_j^s, \mathcal{O}_{G_j^s}(2H)) \cong \mathrm{H}^0(\mathbf{P}^1, \mathcal{O}_{\mathbf{P}^1}(2))^{\oplus 6}$$

be a non–zero section vanishing at Z_j. Then $\mathrm{H}^0(W_j, \mathcal{O}_{W_j}(2H) \otimes \mathcal{I}_{Z_j}) \cong r^{-1}(\langle\sigma\rangle)$ where

$$r : \mathrm{H}^0(W_j, \mathcal{O}_{W_j}(2H)) \to \bigoplus_{s=1}^{6} \mathrm{H}^0(G_j^s, \mathcal{O}_{G_j^s}(2H)) \cong \mathrm{H}^0(\mathbf{P}^1, \mathcal{O}_{\mathbf{P}^1}(2))^{\oplus 6}$$

is the restriction map. The assertion now follows from the restriction exact sequence, since

$$\mathrm{h}^0(W_j, \mathcal{O}_{W_j}(2H) \otimes \mathcal{I}_{Z_j}) = 1 + \mathrm{h}^0(W_j, \mathcal{O}_{W_j}(2H - 6F)) = 4.$$

□

Lemma 14. *The rational map φ_j has degree 2 onto its image, and its restriction to any line of the ruling $|F_{W_j}|$ but the six G_j^s, $1 \leqslant s \leqslant 6$, has degree 2 as well.*

Proof. Let $x \in W_j$ be a general point and let F_x be the line of the ruling containing x. One can find a divisor $D \in |\mathcal{O}_{W_j}(2H) \otimes \mathcal{I}_{Z_j}|$ containing x but not containing F_x. Let $x + x'$ be the length two scheme cut out by D on F_x. By an argument similar to the one in the proof of Lemma 13, one has $\dim(|\mathcal{O}_{W_j}(2H) \otimes \mathcal{I}_{Z_j} \otimes \mathcal{I}_{x+x'}|) = 2$. This shows that x and x' are mapped to the same point by φ. Then, considering the sublinear system

$$2D_j + G_j^1 + \cdots + G_j^6 + F_x + \Phi, \quad \Phi \in |\mathcal{O}_{W_j}(F)|,$$

of $\mathfrak{L}^j$, with fixed divisor $2D_j + G_j^1 + \cdots + G_j^6 + F_x$, the assertion follows from the base point freeness of $|\mathcal{O}_{W_j}(F)|$. □

Proof (of Proposition 24). First we prove that $\mathfrak{L}^j$ has no fixed components, hence that the same holds for $\bar{\mathfrak{L}}^j$. Suppose Φ is such a fixed component. By Lemma 14, $\Phi \cdot F = 0$, hence Φ should consist of curves contained in rulings. The argument of the proof of Lemma 13 shows that no such a curve may occur in Φ, a contradiction.

Let $D \in \bar{\mathfrak{L}}^j$ be a general element. By Lemmas 13 and 14, D is irreducible and hyperelliptic, since $D \cdot F = 2$. Moreover $D^2 = 4$ and $p_a(D) = 3$. This implies that D is smooth and that $\bar{\mathfrak{L}}^j$ is base point free. Moreover the image Σ of φ has degree 2. Since $D \cdot \tilde{D}_j = 0$ and $\tilde{D}_j^2 = -4$, the connected divisor $\tilde{D}_j$ is contracted to a double point v of Σ, which is therefore a cone.

Since D is mapped $2:1$ to a general plane section of Σ, which is a conic, we see that $\deg(B) = 8$. Let $\Phi \in |F|_{W_j}$ be general, and ℓ its image via φ, which is a ruling of Σ. The restriction $\varphi|_\Phi : \Phi \to \ell$ is a degree 2 morphism, which is ramified at the intersection point of Φ with D_j. This implies that ℓ meets B at one single point off the vertex v of Σ. Hence B has a unique irreducible component B_0 which meets the general ruling ℓ in one point off v. We claim that $B = B_0$. If not, $B - B_0$ consists of rulings $\ell_1, \ldots, \ell_n$, corresponding to rulings $F_1, \ldots, F_n$, clearly all different from the G_j^s, with $1 \leqslant s \leqslant 6$. Then the restrictions $\varphi|_{F_i} : F_i \to \ell_i$ would be isomorphisms, for $1 \leqslant i \leqslant n$, which is clearly impossible. Hence B is irreducible, rational, sits in $|\mathcal{O}_\Sigma(4)|$. Finally, taking a plane section of Σ consisting of two general rulings, we see that it has only two intersection points with B off v. Hence B has a point of multipilicity 6 at v and the assertion follows. □

Remark 16. Each of the curves $\hat{G}_j^s + I_j^{s\prime} + I_j^{s\prime\prime} \in |F|_{\bar{W}_j}$, for $1 \leqslant s \leqslant 6$, is mapped by $\bar{\varphi}$ to a ruling ℓ_s of Σ, and this ruling has no intersection point with B off v. This implies that v is an ordinary 6–tuple point for B and that the tangent cone to B at v consists of the rulings $\ell_1, \ldots, \ell_6$ of Σ.

Remark 17. Let $S' \to \Delta$ be the image of $\bar{X} \to \Delta$ via the map defined by the linear system $|\mathcal{L}(-2W_j - \sum_s \tilde{Q}_j^s)|$. One has $S'_t \cong S_{t^2}$ for $t \neq 0$, and the new central fibre S'_0 is a double quadratic cone Σ in $\mathbf{P}^3$.

8.4 The Limit Severi Varieties

In this section we describe the regular components of the limit Severi varieties $\mathfrak{V}_{1,\delta}(\bar{X})$ for $1 \leqslant \delta \leqslant 3$. The discussion here parallels the one in Sect. 5.8, therefore we will be sketchy, leaving to the reader most of the straightforward verifications, based on the description of the limit linear system in Sect. 8.2.

Proposition 25 (Limits of 1-nodal curves). *The regular components of the limit Severi variety $\mathfrak{V}_{1,1}(\bar{X})$ are the following (they all appear with multiplicity* 1, *but the ones in* (iii) *which appear with multiplicity* 2*):*

(i) $V(\delta_{\tilde{S}_0} = 1)$, *which is isomorphic to the Kummer quartic surface* $\check{S}_0 \subset |\mathcal{O}_{S_0}(1)| \cong \check{\mathbf{P}}^3$;
(ii) $V(\tilde{Q}_s, \delta_{\tilde{Q}_s} = 1)$, *which is isomorphic to the smooth quadric* $\check{Q}_s \subset |\mathcal{O}_{Q_s}(1)| \cong \check{\mathbf{P}}^3$, *for* $1 \leqslant s \leqslant 16$;
(iii) $V(\tilde{Q}_s, \tau_{E_s,2} = 1)$, *which is isomorphic to a quadric cone in* $|\mathcal{O}_{Q_s}(1)|$, *for* $1 \leqslant s \leqslant 16$;
(iv) $V(\tilde{Q}_{s'} + \tilde{Q}_{s''}, \delta_{\tilde{Q}_{s'}} = 1)$, *which is isomorphic to* $\check{Q}_s \subset |\mathcal{O}_{Q_s}(1)| \cong \check{\mathbf{P}}^3$, *for* $1 \leqslant s' < s'' \leqslant 16$;
(v) $V(\tilde{Q}_{s'} + \tilde{Q}_{s''} + \tilde{Q}_{s'''}, \delta_{\tilde{Q}_{s'}} = 1)$, *for* $1 \leqslant s', s'', s''' \leqslant 16$ *such that* $\tilde{Q}_{s'}, \tilde{Q}_{s''}, \tilde{Q}_{s'''}$ *are pairwise distinct and do not meet a common* W_j: *it is again isomorphic to* $\check{Q}_s \subset |\mathcal{O}_{Q_s}(1)| \cong \check{\mathbf{P}}^3$;
(vi) $V(2W_j + \tilde{Q}_j^1 + \cdots + \tilde{Q}_j^6, \delta_{W_j} = 1)$, *which is isomorphic to the degree* 14 *surface* $\check{B} \subset |\mathcal{O}_B(1)| \cong \check{\mathbf{P}}^3$, *for* $1 \leqslant j \leqslant 16$.

Corollary 11 (Theorem 2 for $\delta = 1$). *The family* $f : S \to \Delta$ *of general quartic surfaces degenerating to a Kummer surface* S_0 *we started with, with smooth total space* S, *and endowed with the line bundle* $\mathcal{O}_S(1)$, *is* 1*–well behaved, with good model* $\varpi : \bar{X} \to \Delta$. *The limit in* $|\mathcal{O}_{S_0}(1)|$ *of the dual surfaces* $\check{S}_t$, $t \in \Delta^*$, *consists in the union of the dual* $\check{S}_0$ *of* S_0 *(which is again a Kummer surface), plus the* 16 *planes of the* 16_6 *configuration of* $\check{S}_0$, *each counted with multiplicity* 2.

Proof. The push–forward by the morphism $\mathfrak{P}_0''' \to \mathfrak{P}_0 \cong |\mathcal{O}_{S_0}(1)|$ of the regular components of $\mathfrak{V}_{1,1}$ with their respective multiplicities in $\mathfrak{V}_{1,1}^{\mathrm{reg}}$ is $\check{S}_0$ in case (i), $2 \cdot \check{p}_s$ in case (ii), and 0 otherwise. The push-forward of $\mathfrak{V}_{1,1}^{\mathrm{reg}}(\bar{X})$ has thus total degree 36, and is therefore the crude limit Severi variety $\mathfrak{V}_{1,1}^{\mathrm{cr}}(S)$ by Proposition 4. □

Remark 18. **(a)** Similar arguments show that $\varpi : \bar{X} \to \Delta$ is a 1–good model for the degenerations of general quartic surfaces obtained from $\bar{X} \to \Delta$ via the line bundles $\mathcal{L}(-2W_j - \tilde{Q}_j^1 - \cdots - \tilde{Q}_j^6)$ and $\mathcal{L}(-\tilde{Q}_s)$ respectively (see Remarks 17 and 15 for a description of these degenerations).

To see this in the former case, let us consider two general points on a given W_j, and enumerate the regular members of $\mathfrak{V}_{1,1}$ that contain them. There are 2 curves in (i) (indeed, the two points on W_j project to two general points on $D_j \cong \Gamma_j \subset S_0 \subset \mathbf{P}^3$, which span a line $\ell \subset \check{\mathbf{P}}^3$; the limiting curves in S_0 passing through the two original points on W_j correspond to the intersection points of $\check{\ell}$ with $\check{S}_0$; now $\check{\ell}$ meets $\check{S}_0$ with multiplicity 2 at the double point which is the image of Γ_j via the Gauss map, and only the two remaining intersection points are relevant). There are in addition 2 limiting curves in each of the 10 components of type (ii) corresponding to the $\tilde{Q}_s$'s that do not meet W_j, and 14 in the relevant component of type (vi).

In this case, the crude limit Severi variety therefore consists, in the notation of Remark 17, of the degree 14 surface $\check{B}$, plus the plane $\check{v}$ with multiplicity 22 (this has degree 36 as required).

For the degeneration given by $\mathcal{L}(-\tilde{Q}_s)$, the crude limit Severi variety consists, in the notation of Remark 15, of the dual to the smooth quadric Q, plus the dual to the conic Γ with multiplicity 2, plus the 15 planes $\check{p}$ with multiplicity 2, where p ranges among the 15 marked points on the double plane Π.

(b) One can see that $\varpi : \bar{X} \to \Delta$ is not a 1–good model for the degeneration to a union of two double planes obtained via the line bundle $\mathscr{L}(-\tilde{Q}_{s'} - \tilde{Q}_{s''})$ described in Remark 14. In addition (see Step (Ia)) the line bundles $\mathscr{L}(-\tilde{Q}_{s'} - \tilde{Q}_{s''} - \tilde{Q}_{s'''})$, though corresponding to 3–dimensional components of the limit linear system, do not provide suitable degenerations of surfaces. Despite all this, it seems plausible that one can obtain a good model by making further modifications of $\bar{X} \to \Delta$. The first thing to do would be to blow–up the curves $C_{s's''s'''}$.

Proposition 26 (Limits of 2-nodal curves). *The regular components of the limit Severi variety $\mathfrak{V}_{1,2}(\bar{X})$ are the following (they all appear with multiplicity* 1*, except the ones in* (ii) *appearing with multiplicity* 2*):*

(i) $V\big(\tilde{Q}_{s'} + \tilde{Q}_{s''}, \delta_{\tilde{Q}_{s'}} = \delta_{\tilde{Q}_{s''}} = 1\big)$ *for* $s' \neq s''$*, proper transform of the intersection of two smooth quadrics in* $\mathfrak{L}_{s's''}$*;*
(ii) $V\big(\tilde{Q}_{s'} + \tilde{Q}_{s''}, \delta_{\tilde{Q}_{s'}} = 1, \tau_{E_{s''},2} = 1\big)$ *for* $s' \neq s''$*, proper transform of the intersection of a smooth quadric and a quadric cone in* $\mathfrak{L}_{s's''}$*;*
(iii) $V\big(\tilde{Q}_{s'} + \tilde{Q}_{s''} + \tilde{Q}_{s'''}, \delta_{\tilde{Q}_{s'}} = \delta_{\tilde{Q}_{s''}} = 1\big)$ *for* $1 \leqslant s', s'', s''' \leqslant 16$ *such that* $\tilde{Q}_{s'}, \tilde{Q}_{s''}, \tilde{Q}_{s'''}$ *are pairwise distinct and do not meet a common* W_j*, proper transform of the intersection of two smooth quadrics in* $\mathfrak{L}_{s's''s'''}$*;*
(iv) $V\big(2W_j + \tilde{Q}_j^1 + \cdots + \tilde{Q}_j^6, \delta_{W_j} = 2\big)$ *for each* $j \in \{1, \ldots, 16\}$*, proper transform of a degree* 60 *curve in* $\mathfrak{L}^j$*.*

Proof. Again, one checks that the components listed in the above statement are the only ones provided by Proposition 1, taking the following points into account:

(a) The condition $\delta_{\tilde{S}_0} = 2$ is impossible to fulfil, because there is no plane of $\mathbf{P}^3$ tangent to S_0 at exactly two points (see Proposition 20);
(b) The condition $\delta_{\tilde{S}_0} = \delta_{\tilde{Q}_i} = 1$ is also impossible to fulfil, because there is no plane in $\mathbf{P}^3$ tangent to S_0 at exactly one point, and passing through one of its double points. Indeed, let p_i be a double point of S_0, the dual plane $\check{p}_i$ is everywhere tangent to $\check{S}_0$ along the contact conic Gauss image of E_i;
(c) The condition $\delta_{\tilde{Q}_s} = \tau_{E_s,2} = 1$ imposes to a member of $|H|_{\tilde{Q}_s}$ to be the sum of two rulings intersecting at a point on E_s, and such a curve does not belong to the limit Severi variety:
(d) The condition $\tau_{E_{s'},2} = \tau_{E_{s''},2} = 1$ imposes to contain one of the two curves D_j intersecting both $E_{s'}$ and $E_{s''}$, which violates condition (i) of Definition 1.

□

Remark 19. As in Remark 13, we can enumerate the 480 limits of 2–nodal curves passing through a general point in certain irreducible components of $\bar{X}_0$:

(a) For a general point on $\tilde{S}_0$, we find 4 limit curves in each of the $\binom{16}{2} = 120$ components in (i) of Proposition 26;

(b) For a general point on a given W_j, we find 60 limit curves in the appropriate component in (iv), and 4 in each of the $\binom{16}{2} - \binom{6}{2} = 105$ different components of type (i) such that $\tilde{Q}_{s'}$ and $\tilde{Q}_{s''}$ do not both meet W_j.

This shows that $\bar{X} \to \Delta$ is a 2–good model for the degenerations of quartics corresponding to the line bundles $\mathscr{L}$ and $\mathscr{L}(-2W_j - \tilde{Q}_j^1 - \cdots - \tilde{Q}_j^6)$. In particular, it implies Corollary 12 below.

Corollary 12 (Theorem 2 for $\delta = 2$). *Same setting as in Corollary 11. The crude limit Severi variety $\mathfrak{V}_{1,2}^{\mathrm{cr}}(S)$ consists of the images in $|\mathscr{O}_{S_0}(1)|$ of the* 120 *irreducible curves listed in case* (a) *of Remark 19. Each of them projects* 4 : 1 *onto a pencil of planes containing two double points of S_0.*

Proposition 27 (Limits of 3-nodal curves). *The family $\bar{X} \to \Delta$ is absolutely 3–good, and the limit Severi variety $\mathfrak{V}_{1,3}$ is reduced, consisting of:*

(i) *Eight distinct points in each $V\big(-\tilde{Q}_{s'} - \tilde{Q}_{s''} - \tilde{Q}_{s'''}, \delta_{\tilde{Q}_{s'}} = \delta_{\tilde{Q}_{s''}} = \delta_{\tilde{Q}_{s'''}} = 1\big)$, where $1 \leqslant s', s'', s''' \leqslant 16$ are such that $\langle p_{s'}, p_{s''}, p_{s'''} \rangle$ is a plane that does not belong to the 16_6 configuration of S_0;*
(ii) *The* 80 *distinct points in each $V\big(2W_j + \tilde{Q}_j^1 + \cdots + \tilde{Q}_j^6, \delta_{W_j} = 3\big)$, corresponding to the triple points of the double curve of $\check{B} \subset |\mathscr{O}_B(1)| \cong \check{\mathbf{P}}^3$ that are also triple points of $\check{B}$.*

Proof. There are 240 unordered triples $\{s', s'', s'''\}$ such that the corresponding double points of S_0 do not lie on a common D_j, so $\mathfrak{V}_{1,3}^{\mathrm{reg}}$ has degree 3,200, which fits with Proposition 4. □

Corollary 13 (Theorem 2 for $\delta = 3$). *Same setting as in Corollary 11. The crude limit Severi variety $\mathfrak{V}_{1,3}^{\mathrm{cr}}(S) \subset |\mathscr{O}_{S_0}(1)|$ consists of:*

(i) *The* 240 *points corresponding to a plane through three nodes of S_0, but not member of the 16_6 configuration, each counted with multiplicity* 8*;*
(ii) *The* 16 *points corresponding to a member of the 16_6 configuration, each counted with multiplicity* 80.

9 Plane Quartics Curves Through Points in Special Position

In this section we prove the key result needed for the proof of Theorem 3, itself given in Sect. 5.8. We believe this result, independently predicted with tropical methods by E. Brugallé and G. Mikhalkin (private communication), is interesting on its own. Its proof shows once again the usefulness of constructing (relative) good models.

The general framework is the same as that of Sects. 5 and 8, and we are going to be sketchy here.

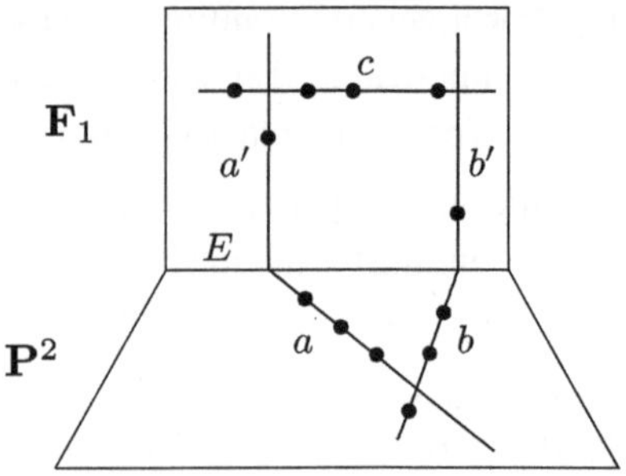

Fig. 10 Degeneration of base points on a *triangle*

9.1 The Degeneration and Its Good Model

We start with the trivial family $f : S := \mathbf{P}^2 \times \Delta \to \Delta$, together with flatly varying data for $t \in \Delta$ of three independent lines a_t, b_t, c_t lying in S_t, and of a 0-dimensional scheme Z_t of degree 12 cut out on $a_t + b_t + c_t$ by a quartic curve Γ_t in S_t, which is general for $t \in \Delta^*$. We denote by $\mathscr{O}_S(1)$ the pull–back line bundle of $\mathscr{O}_{\mathbf{P}^2}(1)$ via the projection $S \to \mathbf{P}^2$.

We blow–up S along the line c_0. This produces a new family $Y \to \Delta$, the central fibre Y_0 of which is the transverse union of a plane P (the proper transform of S_0, which we may identify with P) and of an $\mathbf{F}_1$ surface W (the exceptional divisor). The curve $E := P \cap W$ is the line c_0 in P, and the (-1)–section in W. The limit on Y_0 of the three lines a_t, b_t, c_t on the fibre $Y_t \cong \mathbf{P}^2$, for $t \in \Delta^*$, consists of:

(i) Two general lines a, b in P plus the curves $a', b' \in |F|_W$ matching them on E;
(ii) A curve $c \in |H|_W = |F + E|_W$ on W.

We denote by $\mathscr{O}_Y(1)$ the pull–back of $\mathscr{O}_S(1)$ and we set $\mathscr{L}^\natural = \mathscr{O}_Y(4) \otimes \mathscr{O}_Y(-W)$. One has $\mathscr{L}_t^\natural \cong \mathscr{O}_{\mathbf{P}^2}(4)$ for $t \in \Delta^*$, whereas $\mathscr{L}_0^\natural$ restrict to $\mathscr{O}_P(3H)$ and $\mathscr{O}_W(4F + E) \cong \mathscr{O}_W(4H - 3E)$ respectively. We may assume that the quartic curve $\Gamma_t \in |\mathscr{L}_t^\natural|$ cutting Z_t on $a_t + b_t + c_t$ for $t \in \Delta^*$ tends, for $t \to 0$, to a general curve $\Gamma_0 \subset |\mathscr{L}_0^\natural|$. Then $\Gamma_0 = \Gamma_P + \Gamma_W$, where Γ_P is a general cubic in P and $\Gamma_Q \in |4H - 3E|_W$, with Γ_P and Γ_W matching along E. Accordingly $Z_0 = Z_P + Z_W$, where Z_P has length 6 consisting of 3 points on a and 3 on b, and Z_W consists of 1 point on both a' and b', and 4 points on c (see Fig. 10).

Next we consider the blow–up $\varepsilon : X \to Y$ along the curve Z in Y described by Z_t, for $t \in \Delta$, and thus obtain a new family $\pi : X \to \Delta$, where each X_t is the blow–up of Y_t along Z_t. We call E_Z the exceptional divisor of ε. The fibre of $\varepsilon|_{E_Z} : E_Z \to \Delta$ at $t \in \Delta$ consists of the 12 (-1)–curves of the blow–up of Y_t at Z_t. The central fibre X_0 is the transverse union of $\tilde{P}$ and $\tilde{W}$, respectively the blow–ups of P and W along Z_P and Z_W; we denote by E_P and E_W the corresponding exceptional divisors.

We let $\mathscr{L} := \varepsilon^* \mathscr{L}^\natural \otimes \mathscr{O}_X(-E_Z)$. Recall from Sect. 5.4 that the fibre of $\mathbf{P}(\pi_*(\mathscr{L}))$ over $t \in \Delta^*$ has dimension 3. We will see that $X \to \Delta$, endowed with $\mathscr{L}$, is well behaved and we will describe the crude limit Severi variety $\mathfrak{V}_\delta^{\mathrm{cr}}$ for $1 \leqslant \delta \leqslant 3$. This analysis will prove Theorem 3.

Remark 20. We shall need a detailed description of the linear system $|\mathscr{L}_0|$. The vector space $\mathrm{H}^0(X_0, \mathscr{L}_0)$ is the subspace of $\mathrm{H}^0(\tilde{W}, \mathscr{O}_{\tilde{W}}(4H - 3E - E_W)) \times \mathrm{H}^0(\tilde{P}, \mathscr{O}_{\tilde{P}}(3H - E_P))$ which is the fibred product corresponding to the Cartesian diagram

$$\begin{array}{ccc} \mathrm{H}^0(X_0,\mathscr{L}_0) & \longrightarrow & \mathrm{H}^0(\tilde{P},\mathscr{O}_{\tilde{P}}(3H - E_P)) \\ \downarrow & & \downarrow{\scriptstyle r_P} \\ \mathrm{H}^0(\tilde{W}, \mathscr{O}_{\tilde{W}}(4H - 3E - E_W)) & \xrightarrow[r_W]{} & \mathrm{H}^0(E,\mathscr{L}|_E) \cong \mathrm{H}^0(\mathbf{P}^1,\mathscr{O}_{\mathbf{P}^1}(3)) \end{array} \qquad (29)$$

where r_P, r_W are restriction maps. The map r_W is injective, whereas r_P has a 1-dimensional kernel generated by a section s vanishing on the proper transforms of $a + b + c$. Since $\mathrm{h}^0(X_0, \mathscr{L}_0) \geqslant 4$ by semicontinuity, one has $\mathrm{Im}(r_P) = \mathrm{Im}(r_W)$, and therefore $\mathrm{H}^0(X_0, \mathscr{L}_0) \cong \mathrm{H}^0(\tilde{P}, \mathscr{O}_{\tilde{P}}(3H - E_P))$ has also dimension 4. Geometrically, for a general curve $C_P \in |3H - E_P|$, there is a unique curve $C_W \in |4H-3E-E_W|$ matching it along E and $C_P+C_W \in |\mathscr{L}_0|$. On the other hand $(0, s) \in \mathrm{H}^0(X_0, \mathscr{L}_0)$ is the only non–trivial section (up to a constant) identically vanishing on a component of the central fibre (namely $\tilde{W}$), and $\mathrm{H}^0(X_0, \mathscr{L}_0)/(s) \cong \mathrm{H}^0(\tilde{W}, \mathscr{O}_{\tilde{W}}(4H-3E-E_W))$. Therefore, if we denote by D the point corresponding to $(0, s)$ in $|\mathscr{L}_0|$, a line through D parametrizes the pencil consisting of a fixed divisor in $|4H - 3E - E_W|$ on $\tilde{W}$ plus all divisors in $|3H - E_P|$ matching it on E.

We will denote by $\mathfrak{R}$ the g^2_3 on E given by $|\mathrm{Im}(r_P)| = |\mathrm{Im}(r_W)|$.

To get a good model, we first blow–up the proper transform of a in $\tilde{P}$, and then we blow–up the proper transform of b on the strict transform of $\tilde{P}$. We thus obtain a new family $\varpi : \bar{X} \to \Delta$. The general fibre $\bar{X}_t$, $t \in \Delta^*$, is isomorphic to X_t. The central fibre $\bar{X}_0$ has four components (see Fig. 11):

(i) The proper transform of $\tilde{P}$, which is isomorphic to $\tilde{P}$;
(ii) The proper transform $\bar{W}$ of $\tilde{W}$, which is isomorphic to the blow–up of $\tilde{W}$ at the two points $a \cap E, b \cap E$, with exceptional divisors E_a and E_b;
(iii) The exceptional divisor W_b of the last blow–up, which is isomorphic to $\mathbf{F}_0$;
(iv) The proper transform W_a of the exceptional divisor over a, which is the blow–up of an $\mathbf{F}_0$–surface, at the point corresponding to $a \cap b$ (which is a general point of $\mathbf{F}_0$) with exceptional divisor E_{ab}.

As usual, we go on calling $\mathscr{L}$ the pull–back to $\bar{X}$ of the line bundle $\mathscr{L}$ on X.

9.2 The Limit Linear System

We shall now describe the limit linear system $\mathfrak{L}$ associated to $\mathscr{L}$. As usual, we start with $\mathfrak{P} := \mathbf{P}(\varpi_*(\mathscr{L}))$, and we consider the blow–up $\mathfrak{P}' \to \mathfrak{P}$ at the point $D \in \mathfrak{P}_0 \cong |\mathscr{L}_0|$. The central fibre of $\mathfrak{P}' \to \Delta$ is, as we will see, the limit linear

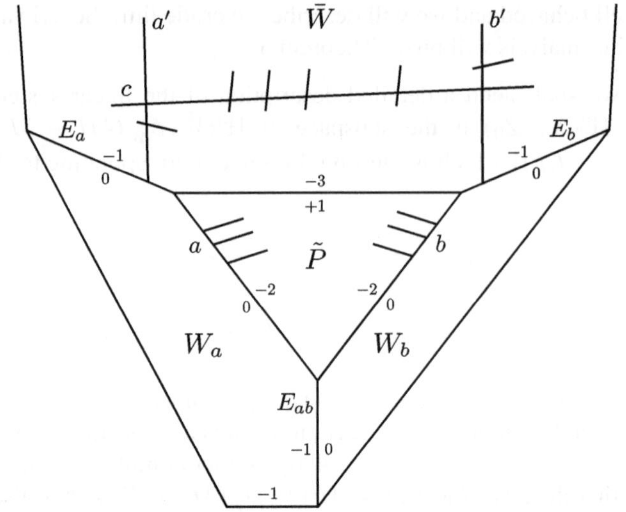

Fig. 11 Good model for plane quartics through 12 points

system $\mathfrak{L}$. It consists of only two components: the proper transform $\mathfrak{L}_1$ of $|\mathscr{L}_0|$ and the exceptional divisor $\mathfrak{L}_2 \cong \mathbf{P}^3$. Let us describe these two components in terms of twisted linear systems on the central fibre.

Since the map r_W in (29) is injective, it is clear that $\mathfrak{L}_2 \cong |\mathscr{L}_0(-\bar{W} - W_a - W_b)|$. The line bundle $\mathscr{L}_0(-\bar{W} - W_a - W_b)$ is trivial on $\tilde{P}$ and restricts to $\mathscr{O}_{\bar{W}}(4H - 2E - 3E_a - 3E_b - E_W)$, $\mathscr{O}_{W_a}(H - E_{ab})$, $\mathscr{O}_{W_b}(H)$ on $\bar{W}$, W_a, W_b respectively. Once chosen $C_W \in \left|\mathscr{O}_{\bar{W}}(4H - 2E - 3E_a - 3E_b - E_W)\right|$, there is only one possible choice of two curves C_a and C_b in $\left|\mathscr{O}_{W_a}(H - E_{ab})\right|$ and $\left|\mathscr{O}_{W_b}(H)\right|$ respectively, that match with C_W along E_a, E_b respectively. They automatically match along E_{ab}.

In conclusion, by mapping $\bar{W}$ to $\mathbf{P}^2$ (via $|H|_{\bar{W}}|$), we have:

Proposition 28. *The component $\mathfrak{L}_2 \cong |\mathscr{L}_0(-\bar{W} - W_a - W_b)|$ of $\mathfrak{P}'_0$ is isomorphic to a 3 dimensional linear system of plane quartics with an imposed double point x, prescribed tangent lines t_1, t_2 at x, and six further base points, two of which general on t_1, t_2 and the remaining four on a general line.*

To identify $\mathfrak{L}_1$ as the blow–up of $|\mathscr{L}_0|$ at D, we take into account Lemma 3, which tells us that the exceptional divisor $\mathfrak{E} \subset \mathfrak{L}_1$ identifies with $\mathfrak{R}$. Since $\mathfrak{E} = \mathfrak{L}_1 \cap \mathfrak{L}_2$, the linear system $\mathfrak{E}$ identifies with a sublinear system of codimension 1 in $\mathfrak{L}_2$, namely that of curves

$$a + b + E + C, \quad C \in \left|4H - 3E - 3E_a - 3E_b - E_W\right|_W$$

(in the setting of Proposition 28, C corresponds to a quartic plane curve with a triple point at x passing through the six simple base points).

It follows from this analysis that $\mathfrak{L}$ is the limit as $t \to 0$ of the linear systems $|\mathscr{L}_t|$, $t \in \Delta^*$, in the sense of Sect. 3.2.

9.3 The Limit Severi Varieties

We will use the notion of **n**–degree introduced in Definition 4. However we will restrict our attention to the case in which we fix 1 or 2 points only on $\tilde{P}$. Hence, if we agree to set $\tilde{P} = Q_1$, then we call P–degree of a component V of $\mathfrak{V}_\delta$ its **n**–degree with $\mathbf{n} = (3-\delta, 0, 0, 0)$; we denote it by $\deg_P(V)$.

Proposition 29 (Limits of 1-nodal curves). *The regular components of the limit Severi variety $\mathfrak{V}_1(\bar{X}, \mathscr{L})$ are the following, all appearing with multiplicity* 1, *except* (iii), *which has multiplicity* 2*:*

(i) $V(\delta_{\tilde{P}} = 1)$, *with P–degree* 9*;*
(ii) $V(\delta_{\bar{W}} = 1)$, *with P–degree* 4*;*
(iii) $V(\tau_{E,2} = 1)$, *with P–degree* 4*;*
(iv) $V(\bar{W} + W_a + W_b, \delta_{\bar{W}} = 1)$, *with P–degree* 0.

Proof. The list is an application of Proposition 1. The only things to prove are the degree assertions. Since $\mathfrak{L}_2$ is trivial on $\tilde{P}$, case (iv) is trivial. Case (i) follows from Proposition 5, because the P–degree of $V(\delta_{\tilde{P}} = 1)$ is the degree 9 of the dual surface of the image X_P of $\tilde{P}$ via the linear system $|3H - E_P|$, which is a cubic surface with an A_2 double point (see Proposition 5).

As for (ii), note that nodal curves in $|4H - 3E - E_W|$ on $\bar{W}$ consist of a ruling in $|F|$ plus a curve C in $|4H - F - 3E - E_W|$. If F does not intersect one of the four exceptional curves in E_W meeting c_0, then $C = c_0 + a' + b'$ and the matching curve on $\tilde{P}$ contains the proper transform of a and b, which is not allowed. So F has to contain one of the four exceptional curves in E_W meeting c_0. This gives rise to four pencils of singular curves in $|4H - 3E - E_W|$, which produce (see Remark 20) four 2–dimensional linear subsystems in $|\mathscr{L}_0|$, and this implies the degree assertion.

The degree assertion in (iii) follows from the fact that a g^1_3 on E has 4 ramification points. □

Proposition 30 (Limits of 2-nodal curves). *The regular components of the limit Severi variety $\mathfrak{V}_2(\bar{X}, \mathscr{L})$ are the following, all appearing with multiplicity* 1, *except* (iv) *and* (v), *which have multiplicity* 2, *and* (vi), *which has multiplicity* 3*:*

(i) $V(\delta_{\tilde{P}} = 2)$, *with P-degree* 9*;*
(ii) $V(\delta_{\bar{W}} = 2)$, *with P-degree* 6*;*
(iii) $V(\delta_{\tilde{P}} = \delta_{\bar{W}} = 1)$, *with P-degree* 36*;*
(iv) $V(\delta_{\tilde{P}} = \tau_{E,2} = 1)$, *with P-degree* 28*;*
(v) $V(\delta_{\bar{W}} = \tau_{E,2} = 1)$, *with P-degree* 8*;*
(vi) $V(\tau_{E,3} = 1)$, *with P-degree* 3*;*
(vii) $V(\bar{W} + W_a + W_b, \delta_{\bar{W}} = 2)$, *with P–degree* 0.

Proof. Again, the list is an immediate application of Proposition 1, and the only things to prove are the degree assertions. Once more case (vii) is clear.

In case (i) the degree equals the number of lines on X_P (the cubic surface image of $\tilde{P}$), that do not contain the double point; this is 9.

In case (ii), we have to consider the binodal curves in $|4H - 3E - E_W|$ not containing E. Such curves split into a sum $\Phi_1 + \Phi_2 + C$, where Φ_1 and Φ_2 are the strict transforms of two curves in $|F|_W$. They are uniquely determined by the choice of two curves in E_W meeting c_0: these fix the two rulings in $|F|$ containing them, and there is a unique curve in $|2H - E|$ containing the remaining curves in E_W. This shows that the degree is 6.

Next, the limit curves of type (iii) consist of a nodal cubic in $|3H - E_P|_{\tilde{P}}$ and a nodal curve in $|4H - 3E - E_W|_{\tilde{W}}$; a ruling necessarily splits from the latter curve. Again, the splitting rulings F are the ones containing one of the four curves in E_W meeting c_0. The curves in $|3H - 2E|_{\tilde{W}}$ containing the remaining curves in E_W, fill up a pencil. Let F_0 be one of these four rulings. The number of nodal curves in $|3H - E_P|_{\tilde{P}}$ passing through the base point $F_0 \cap E$ and through a fixed general point on $\tilde{P}$ equals the degree of the dual surface of X_P, which is 9. For each such curve, there is a unique curve in the aforementioned pencil on $\tilde{W}$ matching it. This shows that the degree is 36.

The general limit curve of type (iv) can be identified with the general plane of $\mathbf{P}^3 = |3H - E_P|^{\vee}_{\tilde{P}}$ which is tangent to both X_P and the curve C_E (image of E in X_P), at different points. The required degree is the number of such planes passing through a general point p of X_P. The planes in question are parametrized in $\check{\mathbf{P}}^3$ by a component Γ_1 of $\check{X}_P \cap \check{C}_E$: one needs to remove from $\check{X}_P \cap \check{C}_E$ the component Γ_2, the general point of which corresponds to a plane which is tangent to X_P at a general point of C_E. The latter appears with multiplicity 2 in $\check{X}_P \cap \check{C}_E$ by Lemma 5. Moreover, $\check{X}_P$ and $\check{C}_E$ have respective degrees 9 and 4 by Proposition 5. Thus we have

$$\deg_P\big(V(\delta_{\tilde{P}} = \tau_{E,2} = 1)\big) = 36 - 2\deg(\Gamma_2).$$

To compute $\deg(\Gamma_2)$, take a general point $q = (q_0 : \ldots : q_3) \in \mathbf{P}^3$, and let $P_q(X_P)$ be the *first polar* of X_P with respect to q, i.e. the surface of homogeneous equation

$$q_0 \frac{\partial f}{\partial x_0} + \cdots + q_0 \frac{\partial f}{\partial x_3} = 0,$$

where $f = 0$ is the homogeneous equation of X_P. The number of planes containing q and tangent to X_P at a point of C_E is then equal to the number of points of $P_q(X_P) \cap C_E$, distinct from the singular point v of X_P. A local computation, which can be left to the reader, shows that v appears with multiplicity 2 in $P_p(X_P) \cap C_E$, which shows that $\deg(\Gamma_2) = 4$, whence $\deg_P\big(V(\delta_{\tilde{P}} = \tau_{E,2} = 1)\big) = 28$.

In case (v), we have to determine the curves in $|4H - 3E - E_W|$ with one node (so that some ruling splits) that are also tangent to E. As usual, the splitting rulings are the one containing one of the four curves in E_W meeting c_0. Inside the residual pencil there are two tangent curves at E. This yields the degree 8 assertion.

Finally, in case (vi), the degree equals the number of flexes of C_E, which is a nodal plane cubic: this is 3. □

Proposition 31 (Limits of 3-nodal curves). *The regular components of the limit Severi variety $\mathfrak{V}_3(\bar{X}, \mathscr{L})$ are the following 0–dimensional varieties, all appearing with multiplicity* 1, *except the ones in* (iv) *and* (v) *appearing with multiplicity* 2, *and* (vi) *with multiplicity* 3*:*

(i) $V(\delta_{\tilde{P}} = 3)$, *which consists of* 6 *points;*
(ii) $V(\delta_{\tilde{P}} = 2, \delta_{\bar{W}} = 1)$, *which consists of* 36 *points;*
(iii) $V(\delta_{\tilde{P}} = 1, \delta_{\bar{W}} = 2)$, *which consists of* 54 *points;*
(iv) $V(\delta_{\tilde{P}} = 2, \tau_{E,2} = 1)$, *which consists of* 18 *points;*
(v) $V(\delta_{\tilde{P}} = \delta_{\bar{W}} = \tau_{E,2} = 1)$, *which consists of* 56 *points;*
(vi) $V(\delta_{\tilde{P}} = \tau_{E,3} = 1)$, *which consists of* 18 *points;*
(vii) $V(\bar{W} + W_a + W_b, \delta_{\bar{W}} = 3)$, *which consists of* 6 *points.*

In the course of the proof, we will need the following lemma.

Lemma 15. *Let p, q be general points on E.*

(i) *The pencil $\mathfrak{l} \subset |3H - E_P|$ of curves containing q, and tangent to E at p, contains exactly* 7 *irreducible nodal curves not singular at p.*
(ii) *The pencil $\mathfrak{m} \subset |3H - E_P|$ of curves with a contact of order* 3 *with E at p contains exactly* 6 *irreducible nodal curves not singular at p.*

Proof. First note that $\mathfrak{l}$ and $\mathfrak{m}$ are indeed pencils by Remark 20. Let $P_{pq} \to \tilde{P}$ be the blow-up at p and q, with exceptional curves E_p and E_q above p and q respectively. Let $P'_{pq} \to P_{pq}$ be the blow-up at the point $E \cap E_p$, with exceptional divisor E'_p. Then $\mathfrak{l}$ pulls back to the linear system $|3H - E_P - E_p - E_q - 2E'_p|$, which induces an elliptic fibration $P'_{pq} \to \mathbf{P}^1$, with singular fibres in number of 12 (each counted with its multiplicity) by Lemma 4. Among them are: (i) the proper transform of $a + b + E$, which has 3 nodes, hence multiplicity 3 as a singular fibre; (ii) the unique curve of $\mathfrak{l}$ containing the (−2)-curve E_p, which has 2 nodes along E_p, hence multiplicity 2 as a singular fibre. The remaining seven singular fibres are the ones we want to count.

The proof of (ii) is similar and can be left to the reader. □

Proof (of Proposition 31). There is no member of $\mathfrak{L}_1$ with 3 nodes on $\bar{W}$, because every such curve contains one of the curves a', b', c_0.

There is no member of $\mathfrak{L}_1$ with two nodes on $\bar{W}$ and a tacnode on E either. Indeed, the component on $\bar{W}$ of such a curve would be the proper transform of a curve of W consisting of two rulings plus a curve in $|2H - E|$, altogether containing Z_W. Each of the two lines passes through one of the points of Z_W on c_0. The curve in $|2H - E|$ must contain the remaining points of Z_W, hence it is uniquely determined and cannot be tangent to E.

Then the list covers all remaining possible cases, and we only have to prove the assertion about the cardinality of the various sets.

The limiting curves of type (i) are in one-to-one correspondence with the unordered triples of lines distinct from a and b in P, the union of which contains the six points of Z_P. There are 6 such triples.

The limiting curves of type (ii) consist of the proper transform C_P in $\tilde{P}$ of the union of a conic and a line on P containing Z_P, plus the union C_W of the proper transforms in $\bar{W}$ of a curve in $|F|$ and one in $|3H - 2E|$ altogether containing Z_W, with C_P and C_W matching along E. We have 9 possible pencils for C_P, corresponding to the choice of two points on Z_P, one on a and one on b; each such pencil determines by restriction on E a line $\mathfrak{l} \subset \mathfrak{R}$. There are 4 possible pencils for C_W, corresponding to the choice of one of the points of Z_W on c_0: there is a unique ruling containing this point, and a pencil of curves in $|3H - 2E|$ containing the five remaining points in Z_W; each such pencil defines a line $\mathfrak{m} \subset \mathfrak{R}$. For each of the above choices, the lines $\mathfrak{l}$ and $\mathfrak{m}$ intersect at one point, whence the order 36.

We know from the proof of Proposition 30 that there are six 2–nodal curves in $|4H - 3E - E_W|$. For each such curve, there is a pencil of matching curves in $|3H - E_P|$. This pencil contains $\deg(\check{X}_P) = 9$ nodal curves, whence the number 54 of limiting curves of type (iii).

The component on $\tilde{P}$ of a limiting curve of type (iv) is the proper transform of the union of a conic and a line on P, containing E_P. As above, there are 9 possible choices for the line. For each such choice, there is a pencil of conics containing the 4 points of Z_P not on the line. This pencil cuts out a g^1_2 on E, and therefore contains 2 curves tangent to E. It follows that there are 18 limiting curves of type (iv).

The component on $\bar{W}$ of a limiting curve of type (v) is the proper transform of a ruling of W plus a curve in $|3H - 2E|$ tangent to E, altogether passing through Z_W. The line necessarily contains one of the four points of Z_W on c_0. There is then a pencil of curves in $|3H - 2E|$ containing the five remaining points of Z_W. It cuts out a g^1_2 on E, hence contains 2 curves tangent to E. For any such curve C_W on $\bar{W}$, there is a pencil of curves on the $\tilde{P}$–side matching it. By Lemma 15, this pencil contains 7 curves, the union of which with C_W is a limiting curve of type (v). This proves that there are 56 such limiting curves.

As for (vi), there are 3 members of $\mathfrak{R}$ that are triple points (see the proof of Proposition 30). Each of them determines a pencil of curves on the $\tilde{P}$–side, which contains six 1-nodal curves by Lemma 15. This implies that there are 18 limiting curves of type (vi).

Finally we have to count the members of $V(\bar{W} + W_a + W_b, \delta_{\bar{W}} = 3)$. They are in one–to–one correspondence with their components on $\bar{W}$, which decompose into the proper transform of unions $C_a \cup C_b$ of two curves $C_a \in |2H - E - 2E_a - E_b|$ and $C_b \in |2H - E - E_a - 2E_b|$, altogether containing Z_W. The curves C_a, C_b must contain the two base points on b', a' respectively. We conclude that each limiting curve of type (vii) corresponds to a partition of the 4 points of Z_W on c_0 in two disjoint sets of two points, and the assertion follows. □

In conclusion, the following is an immediate consequence of Propositions 29–31, together with the formula (2).

Corollary 14 (preliminary version of Theorem 3). *Let a, b, c be three independent lines in the projective plane, and Z be a degree 12 divisor on $a + b + c$ cut out by a general quartic curve. We consider the 3–dimensional sub–linear system $\mathscr{V}$ of $|\mathscr{O}_{\mathbf{P}^2}(4)|$ parametrizing curves containing Z, and we let, for $1 \leqslant \delta \leqslant 3$, $\mathscr{V}_\delta$ be the Zariski closure in $\mathscr{V}$ of the codimension δ locally closed subset parametrizing irreducible δ–nodal curves. One has*

$$\deg(\mathscr{V}_1) \geqslant 21, \quad \deg(\mathscr{V}_2) \geqslant 132, \quad \textit{and} \quad \deg(\mathscr{V}_3) \geqslant 304. \qquad (30)$$

Remark 21. **(a) (Theorem 3)** The three inequalities in (30) above are actually equalities. This is proved in Sect. 5.8, by using both (30) and the degrees of the Severi varieties of a general quartic surface, given by Proposition 4.

Incidentally, this proves that $\varpi : \bar{X} \to \Delta$ is a good model for the family $\hat{f} : \hat{S} \to \Delta$ obtained by blowing–up $S = \mathbf{P}^2 \times \Delta$ along Z, and endowed with the appropriate subline bundle of $\mathscr{O}_{\hat{S}}(1)$.

(b) In particular, we have $\mathfrak{V}_3 = \mathfrak{V}_3^{\text{reg}}$. It then follows from Remark 5 that the relative Severi variety $V_3(\bar{X}, \mathscr{L})$ is smooth at the points of $\mathfrak{V}_3$. This implies that the general fibre of $V_3(\bar{X}, \mathscr{L})$ is reduced. Therefore, in the setting of Corollary 14, if $a + b + c$ and Z are sufficiently general, then $\mathscr{V}_3$ consists of 304 distinct points.

10 Application to the Irreducibility of Severi Varieties and to the Monodromy Action

Set $\mathscr{B} = |\mathscr{O}_{\mathbf{P}^3}(4)|$. We have the *universal family* $p : \mathscr{P} \to \mathscr{B}$, such that the fibre of p over $S \in \mathscr{B}$ is the linear system $|\mathscr{O}_S(1)|$. The variety $\mathscr{P}$ is a component of the *flag Hilbert scheme*, namely the one parametrizing pairs (C, S), where C is a plane quartic curve in $\mathbf{P}^3$ and $S \in \mathscr{B}$ contains C. So $\mathscr{P} \subset \mathscr{B} \times \mathscr{W}$, where $\mathscr{W}$ is the component of the Hilbert scheme of curves in $\mathbf{P}^3$ whose general point corresponds to a plane quartic. The map p is the projection to the first factor; we let q be the projection to the second factor.

Denote by $\mathscr{U} \subset \mathscr{B}$ the open subset parametrizing smooth surfaces, and set $\mathscr{P}_{\mathscr{U}} = p^{-1}(\mathscr{U})$. Inside $\mathscr{P}_{\mathscr{U}}$ we have the *universal Severi varieties* $\mathscr{V}_\delta^\circ$, $1 \leqslant \delta \leqslant 3$, such that for all $S \in \mathscr{U}$, the fibre of $\mathscr{V}_\delta^\circ$ over S is the Severi variety $V_\delta(S, \mathscr{O}_S(1))$. Since S is a $K3$ surface, we know that for all irreducible components V of $V_\delta(S, \mathscr{O}_S(1))$, we have $\dim(V) = 3 - \delta$, so that all components of $\mathscr{V}_\delta^\circ$ have codimension δ in $\mathscr{P}_{\mathscr{U}}$. We then let $\mathscr{V}_\delta$ be the Zariski closure of $\mathscr{V}_\delta^\circ$ in $\mathscr{P}$; we will call it universal Severi variety as well.

The following is immediate (and it is a special case of a more general result, see [11]):

Proposition 32. *The universal Severi varieties $\mathcal{V}_\delta$ are irreducible for $1 \leqslant \delta \leqslant 3$.*

Proof. It suffices to consider the projection $q : \mathcal{V}_\delta \to \mathcal{W}$, and notice that its image is the irreducible variety whose general point corresponds to a quartic curve with δ nodes (cf. [22, 24]), and that the fibres are all irreducible of the same dimension 20. □

Note that the irreducibility of $\mathcal{V}_1$ also follows from the fact that for all $S \in \mathcal{U}$, we have $V_1(S, \mathcal{O}_S(1)) \cong \check{S}$. To the other extreme, $p : \mathcal{V}_3^\circ \to \mathcal{U}$ is a finite cover of degree 3,200. We will denote by $G_{4,3} \leq \mathfrak{S}_{3,200}$ the *monodromy group of this covering*, which acts transitively because $\mathcal{V}_3$ is irreducible.

10.1 The Irreducibility of the Family of Binodal Plane Sections of a General Quartic Surface

In the middle we have $p : \mathcal{V}_2^\circ \to \mathcal{U}$. Though $\mathcal{V}_2$ is irreducible, we cannot deduce from this that for the general $S \in \mathcal{U}$, the Severi variety $V_2(S, \mathcal{O}_S(1))$ (i.e., the curve of binodal plane sections of S) is irreducible. Though commonly accepted as a known fact, we have not been able to find any proof of this in the current literature. It is the purpose of this paragraph to provide a proof of this fact.

In any event, we have a commutative diagram similar to the one in (22)

$$\begin{array}{ccc} \mathcal{V}_2' & \xrightarrow{\ \nu\ } & \mathcal{V}_2^\circ \\ {\scriptstyle p'}\downarrow & \searrow & \downarrow{\scriptstyle p} \\ \mathcal{U}' & \xrightarrow[\ f\ ]{} & \mathcal{U} \end{array}$$

where ν is the normalization of $\mathcal{V}_2^\circ$, and $f \circ p'$ is the Stein factorization of $p \circ \nu : \mathcal{V}_2' \to \mathcal{U}$. The morphism $f : \mathcal{U}' \to \mathcal{U}$ is finite, of degree h equal to the number of irreducible components of $V_2(S, \mathcal{O}_S(1))$ for general $S \in \mathcal{U}$. The monodromy group of this covering acts transitively. This ensures that, for general $S \in \mathcal{U}$, all irreducible components of $V_2(S, \mathcal{O}_S(1))$ have the same degree, which we denote by n. By Proposition 19, we have $n \geqslant 36$.

Theorem 4. *If $S \subset \mathbf{P}^3$ is a general quartic surface, then the curve $V_2(S, \mathcal{O}_S(1))$ is irreducible.*

Proof. Let S_0 be a general quartic Kummer surface, and $f : \mathcal{S} \to \Delta$ a family of surfaces induced as in Example 1 by a pencil generated by S_0 and a general quartic S_∞. Given two distinct nodes p and q of S_0, we denote by $\mathfrak{l}_{pq}$ the pencil of plane sections of S_0 passing through p and q. Corollary 12 asserts that the union of these lines, each counted with multiplicity 4, is the crude limit Severi variety $\mathfrak{V}_2^{\mathrm{cr}}(\mathcal{S}, \mathcal{O}_{\mathcal{S}}(1))$.

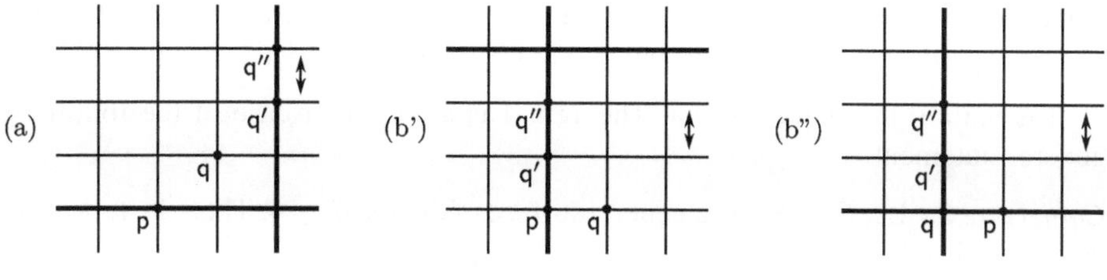

Fig. 12 How to obtain three double points on a double conic

Let Γ_t be an irreducible component of $V_1(S_t, \mathcal{O}_{S_t}(1))$, for $t \in \Delta^*$, and let Γ_0 be its (crude) limit as t tends to 0, which consists of a certain number of (quadruple) curves $\mathfrak{l}_{pq}$. Note that, by Proposition 26, the pull–back of the lines $\mathfrak{l}_{pq}$ to the good limit constructed in Sect. 8 all appear with multiplicity 1 in the limit Severi variety. This yields that, if $\mathfrak{l}$ is an irreducible component of Γ_0, then it cannot be in the limit of an irreducible component Γ_t' of $V_1(S_t, \mathcal{O}_{S_t}(1))$ other than Γ_t.

We shall prove successively the following claims, the last one of which proves the theorem:

(i) Γ_0 contains two curves $\mathfrak{l}_{pq}, \mathfrak{l}_{pq'}$, with $q \neq q'$;
(ii) Γ_0 contains two curves $\mathfrak{l}_{pq}, \mathfrak{l}_{pq'}$, with $q \neq q'$, and p, q, q' on a contact conic D of S_0;
(iii) There is a contact conic D of S_0, such that Γ_0 contains all curves $\mathfrak{l}_{pq}$ with $p, q \in D$;
(iv) Property (iii) holds for every contact conic of S_0;
(v) Γ_0 contains all curves $\mathfrak{l}_{pq}$.

If Γ_0 does not verify (i), then it contains at most 8 curves of type $\mathfrak{l}_{pq}$, a contradiction to $n \geqslant 36$. To prove (ii), we consider two curves $\mathfrak{l}_{pq}$ and $\mathfrak{l}_{pq'}$ contained in Γ_0, and assume that p, q, q' do not lie on a contact conic, otherwise there is nothing to prove. Consider a degeneration of S_0 to a product Kummer surface S, and let p,q,q' be the limits on S of p, q, q' respectively: they are necessarily in one of the three configurations depicted in Fig. 12. In all three cases, we can exchange two horizontal lines in S (as indicated in Fig. 12), thus moving q' to q'', in such a way that p and q remain fixed, and there is a limit in S of contact conics that contains the three points p, q', and q''. Accordingly, there is an element $\gamma \in G_{16,6}$ mapping p, q, q' to p, q, q'' respectively, such that p, q', q'' lie on a contact conic D of S_0. Then $\gamma(\Gamma_0)$ contains $\gamma(\mathfrak{l}_{pq}) = \mathfrak{l}_{pq}$. By the remark preceding the statement of (i)–(v), we have $\gamma(\Gamma_0) = \Gamma_0$. It follows that Γ_0 contains $\mathfrak{l}_{pq'}$ and $\mathfrak{l}_{pq''}$, and therefore satisfies (ii).

Claim (iii) follows from (ii) and the fact that the monodromy acts as $\mathfrak{S}_6$ on the set of nodes lying on D (see Lemma 11). As for (iv), let D' be any other contact conic of S_0. There exists $\gamma \in G_{16,6}$ interchanging D and D' (again by Lemma 11). The action of γ preserves $D \cap D' = \{x, y\}$. We know that Γ_0 contains $\mathfrak{l}_{xy} + \mathfrak{l}_{xy'}$ with $y' \in D$ different from y. Then the same argument as above yields that Γ_0

contains $\mathfrak{l}_{\gamma(x)\gamma(y')}$, where $\gamma(x) \in \{x, y\}$ and $\gamma(y') \in D' - \{x, y\}$. This implies that Γ_0 satisfies (ii) for D', and therefore (iii) holds for D'. Finally (iv) implies (v). □

It is natural to conjecture that Theorem 4 is a particular case of the following general statement:

Conjecture 2. Let $S \subset \mathbf{P}^3$ be a general surface of degree $d \geq 4$. Then the following curves are irreducible:

(i) $V_2(S, \mathscr{O}_S(1))$, the curve of binodal plane sections of S;
(ii) $V_\kappa(S, \mathscr{O}_S(1))$, the curve of cuspidal plane sections of S.

We hope to come back to this in a future work.

10.2 Some Noteworthy Subgroups of $G_{4,3} \leq \mathfrak{S}_{3,200}$

In this section we use the degenerations we studied in Sects. 5 and 8 to give some information on the monodromy group $G_{4,3}$ of $p : \mathscr{V}_3^\circ \to \mathscr{U}$. We will use the following:

Remark 22. Let $f : X \to Y$ be a dominant, generically finite morphism of degree n between projective irreducible varieties, with monodromy group $G \leq \mathfrak{S}_n$. Let $V \subset Y$ be an irreducible codimension 1 subvariety, the generic point of which is a smooth point of Y. Then $f_V := f|_{f^{-1}(V)} : f^{-1}(V) \to V$ is still generically finite, with monodromy group G_V. If V is not contained in the branch locus of f, then $G_V \leq G$.

Suppose to the contrary that V is contained in a component of the branch locus of f. Then $G_V \leq \mathfrak{S}_{n_V}$, with $n_V := \deg f_V < n$, and G_V is no longer a subgroup of G. We can however consider the *local monodromy group* G_V^{loc} *of* f *around* V, i.e. the subgroup of $G \leq \mathfrak{S}_n$ generated by permutations associated to non–trivial loops *turning around* V. Precisely: let U_V be a tubular neighbourhood of V in Y; then G_V^{loc} is the image in G of the subgroup $\pi_1\big(U_V - V\big)$ of $\pi_1(Y - V)$.

There is an epimorphism $G_V^{\mathrm{loc}} \to G_V$, obtained by deforming loops in $U_V - V$ to loops in V. We let H_V^{loc} be the kernel of this epimorphism, so that one has the exact sequence of groups

$$1 \to H_V^{\mathrm{loc}} \to G_V^{\mathrm{loc}} \to G_V \to 1. \tag{31}$$

We first apply this to the degeneration studied in Sect. 5. To this end, we consider the 12–dimensional subvariety $\mathscr{T}$ of $\mathscr{B}$ which is the Zariski closure of the set of fourtuples of distinct planes. Let $f : \tilde{\mathscr{B}}_{\mathrm{tetra}} \to \mathscr{B}$ be the blow–up of $\mathscr{B}$ along $\mathscr{T}$, with exceptional divisor $\tilde{\mathscr{T}}$. The proof of the following lemma (similar to Lemma 3) can be left to the reader:

Lemma 16. *Let* X *be a general point of* $\mathscr{T}$. *Then the fibre of* f *over* X *can be identified with* $|\mathscr{O}_\Lambda(4)|$, *where* $\Lambda = \mathrm{Sing}(X)$.

Thus, for general $X \in \mathscr{T}$, a general point of the fibre of f over X can be identified with a pair (X, D), with $D \in |\mathscr{O}_\Lambda(4)|$ general, where $\Lambda = \mathrm{Sing}(X)$. Consider a family $f : S \to \Delta$ of surfaces in $\mathbf{P}^3$, induced as in Example 1 by a pencil $\mathfrak{l}$ generated by X and a general quartic; then the singular locus of S is a member of $|\mathscr{O}_\Lambda(4)|$, which corresponds to the tangent direction normal to $\mathscr{T}$ defined by $\mathfrak{l}$ in $\mathscr{B}$.

Now the universal family $p : \mathscr{P} \to \mathscr{B}$ can be pulled back to $\tilde{p} : \tilde{\mathscr{P}} \to \tilde{\mathscr{B}}_{\mathrm{tetra}}$, and the analysis of Sect. 5 tells us that we have a generically finite map $\tilde{p} : \tilde{\mathscr{V}}_3 \to \tilde{\mathscr{B}}_{\mathrm{tetra}}$, which restricts to $p : \mathscr{V}_3^\circ \to \mathscr{U}$ over $\mathscr{U}$, and such that $\tilde{\mathscr{T}}$ is in the branch locus of $\tilde{p}$. We let G_{tetra} be the monodromy group of $\tilde{p} : \tilde{\mathscr{V}}_3 \to \tilde{\mathscr{B}}_{\mathrm{tetra}}$ on $\tilde{\mathscr{T}}$, and $G^{\mathrm{loc}}_{\mathrm{tetra}}$, resp. $H^{\mathrm{loc}}_{\mathrm{tetra}}$, be as in (31).

Proposition 33. *Consider a general $(X, D) \in \tilde{\mathscr{B}}_t$. One has:*

(a) $G_{\mathrm{tetra}} \cong \prod_{i=1}^4 G_i$, *where:*

- *(i)* $G_1 \cong \mathfrak{S}_{1{,}024}$ *is the monodromy group of planes containing three points in D, but no edge of X;*
- *(ii)* $G_2 \cong \mathfrak{S}_4 \times \mathfrak{S}_3 \times (\mathfrak{S}_4)^2$ *is the monodromy group of planes containing a vertex of X and two points in D, but no edge of X;*
- *(iii)* $G_3 \cong \mathfrak{S}_6 \times \mathfrak{S}_4$ *is the monodromy group of planes containing an edge of X, and a point in D on the opposite edge of X;*
- *(iv)* $G_4 \cong \mathfrak{S}_4$ *is the monodromy group of faces of X;*

(b) $H^{\mathrm{loc}}_{\mathrm{tetra}} \cong \mathfrak{S}_3 \times G \times H$, *where $G \leq \mathfrak{S}_{16}$ is the monodromy group of bitangent lines to 1–nodal plane quartics as in Proposition 12, and $H \leq \mathfrak{S}_{304}$ is the monodromy group of irreducible trinodal curves in the linear system of quartic curves with 12 base points at a general divisor of $|\mathscr{O}_{a+b+c}(4)|$, with a, b, c three lines not in a pencil (see Sect. 9).*

Proof. The proof follows from Corollary 7. Recall that a group $G \leq \mathfrak{S}_n$ is equal to $\mathfrak{S}_n$, if and only if it contains a transposition and it is doubly transitive. Using this, it is easy to verify the assertions in (ai)–(aiv) (see [21, p.698]). As for (b), the factor $\mathfrak{S}_3$ comes from the fact that the monodromy acts as the full symmetric group on a general line section of the irreducible cubic surface T as in Proposition 13. □

Analogous considerations can be made for the degeneration studied in Sect. 8. In that case, we consider the 18–dimensional subvariety $\mathscr{K}$ of $\mathscr{B}$ which is the Zariski closure of the set $\mathscr{K}^\circ$ of Kummer surfaces. Let $g : \tilde{\mathscr{B}}_{\mathrm{Kum}} \to \mathscr{B}$ be the blow–up along $\mathscr{K}$, with exceptional divisor $\tilde{\mathscr{K}}$. In this case we have:

Lemma 17. *Let $X \in \mathscr{K}$ be a general point, with singular locus N. Then the fibre of g over X can be identified with $|\mathscr{O}_N(4)| \cong \mathbf{P}^{15}$.*

The universal family $p : \mathscr{P} \to \mathscr{B}$ can be pulled back to $\tilde{p} : \tilde{\mathscr{P}} \to \tilde{\mathscr{B}}_{\mathrm{Kum}}$. The analysis of Sect. 8 tells us that we have a map $\tilde{p} : \tilde{\mathscr{V}}_3 \to \tilde{\mathscr{B}}_{\mathrm{Kum}}$, generically finite over $\tilde{\mathscr{K}}$, which is in the branch locus of $\tilde{p}$. We let G_{Kum} be the monodromy group of $\tilde{p}$ on $\tilde{\mathscr{K}}$, and set $G^{\mathrm{loc}}_{\mathrm{Kum}}$ and $H^{\mathrm{loc}}_{\mathrm{Kum}}$ as in (31).

Proposition 34. *One has:*

(a) $G_{\mathrm{Kum}} \cong G_{16,6} \times G'$, *where* G' *is the monodromy group of unordered triples of distinct nodes of a general Kummer surface which do not lie on a contact conic (see Sect. 7.2 for the definition of* $G_{16,6}$*);*

(b) $H^{\mathrm{loc}}_{\mathrm{Kum}} \cong \mathfrak{S}_8 \times G''$, *where* G'' *is the monodromy group of the tritangent planes to a rational curve* B *of degree* 8 *as in the statement of Proposition 24.*

Proof. Part (a) follows right away from Proposition 27. Part (b) also follows, since the monodromy on complete intersections of three general quadrics in $\mathbf{P}^3$ (which gives the multiplicity 8 in (i) of Proposition 27) is clearly the full symmetric group. □

Concerning the group G' appearing in Proposition 34 (a), remember that it acts with at most two orbits on the set of unordered triples of distinct of nodes of a general Kummer surface which do not lie on a contact conic (see Proposition 22 (ii)).

Acknowledgements We thank Erwan Brugallé (who, together with G. Mikhalkin, was able to provide the enumerative numbers of Theorems 1 and 3 using tropical methods), and Concettina Galati for numerous discussions, both enlightening and motivating.

This project profited of various visits of the second author to the first, which have been made possible by the research group GRIFGA, in collaboration between CNRS and INdAM.

The first author is a member of GNSAGA of INdAM and was partly supported by the project "Geometry of Algebraic varieties" funded by the Italian MIUR. The second author is a member of projects CLASS and MACK of the French ANR.

References

1. E. Arbarello, M. Cornalba, P.A. Griffiths, J. Harris, *Geometry of Algebraic Curves. Vol. I.* Grundlehren der Mathematischen Wissenschaften, vol. 267 (Springer, New York, 1985)
2. A. Beauville, *Surfaces Algébriques Complexes.* Number 54 in Astérisque (Société Mathématique de France, Paris, 1978)
3. A. Beauville, Counting rational curves on $K3$ surfaces. Duke Math. J. **97**(1), 99–108 (1999)
4. C. Birkenhake, H. Lange, *Complex Abelian Varieties.* Grundlehren der Mathematischen Wissenschaften, vol. 302, 2nd edn. (Springer, Berlin/Heidelberg, 2004)
5. J.W. Bruce, C.T.C. Wall, On the classification of cubic surfaces. J. Lond. Math. Soc. **19**(2), 245–256 (1979)
6. J. Bryan, N.C. Leung, The enumerative geometry of $K3$ surfaces and modular forms. J. Am. Math. Soc. **13**(2), 371–410 (2000). (electronic)
7. A. Calabri, R. Ferraro, Explicit resolutions of double point singularities of surfaces. Collect. Math. **53**(2), 99–131 (2002)
8. A. Calabri, C. Ciliberto, F. Flamini, R. Miranda, On the K^2 of degenerations of surfaces and the multiple point formula. Ann. Math. **165**, 335–395 (2007)
9. L. Caporaso, J. Harris, Counting plane curves of any genus. Invent. Math. **131**(2), 345–392 (1998)
10. L. Caporaso, J. Harris, Parameter spaces for curves on surfaces and enumeration of rational curves. Compositio Math. **113**(2), 155–208 (1998)
11. C. Ciliberto, T. Dedieu, On universal Severi varieties of low genus $K3$ surfaces. Math. Z. **271**, 953–960 (2012)

12. C. Ciliberto, A. Lopez, R. Miranda, Projective degenerations of K3 surfaces, Gaussian maps, and Fano threefolds. Invent. Math. **114**, 641–667 (1993)
13. T. Dedieu, Severi varieties and self-rational maps of $K3$ surfaces. Int. J. Math. **20**(12), 1455–1477 (2009)
14. I. Dolgachev, *Classical Algebraic Geometry, A Modern View* (Cambridge University Press, Cambridge, 2012)
15. F. Flamini, A.L. Knutsen, G. Pacienza, E. Sernesi, Nodal curves with general moduli on $K3$ surfaces. Commun. Algebra **36**(11), 3955–3971 (2008)
16. R. Friedman, Global smoothings of varieties with normal crossings. Ann. Math. **118**, 75–114 (1983)
17. C. Galati, Degenerating curves and surfaces: first results. Rend. Circ. Mat. Palermo **58**(2), 211–243 (2009)
18. C. Galati and A.L. Knutsen, On the existence of curves with A_k-singularities on $K3$ surfaces, preprint arXiv:1107.4568v4
19. M.R. Gonzalez–Dorrego, $(16,6)$ *Configurations and Geometry of Kummer Surfaces in* $\mathbf{P}^3$. Number 512 in Memoirs of the American Mathematical Society (American Mathematical Society, Providence, 1994)
20. P. Griffiths, J. Harris, On the Noether-Lefschetz theorem and some remarks on codimension-two cycles. Math. Ann. **271**, 31–51 (1985)
21. J. Harris, Galois groups of enumerative problems. Duke Math. J. **46**(4), 685–724 (1979)
22. J. Harris, *On the Severi problem*. Invent. Math. **84**(3), 445–461 (1986)
23. J. Harris, *Algebraic Geometry, A First Course*. Number 133 in Graduate Texts in Mathematics (Springer, New York, 1992)
24. J. Harris, I. Morrison, *Moduli of Curves*. Number 187 in Graduate Texts in Mathematics (Springer, New York, 1998)
25. R.W.H.T. Hudson, *Kummer's Quartic Surface*. Cambridge Mathematical Library (Cambridge University Press, Cambridge/New York, 1990). Revised reprint of the 1905 original
26. T. Keilen, Irreducibility of equisingular families of curves. Trans. Am. Math. Soc. **355**(9), 3485–3512 (2003). (electronic)
27. M. Kemeny, The universal Severi variety of rational curves on $K3$ surfaces. Bull. Lond. Math. Soc. **45**(1), 159–174 (2013)
28. G. Kempf, F.F. Knudsen, D. Mumford, B. Saint–Donat, *Toroidal Embeddings I*. Lecture Notes in Mathematics, vol. 339 (Springer, Berlin/New York, 1973)
29. A. Klemm, D. Maulik, R. Pandharipande, E. Scheidegger, Noether-Lefschetz theory and the Yau-Zaslow conjecture. J. Am. Math. Soc. **23**(4), 1013–1040 (2010)
30. Z. Ran, On nodal plane curves. Invent. Math. **86**(3), 529–534 (1986)
31. Z. Ran, Families of plane curves and their limits: Enriques' conjecture and beyond. Ann. Math. **130**(2), 121–157 (1989)
32. Z. Ran, Enumerative geometry of singular plane curves. Invent. Math. **97**(3), 447–465 (1989)
33. G. Salmon, *A Treatise on the Analytic Geometry of Three Dimensions* (Hodges, Smith, Dublin, 1862)
34. I. Vainsencher, Counting divisors with prescribed singularities. Trans. Am. Math. Soc. **267**(2), 399–422 (1981)
35. S.-T. Yau, E. Zaslow, BPS states, string duality, and nodal curves on $K3$. Nucl. Phys. B **471**(3), 503–512 (1996)

On Images of Weak Fano Manifolds II

Osamu Fujino and Yoshinori Gongyo

Abstract We consider a smooth morphism between smooth complex projective varieties. We give an alternative proof of the following fact: If the anti-canonical divisor of the source space is nef, then so is the anti-canonical divisor of the target space. We do not use mod p reduction arguments.

1 Introduction

We will work over $\mathbb{C}$, the field of complex numbers. The following theorem is the main result of this paper.

Theorem 1 (Main theorem). *Let $f : X \to Y$ be a smooth morphism between smooth projective varieties. Let D be an effective $\mathbb{Q}$-divisor on X such that (X, D) is log canonical,* Supp$D$ *is a simple normal crossing divisor, and* Supp$D$ *is relatively normal crossing over Y. Let Δ be a (not necessarily effective) $\mathbb{Q}$-divisor on Y. Assume that $-(K_X + D) - f^*\Delta$ is nef. Then $-K_Y - \Delta$ is nef.*

By setting $D = 0$ and $\Delta = 0$ in Theorem 1, we obtain the following result, hence a new proof, in characteristic zero, of [5, Corollary 3.15 (a)].

O. Fujino (✉)
Faculty of Science, Department of Mathematics, Kyoto University, Kyoto 606-8502, Japan
e-mail: fujino@math.kyoto-u.ac.jp

Y. Gongyo
Graduate School of Mathematical Sciences, The University of Tokyo, 3-8-1 Komaba, Meguro, Tokyo 153-8914, Japan

Department of Mathematics, Imperial College London, 180 Queen's Gate, London SW7 2AZ, UK
e-mail: gongyo@ms.u-tokyo.ac.jp; y.gongyo@imperial.ac.uk

A. Frühbis-Krüger et al. (eds.), *Algebraic and Complex Geometry*, Springer Proceedings in Mathematics & Statistics 71, DOI 10.1007/978-3-319-05404-9_7,

Corollary 1. *Let $f : X \to Y$ be a smooth morphism between smooth projective varieties. Assume that $-K_X$ is nef. Then $-K_Y$ is nef.*

By setting $D = 0$ and taking for Δ a small ample $\mathbb{Q}$-divisor, we also obtain the following result, which is [10, Corollary 2.9]. Note that Theorem 1 is also a generalization of [8, Theorem 4.8].

Corollary 2 (cf. [10, Corollary 2.9]). *Let $f : X \to Y$ be a smooth morphism between smooth projective varieties. Assume that $-K_X$ is ample. Then $-K_Y$ is ample.*

This last corollary was proved in [10] using mod p reduction arguments. Our proof of Theorem 1 relies instead, as in [4], on a generalization of Viehweg's weak positivity theorem due to Campana [3, Theorem 4.13] which is obtained from the theory of variations of mixed Hodge structure. So our argument is ultimately Hodge-theoretic.

In [8, Theorem 4.1] (see Theorem 2), we also proved a weaker version of Theorem 1 using Kawamata's positivity theorem [8, Theorem 2.2]. We recommend that the readers compare the proof of Theorem 1 with the arguments in [8, Section 4].

By the Lefschetz principle, all the results in this paper hold over any algebraically closed field of characteristic zero. We do not discuss here the case when the characteristic of the base field is positive.

2 Proof of the Main Theorem

In this section, we prove Theorem 1. We closely follow the arguments in [4].

Lemma 1. *Let $f : Z \to C$ be a surjective morphism from a $(d + 1)$-dimensional smooth projective variety Z to a smooth projective curve C. Let B be an ample Cartier divisor on Z such that $R^i f_* \mathcal{O}_Z(kB) = 0$ for every $i > 0$ and $k \geq 1$. Let H be a very ample Cartier divisor on C such that $B^{d+1} < f^*(H - K_C) \cdot B^d$ and $B^{d+1} \leq f^* H \cdot B^d$. Then*

$$(f_* \mathcal{O}_Z(kB))^* \otimes \mathcal{O}_C(lH)$$

is generated by global sections for all $l > k \geq 1$.

Proof. By Grothendieck duality, we have

$$R\mathcal{H}om(Rf_* \mathcal{O}_Z(kB), \omega_C^{\bullet}) \simeq Rf_* R\mathcal{H}om(\mathcal{O}_Z(kB), \omega_Z^{\bullet}),$$

hence we obtain

$$(f_* \mathcal{O}_Z(kB))^* \simeq R^d f_* \mathcal{O}_Z(K_{Z/C} - kB)$$

for $k \geq 1$ and

$$R^i f_* \mathscr{O}_Z(K_{Z/C} - kB) = 0$$

for $k \geq 1$ and $i \neq d$. We note that $f_* \mathscr{O}_Z(kB)$ and its dual $(f_* \mathscr{O}_Z(kB))^*$ are locally free sheaves. Therefore, we have

$$\begin{aligned} &H^1(C, (f_* \mathscr{O}_Z(kB))^* \otimes \mathscr{O}_C((l-1)H)) \\ &\simeq H^1(C, R^d f_* \mathscr{O}_Z(K_{Z/C} - kB) \otimes \mathscr{O}_C((l-1)H)) \\ &\simeq H^{d+1}(Z, \mathscr{O}_Z(K_Z - f^* K_C - kB + f^*(l-1)H)) \end{aligned}$$

for $k \geq 1$. By Serre duality,

$$H^{d+1}(Z, \mathscr{O}_Z(K_Z - f^* K_C - kB + f^*(l-1)H))$$

is dual to

$$H^0(Z, \mathscr{O}_Z(kB + f^* K_C - f^*(l-1)H)).$$

On the other hand, it follows from our assumptions that

$$(kB + f^* K_C - f^*(l-1)H) \cdot B^d < 0$$

if $l - 1 \geq k$. Thus, we obtain

$$H^0(Z, \mathscr{O}_Z(kB + f^* K_C - f^*(l-1)H)) = 0$$

for $l > k$. This means that

$$H^1(C, (f_* \mathscr{O}_Z(kB))^* \otimes \mathscr{O}_C((l-1)H)) = 0$$

for $k \geq 1$ and $l > k$. Therefore, $(f_* \mathscr{O}_Z(kB))^* \otimes \mathscr{O}_C(lH)$ is generated by global sections for $k \geq 1$ and $l > k$.

The following lemma directly follows from [3, Theorem 4.3]. It is essential for the proof of Theorem 1.

Lemma 2. *Let $f : V \to W$ be a surjective morphism between smooth projective varieties with connected fibers. Let Δ be an effective $\mathbb{Q}$-divisor on V such that (V, Δ) is log canonical. Assume that $m\Delta$ is Cartier for some positive integer m. Then $f_* \mathscr{O}_V(m(K_{V/W} + \Delta))$ is weakly positive over some non-empty Zariski open set of W.*

For basic properties of weakly positive sheaves, see [11, Section 2.3]. Although the original proof of [3, Theorem 4.3] depends on Kawamata's difficult result

[9, Theorem 32], the results [6, Theorem 3.9] and [7, Theorem 1.1] are sufficient for the proof of our Lemma 2.

Proof (Proof of Theorem 1). We note that, by Stein factorization, we may assume that f has connected fibers (see [8, Lemma 2.4]). We need to prove that for any finite morphism $\pi : C \to Y$ from a smooth projective curve C, we have $(-\pi^* K_Y - \pi^*\Delta) \cdot C \geq 0$. Let L be an ample Cartier divisor on C. We will prove that for any positive rational number ε, we have $(-\pi^* K_Y - \pi^*\Delta + 2\varepsilon L) \cdot C \geq 0$. We consider the following base change diagram

$$\begin{array}{ccc} Z & \xrightarrow{\ p\ } & X \\ {\scriptstyle g}\downarrow & & \downarrow{\scriptstyle f} \\ C & \xrightarrow[\ \pi\ ]{} & Y \end{array}$$

where $Z = X \times_Y C$. Then $g : Z \to C$ is smooth, Z is smooth, $\mathrm{Supp}(p^*D)$ is relatively normal crossing over C, and $\mathrm{Supp}(p^*D)$ is a simple normal crossing divisor on Z. Let A be a very ample Cartier divisor on X and let δ be a small positive rational number such that $0 < \delta \ll \varepsilon$. Since $-(K_X + D) - f^*\Delta + \delta A$ is ample, we can take an effective $\mathbb{Q}$-divisor F on X such that $-(K_X + D) - f^*\Delta + \delta A \sim_{\mathbb{Q}} F$. Then we have

$$K_{X/Y} + D + F \sim_{\mathbb{Q}} \delta A - f^* K_Y - f^*\Delta.$$

By taking the base change, we obtain

$$K_{Z/C} + p^* D + p^* F \sim_{\mathbb{Q}} \delta p^* A - g^*\pi^* K_Y - g^*\pi^*\Delta.$$

Without loss of generality, we may assume that $\mathrm{Supp}(p^*D + p^*F)$ is a simple normal crossing divisor, p^*D and p^*F have no common irreducible components, and $(Z, p^*D + p^*F)$ is log canonical. Let m be a sufficiently divisible positive integer such that $m\delta$ and $m\varepsilon$ are integers, mp^*D, mp^*F, and $m\Delta$ are Cartier divisors, and

$$m(K_{Z/C} + p^*D + p^*F) \sim m(\delta p^*A - g^*\pi^*K_Y - g^*\pi^*\Delta).$$

We apply Lemma 2 and obtain that

$$g_*\mathscr{O}_Z(m(K_{Z/C} + p^*D + p^*F)) \simeq g_*\mathscr{O}_Z(m(\delta p^*A - g^*\pi^*K_Y - g^*\pi^*\Delta))$$

is weakly positive over some non-empty Zariski open set U of C. Therefore,

$$\mathscr{E}_1 := S^n(g_*\mathscr{O}_Z(m(\delta p^*A - g^*\pi^*K_Y - g^*\pi^*\Delta))) \otimes \mathscr{O}_C(nm\varepsilon L)$$

$$\simeq S^n(g_*\mathscr{O}_Z(m\delta p^*A)) \otimes \mathscr{O}_C(-nm\pi^*K_Y - nm\pi^*\Delta + nm\varepsilon L)$$

is generated by global sections over U for every $n \gg 0$. On the other hand, by Lemma 1, if $m\delta \gg 0$,

$$\begin{aligned}\mathscr{E}_2 &:= \mathscr{O}_C(nm\varepsilon L) \otimes S^n((g_*\mathscr{O}_Z(m\delta p^*A))^*)\\ &\simeq S^n(\mathscr{O}_C(m\varepsilon L) \otimes (g_*\mathscr{O}_Z(m\delta p^*A))^*).\end{aligned}$$

is generated by global sections because $0 < \delta \ll \varepsilon$ and p^*A is ample on Z. Thus there is a homomorphism

$$\alpha : \bigoplus_{\text{finite}} \mathscr{O}_C \to \mathscr{E} := \mathscr{E}_1 \otimes \mathscr{E}_2$$

which is surjective over U. We note that

$$S^n((g_*\mathscr{O}_Z(m\delta p^*A))^*) \simeq (S^n(g_*\mathscr{O}_Z(m\delta p^*A)))^*$$

since $g_*\mathscr{O}_Z(m\delta p^*A)$ is locally free. Therefore, there is a non-trivial trace map

$$S^n(g_*\mathscr{O}_Z(m\delta p^*A)) \otimes S^n((g_*\mathscr{O}_Z(m\delta p^*A))^*) \to \mathscr{O}_C.$$

Hence we have a non-trivial homomorphism

$$\bigoplus_{\text{finite}} \mathscr{O}_C \xrightarrow{\alpha} \mathscr{E} \xrightarrow{\beta} \mathscr{O}_C(-nm\pi^*K_Y - nm\pi^*\Delta + 2nm\varepsilon L),$$

where β is induced by the above trace map. Thus we obtain

$$(-nm\pi^*K_Y - nm\pi^*\Delta + 2nm\varepsilon L) \cdot C = nm(-\pi^*K_Y - \pi^*\Delta + 2\varepsilon L) \cdot C \geq 0.$$

Since ε is an arbitrary positive rational number, we obtain

$$\pi^*(-K_Y - \Delta) \cdot C \geq 0.$$

This means that $-K_Y - \Delta$ is nef on Y.

Remark 1. In Theorem 1, if $-(K_X + D)$ is moreover semi-ample, then we can prove very simply that $-K_Y$ is nef (this is a generalization of [8, Theorem 4.1]). First, by Stein factorization, we may assume that f has connected fibers (see [8, Lemma 2.4]). Next, in the proof of Theorem 1, when $-(K_X + D)$ is semi-ample, we can take $\delta = 0$ and $\Delta = 0$ Then

$$g_*\mathscr{O}_Z(m(K_{Z/C} + p^*D + p^*F)) \simeq \mathscr{O}_C(-m\pi^*K_Y)$$

is weakly positive over some non-empty Zariski open set of C. This means that $-m\pi^*K_Y$ is pseudo-effective. Since C is a smooth projective curve, $-\pi^*K_Y$ is nef. Therefore, $-K_Y$ is nef. In this case, we do not need Lemma 1.

Here we give one more alternative proof of [8, Theorem 4.1], which is implicitly contained in Viehweg's theory of weak positivity [11] and is different from the argument in Remark 1.

Theorem 2 ([8, Theorem 4.1]). *Let $f : X \to Y$ be a smooth morphism between smooth projective varieties. If $-K_X$ is semi-ample, then $-K_Y$ is nef.*

Proof. By Stein factorization, we may assume that f has connected fibers (see [8, Lemma 2.4]). Note that a locally free sheaf $\mathscr{E}$ on Y is nef, equivalently, semi-positive in the sense of Fujita–Kawamata, if and only if $\mathscr{E}$ is weakly positive over Y (see, for example, [11, Proposition 2.9 (e)]). Since f is smooth and $-K_X$ is semi-ample, $f_*\mathscr{O}_X(K_{X/Y} - K_X)$ is locally free and weakly positive over Y (cf. [11, Proposition 2.43]). Therefore, we obtain that $\mathscr{O}_Y(-K_Y) \simeq f_*\mathscr{O}_X(K_{X/Y} - K_X)$ is nef.

The argument in Remark 1 and the proof of Theorem 2 are much simpler than the original proof of [8, Theorem 4.1]. However, that original proof played important roles in [8, Remark 4.2] and the proof of the following result, which completely solves [8, Conjecture 1.3].

Theorem 3 ([2, Theorem 1.3]). *Let $f : X \to Y$ be a smooth morphism between smooth projective varieties. If $-K_X$ is semi-ample, then $-K_Y$ is also semi-ample.*

1 (Analytic method). Sébastien Boucksom pointed out that the following theorem, which is a special case of [1, Theorem 1.2], implies [8, Theorem 4.1] and [10, Corollary 2.9]. Note that a line bundle $\mathscr{L}$ on a compact complex manifold is said to be semi-positive (resp. positive) if $\mathscr{L}$ has a smooth hermitian metric whose curvature form is a semi-positive (resp. positive) $(1, 1)$-form.

Theorem 4 (cf. [1, Theorem 1.2]). *Let $f : X \to Y$ be a smooth morphism from a compact Kähler manifold X to a compact complex manifold Y. If $\mathscr{O}_X(-K_X)$ is semi-positive (resp. positive), then $\mathscr{O}_Y(-K_Y)$ is semi-positive (resp. positive).*

The proof of [1, Theorem 1.2] is analytic and does not use mod p reduction arguments.

We close this paper with a remark on [5]. By modifying the proof of Theorem 1 suitably, we can generalize [5, Corollary 3.14] without any difficulties. We leave the details as an exercise for the readers.

Corollary 3 (cf. [5, Corollary 3.14]). *Let $f : X \to Y$ be a surjective morphism from a smooth projective variety X such that Y is smooth in codimension one. Let D be an effective $\mathbb{Q}$-divisor on X such that* SuppD^{hor}, *where D^{hor} is the horizontal part of D, is a simple normal crossing divisor on X and that (X, D) is log canonical over the generic point of Y. Let Δ be a* not necessarily effective *$\mathbb{Q}$-Cartier $\mathbb{Q}$-divisor on Y.*

- *If* $-(K_X + D) - f^*\Delta$ *is nef, then* $-K_Y - \Delta$ *is generically nef.*
- *If* $-(K_X + D) - f^*\Delta$ *is ample, then* $-K_Y - \Delta$ *is generically ample.*

Acknowledgements The first author was partially supported by the Grant-in-Aid for Young Scientists (A) ♯20684001 from JSPS. The second author was partially supported by the Research Fellowships of the Japan Society for the Promotion of Science for Young Scientists. The authors would like to thank Professor Sébastien Boucksom for informing them of Berndtsson's results [1]. They also would like to thank the Erwin Schrödinger International Institute for Mathematical Physics in Vienna for its hospitality. Finally, they would like to thank the referee for his comments, which helped them revise this paper very much.

References

1. B. Berndtsson, Curvature of vector bundles associated to holomorphic fibrations, Ann. Math. (2) **169**(2), 531–560 (2009)
2. C. Birkar, Y. Chen, Images of manifolds with semi-ample anti-canonical divisor. To appear in J. Algbr. Geom.
3. F. Campana, Orbifolds, special varieties and classification theory, Ann. Inst. Fourier (Grenoble) **54**(3), 499–630 (2004)
4. M. Chen, Q. Zhang, On a question of Demailly–Peternell–Schneider. J. Eur. Math. Soc. **15**, 1853–1858 (2013)
5. O. Debarre, *Higher-Dimensional Algebraic Geometry*. Universitext (Springer, New York, 2001)
6. O. Fujino, Higher direct images of log canonical divisors. J. Differ. Geom. **66**(3), 453–479 (2004)
7. O. Fujino, T. Fujisawa, Variations of mixed Hodge structure and semi-positivity theorems. To appear in Publ. Res. Inst. Math. Sci.
8. O. Fujino, Y. Gongyo, On images of weak Fano manifolds. Math. Z. **270**(1–2), 531–544 (2012)
9. Y. Kawamata, Characterization of abelian varieties. Compos. Math. **43**(2), 253–276 (1981)
10. J. Kollár, Y. Miyaoka, S. Mori, Rational connectedness and boundedness of Fano manifolds. J. Differ. Geom. **36**(3), 765–779 (1992)
11. E. Viehweg, *Quasi-projective Moduli for Polarized Manifolds*. Ergebnisse der Mathematik und ihrer Grenzgebiete (3), vol. 30 (Springer, Berlin, 1995)

The Hyperholomorphic Line Bundle

Nigel Hitchin

Dedicated to Klaus Hulek on the occasion of his 60th birthday

Abstract We study the hyperholomorphic line bundle on a hyperkähler manifold with circle action introduced by A. Haydys, and in particular show how it transforms under a hyperkähler quotient. Applications include ALE spaces and coadjoint orbits.

1 Introduction

In a recent paper [9], A.Haydys introduced a natural line bundle with connection on a hyperkähler manifold with an S^1-action of a certain type. The curvature is of type $(1, 1)$ with respect to all complex structures in the hyperkähler family and for this reason is called *hyperholomorphic*. In [11] a description of this line bundle via a holomorphic bundle on the twistor space was given and in this format calculated for a number of examples of interest to physicists. These are mostly moduli spaces of solutions to gauge-theoretic equations.

In this article we give examples with a more geometrical flavour, in particular on minimal resolutions of Kleinian singularities and cotangent bundles of coadjoint orbits of a compact Lie group. We first approach the subject from the differential-geometric point of view, giving some explicit formulae, and then from the twistor viewpoint, where, as in [11], the holomorphic point of view demonstrates a naturality which is not apparent from the explicit expressions.

In a more general result, which contributes to the examples, we show how the hyperholomorphic bundle descends naturally in a hyperkähler quotient, and for

N. Hitchin (✉)
Mathematical Institute, University of Oxford, Woodstock Road, Oxford OX2 6GG, UK
e-mail: hitchin@maths.ox.ac.uk

A. Frühbis-Krüger et al. (eds.), *Algebraic and Complex Geometry*, Springer Proceedings in Mathematics & Statistics 71, DOI 10.1007/978-3-319-05404-9_8,

the quotient by a linear action on flat space can be identified with a canonical hyperholomorphic line bundle.

2 The Differential Geometric Viewpoint

2.1 The Hyperholomorphic Connection

Let M be a hyperkähler manifold with Kähler forms $\omega_1, \omega_2, \omega_3$ relative to complex structures I, J, K. If the de Rham cohomology class $[\omega_1/2\pi] \in H^2(M, \mathbf{R})$ is in the image of the integral cohomology then there exists a line bundle L and hermitian connection ∇ with curvature ω_1, unique up to tensoring with a flat $U(1)$ bundle. Since ω_1 is of type $(1,1)$ with respect to the complex structure I, L also has a holomorphic structure defined by the $\bar{\partial}$-operator $\nabla^{0,1}$. Given a local holomorphic section s of L, then $\omega_1 = dd^c \log \|s\|^2/2$. Hence, if we multiply the hermitian metric by e^{2f} the curvature of the connection on L compatible with this new structure is

$$F = \omega_1 + dd^c f.$$

Suppose now we have a circle action which fixes ω_1 but acts on the other forms by the transformation $(\omega_2 + i\omega_3) \mapsto e^{i\theta}(\omega_2 + i\omega_3)$. The manifold M must necessarily be noncompact for this. Suppose further that we have chosen a lift of the action to L. This implies in particular the existence of a moment map – a function μ such that $i_X\omega_1 = d\mu$ where X is the vector field generated by the action. Then the result of Haydys [9] (see also [11]) is:

Theorem 1. *The 2-form* $\omega_1 + dd^c\mu$ *is of type* $(1,1)$ *with respect to complex structures* I, J, K.

Thus rescaling the natural metric by $e^{2\mu}$ gives a new connection which defines a holomorphic structure on L relative to all complex structures in the quaternionic family. This is a hyperholomorphic connection, and L is the hyperholomorphic bundle of the title.

There are relatively few hyperkähler metrics which one can write down explicitly but it is instructive to find the line bundle in these cases.

Example. Flat quaternionic space $\mathbf{H}^n$. Writing $\mathbf{H}^n = \mathbf{C}^n \oplus j\mathbf{C}^n$ we have

$$\omega_1 = \frac{i}{2}\sum_i (dz_i \wedge d\bar{z}_i + dw_i \wedge d\bar{w}_i), \qquad \omega_2 + i\omega_3 = \sum_i dz_i \wedge dw_i$$

and the action $(z, w) \mapsto (z, e^{i\theta}w)$ is of the required form. Then

$$F = \omega_1 + dd^c\mu = \frac{i}{2}\sum_i (dz_i \wedge d\bar{z}_i - dw_i \wedge d\bar{w}_i).$$

In the complex structure I this is the trivial holomorphic line bundle with hermitian metric $h = (\|z\|^2 - \|w\|^2)/2$.

In the above we have specified a particular action of the circle on the three Kähler forms $\omega_1, \omega_2, \omega_3$. More generally, if an irreducible hyperkähler manifold M has a circle symmetry group then it acts on the three-dimensional space of covariant constant 2-forms preserving the inner product. The action is either trivial, in which case it is called *triholomorphic*, or it leaves fixed a one-dimensional subspace with an orthogonal complement on which the action is rotation by $n\theta$. The case above is $n = 1$. This occurs for example on the cotangent bundle of a complex manifold where the action is scalar multiplication in a fibre and the symplectic form is the canonical one. In the general case, $\mathbf{Z}_n \subset S^1$ preserves the three Kähler forms and so the quotient $M/\mathbf{Z}_n$ is a hyperkähler orbifold with a circle action as above. The local geometry of the hyperholomorphic bundle is then the same, but the curvature form on M is $F = \omega_1 + ndd^c\mu$.

In what follows we shall also consider flat space as above but with the action $(z, w) \mapsto e^{i\theta}(z, w)$. Then $n = 2$ since $(\omega_2 + i\omega_3) \mapsto e^{2i\theta}(\omega_2 + i\omega_3)$. The moment map $\mu = -(\|z\|^2 + \|w\|^2)/2$ and so $F = \omega_1 + 2dd^c\mu = 0$ and the hyperholomorphic line bundle is trivial as a line bundle with connection. This may seem uninteresting, but in Theorem 4 we shall see how it defines the bundle for a hyperkähler quotient of $\mathbf{H}^n$.

2.2 Hermitian Symmetric Spaces

Biquard and Gauduchon gave in [2] an explicit formula for a hyperkähler metric which, in the complex structure I, is defined on the total space of the cotangent bundle of a hermitian symmetric space G/H. A circle action is given by multiplication of a cotangent vector by a unit complex number and the form $\omega_2 + i\omega_3$ is the canonical symplectic form on the cotangent bundle.

If $p : T^*(G/H) \to G/H$ is the projection and ω is the Kähler form of the symmetric space G/H then the hyperkähler metric is defined by $\omega_1 = p^*\omega + dd^c h$ where, for a cotangent vector v, h is the quartic function on the fibres defined by $h(v) = (f(IR(Iv, v))v, v)$. Here $R(u, v)$ is the curvature tensor of G/H and f is the analytic function

$$f(u) = \frac{1}{u}\left(\sqrt{1+u} - 1 - \log\frac{1+\sqrt{1+u}}{2}\right).$$

This function is applied to $IR(Iv, v)$ which is a non-negative hermitian transformation. In fact since the curvature of a symmetric space is constant we can also view the quadratic map $R(Iv, v)$ from $(\mathfrak{g}/\mathfrak{h})^*$ to $\mathfrak{h} \subset \mathfrak{g}$ as a multiple of the moment map for the isotropy action of H. The strange function $f(u)$ has the property that

$$(uf(u))' = \frac{1}{2u}(\sqrt{1+u} - 1) \tag{1}$$

We first calculate the moment map μ for the circle action. Since the action is purely in the fibres of the cotangent bundle we have

$$i_X\omega_1 = i_X(p^*\omega + dd^c h) = i_X dd^c h.$$

Now the action preserves both h and the complex structure so $(di_X + i_X d)d^c h = \mathscr{L}_X d^c h = d^c(\mathscr{L}_X h) = 0$, which means that $i_X\omega_1 = -d(i_X d^c h)$ and we can take $\mu = -i_X d^c h = (IX)(h)$. The vector field X was generated by $v \mapsto e^{i\theta}v$ so IX is generated by $v \mapsto \lambda^{-1}v$ for $\lambda \in \mathbf{R}^+$. Hence

$$\mu(v) = \frac{\partial}{\partial\lambda}h(\lambda^{-1}v)|_{\lambda=1}.$$

But $h(v) = (f(u)v, v)$ where $u = IR(Iv, v)$ is homogeneous of degree 2 in v and so $\mu(v) = -2(uf'(u)v, v) - 2(f(u)v, v) = -2((uf(u))'v, v)$. Using (1) we see that

$$F = \omega_1 + dd^c\mu = p^*\omega + dd^c k$$

where $k(v) = (g(IR(Iv, v))v, v)$ for the function

$$g(u) = -\frac{1}{u}\left(\log\frac{1+\sqrt{1+u}}{2}\right).$$

This is an explicit formula for the curvature of the hyperholomorphic line bundle (assuming ω is normalized so that $[\omega/2\pi]$ is an integral class).

Note that on the zero-section $v = 0$, F restricts to ω and is S^1-invariant. From [4,5] we can say that this is the unique hyperholomorphic extension to $T^*(G/H)$ of this line bundle with connection on G/H. Later we shall view this in a more natural setting.

2.3 Multi-instanton Metrics

The most concrete examples of hyperkähler metrics are the gravitational multi-instantons of Gibbons and Hawking [6]. These are four-dimensional and in this dimension a hyperholomorphic connection is the same thing as an anti-self-dual one. The general Ansatz for this family of metrics consists of taking a harmonic function V on an open set in $\mathbf{R}^3$, with its flat metric. Writing locally $*dV = d\alpha$ the metric has the form

$$g = V(dx_1^2 + dx_2^2 + dx_3^2) + V^{-1}(d\theta + \alpha)^2.$$

Then $\omega_1 = V\,dx_2 \wedge dx_3 + dx_1 \wedge (d\theta + \alpha)$ is a Kähler form and similarly for ω_2, ω_3.

An example is flat space $\mathbf{C}^2$ with a circle action $(z_1, z_2) \mapsto (e^{i\theta}z_1, e^{-i\theta}z_2)$. (Note that this action is triholomorphic, and so is not of the type we have been considering). The quotient space is $\mathbf{R}^3$ with Euclidean coordinates $x_1 = (|z_1|^2 - |z_2|^2)/2, x_2 + ix_3 = z_1 z_2$ and then the metric has the above form if $V = 1/2r$. The flat space $\mathbf{C}^2\backslash\{0\}$ is here expressed as a principal circle bundle over $\mathbf{R}^3\backslash\{0\}$ and $d\theta + \alpha$ is the connection form for the horizontal distribution defined by metric orthogonality. The curvature of the connection is $d\alpha = *dV$ and the function $V^{-1/2}$ is the length of the vector field Y generated by the action.

The general case has the same principal bundle structure but the flat example shows that a $1/r$ singularity for V does not produce a singularity in the metric: it is simply a fixed point of the circle action on the four-manifold. With this in mind, setting

$$V = \sum_{i=1}^{k+1} \frac{1}{|\mathbf{x} - \mathbf{a_i}|}$$

for distinct points $\mathbf{a_i} \in \mathbf{R}^3$ defines a nonsingular, complete hyperkähler manifold M.

If the points $\mathbf{a}_i$ lie on the x_1-axis then rotation about that axis induces an isometric circle action generating a vector field X. This involves lifting the action on $\mathbf{R}^3$ to the S^1-bundle with connection form α, commuting with the circle action. Such a lifting is defined by a vector field of the form $X = X_H + fY$, where X_H is the horizontal lift of

$$x_2 \frac{\partial}{\partial x_3} - x_3 \frac{\partial}{\partial x_2}$$

and, since $*dV$ is the curvature of the connection, $i_X * dV = df$. Since $\mathscr{L}_X V = 0$ the local existence of such an f is assured. This means

$$df = (x_2 V_2 + x_3 V_3)dx_1 - x_2 V_1 dx_2 - x_3 V_1 dx_3.$$

It follows that, with $\mathbf{a}_i = (a_i, 0, 0)$,

$$f = \sum_{i=1}^{k+1} \frac{x_1 - a_i}{|\mathbf{x} - \mathbf{a_i}|} + c. \tag{2}$$

The Kähler form ω_1 is given by $\omega_1 = V\,dx_2 \wedge dx_3 + dx_1 \wedge (d\theta + \alpha)$. This is the curvature of a connection if its periods lie in $2\pi\mathbf{Z}$. Now the segment $[a_i, a_{i+1}]$ defines a one-parameter family of S^1-orbits which become single points over the end-points and therefore form a 2-sphere in M. The manifold retracts onto a neighbourhood of a chain $[a_1, a_2], [a_2, a_3], \ldots, [a_k, a_{k+1}]$ of k such spheres which therefore generate $H_2(M, \mathbf{Z})$. Integrating ω_1 over the ith sphere gives $2\pi(a_{i+1} - a_i)$

and so for integrality we require $a_{i+1} - a_i$ to be an integer. With these conditions we have, from Haydys's theorem, a hyperholomorphic line bundle which, since the two actions commute, is invariant under the triholomorphic circle action on M.

Kronheimer [12] showed that S^1-invariant instantons on the multi-instanton space became monopoles on $\mathbf{R}^3$ with Dirac singularities at the marked points $\mathbf{a}_i$. Since the hyperholomorphic bundle is invariant we can define it this way by a $U(1)$-monopole: a harmonic function ϕ on $\mathbf{R}^3$ and a connection A such that $F = dA = *d\phi$. The Ansatz is

$$\hat{A} = A - \phi V^{-1}(d\theta + \alpha) \tag{3}$$

where $\hat{A}$ is a local connection form on M. Thus

$$\omega_1 + dd^c\mu = dA - d(\phi V^{-1}) \wedge (d\theta + \alpha) - \phi V^{-1} * dV$$

and taking the interior product with Y we obtain

$$i_Y(\omega_1 + dd^c\mu) = -dx_1 + i_Y dd^c\mu = d(\phi V^{-1}).$$

Since Y is triholomorphic, it preserves I and since it commutes with X it preserves μ so as in the previous section $d(\phi V^{-1}) = -dx_1 - d(i_Y d^c\mu)$ and up to an additive constant,

$$\phi V^{-1} = -x_1 - i_Y d^c\mu = -(x_1 + i_Y(Ii_X\omega_1)) = -(x_1 - g(X,Y))$$

Now $g(X,Y) = V^{-1}f$. therefore

$$\phi = -x_1 V + f = -\sum_{i=1}^{k+1} \frac{a_i}{|\mathbf{x} - \mathbf{a_i}|} + c.$$

Note however that $A \mapsto A + c\alpha, \phi \mapsto \phi + cV$ takes $\hat{A}$ to $A + c\alpha - (\phi + cV)V^{-1}(d\theta + \alpha) = \hat{A} - cd\theta$ and so preserves the anti-self-dual curvature form $d\hat{A}$ (this absorbs the constant ambiguity too). We can therefore also take

$$\phi = \sum_{i=1}^{k} \frac{a_{k+1} - a_i}{|\mathbf{x} - \mathbf{a_i}|} + c$$

and since the coefficients $a_{k+1} - a_i$ are integers, this is a genuine $U(1)$-monopole which satisfies the Dirac quantization condition.

There remains the question of the constant c. This is not in general zero since $k = 1$ is flat space and we have seen the non-zero curvature of the connection in the previous section. Here we have by construction also a circle action which preserves all three Kähler forms so given one lifting of the rotation action on $\mathbf{R}^3$ to M we can compose with a homomorphism to the triholomorphic circle to obtain another. The constant c will then change by $2\pi n, n \in \mathbf{Z}$.

Remark. When $c = 0$ the curvature F is a linear combination of $\mathcal{L}^2$ harmonic forms [8, 14]. In this case if $k = 2m$ and $\mathbf{x}$ lies on the x_1-axis with $a_m \leq x_1 \leq a_{m+1}$ then (2) shows that $f = 0$. Note for future reference that this means that the middle 2-sphere in the chain is point-wise fixed by the circle action.

The complex structure I for the metrics above is the minimal resolution of the Kleinian singularity $xy = z^{k+1}$. There are, thanks to Kronheimer [13], hyperkähler metrics on all such resolutions. These are produced by a finite-dimensional hyperkähler quotient construction and this is semi-explicit – the quotient metric of a subspace of flat space defined by a finite number of quadratic equations – but the hyperholomorphic line bundle is well adapted to the quotient construction.

2.4 Hyperkähler Quotients

The hyperkähler quotient construction of [10] proceeds as follows. Given a hyperkähler manifold with a triholomorphic action of a Lie group G we have, under appropriate conditions, three moment maps ν_1, ν_2, ν_3 corresponding to the three Kähler forms $\omega_1, \omega_2, \omega_3$ and hence a vector-valued moment map $\nu : M \to \mathfrak{g}^* \otimes \mathbf{R}^3$. Then, assuming G acts freely on $\nu^{-1}(0)$, the manifold $\bar{M} = \nu^{-1}(0)/G$ with its quotient metric is hyperkähler.

In our situation we have a distinguished complex structure I preserved by a circle action. The construction can then be viewed in a slightly different way. Firstly $\nu_c = \nu_2 + i\nu_3$ is holomorphic with respect to I and so the zero set $M_0 = \nu_c^{-1}(0)$ is a complex submanifold of M and hence ω_1 restricts to it as a Kähler form. The group G preserves M_0 and ν_3 is the moment map for the restriction of ω_1. Hence the hyperkähler quotient is the symplectic quotient of M_0 by this action.

Theorem 2. *Suppose M has a circle action as in Sect. 2.1, commuting with G, so that the hyperkähler quotient $\bar{M}$ has an induced action. Then the hyperholomorphic line bundle on M descends naturally to the hyperholomorphic line bundle of $\bar{M}$.*

Proof. First recall that for a symplectic manifold (N, ω) with $[\omega/2\pi]$ integral there is a line bundle – the prequantum line bundle – with a unitary connection whose curvature is ω. Given a lift of the action of a group G, the invariant sections on the zero set of the moment map define the prequantum line bundle on the symplectic quotient.

To see this more concretely, let Y be the vertical vector field of the principal $U(1)$-bundle P, X_a the vector field on N given by $a \in \mathfrak{g}$ and $\mu : N \to \mathfrak{g}^*$ the moment map. Then a lift commuting with the $U(1)$-action is defined by $(X_a)_H + \langle \mu, a \rangle Y$ where $(X_a)_H$ is the horizontal lift. An arbitrary section of the line bundle is defined by a function f on P, equivariant under the vertical action, and an invariant section satisfies $((X_a)_H + \langle\mu, a\rangle Y)f = 0$. Thus on $\mu^{-1}(0)$ we have $(X_a)_H f = 0$ which means that the section is covariant constant along the G-orbits. Hence the connection is pulled back from the symplectic quotient $\mu^{-1}(0)/G$.

This is the construction for a symplectic manifold. Now suppose we take our hyperkähler manifold with circle action and commuting triholomorphic G-action with hyperkähler moment map ν. We want to apply the above to $N = M_0 = \nu_c^{-1}(0)$ for the symplectic quotient of M_0 is the hyperkähler quotient of M. Now the circle action does not preserve $\omega_2 + i\omega_3$ but it acts on $d\nu_c = d(\nu_2 + i\nu_3)$ by multiplication by $e^{i\theta}$. If we make a choice of moment map so that the action on ν_c is the same scalar multiplication, then the action will preserve $M_0 = \nu_c^{-1}(0)$. Moreover μ, restricted to M_0, is the moment map for ω_1 restricted to M_0.

The line bundle with hyperholomorphic connection on M, and hence its restriction to M_0, was obtained from the prequantum line bundle by rescaling the hermitian metric by $e^{2\mu}$. By what we have just seen, this descends to $\bar{M}$, the symplectic quotient of $N = M_0$. However, G commutes with the circle action and so μ is G-invariant. It is also the moment map for the induced action on the quotient, and it follows that rescaling the prequantum hermitian metric on $\bar{M}$ gives the hyperholomorphic bundle. □

One other aspect of the quotient is that it comes equipped with a canonical principal G-bundle with a hyperholomorphic connection. Indeed $\nu^{-1}(0)/G = \bar{M}$ and $\nu^{-1}(0)$ is the total space of the principal G-bundle. The induced metric defines an orthogonal subspace in the tangent space to the orbit directions and this is the horizontal space of a connection, which is hyperholomorphic. A differential-geometric proof of this was give in [7] but it can be seen very naturally from the twistor space point of view which we carry out in the next section. In fact, with fewer formulae and more geometry, the hyperholomorphic bundle appears much more naturally using holomorphic techniques.

3 The Twistor Viewpoint

3.1 The Holomorphic Bundle

This section is essentially a review of the construction in [11]. The twistor space Z of a hyperkähler manifold M is the product $Z = M \times S^2$ given the complex structure $(I_{\mathbf{u}}, I)$ where $I_{\mathbf{u}} = u_1 I + u_2 J + u_2 K$ for a unit vector $\mathbf{u} \in \mathbf{R}^3$ and where the second factor is the complex structure of $S^2 = \mathbf{P}^1$. The projection $\pi : Z \to \mathbf{P}^1$ is holomorphic and for each $x \in M$, (x, S^2) is a holomorphic section, a *twistor line*.

The fibre over $\mathbf{u} \in S^2$ is the hyperkähler manifold M with complex structure defined by $\mathbf{u}$ but it also has a holomorphic symplectic form relative to this complex structure. Using an affine coordinate ζ on $\mathbf{P}^1$ where $u_2 + iu_3 = 2\zeta/(1 + |\zeta|^2)$ the complex structures $I, -I$ are defined by $\zeta = 0, \infty$ and the holomorphic symplectic form is $(\omega_2+i\omega_3)+2i\omega_1\zeta+(\omega_2-i\omega_3)\zeta^2$. Globally, this is a twisted relative 2-form ω: a holomorphic section of $\Lambda^2 T^*_{Z/\mathbf{P}^1}(2)$ where the (2) denotes the tensor product with the line bundle $\pi^*\mathcal{O}(2)$, reflecting the quadratic dependence on ζ.

Example. The twistor space for flat $\mathbf{H}^n$ is the total space of the vector bundle $\mathbf{C}^{2n}(1)$ over $\mathbf{P}^1$. This is given by holomorphic coordinates $(v, \xi, \zeta) \in \mathbf{C}^{2n+1}$ on the open set U defined by $\zeta \neq \infty$ and $(\tilde{v}, \tilde{\xi}, \tilde{\zeta})$ for V by $\zeta \neq 0$, with identification $(\tilde{v}, \tilde{\xi}, \tilde{\zeta}) = (v/\zeta, \xi/\zeta, 1/\zeta)$ over $\zeta \in \mathbf{C}^*$. In these coordinates Z is expressed as a C^∞-product by the map $(z, w, \zeta) \mapsto (z + \zeta\bar{w}, w - \zeta\bar{z}, \zeta)$.

If a bundle on M has a hyperholomorphic connection its curvature is of type $(1,1)$ with respect to all complex structures parametrized by ζ and it follows that its pull-back to $Z = M \times S^2$ has a holomorphic structure. Conversely any holomorphic vector bundle on Z which is trivial on the twistor lines (x, S^2) defines a hyperholomorphic connection on a vector bundle over M. This is the hyperkähler version of the Atiyah-Ward result for anti-self-dual connections. For a line bundle the triviality on twistor lines is simply the vanishing of the first Chern class. To get a unitary connection we impose a reality condition. It follows that to describe a hyperholomorphic line bundle on M we simply look for a holomorphic line bundle L_Z on Z determined by the circle action.

Example. In flat space with the action $(z, w) \mapsto (z, e^{i\theta}w)$ one can calculate the line bundle directly. The $(1,0)$-forms on Z for $\zeta \neq \infty$ are spanned by $dz_i + \zeta d\bar{w}_i$, $dw_i - \zeta d\bar{z}_i, d\zeta$ and then with

$$\log h_U = \frac{1}{2}\sum_i z_i\bar{z}_i - w_i\bar{w}_i + \zeta\bar{z}_i\bar{w}_i + \bar{\zeta}z_i w_i$$

we find

$$\bar{\partial}_Z \log h_U = \frac{1}{2}\sum z_i w_i d\bar{\zeta} + z_i d\bar{z}_i - w_i d\bar{w}_i + \bar{\zeta}d(z_i w_i)$$

and hence $\bar{\partial}_Z\partial_Z \log h_U = (\sum_i -dz_i d\bar{z}_i + dw_i d\bar{w}_i)/2$, the curvature of the hyperholomorphic line bundle, on the open set U. Defining $\log h_V = -\log h_U(-1/\bar{\zeta})$ on V, the pair (h_U, h_V) defines a hermitian metric on the line bundle with holomorphic transition function on $U \cap V$

$$\exp(-\sum_i v_i\xi_i/2\zeta).$$

The link between the differential geometric and holomorphic points of view is proved in [11]. In fact the line bundle L_Z is essentially the prequantum line bundle for the family of holomorphic symplectic manifolds defined by Z.

To understand this, and to see where the circle action enters in the construction, first note that since $\omega_2 + i\omega_3$ transforms as $(\omega_2 + i\omega_3) \mapsto e^{i\theta}(\omega_2 + i\omega_3)$, differentiating with respect to θ we have $\omega_2 = \mathcal{L}_X\omega_3 = di_X\omega_3$ and so ω_2 and similarly ω_3 are exact. Thus the 2-form $(\omega_2+i\omega_3)/2i\zeta+\omega_1+(\omega_2-i\omega_3)\zeta/2i$ has the same cohomology class as ω_1 for any ζ and is therefore, given the integrality condition on $[\omega_1/2\pi]$, the curvature of a line bundle on M. In the complex structure at ζ, $(\omega_2+i\omega_3)/2i\zeta+\omega_1+(\omega_2-i\omega_3)\zeta/2i$ is of type $(2,0)$ therefore has no $(0,2)$ part: hence the bundle has a holomorphic structure.

Now observe that the induced circle action on the twistor space generates a holomorphic vector field V on Z. Since the action fixes $\pm I$, V projects to the vector field $i\zeta d/d\zeta$ on $\mathbf{P}^1$ vanishing at $\zeta=0,\infty$. This is a holomorphic section s of $\mathcal{O}(2)$ and so the 2-form we wrote above, $(\omega_2+i\omega_3)/2i\zeta+\omega_1+(\omega_2-i\omega_3)\zeta/2i$ is, on a specific fibre, the restriction of the meromorphic relative differential form $\omega/2is\in\Omega^2_{Z/\mathbf{P}^1}$. It turns out that this relative form is the restriction of a closed meromorphic 2-form F_Z on Z, which is the curvature of a meromorphic connection on the holomorphic line bundle.

Theorem 3 ([11]). *The line bundle L_Z on the twistor space Z admits a meromorphic connection such that*

- *There are simple poles at $\zeta=0,\infty$*
- *The curvature F_Z restricts to*

$$\frac{1}{2i\zeta}(\omega_2+i\omega_3)+\omega_1+\frac{1}{2i}(\omega_2-i\omega_3)\zeta$$

on each fibre over $\mathbf{C}^*\subset\mathbf{P}^1$

- *$i_V F_Z=0$ where V is the vector field generated by the circle action.*

Remark. Suppose the holomorphic vector field V integrates to a $\mathbf{C}^*$-action. Then as F_Z is closed, the last property tells us that this action gives a symplectic isomorphism between any of the holomorphic symplectic manifolds over $\zeta\in\mathbf{C}^*$.

Given that such a connection exists, the line bundle is essentially uniquely determined by the residue of the connection, for given any two such bundles L, L' with the connections as above and with the same residue at $\zeta=0,\infty$, the resulting holomorphic connection on $L'L^*$ would have a curvature which is a holomorphic 2-form. But the normal bundle of a twistor line is $\mathbf{C}^{2n}(1)$ and so $T_Z^*\cong\mathbf{C}^{2n}(-1)\oplus\mathcal{O}(-2)$ on such a line. It follows that there are no holomorphic forms of positive degree on a twistor space since there is a twistor line through each point. Hence the connection on $L'L^*$ is flat and this is in any case the ambiguity in choosing a prequantum connection.

The residue is canonically determined by the data of the action as follows (see [11] for details). Since the connection has a singularity on a divisor of $\mathcal{O}(2)$, its residue will be a section of $T_Z^*(2)$ on that divisor. Now since $T_{\mathbf{P}^1}\cong\mathcal{O}(2)$ the projection $\pi: Z\to\mathbf{P}^1$ gives an exact sequence of bundles:

$$0\to\mathcal{O}\to T_Z^*(2)\to T_{Z/\mathbf{P}^1}^*(2)\to 0$$

and the twisted relative form ω identifies $T_{Z/\mathbf{P}^1}$ with $T^*_{Z/\mathbf{P}^1}(2)$. The resulting extension

$$0 \to \mathcal{O} \to E \to T_{Z/\mathbf{P}^1} \to 0$$

can be identified with $TP/\mathbf{C}^*$ where P is the holomorphic principal bundle of the prequantum line bundle for the real symplectic form ω_1. The vector field V on Z is tangential to the fibres at $\zeta = 0, \infty$ and is X itself. The moment map defines an invariant lift to P and hence a section of $TP/\mathbf{C}^*$. Under the isomorphism above, this is the residue of the connection. If we restrict it as a form to the fibre $\zeta = 0$ it is $i_V(\omega_2 + i\omega_3)/2i$

Examples.

(i) For flat space with the action $(z, w) \mapsto e^{i\theta}(z, w)$ the line bundle L_Z is trivial and the connection with the trivial action is just the meromorphic one-form

$$\frac{1}{2\tilde{\zeta}}\sum_i \tilde{\xi}_i d\tilde{v}_i - \tilde{v}_i d\tilde{\xi}_i = \frac{\zeta}{2}\sum_i \frac{\xi_i}{\zeta} d\frac{v_i}{\zeta} - \frac{v_i}{\zeta} d\frac{\xi_i}{\zeta} = \frac{1}{2\zeta}\sum_i \xi_i dv_i - v_i d\xi_i .$$

With the action $u \mapsto e^{in\theta}u$ it is

$$2\pi i n\frac{d\zeta}{\zeta} + \frac{1}{2\zeta}\sum_i \xi_i dv_i - v_i d\xi_i \tag{1}$$

(ii) Flat space with the other action $(z, w) \mapsto (z, e^{i\theta}w)$ requires local connection forms A_U, A_V such that $A_V = A_U + g_{UV}^{-1} dg_{UV}$. Define

$$A_U = \frac{1}{2\zeta}\sum v_i d\xi_i \qquad A_V = -\frac{1}{2\tilde{\zeta}}\sum_i \tilde{v}_i d\tilde{\xi}_i$$

then on $U \cap V$

$$A_V - A_U = -\frac{\zeta}{2}\sum_i \frac{\xi_i}{\zeta} d\frac{v_i}{\zeta} - \frac{1}{2\zeta}\sum_i v_i d\xi_i = -d\left(\frac{1}{2\zeta}\sum_i v_i\xi_i\right).$$

3.2 *Hyperkähler Quotients*

In the twistor formalism the hyperkähler quotient is a very natural operation: it is just the fibrewise holomorphic symplectic quotient as long as the holomorphic vector fields generated by G integrate to an action of the complexification G^c. Each $a \in \mathfrak{g}$ gives a holomorphic vector field V_a tangential to the fibres of $\pi : Z \to \mathbf{P}^1$ and the three moment maps for $V_a, a \in \mathfrak{g}$ give a complex section

$$\boldsymbol{\nu} = (\nu_2 + i\nu_3) + 2i\nu_1\zeta + (\nu_2 - i\nu_3)\zeta^2$$

of $\mathfrak{g}^c(2)$. The twistor space $\bar{Z}$ of the hyperkähler quotient is then simply $\boldsymbol{\nu}^{-1}(0)/G^c$ where the metric plays a role in determining the stable points for this quotient by a complex group. With this viewpoint the descent of the hyperholomorphic bundle through a quotient is, given Theorem 3, the descent of the prequantum line bundle in a symplectic quotient (it is straightforward to check that the residue descends appropriately).

As we saw in the previous section, a hyperkähler quotient brings with it a canonical hyperholomorphic G-bundle. In fact, in the twistor interpretation, $\boldsymbol{\nu}^{-1}(0)$ is a principal G^c-bundle over the twistor space $\bar{Z} = \boldsymbol{\nu}^{-1}(0)/G^c$ and it satisfies the reality condition to define a hyperholomorphic principal G-bundle over $\bar{M}$. A homomorphism $G \to U(1)$ then defines a hyperholomorphic line bundle and this raises the obvious question about whether, given a circle action, this is the hyperholomorphic bundle of the title.

In fact for a manifold to be a smooth hyperkähler quotient of flat space such homomorphisms must exist. The standard moment map for a linear action is quadratic and the origin lies in $\boldsymbol{\nu}^{-1}(0)$, so for smoothness we must change this by a constant. Equivariance however demands that the constant is an invariant in $\mathfrak{g}^*$: a homomorphism from $\mathfrak{g}$ to $\mathbf{R}$.

Consider flat space $\mathbf{H}^n$ as a right $\mathbf{H}$-module, then $U(n) \subset Sp(n)$ is the subgroup commuting with left multiplication by $e^{i\theta}$: this is a distinguished complex structure I. Let $G \subset U(n)$ and $c \in \mathfrak{g}^*$ be a G-invariant element. If c is integral it corresponds to a homomorphism $\chi : G \to U(1)$. Let ν be the standard quadratic hyperkähler moment map for the linear action, then taking the reduction at $\nu = (c, 0, 0)$, the cohomology class of the Kähler form ω_1 lies in $2\pi H^2(\bar{M}, \mathbf{Z})$. Indeed the integrality for c gives a lift of the G-action to the prequantum line bundle on $\nu_c^{-1}(0)$ which descends.

Theorem 4. *If the hyperkähler quotient $\bar{M}$ of $\mathbf{H}^n$ by G with $\nu = (c, 0, 0)$ is smooth, then the hyperholomorphic line bundle is $\nu^{-1}(c, 0, 0) \times_G \mathbf{C}$ endowed with the canonical connection, where G acts via $\chi : G \to U(1)$.*

Proof. From the twistor point of view the line bundle L_Z on the quotient is defined by the property that local sections are the same as local G^c-invariant sections of the holomorphic line bundle on $\boldsymbol{\nu}^{-1}(0)$. For flat space and the circle action above the latter, as we observed in Sect. 2.1, is a trivial holomorphic bundle but has a non-trivial action defined by χ. Thus on Z the line bundle is associated to the principal G^c-bundle $\boldsymbol{\nu}^{-1}(0)$ by χ. □

Examples.

(i) The simplest example is the cotangent bundle of a complex Grassmannian, one of the Hermitian symmetric spaces of Sect. 2.2. In this case the flat space is $M = V \oplus jV$ for $V = \mathrm{Hom}(\mathbf{C}^k, \mathbf{C}^n)$ and $G = U(n)$ acting in the obvious way. There is just a one-dimensional space of invariant elements in $\mathfrak{g}^*$ and

$H^2(\bar{M}, \mathbf{Z}) \cong \mathbf{Z}$. Notice that -1 acting on the vector space is represented by $-1 \in U(n)$ and hence acts trivially on the quotient. It is therefore $e^{2i\theta}$ which acts effectively on the quotient. Since $e^{i\theta}$ acts on $\omega_2 + i\omega_3$ in flat space by multiplication by $e^{2i\theta}$, on the quotient the induced action is the standard one: in fact the fibre action on the cotangent bundle.

(ii) Taking $M = V \oplus jV$ where $V = \operatorname{End} \mathbf{C}^k \oplus \operatorname{Hom}(\mathbf{C}^k, \mathbf{C}^n)$ and $G = U(k)$ one obtains the moduli space of $U(n)$-instantons on $\mathbf{R}^4$ of charge k or, with a non-zero moment map, the moduli space of noncommutative instantons. For $n = 1$ this is the Hilbert scheme $(\mathbf{C}^2)^{[k]}$ of k points on $\mathbf{C}^2$ and the hyperholomorphic line bundle with complex structure I is defined by the exceptional divisor. The circle action is induced from scalar multiplication on $\mathbf{C}^2$ and so the action on $\omega_2 + i\omega_3$ is multiplication by $e^{2i\theta}$, since on the open set of $(\mathbf{C}^2)^{[k]}$ consisting of the configuration space of $\mathbf{C}^2$ the symplectic form is the sum k copies of $dz \wedge dw$.

(iii) In [13] Kronheimer constructed asymptotically locally Euclidean hyperkähler metrics on minimal resolutions of Kleinian singularities ($\mathbf{C}^2/\Gamma$ for $\Gamma \subset SU(2)$ a finite group) by the quotient construction. The construction is as follows. Let $R = L^2(\Gamma)$ be the regular representation, $\mathbf{C}^2$ the basic representation from $\Gamma \subset SU(2)$ and put $M = (\mathbf{C}^2 \otimes \operatorname{End}(R))^\Gamma$. Since $\operatorname{End}(R)$ has real structure $A \mapsto A^*$ and $SU(2) \cong Sp(1)$ this is a quaternionic vector space and the group $G = U(R)^\Gamma$ acts as quaternionic unitary transformations. The ALE space appears as a hyperkähler quotient of M by the action of G. If $R_0, \ldots, R_k$ are the irreducible representations of Γ, of dimension d_i then

$$R = \bigoplus_i \mathbf{C}^{d_i} \otimes R_i$$

and so $U(R)^\Gamma \cong U(d_0) \times \cdots \times U(d_k)$. From the McKay correspondence each R_i labels a vertex of an extended Dynkin diagram of type A, D, E and then

$$M = \bigoplus_{i \to j} \operatorname{Hom}(\mathbf{C}^{d_i}, \mathbf{C}^{d_j}) \tag{2}$$

the sum taken over the edges of the diagram, once with each orientation. As shown in [13], the invariant subspace of $\mathfrak{g}^*$ can be identified with the Cartan subalgebra of the Lie algebra of type A, D, E as can the cohomology $H^2(\bar{M}, \mathbf{R})$, with $H_2(\bar{M}, \mathbf{Z})$ the root lattice. The case of A_k is the multi-instanton metric of Sect. 2.3, where the chain of 2-spheres constructed explicitly realizes the Dynkin diagram of type A_k.

Here the circle action on the symplectic form of the quotient will be the standard one if there is an element in G which acts as -1. For this, from (2) we need to show that there exist $c_i = \pm 1, 0 \le i \le k$, such that if i, j are joined by an edge of the extended Dynkin diagram then $c_i c_j = -1$. For A_1 this is trivial. Consider the extended Dynkin diagram (for $k > 1$) as a simplicial complex, then this is the same as asking that the $\mathbf{Z}_2$-cocycle associating -1 to each 1-simplex is a coboundary. The diagrams of type D_k, E_6, E_7, E_8 are contractible and so have zero first cohomology so this is true. For A_k the diagram is homeomorphic to a circle and the cohomology class in H^1 vanishes if there is an even number of edges, which is when k is odd.

Now the D, E Dynkin diagrams have a trivalent vertex which, in the presence of our circle action, corresponds to a rational curve of self-intersection -2 which is pointwise fixed, since there cannot be just three fixed points. And, as pointed out in Sect. 2.3, when k is odd, the central curve in the A_k case is fixed.

In these cases, with respect to the complex structure I, we have a rational curve of self-intersection -2 and a neighbourhood of such a curve is biholomorphic to a neighbourhood of the zero section of the cotangent bundle. Moreover the circle action is the standard scalar multiplication in the fibre. Applying [4] this means that the Kronheimer metric with circular symmetry is the unique hyperkähler extension of the induced metric on the distinguished 2-sphere.

3.3 Coadjoint Orbits

The Hermitian symmetric spaces which we considered in Sect. 2.2 are special cases of coadjoint orbits of compact semi-simple Lie groups with their canonical Kähler structure. There is a very natural description of the twistor space of a hyperkähler metric on the cotangent bundle of such a space due originally to Burns [3]. In fact, that paper only asserts the existence of such a metric in a neighbourhood of the zero section, but it was written before hyperkähler quotients, and in particular the infinite-dimensional gauge-theoretic versions, were discovered. Much later, armed with a knowledge of existence theorems for Nahm's equations, Biquard [1] revisited this description being assured of global existence. The action of scalar multiplication in the cotangent fibres by $e^{i\theta}$ gives a circle action and we shall now seek a concrete description of the line bundle L_Z using Burns's approach.

Let G be a semisimple compact Lie group with a bi-invariant metric and $z \in \mathfrak{g}$ be an element with centralizer H. Then in the complex group G^c there are parabolic subgroups P_+, P_- with $P_+ \cap P_- = H^c$. The real (co)adjoint orbit $G/H \cong G^c/P_+ \cong G^c/P_-$, and the complex coadjoint orbit is G^c/H^c.

The Lie algebra $\mathfrak{p}_+ = \mathfrak{h} + \mathfrak{n}_+$ where $\mathfrak{n}_+$ is nilpotent, $z \in \mathfrak{h}$ by definition and we define two complex manifolds

$$Z_0 = G^c \times_{P_+} \{\mathbf{C} \cdot z + \mathfrak{n}_+\} \qquad Z_\infty = G^c \times_{P_-} \{\mathbf{C} \cdot z + \mathfrak{n}_-\}.$$

Since $P_\pm$ fixes z modulo $\mathbf{n}_\pm$, the coefficient of z defines a projection $\pi_0 : Z_0 \to \mathbf{C}$ and similarly for Z_∞. The fibre over 0 is the cotangent bundle $T^*(G^c/P_+) \cong G^c \times_{P_+} \mathfrak{n}_+$ and for $\zeta \neq 0$, $G^c \times_{P_+} \{\zeta z + \mathfrak{n}_+\}$ is an affine bundle over G^c/P_+.

There is another description, however, for $(g, \zeta z + x_+) \mapsto (\mathrm{Ad}\, g(\zeta z + x_+), \zeta)$ identifies the fibre at $\zeta \neq 0$ with the G^c-orbit of ζz. Note that the map $z \mapsto \zeta z$ defines an isomorphism with the orbit of z which is not symplectic for the canonical Kostant-Kirillov form ω_{can} but is for its rescaling $\omega_{\mathrm{can}}/\zeta$.

The twistor space is obtained by identifying Z_0, Z_∞ over $\zeta \in \mathbf{C}^*$ by $(x, \zeta) \mapsto (\zeta^{-2}x, \zeta^{-1})$. Then the two projections define $\pi : Z \to \mathbf{P}^1$ and ω_{can} defines the twisted relative symplectic form.

We define line bundles L_+, L_- over Z_0, Z_∞ by pulling back the prequantum line bundle on $G/H = G^c/P_\pm$ using the projections $p_0 : Z_0 \to G^c/P_+$, $p_\infty : Z_\infty \to G^c/P_-$. Then to define a line bundle L_Z on Z we need an isomorphism between L_+ and L_- over $\mathbf{C}^* \subset \mathbf{P}^1$. But the prequantum line bundle is homogeneous, defined by representations $\chi_\pm : P_\pm \to \mathbf{C}^*$, and these agree on $H^c = P_+ \cap P_-$. This therefore gives an isomorphism $p_+^* L_+ \cong p_-^* L_-$ on $Z_0 \cap Z_\infty \cong G^c/H^c \times \mathbf{C}^*$.

To show that this truly is the twistor version of the hyperholomorphic bundle we may simply note that it does generate a hyperholomorphic line bundle but by the invariance of the construction it is homogeneous on the zero section G/H and hence agrees with a hyperholomorphic bundle there. Invoking [4, 5] once more we see that they are isomorphic everywhere.

References

1. O. Biquard, Twisteurs des orbites coadjointes et métriques hyper-pseudokählériennes. Bull. Soc. Math. Fr. **126**, 79–105 (1998)
2. O. Biquard, P. Gauduchon, Hyperkähler metrics on cotangent bundles of Hermitian symmetric spaces, in *Geometry and Physics* (Aarhus, 1995), ed. by J.E. Andersen et al. Lecture Notes in Pure and Applied Mathematics, vol. 184 (Dekker, New York, 1997), pp. 287–298
3. D. Burns, Some examples of the twistor construction, in *Contributions to Several Complex Variables*, ed. by A. Howard, P-M. Wong. Aspects Mathematics, vol. E9 (Vieweg, Braunschweig, 1986), pp. 51–67
4. B. Feix, Hyperkähler metrics on cotangent bundles. J. Reine Angew. Math. **532**, 33–46 (2001)
5. B. Feix, Hypercomplex manifolds and hyperholomorphic bundles. Math. Proc. Camb. Philos. Soc. **133**, 443–457 (2002)
6. G.W. Gibbons, S.W. Hawking, Gravitational multi-instantons. Phys. Lett. **B78**, 430–432 (1978)
7. T. Gocho, H. Nakajima, Einstein-Hermitian connections on hyper-Kähler quotients. J. Math. Soc. Jpn. **44**, 43–51 (1992)
8. T. Hausel, E. Hunsicker, R. Mazzeo, Hodge cohomology of gravitational instantons. Duke Math. J. **122**, 485–548 (2004)
9. A. Haydys, Hyperkähler and quaternionic Kähler manifolds with S^1-symmetries. J. Geom. Phys. **58**, 293–306 (2008)
10. N.J. Hitchin, A. Karlhede, U. Lindström, M. Roček, Hyperkähler metrics and supersymmetry. Comm. Math. Phys. **108**, 535–589 (1987)
11. N.J. Hitchin, On the hyperkähler/quaternion Kähler correspondence. Commun. Math. Phys. **324**, 77–106 (2013). arXiv 1210.0424
12. P.B. Kronheimer, Monopoles and Taub-NUT metrics, Dissertation, Oxford (1985)
13. P.B. Kronheimer, The construction of ALE spaces as hyper-Kähler quotients. J. Differ. Geom. **29**, 665–683 (1989)
14. P.J. Ruback, The motion of Kaluza-Klein monopoles. Commun. Math. Phys. **107**, 93–102 (1986)

Hodge Numbers for the Cohomology of Calabi-Yau Type Local Systems

Henning Hollborn and Stefan Müller–Stach

Klaus Hulek zum 60. Geburtstag gewidmet

Abstract We determine the Hodge numbers of the cohomology group $H^1_{L^2}(S, \mathbb{V}) = H^1(\bar{S}, j_*\mathbb{V})$ using Higgs cohomology, where the local system $\mathbb{V}$ is induced by a family of Calabi-Yau threefolds over a smooth, quasi-projective curve S. This generalizes previous work to the case of quasi-unipotent, but not necessarily unipotent, local monodromies at infinity. We give applications to Rohde's families of Calabi-Yau 3-folds.

1 Introduction

The first L^2-cohomology group $H^1_{L^2}(S, \mathbb{V}) = H^1(\bar{S}, j_*\mathbb{V})$, where $\mathbb{V}$ is a variation of Hodge structures $\mathbb{V}$ of weight m over a smooth, quasi-projective curve $S = \bar{S} \setminus D \stackrel{j}{\hookrightarrow} \bar{S}$, carries a pure Hodge structure of weight $m + 1$ by [12]. The goal of this paper is to continue the study of its Hodge numbers. We build up on the work done in [2], using the methods of Zucker [12], but in addition the equivalent framework of Higgs bundles from the work of Jost, Yang, and Zuo [7]. In [2] the local monodromies were assumed to be unipotent, but we show that one may skip this assumption, and get similar formulae nevertheless. For simplicity, we will assume that all Hodge numbers of $\mathbb{V}$ are equal to one. Such situations occur for

H. Hollborn • S. Müller–Stach (✉)
Mathematisches Institut der Johannes Gutenberg Universität Mainz, Staudingerweg 9, 55099 Mainz, Germany
e-mail: hollborn@uni-mainz.de; mueller-stach@uni-mainz.de

A. Frühbis-Krüger et al. (eds.), *Algebraic and Complex Geometry*, Springer Proceedings in Mathematics & Statistics 71, DOI 10.1007/978-3-319-05404-9_9,

families of elliptic curves, for the transcendental cohomology of families of K3 surfaces with generic Picard number 19, and for certain families of Calabi-Yau 3-folds.

The case of primary interest will be $m = 3$, i.e., families of Calabi-Yau 3-folds. However, for other applications we will also state results for the cases $m = 1$ and $m = 2$, which go back to work of Stiller [11] and Cox-Zucker [3].

The group $H^1_{L^2}(S, \mathbb{V})$ is of interest in theoretical physics [8], as the presence of codimension two cycles on the total space of a fibration of Calabi-Yau 3-folds implies that its $(2,2)$-Hodge number is non-zero.

The plan of this paper is as follows: After reviewing the basics of L^2-Higgs cohomology, we discuss the cases $m = 1$, $m = 2$ and $m = 3$ separately and state the results in each case, comparing with the existing literature. In case $m = 3$ we extend the results from [2] to the case of non-unipotent monodromies at infinity and complete some tables of Hodge numbers there. In the last section we discuss some examples without maximally unipotent degeneration due to J. C. Rohde [4,9]. These examples are interesting as they contain many CM points in moduli induced by underlying Shimura varieties.

2 The Basic Set-Up: Higgs Cohomology

We consider a smooth, connected, projective family $f : X \longrightarrow S$ of m-dimensional varieties over a smooth quasi-projective curve S. Denote by $\bar{S}$ a smooth compactification of S, and by $\bar{f} : \bar{X} \to \bar{S}$ an extension of f to a flat family over $\bar{S}$. Associated to this situation is a local system $\mathbb{V} = R^m f_* \mathbb{C}$ and the corresponding vector bundle $V := \mathbb{V} \otimes \mathcal{O}_S$ on S. We would like to compute $H^1(\bar{S}, j_* \mathbb{V})$ in terms of the degeneration data of $\bar{f}$.

We denote by T the local monodromy matrix around a point in D at infinity. $\mathbb{V}$ has quasi-unipotent monodromies at all points of $D := \bar{S} \setminus S$. If $\bar{f}$ is semistable in codimension one, then the local monodromies are unipotent. After Deligne, the vector bundle V has a quasi-canonical extension $\bar{V}$ to $\bar{S}$ as a vector bundle together with a logarithmic Gauß-Manin connection

$$\bar{\nabla} : \bar{V} \to \bar{V} \otimes \Omega^1_{\bar{S}}(\log D).$$

In the case of unipotent local monodromies $\bar{V}$ has degree zero, but not in the general case. The Hodge filtration $\bar{V} = F^0 \supset F^1 \supset \cdots \supset F^m \supset F^{m+1} = 0$ also extends to $\bar{S}$ and we define

$$E^{p,m-p} := F^p / F^{p+1}$$

as vector bundles on $\bar{S}$. Let

$$E := \bigoplus_{p=0}^{m} E^{p,m-p}$$

be the associated Higgs bundle with Higgs field

$$\vartheta : E \to E \otimes \Omega^1_{\bar{S}}(\log D),$$

where

$$\vartheta : E^{p,m-p} \to E^{p-1,m-p+1} \otimes \Omega^1_{\bar{S}}(\log D)$$

is induced by $\bar{\nabla}$ and Griffiths transversality. In particular, the Higgs bundle induces a complex of vector bundles

$$E^{\bullet} : E \xrightarrow{\vartheta} E \otimes \Omega^1_{\bar{S}}(\log D).$$

Since $\dim(S) = 1$ here, the usual condition $\vartheta \wedge \vartheta = 0$ is empty, and the complex lives only in degrees 0 and 1. The hypercohomology group $\mathbb{H}^1(E^{\bullet})$ computes $H^1(S, \mathbb{V})$ [7, 12].

If the local monodromy matrix T at some point $P \in D$ is unipotent, then its logarithm $N := \log(T)$ is nilpotent. Any nilpotent endomorphism N of a vector space V_0 satisfying $N^m \neq 0$ and $N^{m+1} = 0$ defines a natural increasing filtration on V_0:

$$0 \subset W_{-m} \subset W_{-m+1} \subset \cdots \subset W_0 \subset W_1 \subset \cdots \subset W_m = V_0,$$

which has the following definition: if $N^{m+1} = 0$ but $N^m \neq 0$, we put

$$W_{m-1} = \mathrm{Ker}(N^m), \quad W_{-m} = \mathrm{Im}(N^m).$$

The further groups W_k for $-m < k \leq m-2$ are inductively constructed by requiring that $N(W_k) = \mathrm{Im}(N) \cap W_{k-2} \subset W_{k-2}$ and

$$N^k : \mathrm{Gr}_k^W(V_0) \to \mathrm{Gr}_{-k}^W(V_0)$$

are isomorphisms. If V_0 is the fiber of $\mathbb{V}$ at a smooth point this filtration is called the *monodromy weight filtration*. Prop. 4.1. of [12] states that in the unipotent case one has a resolution which locally looks like

$$0 \to j_* \mathbb{V} \to [W_0 + t\bar{V}] \xrightarrow{\bar{\nabla}} \frac{dt}{t} \otimes [W_{-2} + t\bar{V}] \to 0.$$

In the quasi-unipotent case with no unipotent part one has on the other hand locally a resolution of the form

$$0 \to j_*\mathbb{V} \to \bar{V} \xrightarrow{\bar{\nabla}} \frac{dt}{t} \otimes \bar{V} \to 0$$

by Prop. 6.9. of [12] and the stalk of $j_*\mathbb{V}$ at $t = 0$ is zero.

Zucker also studies the Hodge filtration on $\bar{V}$. Theorem 11.6 in loc. cit. gives eventually a representation of $H^1(\bar{S}, j_*\mathbb{V})$ and its Hodge components. Instead of this de Rham representation we will switch to the corresponding Higgs version.

We can use the monodromy weight filtration W_* on each fiber E_P, $P \in D$ to define the *L^2-Higgs complex*

$$(\Omega^\bullet_{(2)}(E), \theta) : \Omega^0_{(2)}(E) \subset E, \quad \Omega^1_{(2)}(E) \subset E \otimes \Omega^1_{\bar{S}}(\log D),$$

The sub-sheaves in each degree are defined near $P \in D$ as

$$\Omega^0_{(2)}(E) := W_0 + tE, \quad \Omega^1_{(2)}(E) := (W_{-2} + tE) \otimes \Omega^1_{\bar{S}}(\log D).$$

The notation is such that t is a local parameter with $P = \{t = 0\} \in D$ and the monodromy weight filtration is given by the logarithm $N = \log(T)$. At any point $P \in S$ outside D, the L^2-Higgs complex is just given by the Higgs bundle.

It can be shown [7] that the hypercohomology of the L^2– Higgs complex $(\Omega^\bullet_{(2)}(E), \theta)$ is isomorphic to the L^2-cohomology group

$$H^k_{(2)}(S, E) = H^k(\bar{S}, j_*\mathbb{V}) = \mathbb{H}^k(\Omega^\bullet_{(2)}(E), \theta).$$

In the following sections, we study the local structure of $(\Omega^\bullet_{(2)}, \theta)$ for the case of logarithmic Higgs bundles of type $(1, 1, \ldots, 1, 1)$, so that each summand $E^{p,m-p}$ of E is a line bundle. For the points $P \in D$ one has to distinguish cases corresponding to the possible Jordan normal forms of the endomorphism N. The decomposition

$$E = \bigoplus_{p=0}^{m} E^{p,m-p}$$

induces a decomposition

$$\Omega^\bullet_{(2)}(E) = \bigoplus_{p=0}^{m} \Omega^\bullet_{(2)}(E)^{p,m-p},$$

where

$$\Omega^0_{(2)}(E)^{p,m-p} := \Omega^0_{(2)}(E) \cap E^{p,m-p},$$

$$\Omega^1_{(2)}(E)^{p,m-p} := \Omega^1_{(2)}(E) \cap E^{p-1,m-p+1} \otimes \Omega^1_{\bar{S}}(\log D).$$

The hypercohomology spectral sequence associated to this filtration induces the Hodge structure on $H^k_{(2)}(S, E)$.

3 Elliptic Families

In the case of families of elliptic curves ($m = 1$) we obtain from the previous results:

Theorem 1 (Zucker [12]). *The L^2-Higgs complex for E is given by:*

$$\begin{aligned}
\Omega^0_{(2)}(E)^{1,0} &= E^{1,0}(-I) \\
\Omega^0_{(2)}(E)^{0,1} &= E^{0,1} \\
\Omega^1_{(2)}(E)^{1,0} &= E^{1,0}(II) \otimes \Omega^1_{\bar{S}} \\
\Omega^1_{(2)}(E)^{0,1} &= E^{0,1}(II) \otimes \Omega^1_{\bar{S}}
\end{aligned}$$

Here I is the set of points with unipotent local monodromy (denoted by type I_b in the Kodaira classification of singular fibers), II the set of remaining non-unipotent singular points.

Proof. Elliptic fibrations have either unipotent local monodromy T at points of type I, where the Jordan normal form of T is given by the matrix

$$T = \begin{pmatrix} 1 & 1 \\ 0 & 1 \end{pmatrix}$$

or non-unipotent local monodromies, where T is equivalent to

$$T = \begin{pmatrix} \lambda & 1 \\ 0 & \lambda \end{pmatrix}, \text{ or } T = \begin{pmatrix} \lambda_1 & 0 \\ 0 & \lambda_2 \end{pmatrix}.$$

for some roots of unity $\lambda, \lambda_i \neq 1$. In the first case, Zucker [12, Prop. 4.1.] gives a monodromy weight filtration locally at a point $P = \{t = 0\} \in I$ which looks like $W_0 = W_{-1} = tE^{1,0} \oplus E^{0,1}$ and $W_{-2} = tE$, hence the claim. At a non-unipotent point $P \in II$, [12, Prop. 6.9.] shows the claim as well. □

These observations imply the following well-known theorem.

Theorem 2 (Cox-Zucker [3]). *Assume that $\mathbb{V}$ is irreducible, and that $\vartheta : E^{1,0} \to E^{0,1} \otimes \Omega^1_{\bar{S}}(\log D)$ is a non-zero map with $a + |II| > 0$, where $a := \deg E^{1,0}$. Then the Hodge numbers for the pure Hodge structure of weight 2 on $H^1(\bar{S}, j_*\mathbb{V})$ are*

$$h^{2,0} = h^{0,2} = g - 1 + a + |II|, \; h^{1,1} = 2g - 2 - 2a + |I|.$$

This implies the well-known formula $h^1(j_*\mathbb{V}) = 4g - 4 + |I| + 2|II|$, *see [3, page 39].*

Proof. The Higgs complex is given by

$$\begin{array}{ccc} E^{1,0}(-I) & & E^{0,1} \\ & \searrow^{\vartheta}_{\neq 0} & \\ E^{1,0}(II) \otimes \Omega^1_{\bar{S}} & & E^{0,1}(II) \otimes \Omega^1_{\bar{S}} \end{array}$$

Note that both $\Omega^0_{(2)}(E)^{0,1} = E^{0,1}$ and $\Omega^1_{(2)}(E)^{1,0} = E^{1,0}(II) \otimes \Omega^1_{\bar{S}}$ have neither incoming nor outgoing Higgs differential. By Hodge duality, i.e., $h^{2,0} = h^{0,2}$, we get $h^1(E^{0,1}) = h^0(E^{1,0}(II) \otimes \Omega^1_{\bar{S}})$. Under the assumption $a + |II| > 0$ this gives the formula for $h^{2,0} = h^{0,2}$ by applying Riemann-Roch to the line bundle $E^{1,0}(II) \otimes \Omega^1_{\bar{S}}$. $h^{1,1}$ is h^0 of the cokernel of $\vartheta : \Omega^0_{(2)}(E)^{1,0} \to \Omega^1_{(2)}(E)^{0,1}$, hence the difference of the degrees of both line bundles, from which the rest of the assertion follows. □

Remark 1. The assumptions in the theorem are not independent. The condition that $a + |II| > 0$ is not always satisfied, but in many cases: the parabolic degree of any subbundle $F \subset E$ is defined as

$$\deg_p F := \deg F + \sum_{P \in II} \sum_{0 \le \alpha < 1} \alpha \dim(\mathrm{Gr}_\alpha F_P),$$

where Gr_α is the graded piece of the parabolic filtration corresponding to the monodromy $\exp(2\pi i \alpha)$. For $F = E^{0,1}$, a Higgs subbundle of (E, ϑ) with $\vartheta = 0$, one gets $\deg_p(E^{0,1}) \le \deg_p(E) = 0$ by the Simpson correspondence [6, 10, Prop. 2.1], which implies $\deg_p(E^{1,0}) = -\deg_p(E^{0,1}) \ge 0$. Therefore, if the double sum is not zero, i.e., some $\alpha > 0$ occurs, then $0 \le \deg_p E^{1,0} < a + |II|$, since all Hodge numbers are 1.

Remark 2. In the case $\bar{S} = \mathbb{P}^1$ and $\deg E^{0,1} \le -2$, the proof states that $h^1(E^{0,1}) = h^0((E^{0,1})^\vee \otimes \Omega^1_{\bar{S}}) = h^0(E^{1,0}(II) \otimes \Omega^1_{\bar{S}})$. This implies that $E^{0,1} = (E^{1,0})^{-1}(-II)$.

4 Families of K3 Surfaces

With the previous notation, we consider a smooth projective family of K3 surfaces $f : X \longrightarrow S$ with generic Picard number 19 over a smooth curve S. Associated to this situation is a local system $\mathbb{V} \subset R^2 f_* \mathbb{C}$ of rank 3, given fiberwise by the transcendental cohomology. Let

$$E := E^{2,0} \oplus E^{1,1} \oplus E^{0,2}$$

be the associated Higgs bundle with Higgs field

$$\vartheta : E \to E \otimes \Omega^1_{\bar{S}}(\log D).$$

Now we make the following

Assumption. *Each local monodromy is either unipotent or has no unipotent part.* In other words, there are no mixed cases with non-zero unipotent and non–unipotent pieces. This implies that the Jordan normal forms for the local monodromies are

$$T = \begin{pmatrix} 1 & 1 & 0 \\ 0 & 1 & 1 \\ 0 & 0 & 1 \end{pmatrix}, \begin{pmatrix} 1 & 1 & 0 \\ 0 & 1 & 0 \\ 0 & 0 & 1 \end{pmatrix}, \begin{pmatrix} \lambda & 1 & 0 \\ 0 & \lambda & 1 \\ 0 & 0 & \lambda \end{pmatrix}, \begin{pmatrix} \lambda_1 & 0 & 0 \\ 0 & \lambda_2 & 0 \\ 0 & 0 & \lambda_3 \end{pmatrix} \text{ or } \begin{pmatrix} \lambda_1 & 1 & 0 \\ 0 & \lambda_1 & 0 \\ 0 & 0 & \lambda_2 \end{pmatrix},$$

with $\lambda, \lambda_i \neq 1$ roots of unity.

Lemma 1. *Only the Jordan normal forms*

$$T_I = \begin{pmatrix} 1 & 1 & 0 \\ 0 & 1 & 1 \\ 0 & 0 & 1 \end{pmatrix}, T_{II} = \begin{pmatrix} \lambda & 1 & 0 \\ 0 & \lambda & 1 \\ 0 & 0 & \lambda \end{pmatrix}, T_{II} = \begin{pmatrix} \lambda_1 & 0 & 0 \\ 0 & \lambda_2 & 0 \\ 0 & 0 & \lambda_3 \end{pmatrix} \text{ or } T_{II} = \begin{pmatrix} \lambda_1 & 1 & 0 \\ 0 & \lambda_1 & 0 \\ 0 & 0 & \lambda_2 \end{pmatrix}$$

with $\lambda, \lambda_i \neq 1$ occur. The case I is unipotent, the cases II are strictly quasi–unipotent.

Proof. In the unipotent case, as in [2, p. 11], both maps in the sequence

$$E^{2,0} \xrightarrow{N} E^{1,1} \xrightarrow{N} E^{0,2}$$

are dual to each other. Hence, if $N^2 = 0$, both must be zero, which implies $N = 0$. This excludes the second matrix. □

Theorem 3. *The L^2-Higgs complex for E is given by:*

$$\begin{aligned}
\Omega^0_{(2)}(E)^{2,0} &= E^{2,0}(-I) \\
\Omega^0_{(2)}(E)^{1,1} &= E^{1,1} \\
\Omega^0_{(2)}(E)^{0,2} &= E^{0,2} \\
\Omega^1_{(2)}(E)^{2,0} &= E^{2,0}(II) \otimes \Omega^1_{\bar{S}} \\
\Omega^1_{(2)}(E)^{1,1} &= E^{1,1}(II) \otimes \Omega^1_{\bar{S}} \\
\Omega^1_{(2)}(E)^{0,2} &= E^{0,2}(I + II) \otimes \Omega^1_{\bar{S}}
\end{aligned}$$

Here I is again the set of points with unipotent local monodromy, II the set of remaining non-unipotent singular points.

Proof. The proof is exactly as in the case $m = 1$ using [12, Props. 4.1 and 6.9.]. □

Theorem 4. *Assume that $\mathbb{V}$ is irreducible, and that $\vartheta : E^{2,0} \to E^{1,1} \otimes \Omega^1_{\bar{S}}(\log D)$ as well as $\vartheta : E^{1,1} \to E^{0,2} \otimes \Omega^1_{\bar{S}}(\log D)$ are non-zero maps with $a + |II| > 0$, where $a := \deg E^{2,0}$. Then the Hodge numbers for the pure Hodge structure of weight 3 on $H^1(\bar{S}, j_*\mathbb{V})$ are*

$$h^{3,0} = h^{0,3} = g - 1 + a + |II|,\ h^{2,1} = h^{1,2} = 2g - 2 - a + |I| + \frac{1}{2}|II|.$$

In total, one has $h^1(j_*\mathbb{V}) = 6g - 6 + 2|I| + 3|II|$, which agrees with [2, Prop 3.6.].

Proof. The Higgs complex is given by

$$\begin{array}{ccccc}
E^{2,0}(-I) & & E^{1,1} & & E^{0,2} \\
& \overset{\vartheta}{\underset{\neq 0}{\searrow}} & & \overset{\vartheta}{\underset{\neq 0}{\searrow}} & \\
E^{2,0}(II) \otimes \Omega^1_{\bar{S}} & & E^{1,1}(II) \otimes \Omega^1_{\bar{S}} & & E^{0,2}(I+II) \otimes \Omega^1_{\bar{S}}
\end{array}$$

Note that both $\Omega^0_{(2)}(E)^{0,2} = E^{0,2}$ and $\Omega^1_{(2)}(E)^{2,0} = E^{2,0}(+II) \otimes \Omega^1_{\bar{S}}$ have neither incoming nor outgoing Higgs differential. Hodge duality, i.e., $h^{3,0} = h^{0,3}$, implies $h^0(E^{2,0}(II) \otimes \Omega^1_{\bar{S}}) = h^1(E^{0,2})$. Riemann-Roch applied to $E^{2,0}(II)$ then gives the formula for $h^{3,0} = h^{0,3}$ under the assumption $a + |II| > 0$.

The space $H^{2,1}$ is represented as global sections of the cokernel of the map

$$\Omega^0_{(2)}(E)^{2,0} \xrightarrow{\theta} \Omega^1_{(2)}(E)^{1,1},$$

hence we have to count the zeros of a map of line bundles

$$E^{2,0}(-I) \longrightarrow E^{1,1}(+II) \otimes \Omega^1_{\bar{S}}.$$

This number is given by the difference in degrees of the line bundles, so

$$h^{2,1} = h^{1,2} = \deg E^{1,1}(+II) \otimes \Omega^1_{\bar{S}} - \deg E^{2,0}(-I) = 2g - 2 + \deg E^{1,1} - a + |I| + |II|.$$

It is not true that $\deg E^{1,1} = 0$ in the non–unipotent case. Indeed let $b := \deg E^{1,1}$. We thus obtain $h^{2,1} = 2g - 2 + b - a + |I| + |II|$.

Now we use a checking sum: By [2, Prop 3.6.] we know that

$$h^1(j_*\mathbb{V}) = h^{3,0} + h^{2,1} + h^{1,2} + h^{0,3} = 6g - 6 + 2|I| + 3|II|,$$

since by our assumption non–unipotent local monodromies have zero invariant subspace. This implies that $b = -\frac{1}{2}|II|$. □

Remark 3. Condition $a + |II| > 0$ again follows in many cases, see Remark 1. Assume that $\bar{S} = \mathbb{P}^1$ and that $\deg E^{0,2} \leq -2$. The proof states that $h^1(E^{0,2}) = h^0((E^{0,2})^\vee \otimes \Omega^1_{\bar{S}}) = h^0(E^{2,0}(II) \otimes \Omega^1_{\bar{S}})$. This implies that $E^{0,2} = (E^{2,0})^{-1}(-II)$.

5 Families of Calabi–Yau 3–Folds

We consider a smooth projective family of Calabi–Yau 3–folds $f : X \longrightarrow S$ over a smooth curve S as in [2]. Assume that $\bar{S}$ is a smooth compactification and consider a real VHS $\mathbb{V} \subset R^3 f_* \mathbb{C}$ of rank 4 with Hodge numbers $(1, 1, 1, 1)$. We use the previous notation and make again the assumption that each local monodromy is either unipotent or has no unipotent part.

This implies that the Jordan forms for the local monodromies T are

$$\begin{pmatrix}1&0&0&0\\0&1&1&0\\0&0&1&0\\0&0&0&1\end{pmatrix}, \begin{pmatrix}1&1&0&0\\0&1&0&0\\0&0&1&1\\0&0&0&1\end{pmatrix}, \begin{pmatrix}1&1&0&0\\0&1&1&0\\0&0&1&0\\0&0&0&1\end{pmatrix}, \begin{pmatrix}1&1&0&0\\0&1&1&0\\0&0&1&1\\0&0&0&1\end{pmatrix}, \begin{pmatrix}\lambda_1&0&0&0\\0&\lambda_2&0&0\\0&0&\lambda_3&0\\0&0&0&\lambda_4\end{pmatrix},$$

$$\begin{pmatrix}\lambda_1&0&0&0\\0&\lambda_2&1&0\\0&0&\lambda_2&0\\0&0&0&\lambda_3\end{pmatrix}, \begin{pmatrix}\lambda_1&1&0&0\\0&\lambda_1&0&0\\0&0&\lambda_2&1\\0&0&0&\lambda_2\end{pmatrix}, \begin{pmatrix}\lambda_1&1&0&0\\0&\lambda_1&1&0\\0&0&\lambda_1&0\\0&0&0&\lambda_2\end{pmatrix} \text{ or } \begin{pmatrix}\lambda&1&0&0\\0&\lambda&1&0\\0&0&\lambda&1\\0&0&0&\lambda\end{pmatrix},$$

with $\lambda, \lambda_i \neq 1$ roots of unity.

Lemma 2. *Only the Jordan normal forms*

$$T_I = \begin{pmatrix}1&0&0&0\\0&1&1&0\\0&0&1&0\\0&0&0&1\end{pmatrix}, T_{II} = \begin{pmatrix}1&1&0&0\\0&1&0&0\\0&0&1&1\\0&0&0&1\end{pmatrix}, T_{III} = \begin{pmatrix}1&1&0&0\\0&1&1&0\\0&0&1&1\\0&0&0&1\end{pmatrix},$$

$$T_{IV} = \begin{pmatrix}\lambda_1&0&0&0\\0&\lambda_2&0&0\\0&0&\lambda_3&0\\0&0&0&\lambda_4\end{pmatrix}, T_{IV} = \begin{pmatrix}\lambda_1&0&0&0\\0&\lambda_2&1&0\\0&0&\lambda_2&0\\0&0&0&\lambda_3\end{pmatrix}, T_{IV} = \begin{pmatrix}\lambda_1&1&0&0\\0&\lambda_1&0&0\\0&0&\lambda_2&1\\0&0&0&\lambda_2\end{pmatrix},$$

$$T_{IV} = \begin{pmatrix}\lambda_1&1&0&0\\0&\lambda_1&1&0\\0&0&\lambda_1&0\\0&0&0&\lambda_2\end{pmatrix} \text{ or } T_{IV} = \begin{pmatrix}\lambda&1&0&0\\0&\lambda&1&0\\0&0&\lambda&1\\0&0&0&\lambda\end{pmatrix}$$

with $\lambda, \lambda_i \neq 1$ occur. The cases I, II and III are unipotent, the cases IV are strictly quasi–unipotent.

Proof. See the discussion of normal forms in [2, Sect. 1]. □

Theorem 5. *The L^2-Higgs complex for E is given by:*

$$\begin{aligned}
\Omega^0_{(2)}(E)^{3,0} &= E^{3,0}(-II - III)\\
\Omega^0_{(2)}(E)^{2,1} &= E^{2,1}(-I - III)\\
\Omega^0_{(2)}(E)^{1,2} &= E^{1,2}(-II)\\
\Omega^0_{(2)}(E)^{0,3} &= E^{0,3}\\
\Omega^1_{(2)}(E)^{3,0} &= E^{3,0}(IV) \otimes \Omega^1_{\bar{S}}\\
\Omega^1_{(2)}(E)^{2,1} &= E^{2,1}(IV) \otimes \Omega^1_{\bar{S}}\\
\Omega^1_{(2)}(E)^{1,2} &= E^{1,2}(IV) \otimes \Omega^1_{\bar{S}}\\
\Omega^1_{(2)}(E)^{0,3} &= E^{0,3}(III + IV) \otimes \Omega^1_{\bar{S}}
\end{aligned}$$

Here I, II, III are again the sets of points with unipotent local monodromy, IV the set of remaining non-unipotent singular points.

Proof. The proof is exactly as in the cases $m = 1$ and $m = 2$ using [12, Props. 4.1 and 6.9.]. □

In summary, we get the following result, which agrees with [2, Prop 3.6.] in the unipotent case.

Theorem 6. *Assume that $\mathbb{V}$ is irreducible, and that $\vartheta : E^{3,0} \to E^{2,1} \otimes \Omega^1_{\bar{S}}(\log D)$ as well as $\vartheta : E^{2,1} \to E^{1,2} \otimes \Omega^1_{\bar{S}}(\log D)$ and $\vartheta : E^{1,2} \to E^{0,3} \otimes \Omega^1_{\bar{S}}(\log D)$ are non-zero maps with $a + |IV| > 0$, where $a := \deg E^{3,0}$ and $b := \deg E^{2,1}$. Then the Hodge numbers for the pure Hodge structure of weight 4 on $H^1(\bar{S}, j_*\mathbb{V})$ are*

$$h^{4,0}=h^{0,4} = g - 1 + a + |IV|,\ h^{3,1} = h^{1,3} = 2g - 2 + b - a + |II| + |III| + |IV|,$$
$$h^{2,2} = |I| + |III| - 2b + 2g - 2.$$

In total, one has

$$h^1(j_*\mathbb{V}) = 8g - 8 + |I| + 2|II| + 3|III| + 4|IV|.$$

Proof. The Higgs complex is given by

$$\begin{array}{cccccccc}
E^{3,0}(-II-III) & & E^{2,1}(-I-III) & & E^{1,2}(-II) & & E^{0,3} & \\
& \searrow^{\vartheta}_{\neq 0} & & \searrow^{\vartheta}_{\neq 0} & & \searrow^{\vartheta}_{\neq 0} & & \\
E^{3,0}(IV)\otimes\Omega^1_{\bar{S}} & & E^{2,1}(IV)\otimes\Omega^1_{\bar{S}} & & E^{1,2}(IV)\otimes\Omega^1_{\bar{S}} & & E^{0,3}(III+IV)\otimes\Omega^1_{\bar{S}} &
\end{array}$$

Note that both $\Omega^0_{(2)}(E)^{0,3} = E^{0,3}$ and $\Omega^1_{(2)}(E)^{3,0} = E^{3,0}(IV) \otimes \Omega^1_{\bar{S}}$ have neither incoming nor outgoing Higgs differential. Hodge duality, i.e., $h^{4,0} = h^{0,4}$, implies $h^0(E^{3,0}(IV) \otimes \Omega^1_{\bar{S}}) = h^1(E^{0,3})$. Riemann-Roch applied to $E^{3,0}(IV)$ then gives the formula for $h^{4,0} = h^{0,4}$ under the assumption $a + |IV| > 0$.

As in [2] the space $H^{3,1}$ is represented as global sections of the cokernel of the map $\Omega^0_{(2)}(E)^{3,0} \xrightarrow{\theta} \Omega^1_{(2)}(E)^{2,1}$, hence we have to count the zeros of a map of line bundles

$$E^{3,0}(-II - III) \longrightarrow E^{2,1}(IV) \otimes \Omega^1_{\bar{S}}.$$

This number is therefore given by the difference in degrees of the line bundles, i.e.,

$$h^{3,1} = h^{1,3} = \deg E^{2,1}(IV) \otimes \Omega^1_{\bar{S}} - \deg E^{3,0}(-II-III) = b + 2g - 2 - a + |II| + |III| + |IV|.$$

In a similar way, $H^{2,2}$ is represented as global sections of the cokernel of the map $\Omega^0_{(2)}(E)^{2,1} \xrightarrow{\theta} \Omega^1_{(2)}(E)^{1,2}$, hence we have to count the zeros of the map of line bundles

$$E^{2,1}(-I - III) \longrightarrow E^{1,2}(+IV) \otimes \Omega^1_{\bar{S}}.$$

□

Remark 4. Condition $a + |IV| > 0$ again follows in many cases, see Remark 1. Assume that $\bar{S} = \mathbb{P}^1$ and that $\deg(E^{0,3}) \leq -2$. The proof states that $h^1(E^{0,3}) = h^0((E^{0,3})^\vee \otimes \Omega^1_{\bar{S}}) = h^0(E^{3,0}(IV) \otimes \Omega^1_{\bar{S}})$. This implies that $E^{0,3} = (E^{3,0})^{-1}(-IV)$. Hence, if $a' := -\deg E^{0,3}$, one has $a' = a + |IV|$.

It is not clear that $\deg E^{1,2} = -\deg E^{2,1}$. Indeed let $b' := -\deg E^{1,2}$. We obtain $h^{2,2} = |I| + |III| + |IV| - b - b' + 2g - 2$.

Now we use a checking sum: By [2, Prop 3.6.] we know that

$$h^1(j_*\mathbb{V}) = h^{4,0} + h^{3,1} + h^{2,2} + h^{1,3} + h^{0,4} = 8g - 8 + |I| + 2|II| + 3|III| + 4|IV|,$$

since by our assumption non–unipotent local monodromies have zero invariant subspace. This implies that $b' = b + |IV|$.

Using the formulas obtained above, one can revisit the tables for Hodge numbers in [2] and add the degrees a and b of the Hodge bundles (see table). In the table, e is the degree of a covering map $\mathbb{P}^1 \to \mathbb{P}^1$ of the form $z \mapsto z^e$ ramified in 0 and ∞. The numbering follows the database (Almkvist G, van Enckevort C, van Straten D, Zudilin W, Tables of Calabi-Yau equations, arXiv:math/0507430, unpublished).

In the following sections, we need in addition the following upper bound for a from the work of Jost and Zuo:

Theorem 7 ([6, Theorem 1]).

$$\deg E^{3,0} \leq \left(\frac{1}{2}(h^{2,1} - h_0^{2,1}) + (h^{3,0} - h_0^{3,0})\right)(2g - 2 + \sharp D),$$

where a subscript 0 *denotes the kernel of* ϑ. *More generally, if* $\mathbb{V}$ *is a real VHS of odd weight* $k = 2l + 1 \geq 1$, *then one has*

$$\deg E^{k,0} \leq \left(\frac{1}{2}(h^{k-l,l} - h_0^{k-l,l}) + \sum_{j=0}^{l-1}(h^{k-j,j} - h_0^{k-j,j})\right)(2g - 2 + \sharp D).$$

If we assume that all maps ϑ are non-zero (except the one on $E^{0,3}$), and all ranks $h^{p,q}$ are 1 as in our case, then the inequality simply becomes:

$$\deg E^{3,0} \leq \frac{3}{2}(2g - 2 + \sharp D).$$

In the case $\bar{S} = \mathbb{P}^1$ we therefore obtain $\deg E^{3,0} \leq \frac{3}{2}(\sharp D - 2)$. In the case of 3 singular points, we get $\deg E^{3,0} \leq \frac{3}{2}$, hence $a = \deg E^{3,0} \leq 1$.

6 Rohde's Example

In [4, 9] one finds examples of one-dimensional families $f : X \to S$ of certain Calabi-Yau 3-folds. Their construction is induced by a Borcea-Voisin method, i.e., is obtained from a product of a fixed elliptic curve E and a K3 surface S_λ by application of certain automorphisms. To describe the underlying VHS, in section 2 of [4] a family of genus two Picard curves C_λ is constructed, given by a triple covering $C_\lambda \to \mathbb{P}^1$, and thus coming with an automorphism ξ of order three. The cohomology $H^1(C_\lambda, \mathbb{Q})$ has an eigenspace decomposition according to the eigenvalues ξ and $\bar{\xi} = \xi^2$ and it is strongly related to the cohomology of the fibers of f. Namely, one has

$$H^{3,0}(X_\lambda, \mathbb{Q}) = H^{1,0}(C_\lambda)_{\bar{\xi}},\ H^{2,1}(X_\lambda, \mathbb{Q}) = H^{0,1}(C_\lambda)_{\bar{\xi}},$$
$$H^{1,2}(X_\lambda, \mathbb{Q}) = H^{1,0}(C_\lambda)_{\xi},\ H^{0,3}(X_\lambda, \mathbb{Q}) = H^{0,1}(C_\lambda)_{\xi}.$$

Furthermore, the family C_λ is induced from a Shimura family, see [4]. As a consequence, the Higgs map ϑ induces non-zero maps

$$\vartheta : E^{3,0} \longrightarrow E^{2,1} \otimes \Omega^1_{\bar{S}}(\log D),\ \vartheta : E^{1,2} \longrightarrow E^{0,3} \otimes \Omega^1_{\bar{S}}(\log D)$$

#	Model	T_∞	e	$h^1(j_*\mathbb{V})$	$h^{4,0}$	$h^{3,1}$	$h^{2,2}$	a	b
1	$\mathbb{P}^4[5]$	*IV*	1	0	0	0	0	0	0
			2	1	0	0	1	0	0
			5	0	0	0	0	1	2
			10	1	1	1	1	2	4
2	$\mathbb{P}(1,1,1,2,5)[10]$	*IV*	1	0	0	0	0	0	0
			2	1	0	0	1	0	0
			5	4	0	$0,1,2$	$4,2,0$	0	$0,1,2$
			10	5	0	$0,1,2$	$5,3,1$	1	$2,3,4$
3	$\mathbb{P}^7[2,2,2,2]$	*III*	1	0	0	0	0	0	0
			2	0	0	0	0	1	1
			$2k$	$2k-2$	$k-1$	0	0	k	k
4	$\mathbb{P}^5[3,3]$	*II*	1	0	0	0	0	0	0
			2	1	0	0	1	0	0
			3	0	0	0	0	1	1
			6	3	1	0	1	2	2
5	$\mathbb{P}^6[2,2,3]$	*I*	1	0	0	0	0	0	0
			6	2	0	0 or 1	2 or 0	1	2 or 3
6	$\mathbb{P}^5[2,4]$	*I*	1	0	0	0	0	0	0
			4	0	0	0	0	1	2
			8	4	1	1	0	2	4
7	$\mathbb{P}(1,1,1,1,4)[8]$	*IV*	1	0	0	0	0	0	0
			2	1	0	0	1	0	0
			4	3	0	0 or 1	3 or 1	0	0 or 1
			8	3	0	0 or 1	3 or 1	1	2 or 3
8	$\mathbb{P}(1,1,1,1,2)[6]$	*IV*	1	0	0	0	0	0	0
			2	1	0	0	1	0	0
			6	1	0	0	1	1	2
9	$\mathbb{P}(1,1,1,1,4,6)[2,12]$	*IV*	1	0	0	0	0	0	0
			2	1	0	0	1	0	0
			3	2	0	0 or 1	2 or 0	0	0 or 1
			4	3	0	0 or 1	3 or 1	0	0 or 1
			6	5	0	0,1,2	5,3,1	0	0,1,2
			12	7	0	0,1,2,3	7,5,3,1	1	2,3,4,5
10	$\mathbb{P}(1,1,1,1,2,2)[4,4]$	*II*	1	0	0	0	0	0	0
			2	1	0	0	1	0	0
			4	1	0	0	1	1	1
			8	5	1	0	3	2	2
11	$\mathbb{P}(1,1,1,2,2,3)[4,6]$	*IV*	1	0	0	0	0	0	0
			2	1	0	0	1	0	0
			12	7	0	$0,1,2,3$	$7,5,3,1$	1	$2,3,4,5$
12	$\mathbb{P}(1,1,1,1,1,2)[3,4]$	*IV*	1	0	0	0	0	0	0
			2	1	0	0	1	0	0
			3	2	0	0 or 1	2 or 0	0	0 or 1
			12	7	0	$0,1,2,3$	$7,5,3,1$	1	$2,3,4,5$

(continued)

#	Model	T_∞	e	$h^1(j_*\mathbb{V})$	$h^{4,0}$	$h^{3,1}$	$h^{2,2}$	a	b
13	$\mathbb{P}(1,1,2,2,3,3)[6,6]$	II	1	0	0	0	0	0	0
			2	1	0	0	1	0	0
			3	2	0	0 or 1	2 or 0	0	0 or 1
			6	3	0	0 or 1	3 or 1	1	1 or 2
14	$\mathbb{P}(1,1,1,1,1,3)[2,6]$	I	1	0	0	0	0	0	0
			3	2	0	0 or 1	2 or 0	0	0 or 1
			6	2	0	0 or 1	2 or 0	1	2 or 3

induced by the corresponding Higgs fields for the family C_λ, and the zero morphism

$$\vartheta : E^{2,1} \xrightarrow{0} E^{1,2} \otimes \Omega^1_{\bar{S}}(\log D),$$

by noting that Higgs fields respect eigenspace decompositions.

In this case one knows a little bit more about a and b: One has $\bar{S} = \mathbb{P}^1$ and $\sharp D = 3$ singular points, one of them of type IV. Hence $|IV| = 1$ and $|II| = 2$ in our case. This follows from [4, Sec. 2] from the fact that the resulting Picard-Fuchs equation is a classical hypergeometric equation with singularities at $0, 1, \infty$. Let $F = F^{1,0} \oplus F^{0,1}$ be the Higgs bundle associated to the variation of the genus two curves C_λ. Then F decomposes according to eigenspaces, i.e., $F = F_\xi \oplus F_{\bar{\xi}}$. Due to the existence of non-unipotent points, $F^{1,0}_\square$ and $F^{0,1}_\square$ for $\square \in \{\xi, \bar{\xi}\}$ are not dual to each other. One has:

Lemma 3. *In Rohde's example, each rank two Higgs bundle* $F_\square$ *has a maximal Higgs field, i.e.,*

$$\vartheta : F^{1,0}_\square \xrightarrow{\cong} F^{0,1}_\square \otimes \Omega^1_{\mathbb{P}^1}(\log D).$$

is an isomorphism. Furthermore, $\deg F^{1,0}_\square = 0$ *and* $\deg F^{0,1}_\square = -1$.

Proof. Theorem 7, i.e., the Arakelov inequality of Jost and Zuo [6, Thm. 1], implies that $\deg F^{1,0}_\square \leq \frac{1}{2}$, hence $\deg F^{1,0}_\square \leq 0$. On the other hand, one has $\deg F^{1,0}_\square \geq 0$: Consider the local system $\mathbb{W}_\square$ corresponding to $F_\square$. It satisfies $h^2(\mathbb{P}^1, j_*\mathbb{W}_\square) = h^0(\mathbb{P}^1, j_*\mathbb{W}_\square) = 0$ by the argument of [2, Prop 3.6.]. The Higgs complex for $F_\square$ is given by

$$\begin{array}{ccc} F^{1,0}_\square(-I) & & F^{0,1}_\square \\ & \searrow^{\vartheta}_{\neq 0} & \\ F^{1,0}_\square(II) \otimes \Omega^1_{\mathbb{P}^1} & & F^{0,1}_\square(II) \otimes \Omega^1_{\mathbb{P}^1} \end{array}$$

as in the proof of Theorem 2. Therefore, $H^1(\mathbb{P}^1, F^{1,0}_\square(II) \otimes \Omega^1_{\mathbb{P}^1})$ is the direct summand of Hodge type $(2,1)$ inside $H^2(\mathbb{P}^1, j_*\mathbb{W}_\square) = 0$. Since $|II| = 1$, we obtain $0 = h^1(\mathbb{P}^1, F_\square^{1,0}(1) \otimes \Omega^1_{\mathbb{P}^1}) = h^0(\mathbb{P}^1, (F_\square^{1,0})^{-1}(-1))$. Therefore $F_\square^{1,0} = \mathcal{O}_{\mathbb{P}^1}$. In a similar way, $H^0(F_\square^{0,1})$ contributes to $H^0(\mathbb{P}^1, j_*\mathbb{W}_\square) = 0$, therefore $\deg F_\square^{0,1} < 0$. Since ϑ is a non-zero map, and $\deg(\Omega^1_{\mathbb{P}^1}(\log D)) = 1$, we get that $\deg F_\square^{0,1} = -1$ and $F_\square^{0,1} = \mathcal{O}_{\mathbb{P}^1}(-1)$. □

Corollary 1. *It follows that for the Higgs bundle E one has $a = 0$ and $b = -1$. Furthermore, the Higgs maps $\vartheta : E^{3,0} \longrightarrow E^{2,1} \otimes \Omega^1_{\mathbb{P}^1}(\log D)$, and $\vartheta : E^{1,2} \longrightarrow E^{0,3} \otimes \Omega^1_{\mathbb{P}^1}(\log D)$ are both isomorphisms.*

Note that the identities for $a' = a + |IV|$, $b' = b + |IV|$ in Remark 4 still hold. The properties of E we have shown are summarized in the following definition.

Definition 1. A logarithmic Higgs bundle $E = E^{3,0} \oplus E^{2,1} \oplus E^{1,2} \oplus E^{0,3}$ of weight $m = 3$ and rank 4 on $\bar{S}$ is called decomposed, if $\vartheta : E^{3,0} \to E^{2,1} \otimes \Omega^1_{\bar{S}}(\log D)$ and $\vartheta : E^{1,2} \to E^{0,3} \otimes \Omega^1_{\bar{S}}(\log D)$ are isomorphisms, and $\vartheta : E^{2,1} \to E^{1,2} \otimes \Omega^1_{\bar{S}}(\log D)$ is the zero map.

Theorem 8. *The L^2-Higgs cohomology of a decomposed Higgs bundle $E = E^{3,0} \oplus E^{2,1} \oplus E^{1,2} \oplus E^{0,3}$ of weight $m = 3$ and rank 4 with $a + |IV| > 0$ is described as follows:*

$$h^1_{L^2}(S, \mathbb{V})^{(4,0)} = h^0(\bar{S}, E^{3,0}(IV) \otimes \Omega^1_{\bar{S}}) = g - 1 + a + |IV|,\ h^1_{L^2}(S, \mathbb{V})^{(3,1)} = h^1_{L^2}(S, \mathbb{V})^{(1,3)} = 0,$$

$$h^1_{L^2}(S, \mathbb{V})^{(2,2)} = h^0(\bar{S}, E^{1,2}(IV) \otimes \Omega^1_{\bar{S}}) \oplus h^1(\bar{S}, E^{2,1}(-I - III)) = 2h^0(\bar{S}, E^{1,2}(IV) \otimes \Omega^1_{\bar{S}}).$$

The assumptions imply that $|I| = |III| = 0$ and $a = b + 2g - 2 + \sharp D$.

Proof. We use the same notations for the L^2-Higgs complex $\Omega^0_{(2)}(E) \xrightarrow{\vartheta} \Omega^1_{(2)}(E)$ as above. The symmetry of decomposed Higgs bundles implies that $|I| = |III| = 0$, since such degenerations cannot occur. As E is decomposed, also the arrow $\Omega^0_{(2)}(E)^{2,1} \to \Omega^1_{(2)}(E)^{1,2}$ is still zero. Also the two non-zero arrows in the following diagram remain isomorphisms (which implies again that $|I| = 0$):

$$\begin{array}{ccccccc}
E^{3,0}(-II-III) & & E^{2,1}(-I-III) & & E^{1,2}(-II) & & E^{0,3} \\
& \underset{\cong}{\overset{\vartheta}{\searrow}} & & \underset{0}{\overset{\vartheta}{\searrow}} & & \underset{\cong}{\overset{\vartheta}{\searrow}} & \\
E^{3,0}(IV) \otimes \Omega^1_{\bar{S}} & & E^{2,1}(IV) \otimes \Omega^1_{\bar{S}} & & E^{1,2}(IV) \otimes \Omega^1_{\bar{S}} & & E^{0,3}(III+IV) \otimes \Omega^1_{\bar{S}}
\end{array}$$

It follows that $a = b + 2g - 2 + |II| + |III| + |IV|$, and by Riemann-Roch, using the assumption $a + |IV| > 0$,

$$h^1_{L^2}(S, \mathbb{V})^{(4,0)} = h^1_{L^2}(S, \mathbb{V})^{(0,4)} = h^0(\bar{S}, E^{3,0}(IV) \otimes \Omega^1_{\bar{S}}) = g - 1 + a + |IV|,$$

$$h^1_{L^2}(S, \mathbb{V})^{(3,1)} = h^1_{L^2}(S, \mathbb{V})^{(1,3)} = 0,$$

$$h^1_{L^2}(S, \mathbb{V})^{(2,2)} = h^0(\bar{S}, E^{1,2}(IV) \otimes \Omega^1_{\bar{S}}) \oplus h^1(\bar{S}, E^{2,1}(-I - III)).$$

In the last line, the two summands are dual to each other, which implies again $|I| = |III| = 0$, and $h^1_{L^2}(S,\mathbb{V})^{(2,2)} = 2h^0(\bar{S}, E^{1,2}(IV) \otimes \Omega^1_{\bar{S}})$. □

Theorem 8 implies:

Corollary 2. *In Rohde's example one has* $h^1_{L^2}(S,\mathbb{V}) = 0$*, consequently all Hodge numbers vanish:*

$$h^1_{L^2}(S,\mathbb{V})^{(4,0)}=h^1_{L^2}(S,\mathbb{V})^{(0,4)}=h^1_{L^2}(S,\mathbb{V})^{(3,1)}=h^1_{L^2}(S,\mathbb{V})^{(1,3)}=h^1_{L^2}(S,\mathbb{V})^{(2,2)}=0.$$

In particular, since $|I| = |III| = 0$ and $|II| = 2$, $|IV| = 1$, the check sum

$$h^1(j_*\mathbb{V}) = h^{4,0}+h^{3,1}+h^{2,2}+h^{1,3}+h^{0,4} = 8g-8+|I|+2|II|+3|III|+4|IV| = 0$$

is correct. Base change maps $e : \mathbb{P}^1 \to \mathbb{P}^1$ with prescribed ramification lead to more families where the theorem can be applied. Details can be found in the forthcoming thesis of Henning Hollborn [5].

Acknowledgements This work builds up on [2] in an essential way, and we would like to thank P. L. del Angel, D. van Straten, and K. Zuo for the previous collaboration. We thank Xuanming Ye for several additional discussions, and the referee for some valuable improvements. This work was supported by DFG Sonderforschungsbereich/Transregio 45.

References

1. G. Almkvist, C. van Enckevort, D. van Straten, W. Zudilin, Tables of Calabi-Yau equations, unpublished. arXiv:math/0507430
2. P.L. del Angel, S. Müller-Stach, D. van Straten, K. Zuo, Hodge classes associated to 1-parameter families of Calabi-Yau 3-folds. Acta Mathematica Vietnamica **35**, 1–16 (2010)
3. D. Cox, S. Zucker, Intersection numbers of sections of elliptic surfaces. Invent. Math. **53**, 1–44 (1979)
4. A. Garbagnati, B. van Geemen, The Picard-Fuchs equation of a family of Calabi-Yau threefolds without maximal unipotent monodromy. Int. Math. Res. Not. **2010**(16), 3134–3143 (2010)
5. H. Hollborn, L^2-Kohomologie von Calabi-Yau-Familien über Kurven, Thesis, University of Mainz, 2014.
6. J. Jost, K. Zuo, Arakelov type inequalities for Hodge bundles over algebraic varieties. J. Algebr. Geom. **11**, 535–546 (2002)
7. J. Jost, Y.-H. Yang, K. Zuo, Cohomologies of unipotent harmonic bundles over noncompact curves. Crelle's J. **609**, 137–159 (2007)
8. D. Morrison, J. Walcher, D-branes and normal functions. Adv. Theor. Math. Phys. **13**, 553–598 (2009)
9. J.-C. Rohde, Maximal automorphisms of Calabi-Yau manifolds versus maximally unipotent monodromy. Manuscripta Mathematica **131**, 459–474 (2010)
10. C. Simpson, Constructing variations of Hodge structure using Yang-Mills theory and applications to uniformization. J. Am. Math. Soc. **1**, 867–918 (1988)
11. P. Stiller, *Automorphic Forms and the Picard Number of an Elliptic Surface*. Aspects of Mathematics, vol. E5 (Vieweg Verlag, Braunschweig, 1984)
12. S. Zucker, Hodge theory with degenerating coefficients. Ann. Math. **109**, 415–476 (1979)

Lagrangian Fibrations of Holomorphic-Symplectic Varieties of $K3^{[n]}$-Type

Eyal Markman

Dedicated to Klaus Hulek on the occasion of his sixtieth birthday.

Abstract Let X be a compact Kähler holomorphic-symplectic manifold, which is deformation equivalent to the Hilbert scheme of length n subschemes of a $K3$ surface. Let $\mathscr{L}$ be a nef line-bundle on X, such that the top power $c_1(\mathscr{L})^{2n}$ vanishes and $c_1(\mathscr{L})$ is primitive. Assume that the two dimensional subspace $H^{2,0}(X) \oplus H^{0,2}(X)$ of $H^2(X, \mathbb{C})$ intersects $H^2(X, \mathbb{Z})$ trivially. We prove that the linear system of $\mathscr{L}$ is base point free and it induces a Lagrangian fibration on X. In particular, the line-bundle $\mathscr{L}$ is effective. A determination of the semi-group of effective divisor classes on X follows, when X is projective. For a generic such pair $(X, \mathscr{L})$, not necessarily projective, we show that X is bimeromorphic to a Tate-Shafarevich twist of a moduli space of stable torsion sheaves, each with pure one dimensional support, on a *projective* $K3$ surface.

Contents

Partially supported by Simons Foundation Collaboration Grant 245840 and by NSA grant H98230-13-1-0239.

E. Markman (✉)
Department of mathematics and statistics, University of Massachusetts, Amherst, MA 01003, USA
e-mail: markman@math.umass.edu

A. Frühbis-Krüger et al. (eds.), *Algebraic and Complex Geometry*, Springer Proceedings in Mathematics & Statistics 71, DOI 10.1007/978-3-319-05404-9_10,

1 Introduction

An *irreducible holomorphic symplectic manifold* is a simply connected compact Kähler manifold such that $H^0(X, \wedge^2 T^*X)$ is generated by an everywhere non-degenerate holomorphic 2-form [4]. A compact Kähler manifold X is said to be of *$K3^{[n]}$-type*, if it is deformation equivalent to the Hilbert scheme $S^{[n]}$ of length n subschemes of a $K3$ surface S. Any manifold of $K3^{[n]}$-type is irreducible holomorphic symplectic [4]. The second integral cohomology of an irreducible holomorphic symplectic manifold X admits a natural symmetric non-degenerate integral bilinear pairing $(\bullet, \bullet)$ of signature $(3, b_2(X) - 3)$, called the *Beauville-Bogomolov-Fujiki pairing*. The Beauville-Bogomolov-Fujiki pairing is monodromy invariant, and is thus an invariant of the deformation class of X.

Definition 1.1. An irreducible holomorphic symplectic manifold X is said to be *special*, if the intersection in $H^2(X, \mathbb{C})$ of $H^2(X, \mathbb{Z})$ and $H^{2,0}(X) \oplus H^{0,2}(X)$ is a non-zero subgroup.

The locus of special periods forms a countable union of real analytic subvarieties of half the dimension in the corresponding moduli space.

Definition 1.2. Let X be a $2n$-dimensional irreducible holomorphic symplectic manifold and $\mathscr{L}$ a line bundle on X. We say that $\mathscr{L}$ *induces a Lagrangian fibration*, if it satisfies the following two conditions.

1. $h^0(X, \mathscr{L}) = n + 1$.
2. The linear system $|\mathscr{L}|$ is base point free, and the generic fiber of the morphism $\pi: X \to |\mathscr{L}|^*$ is a connected Lagrangian subvariety.

A line bundle $\mathscr{L}$ on a holomorphic symplectic manifold X is said to be *nef*, if $c_1(\mathscr{L})$ belongs to the closure in $H^{1,1}(X, \mathbb{R})$ of the Kähler cone of X.

Theorem 1.3. *Let X be an irreducible holomorphic symplectic manifold of $K3^{[n]}$-type and $\mathscr{L}$ a nef line-bundle, such that $c_1(\mathscr{L})$ is primitive and isotropic with respect* *to the Beauville-Bogomolov-Fujiki pairing. Assume that X is non-special. Then the line bundle $\mathscr{L}$ induces a Lagrangian fibration $\pi : X \to |\mathscr{L}|^*$.*

See Theorem 6.3 for a variant of Theorem 1.3 dropping the assumption that $\mathscr{L}$ is nef. Theorem 1.3 is proven in Sect. 6. The proof relies on Verbitsky's Global Torelli Theorem [14, 40], on the determination of the monodromy group of X [21, 22], and on a result of Matsushita that Lagrangian fibrations form an *open* subset in the moduli space of pairs $(X, \mathscr{L})$ [27]. Let us sketch the three main new ingredients in the proof of Theorem 1.3.

(1) We associate to the pair $(X, \mathscr{L})$ in Theorem 1.3 a projective $K3$ surface S with a nef line bundle $\mathscr{B}$ of degree $\frac{2n-2}{d^2}$, where $d := \gcd\{(c_1(\mathscr{L}), \lambda) : \lambda \in H^2(X, \mathbb{Z})\}$. The sub-lattice $c_1(\mathscr{B})^\perp$ orthogonal to $c_1(\mathscr{B})$ in $H^2(S, \mathbb{Z})$ is Hodge-isometric to $c_1(\mathscr{L})^\perp / \mathbb{Z}c_1(\mathscr{L})$. The construction realizes the period domain Ω_{20} of the pairs $(X, \mathscr{L})$ as an affine line bundle over a period domain Ω_{19} of semi-polarized $K3$ surfaces (Sect. 4).
(2) The bundle map $q : \Omega_{20} \to \Omega_{19}$ is invariant with respect to a subgroup Q of the monodromy group (Lemma 5.3). The group Q is isomorphic to $c_1(\mathscr{B})^\perp$. Q acts on the fiber of q over the period of a semi-polarized $K3$ surface $(S, \mathscr{B})$. Similarly, the lattice $c_1(\mathscr{B})^\perp$ projects to a subgroup of $H^{0,2}(S)$, which acts on $H^{0,2}(S)$ by translations. There exists an isomorphism, of the fiber of q with $H^{0,2}(S)$, which is equivariant with respect to the two actions (Lemma 5.4).
(3) The fiber of q over the period of a semi-polarized $K3$ surface $(S, \mathscr{B})$ contains the period of a moduli space of sheaves on S with pure one-dimensional support in the linear system $|\mathscr{B}^d|$ (Sect. 5.1). Each such moduli space of sheaves is known to be a Lagrangian fibration [34].

The assumption that X is non-special in Theorem 1.3 is probably not necessary. Unfortunately, our proof will rely on it. When X is non-special the Q-orbit, of every point in the fiber of q through the period of X, is a dense subset of the fiber (Lemma 5.4). This density will have a central role in this paper due to the following elementary observation.

Observation 1.4. *Let T be a topological space and Q a group acting on T. Assume that the Q-orbit of every point of T is dense in T. Then any nonempty Q-invariant open subset of T must be the whole of T.*

The above observation will be used in an essential way in three different proofs (Theorem 6.1, Proposition 7.7, and Theorem 7.11).

The statement of the next result requires the notion of a Tate-Shafarevich twist, which we now recall. Let M be a complex manifold and $\pi : M \to B$ a proper map with connected fibers of pure dimension n. Assume that the generic fiber of π is a smooth abelian variety. Let $\{U_i\}$ be an open covering of B in the analytic topology. Set $U_{ij} := U_i \cap U_j$ and $M_{ij} := \pi^{-1}(U_{ij})$. Assume given a 1-co-cycle g_{ij} of automorphisms of M_{ij}, satisfying $\pi \circ g_{ij} = \pi$, and acting by translations on the smooth fibers of π. We can re-glue the open covering $\{M_i\}$ of M using the co-cycle $\{g_{ij}\}$ to get a complex manifold M' and a proper map $\pi' : M' \to B$, whose fibers are isomorphic to those of π. We refer to (M', π') as the *Tate-Shafarevich twist* of (M, π) associated to the co-cycle $\{g_{ij}\}$. Tate-Shafarevich twists are standard in the study of elliptic fibrations [10, 17].

Let $\mathscr{L}$ be a semi-ample line bundle on a $K3$ surface S with an indivisible class $c_1(\mathscr{L})$. Given an ample line bundle H on S and an integer χ, denote by $M_H(0, \mathscr{L}^d, \chi)$ the moduli space of H-stable coherent sheaves on S of rank zero, determinant $\mathscr{L}^d$, and Euler characteristic χ. Assume that d and χ are relatively prime. For a generic polarization H, the moduli space $M_H(0, \mathscr{L}^d, \chi)$ is smooth and projective and it admits a Lagrangian fibration over the linear system $|\mathscr{L}^d|$ [34].

Let X be an irreducible holomorphic symplectic manifold of $K3^{[n]}$-type and $\pi : X \to \mathbb{P}^n$ a Lagrangian fibration. Set $\alpha := \pi^* c_1(\mathscr{O}_{\mathbb{P}^n}(1))$. The *divisibility* of $(\alpha, \bullet)$ is the positive integer $d := \gcd\{(\alpha, \lambda) \ : \ \lambda \in H^2(X, \mathbb{Z})\}$. The integer d^2 divides $n-1$ (Lemma 2.5).

Theorem 1.5. *Assume that X is non-special and the intersection $H^{1,1}(X, \mathbb{Z}) \cap \alpha^{\perp}$ is $\mathbb{Z}\alpha$. There exists a $K3$ surface S, a semi-ample line bundle $\mathscr{L}$ on S of degree $\frac{2n-2}{d^2}$ with an indivisible class $c_1(\mathscr{L})$, an integer χ relatively prime to d, and a polarization H on S, such that X is bimeromorphic to a Tate-Shafarevich twist of the Lagrangian fibration $M_H(0, \mathscr{L}^d, \chi) \to |\mathscr{L}^d|$.*

Theorem 1.5 is proven in Sect. 7. The semi-polarized $K3$ surface $(S, \mathscr{L})$ in Theorem 1.5 is the one mentioned already above, which is associated to (X, α) in Sect. 4.1. The equality $H^{1,1}(X, \mathbb{Z}) \cap \alpha^{\perp} = \mathbb{Z}\alpha$ is equivalent to the statement that $\mathrm{Pic}(S)$ is cyclic generated by $\mathscr{L}$. This condition is relaxed in Theorem 7.13, which strengthens Theorem 1.5.

A reduced and irreducible divisor on X is called *prime exceptional*, if it has negative Beauville-Bogomolov-Fujiki degree. A divisor D on X is called *movable*, if the base locus of the linear system $|D|$ has co-dimension ≥ 2 in X. The *movable cone* $\mathscr{MV}_X$ of X is the cone in $N^1(X) := H^{1,1}(X, \mathbb{Z}) \otimes_{\mathbb{Z}} \mathbb{R}$ generated by classes of movable divisors. Assume that X is a projective irreducible holomorphic symplectic manifold of $K3^{[n]}$-type and let $h \in N^1(X)$ be an ample class. Denote by $\mathscr{P}ex_X \subset H^{1,1}(X, \mathbb{Z})$ the set of classes of prime exceptional divisors. The set $\mathscr{P}ex_X$ is determined in [24, Theorem 1.11 and Sec. 1.5]. The closure of the movable cone in $N^1(X)$ is determined as follows:

$$\overline{\mathscr{MV}}_X = \{c \in N^1(X) : (c, c) \geq 0, \ (c, h) \geq 0, \ \text{and } (c, e) \geq 0, \ \text{for all } e \in \mathscr{P}ex_X\},$$

by a result of Boucksom [6, 23, Prop. 5.6 and Lemma 6.22].[1]

Corollary 1.6. *Let X be a projective irreducible holomorphic symplectic manifold of $K3^{[n]}$-type. The semi-group of effective divisor classes on X is generated by*

[1] Prop. 5.6 and Lemma 6.22 in the last reference [23]. The same convention will be used throughout the paper for all citations with multiple references.

the classes of prime exceptional divisors and integral points in the closure of the movable cone in $N^1(X)$.

Corollary 1.6 was shown to follow from Theorem 1.3 in [23, Paragraph following Question 10.11].

We classify the deformation types of pairs $(X, \mathcal{L})$, consisting of an irreducible holomorphic symplectic manifold X of $K3^{[n]}$-type, $n \geq 2$, and a line bundle $\mathcal{L}$ on X with a primitive and isotropic first Chern class, such that $(c_1(\mathcal{L}), \kappa) > 0$, for some Kähler class κ. The following proposition is proven in Sect. 4.3, using monodromy invariants introduced in Lemma 2.5.

Proposition 1.7. *Let d be a positive integer, such that d^2 divides $n-1$. If $1 \leq d \leq 4$, then there exists a unique deformation type of pairs $(X, \mathcal{L})$, with $c_1(\mathcal{L})$ primitive and isotropic, such that $(c_1(\mathcal{L}), \bullet)$ has divisibility d. For $d \geq 5$, let $\nu(d)$ be half the number of multiplicative units in the ring $\mathbb{Z}/d\mathbb{Z}$. Then there are $\nu(d)$ deformation types of pairs $(X, \mathcal{L})$ as above, with $(c_1(\mathcal{L}), \bullet)$ of divisibility d.*

A *generalized Kummer variety* of dimension $2n$ is the fiber of the Albanese map $S^{[n+1]} \to S$ from the Hilbert scheme of length n subschemes of an abelian surface S to S itself [4]. We expect all of the above results to have analogues for X an irreducible holomorphic-symplectic manifold deformation equivalent to a generalized Kummer variety. Yoshioka proved Theorem 1.3 for those X associated to a moduli space of sheaves on an abelian surface [43]. Let the pair $(X, \mathcal{L})$ consist of X, deformation equivalent to a generalized Kummer, and a line bundle $\mathcal{L}$ with a primitive and isotropic first Chern class. The basic construction of Sect. 4.1 associates to the pair $(X, \mathcal{L})$, with $\dim(X) = 2n$, $n \geq 2$, and with $(c_1(\mathcal{L}), \bullet)$ of divisibility d, two dual pairs (S_1, α_1) and (S_2, α_2), each consisting of an abelian surface S_i and a class α_i in the Neron-Severi group of S_i of self intersection $\frac{2n+2}{d^2}$, such that $S_2 \cong S_1^*$ and the natural isometry $H^2(S_1, \mathbb{Z}) \cong H^2(S_2, \mathbb{Z})$ maps α_1 to α_2. A conjectural determination of the monodromy group of generalized Kummer varieties was suggested in the comment after [25, Prop. 4.8]. Assuming that the monodromy group is as conjectured, we expect that the proofs of all the results above can be adapted to this deformation type.

A version of Theorem 1.3 has been conjectured for irreducible holomorphic symplectic manifolds of all deformation types [5, 26, 39, Conjecture 2]. Markushevich, Sawon, and Yoshioka proved a version of Theorem 1.3, when X is the Hilbert scheme of n points on a $K3$ surface and $(c_1(\mathcal{L}), \bullet)$ has divisibility 1 [26, Cor. 4.4] and [39] (the regularity of the fibration, in Sect. 5 of [39], is due to Yoshioka). Bayer and Macri recently proved a strong version of Theorem 1.3 for moduli spaces of sheaves on a projective $K3$ surface [3].

Remark 1.8 (Added in the final revision). Let X_0 be an irreducible holomorphic symplectic manifold and $\mathcal{L}_0$ a nef line bundle on X_0, such that $c_1(\mathcal{L}_0)$ is primitive and isotropic with respect to the Beauville-Bogomolov-Fujiki pairing. Matsushita proved that if $\mathcal{L}_0$ induces a Lagrangian fibration, then so does $\mathcal{L}$ for every pair $(X, \mathcal{L})$ deformation equivalent to $(X_0, \mathcal{L}_0)$, with X irreducible holomorpic symplectic and $\mathscr{L}$ nef (preprint posted very recently [28], announced earlier in his talk [31]). It follows that Theorem 1.3 above holds also without the assumption that X is non-special, since a pair $(X, \mathscr{L})$ with X special is a deformation of a pair $(X_0, \mathscr{L}_0)$ with X_0 non-special. In fact, this stronger version of Theorem 1.3, dropping the non-speciality, follows already from the combination of Matsushita's result and Example 3.1 below, since Example 3.1 exhibits a pair $(X_0, \mathscr{L}_0)$, with a line bundle $\mathscr{L}_0$ inducing a Lagrangian fibration, in each deformation class of pairs $(X, \mathscr{L})$ with X of $K3^{[n]}$-type and $c_1(\mathscr{L})$ primitive, isotropic, and on the boundary of the positive cone. Matsushita's result does not seem to provide an alternative proof of Theorem 1.5 and the only proof we know is presented in Sect. 7 and relies on the preceding sections.

2 Classification of Primitive-Isotropic Classes

A lattice, in this note, is a finitely generated free abelian group with a symmetric bilinear pairing $(\bullet, \bullet) : L \otimes_{\mathbb{Z}} L \to \mathbb{Z}$. The pairing may be degenerate. The isometry group $O(L)$ is the group of automorphisms of L preserving the bilinear pairing.

Definition 2.1. Two pairs (L_i, v_i), $i = 1, 2$, each consisting of a lattice L_i and an element $v_i \in L_i$, are said to be *isometric*, if there exists an isometry $g : L_1 \to L_2$, such that $g(v_1) = v_2$.

Let X be an irreducible holomorphic symplectic manifold of $K3^{[n]}$-type, $n \geq 2$. Set $\Lambda := H^2(X, \mathbb{Z})$. We will refer to Λ as the *$K3^{[n]}$-lattice.* Let $\tilde{\Lambda}$ be the Mukai lattice, i.e., the orthogonal direct sum of two copies of the negative definite $E_8(-1)$ lattice and four copies of the even unimodular rank two lattice with signature $(1, -1)$.

Theorem 2.2 ([22], Theorem 1.10). *X comes with a natural $O(\tilde{\Lambda})$-orbit ι_X of primitive isometric embeddings $\iota : H^2(X, \mathbb{Z}) \hookrightarrow \tilde{\Lambda}$.*

Choose a primitive isometric embedding $\iota : \Lambda \hookrightarrow \tilde{\Lambda}$ in the canonical $O(\tilde{\Lambda})$-orbit ι_X provided by Theorem 2.2. Choose a generator $v \in \tilde{\Lambda}$ of the rank 1 sub-lattice orthogonal to $\iota(\Lambda)$. We say that an isometry $g \in O(\Lambda)$ *stabilizes* the $O(\tilde{\Lambda})$-orbit ι_X, if given a representative isometric embedding ι in the orbit ι_X, there exists an isometry $\tilde{g} \in O(\tilde{\Lambda})$ satisfying $\iota \circ g = \tilde{g} \circ \iota$. Note that $\tilde{g}$ necessarily maps v to $\pm v$.

Set $\Lambda_{\mathbb{R}} := \Lambda \otimes_{\mathbb{Z}} \mathbb{R}$. Let $\mathscr{C} \subset \Lambda_{\mathbb{R}}$ be the positive cone $\{x \in \Lambda_{\mathbb{R}} : (x, x) > 0\}$. Then $H^2(\tilde{\mathscr{C}}, \mathbb{Z})$ is isomorphic to $\mathbb{Z}$ and is a natural character of the isometry group $O(\Lambda)$ [23, Lemma 4.1]. Denote by $O^+(\Lambda)$ the kernel of this orientation character. Isometries in $O^+(\Lambda)$ are said to be *orientation preserving.*

Definition 2.3. Let X, X_1, and X_2 be irreducible holomorphic symplectic manifolds. An isometry $g : H^2(X_1, \mathbb{Z}) \to H^2(X_2, \mathbb{Z})$ is a *parallel transport operator,* if there exists a family $\pi : \mathscr{X} \to B$ (which may depend on g) of irreducible holomorphic symplectic manifolds, points b_1 and b_2 in B, isomorphisms $X_i \cong \mathscr{X}_{b_i}$, where $\mathscr{X}_{b_i}$ is the fiber over b_i, $i = 1, 2$, and a continuous path γ from b_1 to b_2, such that parallel transport along γ in the local system $R^2\pi_*\mathbb{Z}$ induces the isometry g. When $X = X_1 = X_2$, we call g a *monodromy operator*. The *monodromy group* $Mon^2(X)$ of X is the subgroup, of the isometry group of $H^2(X, \mathbb{Z})$, generated by monodromy operators.

Theorem 2.4 ([22], Theorem 1.2 and Lemma 4.2). *The subgroup $Mon^2(X)$ of $O(\Lambda)$ consists of orientation preserving isometries stabilizing the orbit ι_X.*

Given a lattice L, let $I_n(L) \subset L$ be the subset of primitive classes v with $(v, v) = 2n-2$. Notice that the orbit set $I_n(L)/O(L)$ parametrizes the set of isometry classes of pairs (L', v'), such that L' is isometric to L and v' is a primitive class in L' with $(v', v') = 2n - 2$ [23, Lemma 9.14].

Let n be an integer ≥ 2, let Λ be the $K3^{[n]}$-lattice, and let $\alpha \in \Lambda$ be a primitive isotropic class. Let $\mathrm{div}(\alpha, \bullet)$ be the largest positive integer, such that $(\alpha, \bullet)/\mathrm{div}(\alpha, \bullet)$ is an integral class of Λ^*. Set $d := \mathrm{div}(\alpha, \bullet)$ and

$$\beta := \iota(\alpha).$$

Let $L \subset \tilde{\Lambda}$ be the saturation[2] of $\mathrm{span}_{\mathbb{Z}}\{\beta, v\}$. Clearly, the isometry class of (L, v) depends only on α and the $O(\tilde{\Lambda})$-orbit of ι. Consequently, the isometry class of (L, v) depends only on α, as the $O(\tilde{\Lambda})$-orbit ι_X of ι is natural, by Theorem 2.2. We denote by $[L, v](\alpha)$ the isometry class of the pair (L, v) associated to α.

Lemma 2.5. *(1) d^2 divides $n - 1$.*

(2) L is isometric to the lattice $L_{n,d}$ with Gram matrix $\frac{2n-2}{d^2}\begin{pmatrix} 1 & 0 \\ 0 & 0 \end{pmatrix}$.

(3) Let $d \geq 1$ be an integer, such that d^2 divides $n - 1$. The map $\alpha \mapsto [L, v](\alpha)$ induces a one-to-one correspondence between the set of $Mon^2(X)$-orbits, of primitive isotropic classes α with $\mathrm{div}(\alpha, \bullet) = d$, and the set of isometry classes $I_n(L_{n,d})/O(L_{n,d})$.

(4) There exists an integer b, such that $(\beta - bv)/d$ is an integral class of L. The isometry class $[L, v](\alpha)$ is represented by $(L_{n,d}, (d, b))$, for any such integer b.

Proof. Part (1): There exists a class $\delta \in \Lambda$, such that $(\delta, \delta) = 2 - 2n$ and the sub-lattice $\delta^{\perp}_{\Lambda}$ of Λ, orthogonal to δ, is a unimodular lattice isometric to the $K3$-lattice. The sub-lattice $[\iota(\delta^{\perp}_{\Lambda})]^{\perp}_{\tilde{\Lambda}}$ of $\tilde{\Lambda}$, which is the saturation of $\mathrm{span}\{\iota(\delta), v\}$, is unimodular, hence isometric to the unimodular hyperbolic plane U with Gram matrix $\begin{pmatrix} 0 & -1 \\ -1 & 0 \end{pmatrix}$. We may further assume that $v = (1, 1-n)$ and $\iota(\delta) = (1, n-1)$, under this isomorphism. If X is the Hilbert scheme $S^{[n]}$ of a $K3$-surface and δ is half the class of the big diagonal, then δ satisfies the above properties. Write

---

[2] The *saturation* of a sublattice L' of Λ is the maximal sublattice L of Λ, of the same rank as L', which contains L'.

$\alpha = a\xi + b\delta$, where ξ is a primitive class of the $K3$-lattice $\delta^{\perp}_{\tilde{\Lambda}}$, $a > 0$, and $\gcd(a,b) = 1$. We get

$$0 = (\alpha,\alpha) = a^2(\xi,\xi) - (2n-2)b^2,$$

and (ξ,ξ) is even. Hence, a^2 divides $n-1$. Furthermore, $\mathrm{div}(\delta,\bullet) = 2n-2$, $\mathrm{div}(\xi,\bullet) = 1$, since $\delta^{\perp}_{\tilde{\Lambda}}$ is unimodular, and $\mathrm{div}(\alpha,\bullet) = \gcd(\mathrm{div}(a\xi,\bullet),\mathrm{div}(b\delta,\bullet)) = \gcd(a,(2n-2)b) = a$. Thus, $a = d := \mathrm{div}(\alpha,\bullet)$.

Part (2): Note that $\iota(\delta) - v = (2n-2)e$, where e is a primitive isotropic class of $\tilde{\Lambda}$. Set $\gamma := \frac{1}{d}(\beta - bv) = \iota(\xi) + \frac{b(2n-2)}{d}e$. We claim that the lattice $L := \mathrm{span}_{\mathbb{Z}}\{v,\gamma\}$ is saturated in $\tilde{\Lambda}$. Indeed, choose $\eta \in \delta^{\perp}_{\tilde{\Lambda}}$, such that $(\xi,\eta) = 1$. Then $\begin{pmatrix}(v,e) & (v,\eta)\\ (\gamma,e) & (\gamma,\eta)\end{pmatrix} = \begin{pmatrix}-1 & 0\\ 0 & 1\end{pmatrix}$.

Let G be the Gram matrix of L in the basis $\{v,\gamma\}$. Then

$$G = \frac{2n-2}{d^2}\begin{pmatrix} d^2 & -bd\\ -bd & b^2\end{pmatrix} = \frac{2n-2}{d^2}\begin{pmatrix} d\\ -b\end{pmatrix}\begin{pmatrix} d & -b\end{pmatrix}.$$

Choose a 2×2 invertible matrix A, with integer coefficients, such that $A\begin{pmatrix} d\\ -b\end{pmatrix} = \begin{pmatrix} 1\\ 0\end{pmatrix}$. Then AGA^t is the Gram matrix of $L_{n,d}$.

Part (3): Assume given two primitive isotropic classes α_1 and α_2 in $\Lambda := H^2(X,\mathbb{Z})$ and let (L_i,v_i) be the pair associated to α_i as above, for $i = 1,2$. In other words, $\iota_i : \Lambda \hookrightarrow \tilde{\Lambda}$ is a primitive embedding in the orbit ι_X, v_i generates the sub-lattice of $\tilde{\Lambda}$ orthogonal to the image of ι_i, and L_i is the saturation of $\mathrm{span}_{\mathbb{Z}}\{\iota(\alpha_i),v_i\}$.

Let us check that the map $\alpha \mapsto [L,v](\alpha)$ is constant on $Mon^2(X)$-orbits. Assume that there exists an element $\mu \in Mon^2(X)$, such that $\mu(\alpha_1) = \alpha_2$. Then there exists an isometry $\tilde{\mu} \in O(\tilde{\Lambda})$, satisfying $\tilde{\mu}\circ\iota_1 = \iota_2\circ\mu$, by Theorem 2.4. We get that $\tilde{\mu}(L_1) = L_2$ and $\tilde{\mu}(v_1) = v_2$, or $\tilde{\mu}(v_1) = -v_2$. So, the isometry $\tilde{\mu}$ or $-\tilde{\mu}$ from L_1 onto L_2 provides an isometry of the pairs (L_i,v_i), $i = 1,2$.

We show next that the map $\alpha \mapsto [L,v](\alpha)$ is injective, i.e., that the isometry class of the pair (L,v) determines the $Mon^2(X)$-orbit of α. Assume that there exists as isometry $f : L_1 \to L_2$, such that $f(v_1) = v_2$. Then there exists an isometry $\tilde{f} \in O(\tilde{\Lambda})$, such that $\tilde{f}(L_1) = L_2$ and the restriction of $\tilde{f}$ to L_1 is f, by ([36], Proposition 1.17.1 and Theorem 1.14.4, see also [21], Lemma 8.1 for more details). In particular, $\tilde{f}(v_1) = v_2$. There exists a unique isometry $h \in O(\Lambda)$ satisfying $\iota_2\circ h = \tilde{f}\circ\iota_1$. There exists an isometry $\phi \in O(\tilde{\Lambda})$, such that $\phi\circ\iota_2 = \iota_1$, since both ι_i belong to the same $O(\tilde{\Lambda})$-orbit ι_X. We get the equality $\iota_1\circ h = \phi\circ\iota_2\circ h = (\phi\circ\tilde{f})\circ\iota_1$. If h is orientation preserving, then h belongs to $Mon^2(X)$, otherwise, $-h$ does, by Theorem 2.4. Let $\mu = h$, if it is orientation preserving. Otherwise, set $\mu := -h$. Then μ is a monodromy operator and $\iota_2(\mu(\alpha_1)) = \pm\iota_2(h(\alpha_1)) = \pm\tilde{f}(\iota_1(\alpha_1))$. The class $\iota_1(\alpha_1)$ spans the null space of L_1, and $\tilde{f}$ restricts to an isometry from L_1 to L_2. Hence, $\iota_2(\mu(\alpha_1))$ spans the null space of L_2. Hence, $\mu(\alpha_1) = \pm\alpha_2$.

Finally we show that α_2 and $-\alpha_2$ belong to the same $Mon^2(X)$-orbit. There exists an element $\tau \in \Lambda$ satisfying $(\tau,\tau) = 2$, and $(\tau,\alpha_2) = 0$. The isometry $\rho_\tau \in O(\Lambda)$, given by $\rho_\tau(\lambda) = -\lambda + (\lambda,\tau)\tau$, belongs to $Mon^2(X)$, by ([21], Corollary 1.8), and it sends α_2 to $-\alpha_2$.

It remains to prove that the map $\alpha \mapsto [L,v](\alpha)$ is surjective. Assume given a primitive class $v \in L_{n,d}$ with $(v,v) = 2n-2$. There exists a primitive isometric embedding $f : L_{n,d} \hookrightarrow \tilde{\Lambda}$, by ([36], Proposition 1.17.1). The lattice $f(v)^{\perp}_{\tilde{\Lambda}}$, orthogonal to $f(v)$ in $\tilde{\Lambda}$, is isometric to the $K3^{[n]}$-lattice Λ. Choose such an isometry $h : f(v)^{\perp}_{\tilde{\Lambda}} \to \Lambda$, with the property that $h^{-1} : \Lambda \hookrightarrow \tilde{\Lambda}$ belongs to the $O(\tilde{\Lambda})$-orbit ι_X. Such a choice exists, since $O(\Lambda)$ acts transitively on the orbit space $O(\Lambda,\tilde{\Lambda})/O(\tilde{\Lambda})$, by ([22], Lemma 4.3). Above, $O(\Lambda,\tilde{\Lambda})$ denotes the set of primitive isometric embeddings of Λ in $\tilde{\Lambda}$. Denote by $\beta \in L_{n,d}$ a generator of the null space of $L_{n,d}$. Set $\alpha := h(f(\beta))$. Then α is a class in Λ, such that $[L,v](\alpha)$ is represented by $(L_{n,d}, v)$.

Part (4): The existence of such an integer b was established in the course of proving part (1). The rest of the statement follows from Lemma 2.6. □

If $d = 2$, set $\nu(d) := 1$. If $d > 2$, let $\nu(d)$ be half the number of multiplicative units in the ring $\mathbb{Z}/d\mathbb{Z}$.

Lemma 2.6. *A vector $(x,y) \in L_{n,d}$ is primitive of degree $2n-2$, if and only if $|x| = d$ and $\gcd(d,y) = 1$. Two primitive vectors (d,y), (d,z) belong to the same $O(L_{n,d})$-orbit, if and only if $y \equiv z$ modulo d, or $y \equiv -z$ modulo d. Consequently, $\nu(d)$ is equal to the number of $O(L_{n,d})$-orbits of primitive vectors in $L_{n,d}$ of degree $2n-2$.*

Proof. The isometry group of $L_{n,d}$ consists of matrices of the form $\begin{pmatrix} \pm 1 & 0 \\ c & \pm 1 \end{pmatrix}$. The orbit $O(L_{n,d})(d,y)$ consists of vectors of the form $(\pm d, cd \pm y)$. Consequently, the number of $O(L_{n,d})$-orbits of primitive vectors in $L_{n,d}$ of degree $2n-2$ is equal to the number of orbits in $\{y : 0 < y < d \text{ and } \gcd(y,d) = 1\}$ under the action $y \mapsto d - y$. The latter number is $\nu(d)$. □

3 An Example of a Lagrangian Fibration for Each Value of the Monodromy Invariants

Let S be a projective $K3$ surface, $K(S)$ its topological K-group, generated by classes of complex vector bundles, and $H^*(S,\mathbb{Z})$ its integral cohomology ring. Let $td_S := 1 + \frac{c_2(S)}{12}$ be the Todd class of S and $\sqrt{td_S} := 1 + \frac{c_2(S)}{24}$ its square root. The homomorphism $v : K(S) \to H^*(S,\mathbb{Z})$, given by $v(x) = ch(x)\sqrt{td_S}$ is an isomorphism of free abelian groups. Given a coherent sheaf E on S, the class $v(E)$ is called the *Mukai vector* of E. Given integers r and s and a class $c \in H^2(S,\mathbb{Z})$, we will denote by (r,c,s) the class of $H^*(S,\mathbb{Z})$, whose graded summand in $H^0(S,\mathbb{Z})$ is r times the class Poincare dual to S, its graded summand in $H^2(S,\mathbb{Z})$ is c, and its graded summand in $H^4(S,\mathbb{Z})$ is s times the class Poincare dual to a point. We endow $H^*(S,\mathbb{Z})$ with the *Mukai pairing*

$$((r,c,s),(r',c',s')) \ := \ (c,c') - rs' - r's,$$

where $(c,c') := \int_S c \cup c'$. Then $(v(x),v(y)) = -\chi(x \otimes y)$, where $\chi : K(S) \to \mathbb{Z}$ is the Euler characteristic [35]. $H^*(S,\mathbb{Z})$, endowed with the Mukai pairing, is called the *Mukai lattice*. The Mukai lattice is an even unimodular lattice of rank 24, which is isometric to the orthogonal direct sum of two copies of the negative definite $E_8(-1)$ lattice and four copies of the even unimodular rank 2 hyperbolic lattice U.

Let $v \in K(S)$ be the class with Mukai vector $(0, d\xi, s)$ in $H^*(S,\mathbb{Z})$, such that ξ a primitive effective class in $H^{1,1}(S,\mathbb{Z})$, $(\xi,\xi) > 0$, d is a positive integer, and $\gcd(d,s) = 1$. There is a system of hyperplanes in the ample cone of S, called v-walls, that is countable but locally finite [15, Ch. 4C]. An ample class is called *v-generic*, if it does not belong to any v-wall. Choose a v-generic ample class H. Let $M_H(v)$ be the moduli space of H-stable sheaves on the $K3$ surface S with class v. $M_H(v)$ is a smooth projective irreducible holomorphic symplectic variety of $K3^{[n]}$-type, with $n = \frac{(v,v)+2}{2} = \frac{d^2(\xi,\xi)+2}{2}$. This is a special case of a result, which is due to several people, including Huybrechts, Mukai, O'Grady [38], and Yoshioka [44]. It can be found in its final form in [44].

Over $S \times M_H(v)$ there exists a universal sheaf $\mathscr{F}$, possibly twisted with respect to a non-trivial Brauer class pulled-back from $M_H(v)$. Associated to $\mathscr{F}$ is a class $[\mathscr{F}]$ in $K(S \times M_H(v))$ ([20], Definition 26). Let π_i be the projection from $S \times M_H(v)$ onto the i-th factor. Denote by $v^\perp$ the sub-lattice in $H^*(S,\mathbb{Z})$ orthogonal to v. The second integral cohomology $H^2(M_H(v),\mathbb{Z})$, its Hodge structure, and its Beauville-Bogomolov-Fujiki pairing, are all described by Mukai's Hodge-isometry

$$\theta \, : \, v^\perp \ \longrightarrow \ H^2(M_H(v),\mathbb{Z}), \tag{3.1}$$

given by $\theta(x) := c_1\left(\pi_{2_!}\{\pi_1^!(x^\vee) \otimes [\mathscr{F}]\}\right)$ (see [44]).

We provide next an example of a moduli space $M_H(v)$ and a primitive isotropic class $\alpha \in H^{1,1}(M_H(v),\mathbb{Z})$, such that $[L,v](\alpha)$ is represented by $(L_{n,d},(d,b))$, for every integer $n \geq 2$, for every positive integer d, such that d^2 divides $n-1$, and for every integer b satisfying $\gcd(b,d) = 1$.

Example 3.1. Let d be a positive integer, such that d^2 divides $n-1$. Let S be a $K3$ surface with a nef line bundle $\mathscr{L}$ of degree $\frac{2n-2}{d^2}$. Let λ be the class $c_1(\mathscr{L})$ in $H^2(S,\mathbb{Z})$. Fix an integer b satisfying $\gcd(b,d) = 1$. Set $v := (0, d\lambda, s)$, where s is an integer satisfying $sb = 1$ (modulo d). Then v is a primitive Mukai vector and $(v,v) = 2n-2$. Choose a v-generic ample line bundle H. A sheaf F of class v is H-stable, if and only if it is H-semi-stable. The moduli space $M_H(v)$, of H-stable sheaves of class v, is smooth, projective, holomorphic symplectic, and of $K3^{[n]}$-type. Set $\alpha := \theta((0,0,1))$. Let $\iota : H^2(M_H(v),\mathbb{Z}) \to H^*(S,\mathbb{Z})$ be the composition of θ^{-1} with the inclusion of $v^\perp$ into $H^*(S,\mathbb{Z})$. A Mukai vector (r,c,t) belongs to $v^\perp$, if and only if $rs = d(c, \lambda)$. It follows that d divides r, since $\gcd(d, s) = 1$. Thus, $\mathrm{div}(\alpha, \bullet) = d$. Now

$$\iota(\alpha) - bv = (0, -bd\lambda, 1 - bs)$$

is divisible by d, by our assumption on s. Hence, the monodromy invariant $[L, v](\alpha)$ is equal to the isometry class of $(L_{n,d}, (d, b))$, by Lemma 2.5. The cohomology $H^1(S, \mathcal{L}^d)$ vanishes, since $\mathcal{L}$ is a nef divisor of positive degree [32, Prop. 1]. Thus, the vector space $H^0(S, \mathcal{L}^d)$ has dimension $\chi(\mathcal{L}^d) = n + 1$. The support morphism $\pi : M_H(v) \to |\mathcal{L}^d|$ realizes $M_H(v)$ as a completely integrable system. The equality $\pi^* c_1(\mathcal{O}_{|\mathcal{L}^d|}(1)) = \alpha$ is easily verified.

4 Period Domains and Period Maps

4.1 A Projective $K3$ Surface Associated to an Isotropic Class

Let X be an irreducible holomorphic symplectic manifold of $K3^{[n]}$-type, $n \geq 2$. Assume that there exists a non-zero primitive isotropic class $\alpha \in H^{1,1}(X, \mathbb{Z})$. Let $\tilde{\Lambda}$ be the Mukai lattice. Choose a primitive isometric embedding $\iota : H^2(X, \mathbb{Z}) \to \tilde{\Lambda}$ in the canonical $O(\tilde{\Lambda})$-orbit ι_X of Theorem 2.2. Set $\tilde{\Lambda}_{\mathbb{C}} := \tilde{\Lambda} \otimes_{\mathbb{Z}} \mathbb{C}$. Endow $\tilde{\Lambda}_{\mathbb{C}}$ with the weight 2 Hodge structure, so that $\tilde{\Lambda}_{\mathbb{C}}^{2,0} = \iota(H^{2,0}(X))$. Set $\beta := \iota(\alpha)$. Then β belongs to $\tilde{\Lambda}_{\mathbb{C}}^{1,1}$. Set

$$\Lambda_{k3} \;:=\; \beta^{\perp}_{\tilde{\Lambda}}/\mathbb{Z}\beta$$

and endow Λ_{k3} with the induced Hodge structure. Let U be the even unimodular rank 2 lattice of signature $(1, 1)$, and $E_8(-1)$ the negative definite E_8 lattice. Then Λ_{k3} is isometric to the $K3$ lattice, which is the orthogonal direct sum of two copies of $E_8(-1)$ and three copies of U. Indeed, this is clear if β is a class in a direct summand of $\tilde{\Lambda}$ isometric to U. It follows in general, since the isometry group of $\tilde{\Lambda}$ acts transitively on the set of primitive isotropic classes in $\tilde{\Lambda}$. The induced Hodge structure on Λ_{k3} is the weight 2 Hodge structure of some $K3$ surface $S(\alpha)$, by the surjectivity of the period map.

Let v be a generator of the rank 1 sub-lattice of $\tilde{\Lambda}$ orthogonal to the image of ι. Then v is of Hodge-type $(1, 1)$. Set $\Lambda := H^2(X, \mathbb{Z})$. Then $v^{\perp}$ is isometric to Λ. We claim that $(v, v) = 2n - 2$. Indeed, the pairing induces an isomorphism of the two discriminant groups $(\mathbb{Z}v)^*/\mathbb{Z}v$ and Λ^*/Λ, since $\mathbb{Z}v$ and Λ are a pair of primitive sublattices, which are orthogonal complements in the unimodular lattice $\tilde{\Lambda}$. We conclude that the order $|(v, v)|$ of $(\mathbb{Z}v)^*/\mathbb{Z}v$ is equal to the order $2n - 2$ of Λ^*/Λ. Finally, $(v, v) > 0$, by comparing the signatures of Λ and $\tilde{\Lambda}$.

Let $\bar{v}$ be the coset $v + \mathbb{Z}\beta$ in Λ_{k3}. Then $\bar{v}$ is of Hodge-type $(1, 1)$ and $(\bar{v}, \bar{v}) = 2n - 2$. Hence $S(\alpha)$ is a projective $K3$ surface (even if X is not projective). We may further choose the Hodge isometry $\eta : H^2(S(\alpha),\mathbb{Z}) \to \Lambda_{k3}$, so that that $\bar{v}$ corresponds to a class in the positive cone of $S(\alpha)$, possibly after replacing v by $-v$. We may further assume that $\bar{v}$ corresponds to a nef class of $S(\alpha)$, possibly after replacing η with $\eta \circ w$, where w is an element of the subgroup $W \subset O^+(H^2(S(\alpha),\mathbb{Z}))$, generated by reflections by classes of smooth rational curves on $S(\alpha)$ [19, Prop. 1.9].

4.2 A Period Domain as an Affine Line Bundle Over Another

Keep the notation of Sect. 4.1. Set $\Lambda := H^2(X,\mathbb{Z})$. Set $d := \mathrm{div}(\alpha,\bullet)$. Let $\alpha^{\perp}_{\Lambda}$ be the (degenerate) lattice orthogonal to α in Λ. Set $Q_\alpha := \alpha^{\perp}_{\Lambda}/\mathbb{Z}\alpha$.

Lemma 4.1. *Q_α is isometric to the sub-lattice $\bar{v}^{\perp}$ of Λ_{k3} and both are isometric to the orthogonal direct sum*

$$E_8(-1) \oplus E_8(-1) \oplus U \oplus U \oplus \mathbb{Z}\lambda,$$

where $(\lambda,\lambda) = \frac{2-2n}{d^2}$.

Proof. The $K3$ lattice $\Lambda_{k3} := [\beta^{\perp}_{\tilde{\Lambda}}]/\mathbb{Z}\beta$ is isometric to $E_8(-1) \oplus E_8(-1) \oplus U \oplus U \oplus U$. Let L be the saturation of $\mathrm{span}_{\mathbb{Z}}\{v,\beta\}$ in $\tilde{\Lambda}$. Then L is contained in $\beta^{\perp}_{\tilde{\Lambda}}$ and the image of L in Λ_{k3} is spanned by a class ξ of self-intersection $\frac{2n-2}{d^2}$, such that $\bar{v} = d\xi$, by Lemma 2.5.

It remains to prove that Q_α is isometric to $\xi^{\perp}_{\Lambda_{k3}}$. Consider the following commutative diagram.

$$\begin{array}{ccccccc}
0 \to & \mathbb{Z}\beta & \to & \beta^{\perp}_{\tilde{\Lambda}} & \to & \Lambda_{k3} & \to 0 \\
 & =\uparrow & & \uparrow & & \uparrow j & \\
0 \to & \mathbb{Z}\beta & \to & L^{\perp}_{\tilde{\Lambda}} & \to & L^{\perp}_{\tilde{\Lambda}}/\mathbb{Z}\beta & \to 0 \\
 & \cong\uparrow & & \cong\uparrow \iota & & \uparrow \bar{\iota} & \\
0 \to & \mathbb{Z}\alpha & \to & \alpha^{\perp}_{\Lambda} & \to & Q_\alpha & \to 0.
\end{array}$$

The lower vertical arrow $\bar{\iota}$ in the rightmost column is evidently an isomorphism. The image of the upper one j is precisely $\xi^{\perp}_{\Lambda_{k3}}$. □

Let Ω_Λ be the period domain

$$\Omega_\Lambda := \{\ell \in \mathbb{P}[H^2(X,\mathbb{C})] \ : \ (\ell,\ell) = 0 \text{ and } (\ell,\bar{\ell}) > 0\}. \tag{4.1}$$

Set

$$\Omega_{\alpha^\perp} := \{\ell \in \Omega_\Lambda \ : \ (\ell,\alpha) = 0\}. \tag{4.2}$$

Then $\Omega_{\alpha^\perp}$ is an affine line-bundle over the period domain

$$\Omega_{Q_\alpha} := \{\ell \in \mathbb{P}[Q_\alpha \otimes_{\mathbb{Z}} \mathbb{C}] \; : \; (\ell,\ell) = 0 \text{ and } (\ell,\bar{\ell}) > 0\}.$$

Given a point of Ω_{Q_α}, corresponding to a one-dimensional subspace ℓ of $Q_\alpha \otimes_{\mathbb{Z}} \mathbb{C}$, we get a two dimensional subspace V_ℓ of $H^2(X,\mathbb{C})$ orthogonal to α and containing α. The line in $\Omega_{\alpha^\perp}$, over the point ℓ of Ω_{Q_α}, is $\mathbb{P}[V_\ell]\backslash\{\mathbb{P}[\mathbb{C}\alpha]\}$. Denote by

$$q \; : \; \Omega_{\alpha^\perp} \;\rightarrow\; \Omega_{Q_\alpha} \tag{4.3}$$

the bundle map. A *semi-polarized* $K3$ surface of degree k is a pair consisting of a $K3$ surface together with a nef line bundle of degree k (also known as weak algebraic polarization of degree k in [33, Section 5]). Note that each component of Ω_{Q_α} is isomorphic to the period domain of the moduli space of semi-polarized $K3$ surfaces of degree $\frac{2n-2}{d^2}$.

Definition 4.2. Fibers of q will be called *Tate-Shafarevich lines* for reasons that will become apparent in Sect. 7.

Tate-Shafarevich lines are limits of twistor lines, as will be explained in Remark 4.6.

4.3 The Period Map

Given a period $\ell \in \Omega_\Lambda$, set $\Lambda^{1,1}(\ell,\mathbb{Z}) := \{\lambda \in \Lambda \; : \; (\lambda,\ell) = 0\}$. Define $Q_\alpha^{1,1}(q(\ell),\mathbb{Z})$ similarly. We get the short exact sequence

$$0 \rightarrow \mathbb{Z}\alpha \rightarrow \left[\alpha^\perp \cap \Lambda^{1,1}(\ell,\mathbb{Z})\right] \rightarrow Q_\alpha^{1,1}(q(\ell),\mathbb{Z}) \rightarrow 0.$$

$\Omega_{\alpha^\perp}$ has two connected components, since Ω_{Q_α} has two connected components. Indeed, Q_α has signature $(2, b_2(X) - 4)$, and a period ℓ comes with an oriented positive definite plane $[\ell \oplus \bar{\ell}] \cap [\Lambda_{\mathbb{R}}]$, which, in turn, determines the orientation of the positive cone in $Q_\alpha \otimes_{\mathbb{Z}} \mathbb{R}$.

The positive cone $\tilde{\mathscr{C}}_\Lambda$ in $\Lambda_{\mathbb{R}}$ is the cone

$$\tilde{\mathscr{C}}_\Lambda := \{x \in \Lambda_{\mathbb{R}} \; : \; (x,x) > 0\}. \tag{4.4}$$

The cohomology group $H^2(\tilde{\mathscr{C}}_\Lambda,\mathbb{Z})$ is isomorphic to $\mathbb{Z}$ and an *orientation* of $\tilde{\mathscr{C}}_\Lambda$ is the choice of one of the two generator of $H^2(\tilde{\mathscr{C}}_\Lambda,\mathbb{Z})$. An orientation of $\tilde{\mathscr{C}}_\Lambda$ determines an orientation of every positive definite three dimensional subspace of $\Lambda_{\mathbb{R}}$ [23, Lemma 4.1]. A choice of an orientation of $\tilde{\mathscr{C}}_\Lambda$ determines a choice of a component of $\Omega_{\alpha^\perp}$ as follows. A period $\ell \in \Omega_\Lambda$ determines the subspace $\Lambda^{1,1}(\ell,\mathbb{R})$ and the cone $\mathscr{C}'_\ell := \{x \in \Lambda^{1,1}(\ell,\mathbb{R}) \; : \; (x,x) > 0\}$ in $\Lambda^{1,1}(\ell,\mathbb{R})$ has two connected components. A choice of a connected component of $\mathscr{C}'_\ell$ is equivalent to a choice of an orientation of the positive cone of $\tilde{\mathscr{C}}_\Lambda$. Indeed, a non-zero element $\sigma \in \ell$ and an element $\omega \in \mathscr{C}'_\ell$ determine a basis $\{Re(\sigma), Im(\sigma), \omega\}$, hence an orientation, of a positive definite three dimensional subspace of $\Lambda_{\mathbb{R}}$, and the corresponding orientation of $\tilde{\mathscr{C}}_\Lambda$ is independent of the choice of σ and ω. Thus, the choice of the orientation of the positive cone $\tilde{\mathscr{C}}_\Lambda$ determines a connected component $\mathscr{C}_\ell$ of $\mathscr{C}'_\ell$, called the *positive cone* (for the orientation). If ℓ belongs to $\Omega_{\alpha^\perp}$, then the class α belongs to $\Lambda^{1,1}(\ell,\mathbb{R})$ and α is in the closure of precisely one of the two connected components of $\mathscr{C}'_\ell$. The connected component of $\Omega_{\alpha^\perp}$, compatible with the chosen orientation of $\tilde{\mathscr{C}}_\Lambda$, is the one for which α belongs to the boundary of the positive cone $\mathscr{C}_\ell$ for the chosen orientation.

A *marked pair* (Y,ψ) consists of an irreducible holomorphic symplectic manifold Y and an isometry ψ from $H^2(Y,\mathbb{Z})$ onto a fixed lattice. The moduli space of isomorphism classes of marked pairs is a non-Hausdorff complex manifold [13]. Let $\mathfrak{M}^0_\Lambda$ be a connected component of the moduli space of marked pairs of $K3^{[n]}$-type, where the fixed lattice is Λ. The *period map*

$$P_0 : \mathfrak{M}^0_\Lambda \to \Omega_\Lambda$$

sends a marked pair (Y,ψ) to the point $\psi(H^{2,0}(Y))$ of Ω_Λ. P_0 is a holomorphic map and a local homeomorphism [4]. The positive cone $\mathscr{C}_Y$ is the connected component of the cone $\{x \in H^{1,1}(Y,\mathbb{R}) \ : \ (x,x) > 0\}$ containing the Kähler cone. Hence, the positive cone in $H^2(Y,\mathbb{R})$ comes with a canonical orientation and the marking ψ determines an orientation of the positive cone in $\tilde{\mathscr{C}}_\Lambda$. We conclude that $\mathfrak{M}^0_\Lambda$ determines an orientation of the positive cone $\tilde{\mathscr{C}}_\Lambda$ [23, Sec. 4]. Let

$$\Omega^+_{\alpha^\perp} \tag{4.5}$$

be the connected component of $\Omega_{\alpha^\perp}$, inducing the same orientation of $\tilde{\mathscr{C}}_\Lambda$ as $\mathfrak{M}^0_\Lambda$. Let

$$\mathfrak{M}^0_{\alpha^\perp} \tag{4.6}$$

be the inverse image $P_0^{-1}(\Omega^+_{\alpha^\perp})$.

Theorem 4.3 (The Global Torelli Theorem [14, 40]). *The period map P_0 : $\mathfrak{M}^0_\Lambda \to \Omega_\Lambda$ is surjective. Any two points in the same fiber of P_0 are inseparable. If (X_1,η_1) and (X_2,η_2) correspond to two inseparable points in $\mathfrak{M}^0_\Lambda$, then X_1 and X_2 are bimeromorphic. If the Kähler cone of X is equal to its positive cone and (X,η) corresponds to a point of $\mathfrak{M}^0_\Lambda$, then this point is separated.*

Lemma 4.4. $\mathfrak{M}^0_{\alpha^\perp}$ *is path-connected.*

Proof. The statement follows from the Global Torelli Theorem 4.3 and the fact that $\Omega^+_{\alpha^\perp}$ is connected. The proof is similar to that of [24, Proposition 5.11]. □

Proposition 4.5. *Let X_1 and X_2 be two irreducible holomorphic symplectic manifolds of $K3^{[n]}$-type and $\eta_j : H^2(X_j, \mathbb{Z}) \to \Lambda$, $j = 1, 2$, isometries. The marked pairs (X_1, η_1) and (X_2, η_2) belong to the same connected moduli space $\mathfrak{M}^0_{\alpha^\perp}$, provided the following conditions hold.*

(1) The $O(\tilde{\Lambda})$ orbits $\iota_{X_j} \circ \eta_j^{-1}$, $j = 1, 2$, are equal. Above ι_{X_j} is the canonical $O(\tilde{\Lambda})$-orbit of primitive isometric embeddings of $H^2(X_j, \mathbb{Z})$ into $\tilde{\Lambda}$ mentioned in Theorem 2.2.
(2) $\eta_2^{-1} \circ \eta_1 : H^2(X_1, \mathbb{Z}) \to H^2(X_2, \mathbb{Z})$ is orientation preserving.
(3) $\eta_j^{-1}(\alpha)$ is of Hodge type $(1, 1)$ and it belongs to the boundary of the positive cone $\mathscr{C}_{X_j}$ in $H^{1,1}(X_j, \mathbb{R})$, for $j = 1, 2$.

Proof. Conditions 1 and 2 imply that $\eta_2^{-1} \circ \eta_1$ is a parallel-transport operator, by Theorem 2.4. Hence, the two marked pairs belong to the same connected component $\mathfrak{M}^0_\Lambda$ of $\mathfrak{M}_\Lambda$. Condition 3 implies that both belong to $\mathfrak{M}^0_{\alpha^\perp}$, and the latter is connected, by Lemma 4.4. □

Proof (of Proposition 1.7). Lemma 2.5 introduced the monodromy invariant $[L, v](c_1(\mathscr{L}))$ of the pair $(X, \mathscr{L})$. The claimed number of deformation types in the statement of the proposition is equal to the number of values of the monodromy invariant $[L, v](\bullet)$ for fixed n and d, by Lemma 2.6. Assume given another pair $(X', \mathscr{L}')$ as above, such that the monodromy invariants $[L, v](c_1(\mathscr{L}'))$ and $[L, v](c_1(\mathscr{L}))$ are equal. Choose a parallel transport operator $g : H^2(X', \mathbb{Z}) \to H^2(X, \mathbb{Z})$. We do not assume that $g(c_1(\mathscr{L}'))$ is of Hodge type $(1, 1)$. Set $\alpha := c_1(\mathscr{L})$ and $\alpha' := c_1(\mathscr{L}')$. The monodromy invariant $[L, v](g(\alpha'))$ is equal to $[L, v](\alpha')$ and hence also to $[L, v](\alpha)$. Hence, there exists a monodromy operator $f \in Mon^2(X)$, such that $fg(\alpha') = \alpha$, by Lemma 2.5. Choose a marking $\eta : H^2(X, \mathbb{Z}) \to \Lambda$. Then $\eta' := \eta \circ f \circ g$ is a marking of X' satisfying $\eta(\alpha) = \eta'(\alpha')$. Hence, the triples (X, α, η) and (X', α', η') both belong to the moduli space $\mathfrak{M}^0_{\eta(\alpha)^\perp}$, by Proposition 4.5. $\mathfrak{M}^0_{\eta(\alpha)^\perp}$ is connected, by Lemma 4.4. Hence, $(X, \mathscr{L})$ and $(X', \mathscr{L}')$ are deformation equivalent. □

Remark 4.6. Tate-Shafarevich lines (Definition 4.2) are limits of twistor lines in the following sense. Let ℓ be a point of Ω_Λ and ω a class in the positive cone $\mathscr{C}_\ell$ in $\Lambda^{1,1}(\ell, \mathbb{R})$. Assume that ω is not orthogonal to any class in $\Lambda^{1,1}(\ell, \mathbb{Z})$. Then there exists a marked pair (X, η) in each connected component $\mathfrak{M}^0_\Lambda$ of the moduli space of marked pairs, such that $P(X, \eta) = \ell$ and $\eta^{-1}(\omega)$ is a Kähler class of X [13, Cor. 5.7]. Set $W' := \ell \oplus \bar{\ell} \oplus \mathbb{C}\omega$. $\mathbb{P}(W') \cap \Omega_\Lambda$ is a *twistor line* for (X, η); it admits a canonical lift to a smooth rational curve in $\mathfrak{M}^0_\Lambda$ containing the point (X, η) [13, Cor. 5.8]. This lift corresponds to an action of the quaternions $\mathbb{H}$ on the real tangent bundle of the differentiable manifold X, such that the unit quaternions act as integrable complex structures, one of which is the complex structure of X. Let $\alpha \in \Lambda$ be the primitive isotropic class as above. Assume that ℓ belongs to $\Omega^+_{\alpha^\perp}$. Consider the three dimensional subspace $W := \ell \oplus \bar{\ell} \oplus \mathbb{C}\alpha$ of $H^2(X, \mathbb{C})$. Then W is a limit of a sequence of three dimensional subspaces W_i', associated to some sequence of classes ω_i as above, since α belongs to the boundary of the positive cone $\mathscr{C}_\ell$. Now W is contained in $\alpha^\perp$, and so $\mathbb{P}(W) \cap \Omega_{\alpha^\perp} = \mathbb{P}(W) \cap \Omega_\Lambda$. In this degenerate case, the conic $\mathbb{P}(W) \cap \Omega_\Lambda$ consists of two irreducible components, the Tate-Shafarevich line $\mathbb{P}[\ell \oplus \mathbb{C}\alpha] \setminus \{\mathbb{P}[\mathbb{C}\alpha]\}$ in $\Omega^+_{\alpha^\perp}$ and the line $\mathbb{P}[\bar{\ell} \oplus \mathbb{C}\alpha] \setminus \{\mathbb{P}[\mathbb{C}\alpha]\}$ in the other connected component $\Omega^-_{\alpha^\perp}$ of $\Omega_{\alpha^\perp}$. Theorem 7.11 will provide a lift of a generic Tate-Shafarevich line in the period domain to a line in the moduli space of marked pairs.

A summary of notation related to lattices and period domains

U	The rank 2 even unimodular lattice of signature $(1, 1)$
$E_8(-1)$	The root lattice of type E_8 with a negative definite pairing
$\widetilde{\Lambda}$	The Mukai lattice; the orthogonal direct sum $U^{\oplus 4} \oplus E_8(-1)^{\oplus 2}$
Λ	The $K3^{[n]}$-lattice; the orthogonal direct sum $U^{\oplus 3} \oplus E_8(-1)^{\oplus 2} \oplus \langle 2-2n \rangle$, where $\langle 2-2n \rangle$ is the rank 1 lattice generated by a class of self-intersection $2-2n$
α	A primitive isotropic class in Λ
Q_α	The subquotient $\alpha^\perp/\mathbb{Z}\alpha$
ι	A primitive embedding of Λ in $\widetilde{\Lambda}$
β	The primitive isotropic class $\iota(\alpha)$ in $\widetilde{\Lambda}$
Λ_{k3}	The subquotient $\beta^\perp/\mathbb{Z}\beta$, which is isomorphic to the $K3$ lattice $U^{\oplus 3} \oplus E_8(-1)^{\oplus 2}$
v	A generator of the rank 1 sublattice of $\widetilde{\Lambda}$ orthogonal to $\iota(\Lambda)$
$\bar{v}$	The coset $v + \mathbb{Z}\beta$ in Λ_{k3}
d	The divisibility of $(\alpha, \bullet)$ in Λ^*; $d := \gcd\{(\alpha, \lambda) : \lambda \in \Lambda\}$
ξ	The integral element $(1/d)\bar{v}$ of Λ_{k3}. We have $(\xi, \xi) = \frac{2n-2}{d^2}$
Ω_Λ	The period domain given in (4.1)
$\widetilde{\mathscr{C}}_\Lambda$	The positive cone given in (4.4)
Ω^+_Λ	The connected component of Ω_Λ determined by the orientation of $\widetilde{\mathscr{C}}_\Lambda$
$\Omega_{\alpha^\perp}$	The hyperplane section of Ω_Λ given in (4.2)
$\Omega^+_{\alpha^\perp}$	The connected component of $\Omega_{\alpha^\perp}$ given in (4.5)
Ω_{Q_α}	The period domain of the lattice Q_α
q	The fibration $q : \Omega_{\alpha^\perp} \to \Omega_{Q_\alpha}$ by Tate-Shafarevich lines given in (4.3)
$\mathfrak{M}^0_\Lambda$	A connected component of the moduli space of marked pairs
P_0	The period map $P_0 : \mathfrak{M}^0_\Lambda \to \Omega^+_\Lambda$
$\mathfrak{M}^0_{\alpha^\perp}$	The inverse image of $\Omega^+_{\alpha^\perp}$ in $\mathfrak{M}^0_\Lambda$ via P_0
$[L, v](\alpha)$	The monodromy invariant associated to the class α in Lemma 2.5 (4)

5 Density of Periods of Relative Compactified Jacobians

We keep the notation of Sect. 4. In Sect. 5.1 we construct a section $\tau : \Omega^+_{Q_\alpha} \to \Omega^+_{\alpha^\perp}$, given in (5.2), of the fibration $q : \Omega^+_{\alpha^\perp} \to \Omega^+_{Q_\alpha}$ by Tate-Shafarevich lines. We then show that τ maps a period $\underline{\ell}$, of a semi-polarized $K3$ surface $(S, \mathscr{B})$ in the period domain $\Omega^+_{Q_\alpha}$, to the period $\tau(\underline{\ell})$ of a moduli space M of sheaves on S

with pure one-dimensional support in the linear system $|\mathscr{B}^d|$. The moduli space M admits a Lagrangian fibration over $|\mathscr{B}^d|$. In Sect. 5.2 we construct an injective homomorphism $g : Q_\alpha \to O(\Lambda)$, whose image is contained in the subgroup of the monodromy group which stabilizes α. We get an action of Q_α on the period domain $\Omega^+_{\alpha^\perp}$, which lifts to an action on connected components $\mathfrak{M}^0_{\alpha^\perp}$ of the moduli space of marked pairs given in Eq. (4.6). We then show that the fibration q by Tate-Shafarevich lines is $g(Q_\alpha)$-invariant. In Sect. 5.3 we prove that the $g(Q_\alpha)$-orbit of every point in a non-special Tate-Shafarevich line is dense in that line. Consequently, the non-special Tate-Shafarevich line $q^{-1}(\underline{\ell})$ contains the dense orbit $g(Q_\alpha)\tau(\underline{\ell})$ of periods of marked pairs in $\mathfrak{M}^0_{\alpha^\perp}$ admitting a Lagrangian fibration.

Conventions: The discussion in the current Sect. 5 concerns only period domains, so we are free to choose the embedding ι. When we consider in subsequent sections a component $\mathfrak{M}^0_\Lambda$ of the moduli space of marked pairs (X, η) of $K3^{[n]}$-type, together with such an embedding $\iota : \Lambda \to \tilde{\Lambda}$, we will always assume that ι is chosen so that $\iota \circ \eta$ belongs to the canonical $O(\tilde{\Lambda})$-orbit ι_X of Theorem 2.2, for all (X, η) in $\mathfrak{M}^0_\Lambda$. We choose the orientation of the positive cone $\tilde{\mathscr{C}}_\Lambda$ of Λ, so that α belongs to the boundary of the positive cone in $\Lambda^{1,1}(\ell, \mathbb{R})$, for every $\ell \in \Omega^+_{\alpha^\perp}$. We choose the orientation of the positive cone $\tilde{\mathscr{C}}_{\Lambda_{k3}}$, so that $\bar{v}$ belongs to the positive cone in $\Lambda^{1,1}_{k3}(\underline{\ell}, \mathbb{R})$, for every $\underline{\ell} \in \Omega^+_{\bar{v}^\perp}$. Note that the composition $\alpha^\perp_\Lambda \stackrel{\iota}{\to} \beta^\perp_{\tilde{\Lambda}} \to \Lambda_{k3}$ induces an isometry from $Q_\alpha := \alpha^\perp_\Lambda / \mathbb{Z}\alpha$ onto $\bar{v}^\perp_{\Lambda_{k3}}$, by Lemma 4.1. The choice of orientation of the positive cone of Λ_{k3} determines an orientation of the positive cone of Q_α.

5.1 A Period of a Lagrangian Fibration in Each Tate-Shafarevich Line

Choose a class γ in $\tilde{\Lambda}$ satisfying $(\gamma, \beta) = -1$ and $(\gamma, \gamma) = 0$. Note that β and γ span a unimodular sub-lattice of $\tilde{\Lambda}$ of signature $(1, 1)$. We construct next a section of the affine bundle $q : \Omega_{\alpha^\perp} \to \Omega_{Q_\alpha}$, given in Eq. (4.3), in terms of γ. We have the following split short exact sequence.

$$\begin{array}{c} 0 \longrightarrow \mathbb{Z}\beta \longrightarrow \beta^\perp_{\tilde{\Lambda}} \stackrel{j}{\longrightarrow} \Lambda_{k3} \longrightarrow 0. \\ \sigma_\gamma : \beta^\perp_{\tilde{\Lambda}} \to \mathbb{Z}\beta, \quad \tilde{\tau}_\gamma : \Lambda_{k3} \to \beta^\perp_{\tilde{\Lambda}} \end{array} \tag{5.1}$$

Above, $\sigma_\gamma(x) = -(x, \gamma)\beta$, and $\tilde{\tau}_\gamma(y) = \tilde{y} + (\tilde{y}, \gamma)\beta$, where $\tilde{y}$ is any element of $\beta^\perp_{\tilde{\Lambda}}$ satisfying $j(\tilde{y}) = y$. One sees that $\tilde{\tau}_\gamma$ is well defined as follows. If $\tilde{y}_1$ and $\tilde{y}_2$ satisfy $j(\tilde{y}_k) = y$, then the difference $[\tilde{y}_1 + (\tilde{y}_1, \gamma)\beta] - [\tilde{y}_2 + (\tilde{y}_2, \gamma)\beta]$ belongs to the kernel of j and is sent to 0 via σ_γ, so the difference is equal to 0. Note that $\tilde{\tau}_\gamma$ is an isometric embedding and its image is precisely $\{\beta, \gamma\}^\perp_{\tilde{\Lambda}}$.

We regard $\Omega^+_{Q_\alpha}$ as the period domain for semi-polarized $K3$ surfaces, with a nef line bundle of degree $\frac{2n-2}{d^2}$, via the isomorphism $\bar{v}^\perp_{\Lambda_{k3}} \cong Q_\alpha$ of Lemma 4.1. The homomorphism $\iota^{-1} \circ \tilde{\tau}_\gamma$ induces an isometric embedding of Q_α in $\alpha^\perp_\Lambda$. We get a section

$$\tau_\gamma \,:\, \Omega^+_{Q_\alpha} \to \Omega^+_{\alpha^\perp} \tag{5.2}$$

of $q : \Omega^+_{\alpha^\perp} \to \Omega^+_{Q_\alpha}$. Following is an explicit description of τ_γ. Let $\underline{\ell}$ be a period in $\Omega^+_{Q_\alpha}$. Choose a period ℓ in $\Omega^+_{\alpha^\perp}$ satisfying $q(\ell) = \underline{\ell}$. Let x be a non-zero element of the line ℓ in $\alpha^\perp_\Lambda \otimes_{\mathbb{Z}} \mathbb{C}$. Then

$$\tau_\gamma(\underline{\ell}) = \mathrm{span}_{\mathbb{C}}\{x + (\iota(x), \gamma)\alpha\}. \tag{5.3}$$

We see that γ belongs to $\tilde{\Lambda}^{1,1}(\tau_\gamma(\underline{\ell}))$, for every $\underline{\ell}$ in $\Omega^+_{Q_\alpha}$.

Fix a period $\underline{\ell}$ in $\Omega^+_{Q_\alpha}$. We construct next a marked pair $(M_H(u), \eta_1)$ with period $\tau_\gamma(\underline{\ell})$, such that $\eta_1^{-1}(\alpha)$ induces a Lagrangian fibration. Let S be a $K3$ surface and $\eta : H^2(S,\mathbb{Z}) \to \Lambda_{k3}$ a marking, such that the period $\eta(H^{2,0}(S))$ is $\underline{\ell}$. Such a marked pair (S,η) exists, by the surjectivity of the period map. Extend η to the Hodge isometry

$$\tilde{\eta} \,:\, H^*(S,\mathbb{Z}) \to \tilde{\Lambda},$$

given by $\tilde{\eta}((0,0,1)) = \beta$, $\tilde{\eta}((1,0,0)) = \gamma$, and $\tilde{\eta}$ restricts to $H^2(S,\mathbb{Z})$ as $\tilde{\tau}_\gamma \circ \eta$. We have the equality $v = \sigma_\gamma(v) + \tilde{\tau}_\gamma(\bar{v}) = -(\gamma, v)\beta + \tilde{\tau}_\gamma(\bar{v})$. Set $a := -(\gamma, v)$ and $u := (0, \eta^{-1}(\bar{v}), a)$. Then $\tilde{\eta}(u) = v$. We may choose the marking η so that the class $\eta^{-1}(\bar{v})$ is nef, possibly after replacing η by $\pm\eta \circ w$, where w is an element of the group of isometries of $H^2(S,\mathbb{Z})$, generated by reflections by -2 curves [2, Ch. VIII Prop. 3.9]. Choose a u-generic polarization H of S. Then $M_H(u)$ is a projective irreducible holomorphic symplectic manifold. Let

$$\theta : u^\perp \to H^2(M_H(u), \mathbb{Z})$$

be Mukai's isometry, given in Eq. (3.1). We get the commutative diagram:

$$\begin{array}{ccccc}
\Lambda & \xrightarrow[\cong]{\iota} & v^\perp & \xrightarrow{\subset} & \tilde{\Lambda} \\
\uparrow{\scriptstyle \eta_1} & & \uparrow{\scriptstyle \eta_2} & & \uparrow{\scriptstyle \tilde{\eta}} \\
H^2(M_H(u),\mathbb{Z}) & \xrightarrow{\theta^{-1}} & u^\perp & \xrightarrow{\subset} & H^*(S,\mathbb{Z}),
\end{array} \tag{5.4}$$

where η_2 is the restriction of $\tilde{\eta}$ and $\eta_1 = \iota^{-1} \circ \eta_2 \circ \theta^{-1}$. Note that $\eta_1(\theta(0,0,1)) = \alpha$. Let L be the saturation in $H^*(S,\mathbb{Z})$ of the sub-lattice spanned by $(0,0,1)$ and u. Let b be an integer satisfying $ab \equiv 1$ (modulo d). The monodromy invariant

$[L, u](\theta(0,0,1))$ of Lemma 2.5 is the isometry class of the pair $(L_{n,d}, (d,b))$, by the commutativity of the above diagram. Furthermore, η_1 is a Hodge isometry with respect to the Hodge structure on Λ induced by $\tau_\gamma(\underline{\ell})$. In particular, $(M_H(u), \eta_1)$ is a marked pair with period $\tau_\gamma(\underline{\ell})$. Example 3.1 exhibits $\theta(0,0,1)$ as the class $\pi^* c_1(\mathcal{O}_{|\mathcal{L}^d|}(1))$, for a Lagrangian fibration $\pi : M_H(u) \to |\mathcal{L}^d|$, where $\mathcal{L}$ is the line bundle over S with class $\eta^{-1}(\xi)$.

Remark 5.1. The isometry η_1 is compatible with the orientations of the positive cones, the canonical one of $H^2(M_H(u), \mathbb{Z})$ and the chosen one of Λ. Indeed, it maps the class $\theta(0,0,1)$, on the boundary of the positive cone of $H^{1,1}(M_H(u), \mathbb{R})$, to the class α on the boundary of the positive cone of $\Lambda^{1,1}(\tau_\gamma(\underline{\ell}), \mathbb{R})$. The composition $\tilde{\eta} \circ \theta^{-1}$ in Diagram (5.4) belongs to the canonical orbit $\iota_{M_H(u)}$ of Theorem 2.2, by [22, Theorem 1.14]. The commutativity of the Diagram implies that the isometric embedding $\iota \circ \eta_1$ also belongs to the orbit $\iota_{M_H(u)}$.

5.2 *Monodromy Equivariance of the Fibration by Tate-Shafarevich Lines*

Denote by $O(\tilde{\Lambda})^+_{\beta,v}$ the subgroup of $O(\tilde{\Lambda})^+$ stabilizing both β and v. Following is a natural homomorphism

$$h : O(\tilde{\Lambda})^+_{\beta,v} \to O(\Lambda_{k3})_{\bar{v}}. \tag{5.5}$$

If ψ belongs to $O(\tilde{\Lambda})^+_{\beta,v}$, then $\psi(\beta) = \beta$ and $\beta^{\perp}_{\tilde{\Lambda}}$ is ψ-invariant. Thus ψ induces an isometry $h(\psi)$ of $\Lambda_{k3} := \beta^{\perp}_{\tilde{\Lambda}}/\mathbb{Z}\beta$. We construct next a large subgroup in the kernel of h.

Given an element z of $\tilde{\Lambda}$, orthogonal to β and v, define the map $\tilde{g}_z : \tilde{\Lambda} \to \tilde{\Lambda}$ by

$$\tilde{g}_z(x) := x - (x,\beta)z + \left[(x,z) - \frac{1}{2}(x,\beta)(z,z)\right]\beta.$$

Lemma 5.2. *The map $\tilde{g}_z$ is the unique isometry in $O(\tilde{\Lambda})_{\beta,v}$, which sends γ to an element of $\tilde{\Lambda}$ congruent to $\gamma + z$ modulo $\mathbb{Z}\beta$ and belongs to the kernel of h. The isometry $\tilde{g}_z$ is orientation preserving.*

Proof. We first define an isometry f with the above property, then prove its uniqueness, and finally prove that it is equal to $\tilde{g}_z$. Set $\gamma_1 := \gamma + z + \left[(\gamma, z) + \frac{1}{2}(z,z)\right]\beta$. Then $(\gamma_1, \gamma_1) = 0$, $(\gamma_1, \beta) = -1$, and γ_1 is the unique element of $\tilde{\Lambda}$ satisfying the above equalities and congruent to $\gamma + z$ modulo $\mathbb{Z}\beta$. Define $\tilde{\sigma}_\gamma : \tilde{\Lambda} \to \mathbb{Z}\beta + \mathbb{Z}\gamma$ by $\tilde{\sigma}_\gamma(x) := -(x,\beta)\gamma - (x,\gamma)\beta$. We get the commutative diagram with split short exact rows:

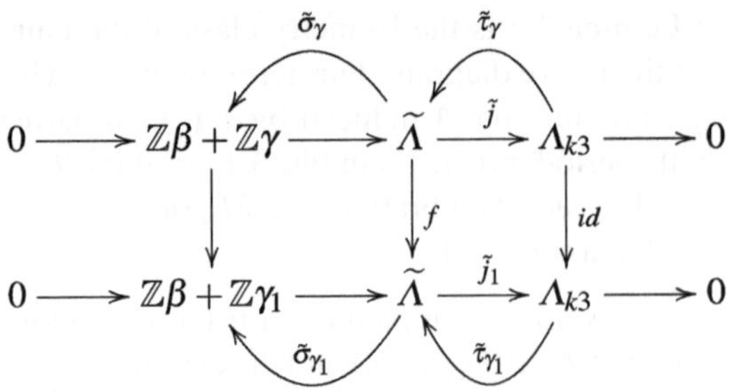

Above $\tilde{\tau}_\gamma$ and j are the homomorphisms given in Eq. (5.1), $\tilde{j}(x) = j(x + (x, \beta)\gamma)$, and $\tilde{\sigma}_{\gamma_1}$, $\tilde{\tau}_{\gamma_1}$, and $\tilde{j}_1$ are defined similarly, replacing γ by γ_1. The map f is defined by $f(\beta) = \beta$, $f(\gamma) = \gamma_1$, and $f(\tilde{\tau}_\gamma(y)) = \tilde{\tau}_{\gamma_1}(y)$. Then f is clearly an isometry.

The isometry f can be extended to an isometry of $\tilde{\Lambda}_{\mathbb{R}}$ and we can continuously deform z to 0 in $\{\beta, v\}^\perp \otimes_{\mathbb{Z}} \mathbb{R}$, resulting in a continuous deformation of f to the identity. Hence, f is orientation preserving.

Note the equalities $\tilde{\sigma}_\gamma(v) = -(v, \gamma)\beta = -(v, \gamma_1)\beta = \tilde{\sigma}_{\gamma_1}(v)$, where the middle one follows from that fact that both z and β are orthogonal to v. We get the equality

$$\tilde{\tau}_\gamma(\bar{v}) = v - \tilde{\sigma}_\gamma(v) = v - \tilde{\sigma}_{\gamma_1}(v) = \tilde{\tau}_{\gamma_1}(\bar{v}).$$

Thus $f(v) = v$ and f belongs to $O(\tilde{\Lambda})^+_{\beta,v}$. Let x be an element of $\beta^\perp$. Then $\tilde{j}(x) = j(x) = \tilde{j}_1(x)$. Set $y := j(x)$. Now $\tilde{\tau}_\gamma(y) \equiv \tilde{\tau}_{\gamma_1}(y)$ modulo $\mathbb{Z}\beta$, by definition of both. Hence, $h(f)$ is the identity isometry of Λ_{k3}.

Let f' be another isometry of $\tilde{\Lambda}$ satisfying the assumptions of the Lemma. Then $f'(\gamma) = \gamma_1$, by the characterization of γ_1 mentioned above. Set $e := f^{-1} \circ f'$. Then $e(\beta) = \beta$, $e(\gamma) = \gamma$, $e(v) = v$, and $h(e) = id$. Given $x \in \beta^\perp$, we get that $e(x) \equiv x$ modulo $\mathbb{Z}\beta$. Now $(e(x), \gamma) = (e(x), e(\gamma)) = (x, \gamma)$. Thus, e restricts to the identity on $\beta^\perp$. We conclude that e is the identity of $\tilde{\Lambda}$, as the latter is spanned by γ and $\beta^\perp$. Thus $f' = f$.

It remains to prove the equality $f = \tilde{g}_z$. We already know that $f(\gamma) = \gamma_1 = \tilde{g}_z(\gamma)$ and $f(\beta) = \beta = \tilde{g}_z(\beta)$. Given $y \in \Lambda_{k3}$, we have

$$\tilde{g}_z(\tilde{\tau}_\gamma(y)) = \tilde{\tau}_\gamma(y) + (\tilde{\tau}_\gamma(y), z)\beta = \tilde{\tau}_{\gamma_1}(y) = f(\tilde{\tau}_\gamma(y)).$$

Hence, $\tilde{g}_z = f$. □

Let

$$\tilde{g} : \alpha_\Lambda^\perp \to O(\tilde{\Lambda})^+_{\beta,v}$$

be the map sending z to $\tilde{g}_{\iota(z)}$. Denote by $Mon^2(\Lambda, \iota)$ the subgroup of $O^+(\Lambda)$ of isometries stabilizing the orbit $O(\tilde{\Lambda})\iota$. Note that $O(\tilde{\Lambda})^+_v$ is conjugated via ι onto $Mon^2(\Lambda, \iota)$, if $n = 2$, and to an index 2 subgroup of $Mon^2(\Lambda, \iota)$, if $n \geq 2$ [21, Lemma 4.10]. Let $Mon^2(\Lambda, \iota)_\alpha$ be the subgroup of $Mon^2(\Lambda, \iota)$ stabilizing α.

Lemma 5.3. *(1) The map $\tilde{g}$ is a group homomorphism with kernel $\mathbb{Z}\alpha$. It thus factors through an injective homomorphism*

$$g : Q_\alpha \to Mon^2(\Lambda, \iota)_\alpha.$$

(2) Let z be an element of $\alpha^\perp_{\tilde{\Lambda}}$ and $[z]$ its coset in Q_α. Then $g_{[z]} : \alpha^\perp \to \alpha^\perp$ sends $x \in \alpha^\perp$ to $x + (x, z)\alpha$.

(3) The map $q : \Omega^+_{\alpha^\perp} \to \Omega^+_{Q_\alpha}$ is $Mon^2(\Lambda, \iota)_\alpha$-equivariant and it is invariant with respect to the image $g(Q_\alpha) \subset Mon^2(\Lambda, \iota)_\alpha$ of g.

(4) The image of $\tilde{g}$ is equal to the kernel of the homomorphism h, given in Eq. (5.5).

Proof. Part (1) follows from the characterization of $\tilde{g}_z$ in Lemma 5.2. Part (2) is straightforward as is the $Mon^2(\Lambda, \iota)_\alpha$-equivariance of q. The $g(Q_\alpha)$-invariance of q follows from part (2). Part (3) is thus proven.

Part (4): The image of $\tilde{g}$ is contained in the kernel of h, by Lemma 5.2. Let $f \in O(\tilde{\Lambda})_{\beta,v}$ belong to the kernel of h. Set $\gamma_1 := f(\gamma)$ and $z := \gamma_1 - \gamma$. Then $(\gamma_1, \beta) = (f(\gamma), \beta) = (f(\gamma), f(\beta)) = (\gamma, \beta)$ and similarly $(\gamma_1, v) = (\gamma, v)$. Hence, $(z, \beta) = 0$ and $(z, v) = 0$. The isometry $\tilde{g}_z$ is thus well defined and it is equal to f, by Lemma 5.2. □

5.3 Density

A period $\underline{\ell}$ in $\Omega_{\Lambda_{k3}}$ is said to be *special*, if it satisfies the condition analogous to the one in Definition 1.1. We identify Ω_{Q_α} as a submanifold of $\Omega_{\Lambda_{k3}}$, via Lemma 4.1. Note that a period $\ell \in \Omega_{\alpha^\perp}$ is special, if and only if the period $q(\ell)$ is.

Lemma 5.4. *1. $g(Q_\alpha)$ has a dense orbit in $q^{-1}(\underline{\ell})$, if and only if $\underline{\ell}$ is non-special.*
2. If $g(Q_\alpha)$ has a dense orbit in $q^{-1}(\underline{\ell})$, then every $g(Q_\alpha)$-orbit in $q^{-1}(\underline{\ell})$ is dense.

Proof. Part 2 follows from the description of the action in Lemma 5.3 part 2. We prove part 1. Fix a period ℓ such that $q(\ell) = \underline{\ell}$ and choose a non-zero element t of the line ℓ in $\alpha^\perp_{\tilde{\Lambda}} \otimes_{\mathbb{Z}} \mathbb{C}$. Then $q^{-1}(\underline{\ell}) = \mathbb{P}[\mathbb{C}\alpha + \mathbb{C}t] \setminus \{\mathbb{P}[\mathbb{C}\alpha]\}$ and $g_{[z]}(a\alpha + t) = (a + (t, z))\alpha + t$, by Lemma 5.3 part 2. The fiber $q^{-1}(\underline{\ell})$ has a dense $g(Q_\alpha)$-orbit, if and only if the image of

$$(t, \bullet) : Q_\alpha \to \mathbb{C} \tag{5.6}$$

is dense in $\mathbb{C}$.

Suppose first that $\underline{\ell}$ is special. Set $V := [\underline{\ell} \oplus \bar{\underline{\ell}}] \cap [Q_\alpha \otimes_{\mathbb{Z}} \mathbb{R}]$. Let λ be a non-zero element in $V \cap Q_\alpha$. There exists an element $t \in \underline{\ell}$, such that $\lambda = t + \bar{t}$. Given an element $z \in Q_\alpha$, then $2Re(z, t) = (z, t) + (z, \bar{t}) = (z, \lambda)$ is an integer. Thus, $Re(z, t)$ belongs to the discrete subgroup $\frac{1}{2}\mathbb{Z}$ of $\mathbb{R}$. Hence, the image of the homomorphism (5.6) is not dense in $\mathbb{C}$.

Assume next that $\underline{\ell}$ is non-special. Denote by $\Theta(\underline{\ell}) \subset Q_\alpha$ the lattice orthogonal to the kernel of the homomorphism (5.6). $\Theta(\underline{\ell})$ is the transcendental lattice of the $K3$-surface with period $\underline{\ell}$. We know that $\Theta(\underline{\ell})$ has rank at least two, and if the rank of $\Theta(\underline{\ell})$ is 2, then the Hodge decomposition is defined over $\mathbb{Q}$ and so $\underline{\ell}$ is special. Thus, the rank of $\Theta(\underline{\ell})$ is at least three. Let $G \subset \Theta(\underline{\ell})$ be a co-rank 1 subgroup. We claim that the image (t, G), of G via the homomorphism (5.6), spans $\mathbb{C}$ as a 2-dimensional real vector space. The latter statement is equivalent to the statement that the image of G in V^*, under the map $z \mapsto (z, \bullet)$ which has real values on V, spans V^*. The equivalence is clear considering the following isomorphisms of two dimensional real vector spaces:

$$\mathbb{C} \xleftarrow{ev_t} \mathrm{Hom}_{\mathbb{C}}(\ell, \mathbb{C}) \xrightarrow{Re} \mathrm{Hom}_{\mathbb{R}}(\ell, \mathbb{R}) \xrightarrow{p^*} \mathrm{Hom}_{\mathbb{R}}(V, \mathbb{R}) = V^*,$$

where ev_t is evaluation at t, Re takes $(z, \bullet)$ to its real part $Re(z, \bullet)$, and p^* is pullback via the projection $p : V \to \ell$ on the $(2,0)$ part. Assume that the image of G in V^* spans a one-dimensional subspace W. Let U be the subspace of V annihilated by W, and hence also by $(z, \bullet)$, $z \in G$. Then the kernel of the homomorphism $\Lambda_{k3} \to U^*$, given by $z \mapsto (z, \bullet)$, has co-rank 1 in Λ_{k3}. It follows that the decomposition $\Lambda_{k3} \otimes_{\mathbb{Z}} \mathbb{R} = U \oplus U^\perp$ is defined over $\mathbb{Q}$. Thus, $U \cap \Lambda_{k3}$ is non-trivial and $\underline{\ell}$ is special. A contradiction. Thus, indeed, the image (t, G) of G spans $\mathbb{C}$. Let $Z \subset \mathbb{C}$ be the image $(t, \Theta(\underline{\ell}))$ of $\Theta(\underline{\ell})$ via the homomorphism (5.6). We have established that Z satisfies the hypothesis of Lemma 5.5 below, which implies that the image of the homomorphism (5.6) is dense in $\mathbb{C}$. □

Lemma 5.5. *Let $Z \subset \mathbb{R}^2$ be a free additive subgroup of rank ≥ 3. Assume that any co-rank 1 subgroup of Z spans $\mathbb{R}^2$ as a real vector space. Then Z is dense in $\mathbb{R}^2$.*

Proof. Let Σ be the set of all bases of $\mathbb{R}^2$, consisting of elements of Z. Given a basis $\beta \in \Sigma$, $\beta = \{z_1, z_2\}$, set $|\beta| = |z_1| + |z_2|$. Set $I := \inf\{|\beta| : \beta \in \Sigma\}$. Note that the closed parallelogram P_β with vertices $\{0, z_1, z_2, z_1 + z_2\}$ has diameter $< |\beta|$. Furthermore, every point of the plane belongs to a translate of P_β by an element of the subset $\mathrm{span}_{\mathbb{Z}}\{z_1, z_2\}$ of Z. Hence, it suffices to prove that $I = 0$.

The proof is by contradiction. Assume that $I > 0$. Let $\beta = \{z_1, z_2\}$ be a basis satisfying $I \leq |\beta| < \frac{12}{11} I$. We may assume, without loss of generality, that $|z_1| \geq |z_2|$.

We prove next that there exists an element $w \in Z$, such that $w = c_1 z_1 + c_2 z_2$, where the coefficients c_i are irrational. Set $r := \mathrm{rank}(Z)$. Let $z_3, \ldots, z_r$ be elements of Z completing $\{z_1, z_2\}$ to a subset, which is linearly independent over $\mathbb{Q}$. Write $z_j = c_{j,1} z_1 + c_{j,2} z_2$, for $3 \leq j \leq r$. Assume that $c_{j,1}$ are rational, for $3 \leq j \leq r$. Then there exists a positive integer N, such that $N c_{j,1}$ are integers, for all $3 \leq j \leq r$. Then

$$\{z_2, N z_3 - N c_{3,1} z_1, \ldots, N z_r - N c_{r,1} z_1\}$$

spans a co-rank 1 subgroup of Z, which lies on $\mathbb{R} z_2$. This contradicts the assumption on Z. Hence, there exists an element $w \in Z$, such that $w = c_1 z_1 + c_2 z_2$, where the coefficient c_1 is irrational. Repeating the above argument for c_2, we get the desired conclusion.

Choose an element w as above. By adding vectors in $\mathrm{span}_{\mathbb{Z}}\{z_1, z_2\}$, and possibly after changing the signs of z_1 or z_2, we may assume that $w = c_1 z_1 + c_2 z_2$, with $0 < c_1 < \frac{1}{2}$ and $0 < c_2 < \frac{1}{2}$. Then w belongs to the parallelogram $\frac{1}{2}P_\beta$ with vertices $\{0, \frac{z_1}{2}, \frac{z_2}{2}, \frac{z_1+z_2}{2}\}$. If c_1 and c_2 are both larger than $\frac{1}{3}$ replace w by z_1+z_2-2w. We may thus assume further, that at least one c_i is $\leq \frac{1}{3}$. In particular, $|w| \leq c_1 |z_1| + c_2 |z_2| < \frac{5}{6} |z_1|$. Consider the new basis $\tilde{\beta} := \{w, z_2\}$ of $\mathbb{R}^2$. Then $\left|\tilde{\beta}\right| = |w| + |z_2| < \frac{5}{6}|z_1| + |z_2| = |\beta| - \frac{1}{6}|z_1| \leq \frac{11}{12}|\beta| < I$. We obtain the desired contradiction. □

Denote by $J_\alpha \subset \Omega_{\alpha^\perp}$ the union of all the $g(Q_\alpha)$ translates of the section τ_γ constructed in Eq. (5.2) above.

$$J_\alpha := \bigcup_{\gamma \in Q_\alpha} g_\gamma\left[\tau_\gamma\left(\Omega^+_{Q_\alpha}\right)\right].$$

One easily checks that $g_{[z]} \circ \tau_\gamma = \tau_\delta$, where $\delta := \gamma + \iota(z) + (\gamma, \iota(z))\beta + \frac{(z,z)}{2}\beta$, for all $z \in \alpha^\perp_\Lambda$, and so J_α is independent of the choice of γ.

Proposition 5.6. *(1) J_α is a dense subset of $\Omega^+_{\alpha^\perp}$.*

(2) If V is a $g(Q_\alpha)$-invariant open subset of $\Omega^+_{\alpha^\perp}$, which contains J_α, then V contains every non-special period in $\Omega^+_{\alpha^\perp}$.

(3) For every $\ell \in J_\alpha$, there exists a marked pair (M, η), consisting of a smooth projective irreducible holomorphic symplectic manifold M of $K3^{[n]}$-type and a marking $\eta : H^2(M, \mathbb{Z}) \to \Lambda$ with period ℓ satisfying the following properties.

(a) The composition $\iota \circ \eta : H^2(M, \mathbb{Z}) \to \tilde{\Lambda}$ belongs to the canonical $O(\tilde{\Lambda})$-orbit ι_M of Theorem 2.2.

(b) There exists a Lagrangian fibration $\pi : M \to \mathbb{P}^n$, such that the class $\eta^{-1}(\alpha)$ is equal to $\pi^* c_1(\mathcal{O}_{\mathbb{P}^n}(1))$.

Proof.

(1) The density of J_α follows from Lemma 5.4.

(2) V intersects every non-special fiber $q^{-1}(\underline{\ell})$ in a non-empty open $g(Q_\alpha)$-equivariant subset of the latter. The complement $q^{-1}(\underline{\ell}) \setminus V$ is thus a closed $g(Q_\alpha)$-equivariant proper subset of the fiber. But any $g(Q_\alpha)$-orbit in the non-special fiber $q^{-1}(\underline{\ell})$ is dense in $q^{-1}(\underline{\ell})$, by Lemma 5.4. Hence, the complement $q^{-1}(\underline{\ell}) \setminus V$ must be empty.

(3) If ℓ_0 belongs to the section $\tau_\gamma\left(\Omega^+_{Q_\alpha}\right)$, then such a pair $(M, \eta) := (M_H(u), \eta_1)$ was constructed in Diagram (5.4) as mentioned in Remark 5.1. If $\ell = g_z(\ell_0)$, $z \in \alpha^\perp_\Lambda$, set $(M, \eta) = (M_H(u), g_z \circ \eta_1)$. □

6 Primitive Isotropic Classes and Lagrangian Fibrations

We prove Theorem 1.3 in this section using the geometry of the moduli space $\mathfrak{M}^0_{\alpha^\perp}$ given in Eq. (4.6). Recall that $\mathfrak{M}^0_{\alpha^\perp}$ is a connected component of the moduli space of marked pairs (X,η) with X of $K3^{[n]}$-type and such that $\eta^{-1}(\alpha)$ is a primitive isotropic class of Hodge type $(1,1)$ in the boundary of the positive cone in $H^{1,1}(X,\mathbb{R})$.

Fix a connected moduli space $\mathfrak{M}^0_{\alpha^\perp}$ as in Eq. (4.6). Denote by $\mathscr{L}_{\eta^{-1}(\alpha)}$ the line bundle on X with $c_1(\mathscr{L}) = \eta^{-1}(\alpha)$. Let V be the subset of $\mathfrak{M}^0_{\alpha^\perp}$ consisting of all pairs (X,η), such that $\mathscr{L}_{\eta^{-1}(\alpha)}$ induces a Lagrangian fibration.

Theorem 6.1. *The image of V via the period map contains every non-special period in $\Omega^+_{\alpha^\perp}$.*

Proof. Let (X,η) be a marked pair in $\mathfrak{M}^0_{\alpha^\perp}$. The property that $\eta^{-1}(\alpha)$ is the first Chern class of a line-bundle $\mathscr{L}$ on X, which induces a Lagrangian fibration $X \to |\mathscr{L}|^*$, is an open property in the moduli space of marked pairs, by a result of Matsushita [27]. V is thus an open subset.

Choose a primitive embedding $\iota : \Lambda \to \tilde{\Lambda}$ with the property that $\iota \circ \eta$ belongs to the canonical $O(\tilde{\Lambda})$-orbit ι_X of Theorem 2.2, for all (X,η) in $\mathfrak{M}^0_{\Lambda}$. Let $Mon^2(\Lambda,\iota)$ and its subgroup $Mon^2(\Lambda,\iota)_\alpha$ be the subgroups of $O^+(\Lambda)$ introduced in Lemma 5.3. The component $\mathfrak{M}^0_{\Lambda}$ of the moduli space of marked pairs is invariant under $Mon^2(\Lambda,\iota)$, by Theorem 2.4. The subset $\mathfrak{M}^0_{\alpha^\perp}$ of $\mathfrak{M}^0_{\Lambda}$ is invariant under the subgroup $Mon^2(\Lambda,\iota)_\alpha$. Hence, the subset V is $Mon^2(\Lambda,\iota)_\alpha$ invariant. The construction in Sect. 5.1 yields a marked pair $(M_H(u),\eta_1)$ with period in the image of the section $\tau_\gamma : \Omega^+_{Q_\alpha} \to \Omega^+_{\alpha^\perp}$, given in Eq. (5.2). Furthermore, the class $\eta_1^{-1}(\alpha)$ induces a Lagrangian fibration of $M_H(u)$. The marked pair $(M_H(u),\eta_1)$ belongs to $\mathfrak{M}^0_{\alpha^\perp}$, by Proposition 4.5 (Remark 5.1 verifies the conditions of Proposition 4.5). Hence, $(M_H(u),\eta_1)$ belongs to V and the image of the section $\tau_\gamma : \Omega^+_{Q_\alpha} \to \Omega^+_{\alpha^\perp}$ is thus contained in the image of V via the period map. The period map P_0 is $Mon^2(\Lambda,\iota)_\alpha$ equivariant and a local homeomorphism, by the Local Torelli Theorem [4]. Hence, the image $P_0(V)$ is an open and $Mon^2(\Lambda,\iota)_\alpha$ invariant subset of $\Omega^+_{\alpha^\perp}$. Any $Mon^2(\Lambda,\iota)_\alpha$ invariant subset, which contains the section $\tau_\gamma(\Omega^+_{Q_\alpha})$, contains also the dense subset J_α of Proposition 5.6. $P_0(V)$ thus contains every non-special period in $\Omega^+_{\alpha^\perp}$, by Proposition 5.6 (2). □

We will need the following criterion of Kawamata for a line bundle to be semi-ample. Let X be a smooth projective variety and D a divisor class on X. Set $\nu(X,D) := \max\{e : D^e \not\equiv 0\}$, where $\equiv$ denotes numerical equivalence. If $D \equiv 0$, we set $\nu(X,D) = 0$. Denote by $\Phi_{kD} : X \dashrightarrow |kD|^*$ the rational map, defined whenever the linear system is non-empty. Set $\kappa(X,D) := \max\{\dim \Phi_{kD}(X) : k > 0\}$, if $|kD|$ is non-empty for some positive integer k, and $\kappa(X,D) := -\infty$, otherwise.

Theorem 6.2 (A special case of [16, Theorem 6.1]). *Let X be a smooth projective variety with a trivial canonical bundle and D a nef divisor. Assume that $\nu(X, D) = \kappa(X, D)$ and $\kappa(X, D) \geq 0$. Then D is semi-ample, i.e., there exists a positive integer k such that the linear system $|kD|$ is base point free.*

An alternate proof of Kawamata's Theorem is provided in [11]. A reduced and irreducible divisor E on X is called *prime-exceptional*, if the class $e \in H^2(X,\mathbb{Z})$ of E satisfies $(e,e) < 0$. Consider the reflection $R_E : H^2(X,\mathbb{Z}) \to H^2(X,\mathbb{Z})$, given by

$$R_E(x) = x - \frac{2(x,e)}{(e,e)}e.$$

It is known that the reflection R_E by the class of a prime exceptional divisor E is a monodromy operator, and in particular an integral isometry [24, Cor. 3.6]. Let $W(X) \subset O(H^2(X,\mathbb{Z}))$ be the subgroup generated by reflections R_E by classes of prime exceptional divisors $E \subset X$. Elements of $W(X)$ preserve the Hodge structure, hence $W(X)$ acts on $H^{1,1}(X,\mathbb{Z})$.

Let $\mathcal{P}ex_X \subset H^{1,1}(X,\mathbb{Z})$ be the set of classes of prime exceptional divisors. The *fundamental exceptional chamber* of the positive cone $\mathcal{C}_X$ is the set

$$\mathcal{F}\mathcal{E}_X := \{a \in \mathcal{C}_X : (a,e) > 0, \text{ for all } e \in \mathcal{P}ex_X\}.$$

The closure of $\mathcal{F}\mathcal{E}_X$ in $\mathcal{C}_X$ is a fundamental domain for the action of $W(X)$ [23, Theorem 6.18]. Let $f : X \dashrightarrow Y$ be a bimeromorphic map to an irreducible holomorphic symplectic manifold Y and $\mathcal{K}_Y$ the Kähler cone of Y. Then $f^*\mathcal{K}_Y$ is an open subset of $\mathcal{F}\mathcal{E}_X$. Furthermore, the union of $f^*\mathcal{K}_Y$, as f and Y vary over all such pairs, is a dense open subset of $\mathcal{F}\mathcal{E}_X$, by a result of Boucksom [6] (see also [23, Theorem 1.5]).

Proof (of Theorem 1.3). Step 1: Keep the notation in the opening paragraph of Sect. 5. Choose a marking $\eta : H^2(X,\mathbb{Z}) \to \Lambda$, such that $\iota \circ \eta$ belongs to the canonical $O(\tilde{\Lambda})$-orbit ι_X. Set $\alpha := \eta(c_1(\mathcal{L}))$. Then (X,η) belongs to a component $\mathfrak{M}^0_{\alpha^\perp}$ of the moduli space of marked pairs of $K3^{[n]}$-type considered in Theorem 6.1. We use here the assumption that $\mathcal{L}$ is nef in order to deduce that $\eta^{-1}(\alpha)$ belongs to the boundary of the positive cone of X, used in Theorem 6.1.

The period $P_0(X,\eta)$ is non-special, by assumption. There exists a marked pair (Y,ψ) in $\mathfrak{M}^0_{\alpha^\perp}$ satisfying $P_0(Y,\psi) = P_0(X,\eta)$, such that the class $\psi^{-1}(\alpha)$ induces a Lagrangian fibration, by Theorem 6.1. The marked pairs (X,η) and (Y,ψ) correspond to inseparable points in the moduli space $\mathfrak{M}^0_{\alpha^\perp}$, by the Global Torelli Theorem 4.3. Hence, there exists an analytic correspondence $Z \subset X \times Y$, $Z = \sum_{i=0}^k Z_i$ in $X \times Y$, of pure dimension $2n$, with the following properties, by results of Huybrechts [13, Theorem 4.3] (see also [23, Sec. 3.2]).

(1) The homomorphism $Z_* : H^*(X,\mathbb{Z}) \to H^*(Y,\mathbb{Z})$ is a Hodge isometry, which is equal to $\psi^{-1} \circ \eta$. The irreducible component Z_0 of the correspondence is the graph of a bimeromorphic map $f : X \dashrightarrow Y$.
(2) The images in X and Y of all other components Z_i, $i > 0$, are of positive co-dimension.

Step 2: We prove next that the line bundle $\mathscr{L}$ over X is semi-ample. We consider separately the projective and non-algebraic cases.

Step 2.1: Assume that X is not projective.[3] We claim that $f_*(c_1(\mathscr{L})) = \psi^{-1}(\alpha)$. The Neron-Severi group $NS(X)$ does not contain any positive class, by Huybrechts projectivity criterion [13]. Hence, the Beauville-Bogomolov-Fujiki pairing restricts to $NS(X)$ as a non-positive pairing with a rank one null sub-lattice spanned by the class $c_1(\mathscr{L})$. Similarly, the Beauville-Bogomolov-Fujiki pairing restricts to $NS(Y)$ with a rank one null space spanned by $\psi^{-1}(\alpha)$. Hence, $f_*(c_1(\mathscr{L})) = \pm\psi^{-1}(\alpha)$. Now $\psi^{-1}(\alpha)$ is semi-ample and hence belongs to the closure of $\mathscr{F}\mathscr{E}_Y$. The class $c_1(\mathscr{L})$ is assumed nef, and hence belongs to the closure of $\mathscr{F}\mathscr{E}_X$. The bimeromorphic map f induces a Hodge-isometry $f_* : H^2(X,\mathbb{Z}) \to H^2(Y,\mathbb{Z})$, which maps $\mathscr{F}\mathscr{E}_X$ onto $\mathscr{F}\mathscr{E}_Y$ [6]. Hence, $f_*(c_1(\mathscr{L}))$ belongs to $\overline{\mathscr{F}\mathscr{E}_Y}$ as well. We conclude the equality $f_*(c_1(\mathscr{L})) = \psi^{-1}(\alpha)$.

Let $\mathscr{L}_2$ be the line bundle with $c_1(\mathscr{L}_2) = \psi^{-1}(\alpha)$. The bimeromorphic map $f : X \dashrightarrow Y$ is holomorphic in co-dimension one, and so induces an isomorphism $f_1 : |\mathscr{L}| \to |\mathscr{L}_2|$ of the two linear systems. Denote by $\Phi_{\mathscr{L}_2} : Y \to |\mathscr{L}_2|^*$ the Lagrangian fibration induced by $\mathscr{L}_2$. We conclude that $|\mathscr{L}|$ is n dimensional and the meromorphic map $\Phi_{\mathscr{L}} : X \dashrightarrow |\mathscr{L}|^*$ is an algebraic reduction of X (see [8]). By definition, an *algebraic reduction* of X is a dominant meromorphic map $\pi : X \dashrightarrow B$ to a normal projective variety B, such that π^* induces an isomorphism of the function fields of meromorphic functions [8]. Only the birational class of B is determined by X. Fibers of the algebraic reduction π are defined via a resolution of indeterminacy, and are closed connected analytic subsets of X. In our case, the generic fiber of $\Phi_{\mathscr{L}}$ is bimeromorphic to the generic fiber of $\Phi_{\mathscr{L}_2}$. The generic fiber of $\Phi_{\mathscr{L}_2}$ is a complex torus, and hence algebraic, by [7, Prop. 2.1]. Hence, the generic fiber of $\Phi_{\mathscr{L}}$ has algebraic dimension n. It follows that the line bundle $\mathscr{L}$ is semi-ample, it is the pullback of an ample line-bundle over B, via a holomorphic reduction map $\pi : X \to B$ which is a regular morphism, by [8, Theorems 1.5 and 3.1].

Step 2.2: When X is projective there exists an element $w \in W(X)$, such that Huybrecht's birational map $f : X \dashrightarrow Y$ satisfies $f^* \circ \psi^{-1} \circ \eta = w$, by [23, Theorem 1.6]. Set $\alpha_X := \eta^{-1}(\alpha)$ and $\alpha_Y := \psi^{-1}(\alpha)$. We get the equality $w(\alpha_X) = f^*(\alpha_Y)$.

Let $\overline{\mathscr{F}\mathscr{E}_X}$ be the closure of the fundamental exceptional chamber $\mathscr{F}\mathscr{E}_X$ in $H^{1,1}(X,\mathbb{R})$. The class α_X is nef, by assumption, and it thus belongs to $\overline{\mathscr{F}\mathscr{E}_X}$. We already know that α_Y is the class of a line bundle, which induces a Lagrangian

[3] I thank K. Oguiso and S. Rollenske for pointing out to me that in the non-algebraic case the result should follow from the above via the results of Ref. [8].

fibration. Hence, $f^*(\alpha_Y)$ belongs to $\overline{\mathcal{FE}}_X$. The class $w(\alpha_X)$ thus belongs to the intersection $w\left(\overline{\mathcal{FE}}_X\right) \cap \overline{\mathcal{FE}}_X$.

Let J be the subset of $\mathcal{P}ex_X$ given by $J = \{e \in \mathcal{P}ex_X \ : \ (e, \alpha_X) = 0\}$. Denote by W_J the subgroup of $W(X)$ generated by reflections R_e, for all $e \in J$. Then W_J is equal to

$$\{w \in W(X) \ : \ w(\alpha_X) \in \overline{\mathcal{FE}}_X\},$$

by a general property of crystalographic hyperbolic reflection groups [12, Lecture 3, Proposition on page 15]. We conclude that $w(\alpha_X) = \alpha_X$ and

$$\alpha_X = f^*(\alpha_Y). \tag{6.1}$$

We are ready to prove[4] that $\mathcal{L}$ is semi-ample. The rational map f is regular in co-dimension one. The map f thus induces an isomorphism $f_m : |\mathcal{L}^m| \to |\mathcal{L}_2^m|$, for every integer m. Hence, $\kappa(X, \mathcal{L}) = \kappa(Y, \mathcal{L}_2) = n$. Any non-zero isotropic divisor class D on a $2n$ dimensional irreducible holomorphic symplectic manifold satisfies $\nu(X, D) = n$, by a result of Verbitsky [41]. Hence, $\nu(X, \mathcal{L}) = n$. The line bundle $\mathcal{L}$ is assumed to be nef. Hence, $\mathcal{L}$ is semi-ample, by Theorem 6.2.

Step 3: We return to the general case, where X may or may not be projective. In both cases we have seen that there exists a positive integer m, such that the linear system $|\mathcal{L}^m|$ is base point free and $\Phi_{\mathcal{L}^m}$ is a regular morphism. Furthermore, the bimeromorphic map $f : X \dashrightarrow Y$ is regular in co-dimension one and thus induces an isomorphism $f_k : |\mathcal{L}^k| \to |\mathcal{L}_2^k|$, for every positive integer k. Denote by $f_k^* : |\mathcal{L}_2^k|^* \to |\mathcal{L}^k|^*$ the transpose of f_k. We get the equality $\Phi_{\mathcal{L}^k} = f_k^* \circ \Phi_{\mathcal{L}_2^k} \circ f$, for all k. Let $V_m : |\mathcal{L}_2|^* \to |\mathcal{L}_2^m|^*$ be the Veronese embedding. We get the equalities

$$V_m \circ (f_1^*)^{-1} \circ \Phi_{\mathcal{L}} = V_m \circ \Phi_{\mathcal{L}_2} \circ f = \Phi_{\mathcal{L}_2^m} \circ f = (f_m^*)^{-1} \circ \Phi_{\mathcal{L}^m}. \tag{6.2}$$

Now, $V_m \circ (f_1^*)^{-1} : |\mathcal{L}|^* \to |\mathcal{L}_2^m|^*$ is a closed immersion and the morphism on the right hand side of (6.2) is regular. Hence, the rational map $\Phi_{\mathcal{L}}$ is a regular morphism. The base locus of the linear system $|\mathcal{L}|$ is thus either empty, or a divisor. The latter is impossible, since f is regular in co-dimension one and $|\mathcal{L}_2|$ is base point free. Hence, $|\mathcal{L}|$ is base point free. □

Let X and $\mathcal{L}$ be as in Theorem 1.3, except that we drop the assumption that $\mathcal{L}$ is nef and assume only that $c_1(\mathcal{L})$ belongs to the boundary of the positive cone. Assume that X is projective.

---

[4] I thank C. Lehn for Ref. [18, Prop. 2.4], used in an earlier argument, and T. Peternell and Y. Kawamata for suggesting the current more direct argument.

Theorem 6.3. *There exists an element $w \in W(X)$, a projective irreducible holomorphic symplectic manifold Y, a birational map $f : X \dashrightarrow Y$, and a Lagrangian fibration $\pi : Y \to \mathbb{P}^n$, such that $w(\mathcal{L}) = f^*\pi^*\mathcal{O}_{\mathbb{P}^n}(1)$.*

Proof. Let (Y, ψ) be the marked pair constructed in Step 1 of the proof of Theorem 1.3. Then Y admits a Lagrangian fibration $\pi : Y \to \mathbb{P}^n$ and the class $\pi^* c_1(\mathcal{O}_{\mathbb{P}^n}(1))$ was denoted α_Y in that proof. In step 2.2 of that proof we showed the existence of a birational map $f : X \dashrightarrow Y$ and an element $w \in W(X)$, such that $w(c_1(\mathcal{L})) = f^*(\alpha_Y)$ (see Equality (6.1)). □

7 Tate-Shafarevich Lines and Twists

7.1 The Geometry of the Universal Curve

Let S be a projective $K3$ surface, d a positive integer, and $\mathcal{L}$ a nef line bundle on S of positive degree, such that the class $c_1(\mathcal{L})$ is indivisible. Set $n := 1 + \frac{d^2 \deg(\mathcal{L})}{2}$. Let $\mathcal{C} \subset S \times |\mathcal{L}^d|$ be the universal curve, π_i the projection from $S \times |\mathcal{L}^d|$ to the i-th factor, $i = 1, 2$, and p_i the restriction of π_i to $\mathcal{C}$. We assume in this section the following assumptions about the line bundle $\mathcal{L}$.

Assumption 7.1. *(1) The linear system $|\mathcal{L}^d|$ is base point free.*
(2) The locus in $|\mathcal{L}^d|$, consisting of divisors which are non-reduced, or reducible having a singularity which is not an ordinary double point, has co-dimension at least 2.

Remark 7.2. Assumption 7.1 holds whenever $\mathrm{Pic}(S)$ is cyclic generated by $\mathcal{L}$. The base point freeness Assumption 7.1 (1) follows from [32, Prop. 1]. Assumption 7.1 (2) is verified as follows. If $a+b = d$, $a \geq 1$, $b \geq 1$, then the image of $|\mathcal{L}^a| \times |\mathcal{L}^b|$ in $|\mathcal{L}^d|$ has co-dimension $2ab\left(\frac{n-1}{d^2}\right) - 1$. The co-dimension is at least two, except in the case $(n, d) = (5, 2)$. In the latter case $|\mathcal{L}| \cong \mathbb{P}^2$, $|\mathcal{L}^2| \cong \mathbb{P}^5$ and the generic curve in the image of $|\mathcal{L}| \times |\mathcal{L}|$ in $|\mathcal{L}^2|$ is the union of two smooth curves of genus 2 meeting transversely at two points. Hence, Assumption 7.1 (2) holds in this case as well.

The morphism $p_1 : \mathcal{C} \to S$ is a projective hyperplane sub-bundle of the trivial bundle over S with fiber $|\mathcal{L}^d|$, by the base point freeness Assumption 7.1 (1). Assumption 7.1 (2) will be used in the proof of Lemma 7.9. Consider the exponential short exact sequence over $\mathcal{C}$

$$0 \to \mathbb{Z} \to \mathcal{O}_{\mathcal{C}} \to \mathcal{O}^*_{\mathcal{C}} \to 0.$$

We get the exact sequence of sheaves of abelian groups over $|\mathcal{L}^d|$

$$0 \longrightarrow R^1 p_{2_*}\mathbb{Z} \longrightarrow R^1 p_{2_*}\mathcal{O}_{\mathcal{C}} \longrightarrow R^1 p_{2_*}\mathcal{O}^*_{\mathcal{C}} \xrightarrow{\deg} R^2 p_{2_*}\mathbb{Z} \longrightarrow 0, \tag{7.1}$$

where we work in the complex analytic category. Note that deg above is surjective, since $R^2 p_{2*}\mathcal{O}_{\mathscr{C}}$ vanishes. Set $Ш := H^1(|\mathscr{L}^d|, R^1 p_{2*}\mathcal{O}^*_{\mathscr{C}})$ and $\widetilde{Ш} := H^1(|\mathscr{L}^d|, R^1 p_{2*}\mathcal{O}_{\mathscr{C}})$. Set $Br'(S) := H^2(S, \mathcal{O}^*_S)$ and $Br'(\mathscr{C}) := H^2(\mathscr{C}, \mathcal{O}^*_{\mathscr{C}})$.

Lemma 7.3. *(1) There is a natural isomorphism*

$$R^1 p_{2*}\mathcal{O}_{\mathscr{C}} \cong T^*|\mathscr{L}^d| \otimes_{\mathbb{C}} H^{2,0}(S)^*.$$

(2) $\widetilde{Ш}$ is naturally isomorphic to $H^{0,2}(\mathscr{C})$. Consequently, $\widetilde{Ш}$ is one dimensional.
(3) $H^2(\mathscr{C}, \mathbb{Z})$ decomposes as a direct sum

$$H^2(\mathscr{C}, \mathbb{Z}) = p_1^* H^2(S, \mathbb{Z}) \oplus p_2^* H^2(|\mathscr{L}^d|, \mathbb{Z}).$$

The groups $H^i(\mathscr{C}, \mathbb{Z})$ vanish for odd i. The Dolbeault cohomologies $H^{p,q}(\mathscr{C})$ vanish, if $|p - q| > 2$.
(4) The pullback homomorphism $p_1^* : H^2(S, \mathcal{O}^*_S) \to H^2(\mathscr{C}, \mathcal{O}^*_{\mathscr{C}})$ is an isomorphism. The Leray spectral sequence yields an isomorphism

$$b : H^2(\mathscr{C}, \mathcal{O}^*_{\mathscr{C}}) \to H^1(|\mathscr{L}^d|, R^1 p_{2*}\mathcal{O}^*_{\mathscr{C}}).$$

Consequently, we have the isomorphisms

$$Br'(S) \xrightarrow[\cong]{p_1^*} Br'(\mathscr{C}) \xrightarrow[\cong]{b} Ш.$$

Let $\mathscr{F}$ be a sheaf of abelian groups over $\mathscr{C}$. Let $F^p H^k(\mathscr{C}, \mathscr{F})$ be the Leray filtration associated to the morphism $p_2 : \mathscr{C} \to |\mathscr{L}^d|$ and $E_\infty^{p,q} := F^p H^{p+q}(\mathscr{C}, \mathscr{F})/F^{p+1} H^{p+q}(\mathscr{C}, \mathscr{F})$ its graded pieces. Recall that the $E_2^{p,q}$ terms are $E_2^{p,q} := H^p(|\mathscr{L}^d|, R^q p_{2*}\mathscr{F})$ and the differential at this step is $d_2 : E_2^{p,q} \to E_2^{p+2,q-1}$.

Proof. (1) We have the isomorphism $\mathcal{O}_{S\times|\mathscr{L}^d|}(\mathscr{C}) \cong \pi_1^*\mathscr{L}^d \otimes \pi_2^*\mathcal{O}_{|\mathscr{L}^d|}(1)$. Apply the functor $R\pi_{2*}$ to the short exact sequence $0 \to \mathcal{O}_{S\times|\mathscr{L}^d|} \to \mathcal{O}_{S\times|\mathscr{L}^d|}(\mathscr{C}) \to \mathcal{O}_{\mathscr{C}}(\mathscr{C}) \to 0$ to obtain the Euler sequence of the tangent bundle.

$$0 \to \mathcal{O}_{|\mathscr{L}^d|} \to H^0(S, \mathscr{L}^d) \otimes_{\mathbb{C}} \mathcal{O}_{|\mathscr{L}^d|}(1) \to T|\mathscr{L}^d| \to 0.$$

Now $\mathcal{O}_{\mathscr{C}}(\mathscr{C}) \otimes_{\mathbb{C}} H^{2,0}(S)$ is isomorphic to the relative dualizing sheaf ω_{p_2}. We get the isomorphisms

$$\begin{aligned} R^1 p_{2*}\mathcal{O}_{\mathscr{C}} &\cong [R^0 p_{2*}\mathcal{O}_{\mathscr{C}}(\mathscr{C}) \otimes_{\mathbb{C}} H^{2,0}(S)]^* \cong [R^0 p_{2*}\mathcal{O}_{\mathscr{C}}(\mathscr{C})]^* \otimes_{\mathbb{C}} H^{2,0}(S)^* \\ &\cong T^*|\mathscr{L}^d| \otimes_{\mathbb{C}} H^{2,0}(S)^*. \end{aligned}$$

(2) $R^2 p_{2*}\mathcal{O}_{\mathscr{C}}$ vanishes, since p_2 has one-dimensional fibers. $H^2(|\mathscr{L}^d|, p_{2*}\mathcal{O}_{\mathscr{C}})$ vanishes, since $p_{2*}\mathcal{O}_{\mathscr{C}} \cong \mathcal{O}_{|\mathscr{L}^d|}$. The latter isomorphism follow from the

fact that p_2 has connected fibers. We conclude that $H^2(\mathcal{C},\mathcal{O}_{\mathcal{C}})$ is isomorphic to the $E_\infty^{1,1}$ graded summand of its Leray filtration. The differential $d_2 : H^1(|\mathcal{L}^d|, R^1p_{2_*}\mathcal{O}_{\mathcal{C}}) \to H^3(|\mathcal{L}^d|, p_{2_*}\mathcal{O}_{\mathcal{C}})$ vanishes, since $H^{0,3}(|\mathcal{L}^d|)$ vanishes. Hence, the $E_2^{1,1}$ term $\widetilde{\text{III}} := H^1(|\mathcal{L}^d|, R^1p_{2_*}\mathcal{O}_{\mathcal{C}})$ is isomorphic to $H^2(\mathcal{C},\mathcal{O}_{\mathcal{C}})$.

(3) The statement is topological and so it suffices to prove it in the case where $\mathrm{Pic}(S)$ is cyclic generated by $\mathcal{L}$. In this case $\mathcal{L}$ is ample, and so the line bundle $\pi_1^*\mathcal{L}^d \otimes \pi_2^*\mathcal{O}_{|\mathcal{L}^d|}(1)$ is ample. The Lefschetz Theorem on Hyperplane sections implies that the restriction homomorphism $H^2(S\times|\mathcal{L}^d|,\mathbb{Z}) \to H^2(\mathcal{C},\mathbb{Z})$ is an isomorphism.

$\mathcal{C}$ is the projectivization of a rank n vector bundle F over S. Hence, $H^*(\mathcal{C},\mathbb{Z})$ is the quotient of $H^*(S,\mathbb{Z})[x]$, with x of degree 2, by the ideal generated by $\sum_{i=0}^{n+1} c_i(F)x^i$. The image of x in $H^*(\mathcal{C},\mathbb{Z})$ corresponds to the class $\bar{x} := c_1(\mathcal{O}_{\mathcal{C}}(1))$ of Hodge type $(1,1)$. In particular, $H^*(\mathcal{C},\mathbb{Z})$ is a free $H^*(S,\mathbb{Z})$-module of rank n generated by $1, \bar{x}, \ldots, \bar{x}^{n-1}$.

(4) The vanishing of $H^3(S,\mathbb{Z})$ and $H^3(\mathcal{C},\mathbb{Z})$ yields the commutative diagram with exact rows:

$$\begin{array}{ccccccccc}
0 & \longrightarrow & H^2(S,\mathbb{Z})/NS(S) & \longrightarrow & H^2(S,\mathcal{O}_S) & \longrightarrow & H^2(S,\mathcal{O}_S^*) & \longrightarrow & 0\\
 & & \downarrow p_1^* & & \downarrow p_1^* \cong & & \downarrow p_1^* & & \\
0 & \longrightarrow & H^2(\mathcal{C},\mathbb{Z})/NS(\mathcal{C}) & \longrightarrow & H^2(\mathcal{C},\mathcal{O}_{\mathcal{C}}) & \longrightarrow & H^2(\mathcal{C},\mathcal{O}_{\mathcal{C}}^*) & \longrightarrow & 0
\end{array}$$

Part (3) of the Lemma implies that the left and middle vertical homomorphism are isomorphisms. It follows that the right vertical homomorphism is an isomorphism as well.

The sheaf $R^2p_{2_*}\mathcal{O}_{\mathcal{C}}^*$ vanishes, by the exactness of $R^2p_{2_*}\mathcal{O}_{\mathcal{C}} \to R^2p_{2_*}\mathcal{O}_{\mathcal{C}}^* \to R^3p_{2_*}\mathbb{Z}$ and the vanishing of the left and right sheaves due to the fact that p_2 has one-dimensional fibers. The sheaf $p_{2_*}\mathcal{O}_{\mathcal{C}}^*$ is isomorphic to $\mathcal{O}_{|\mathcal{L}^d|}^*$, since p_2 has connected complete fibers. Thus, $H^2(\mathcal{C},\mathcal{O}_{\mathcal{C}}^*)$ is isomorphic to the kernel of the differential

$$d_2 : E_2^{1,1} := H^1(|\mathcal{L}^d|, R^1p_{2_*}\mathcal{O}_{\mathcal{C}}^*) \to E_2^{3,0} := H^3(|\mathcal{L}^d|, \mathcal{O}_{|\mathcal{L}^d|}^*). \tag{7.2}$$

We prove next that d_2 vanishes. The co-kernel of d_2 is equal to $F^3H^3(\mathcal{C},\mathcal{O}_{\mathcal{C}}^*)$. Now $F^3H^3(\mathcal{C},\mathcal{O}_{\mathcal{C}}^*)$ is equal to the image of $p_2^* : H^3(|\mathcal{L}^d|, \mathcal{O}_{|\mathcal{L}^d|}^*) \to H^3(\mathcal{C},\mathcal{O}_{\mathcal{C}}^*)$. We have a commutative diagram

$$\begin{array}{ccc}
H^3(\mathcal{C},\mathcal{O}_{\mathcal{C}}^*) & \xrightarrow{\cong} & H^4(\mathcal{C},\mathbb{Z})\\
\uparrow p_2^* & & \uparrow p_2^*\\
H^3(|\mathcal{L}^d|, \mathcal{O}_{|\mathcal{L}^d|}^*) & \xrightarrow{\cong} & H^4(|\mathcal{L}^d|,\mathbb{Z}).
\end{array}$$

The horizontal homomorphisms, induced by the connecting homomorphism of the exponential sequence, are isomorphisms, since $h^{0,3}(\mathscr{C}) = h^{0,3}(|\mathscr{L}^d|) = 0$ and $h^{0,4}(\mathscr{C}) = h^{0,4}(|\mathscr{L}^d|) = 0$. The right vertical homomorphism is injective. We conclude that the left vertical homomorphism is injective. Hence the differential d_2 in (7.2) vanishes and $H^2(\mathscr{C}, \mathscr{O}^*_{\mathscr{C}})$ is isomorphism to $H^1(|\mathscr{L}^d|, R^1 p_{2_*} \mathscr{O}^*_{\mathscr{C}})$, yielding the isomorphism b. □

Let $\Sigma \subset H^2(S, \mathbb{Z})$ be the sub-lattice generated by classes of irreducible components of divisors in the linear system $|\mathscr{L}^d|$. Denote by $\Sigma^\perp$ the sub-lattice of $H^2(S, \mathbb{Z})$ orthogonal to Σ.

Lemma 7.4. *(1) The Leray filtration of $H^2(\mathscr{C}, \mathbb{Z})$ associated to p_2 is identified as follows:*

$$F^2 H^2(\mathscr{C}, \mathbb{Z}) = p_2^* H^2(|\mathscr{L}^d|, \mathbb{Z}),$$
$$F^1 H^2(\mathscr{C}, \mathbb{Z}) = p_2^* H^2(|\mathscr{L}^d|, \mathbb{Z}) \oplus p_1^* \Sigma^\perp.$$

(2) $E_2^{p,q} = E_\infty^{p,q}$, if $(p,q) = (2,0)$, or $(1,1)$. Consequently, we get the following isomorphisms.

$$E_2^{2,0} := H^2(|\mathscr{L}^d|, p_{2_*}\mathbb{Z}) \cong p_2^* H^2(|\mathscr{L}^d|, \mathbb{Z}),$$
$$E_2^{1,1} := H^1(|\mathscr{L}^d|, R^1 p_{2_*}\mathbb{Z}) \cong p_1^* \Sigma^\perp,$$

(3) If the sub-lattice Σ is saturated in $H^2(S, \mathbb{Z})$, then $H^2(|\mathscr{L}^d|, R^1 p_{2_*}\mathbb{Z})$ vanishes.

Proof. (1), (2) The sheaf $p_{2_*}\mathbb{Z}$ is the constant sheaf $\mathbb{Z}$, since p_2 has connected fibers. Then $E_2^{3,0} = H^3(|\mathscr{L}^d|, \mathbb{Z}) = 0$, and so $E_\infty^{1,1} = E_2^{1,1} = H^1(|\mathscr{L}^d|, R^1 p_{2_*}\mathbb{Z})$. $E_2^{2,0} := H^2(|\mathscr{L}^d|, p_{2_*}\mathbb{Z})$ has rank 1 and it maps injectively into $H^2(\mathscr{C}, \mathbb{Z})$, with image equal to $p_2^* H^2(|\mathscr{L}^d|, \mathbb{Z})$. Thus, $E_2^{2,0} = E_\infty^{2,0}$ and $E_\infty^{1,1} := F^1 H^2(\mathscr{C}, \mathbb{Z}) / E_\infty^{2,0}$ is isomorphic to $F^1 H^2(\mathscr{C}, \mathbb{Z}) / p_2^* H^2(|\mathscr{L}^d|, \mathbb{Z})$. Finally, $E_2^{0,2}$ is the kernel of

$$d_2 : H^0(|\mathscr{L}^d|, R^2 p_{2_*}\mathbb{Z}) \to H^2(|\mathscr{L}^d|, R^1 p_{2_*}\mathbb{Z}).$$

Thus, $F^1 H^2(\mathscr{C}, \mathbb{Z})$ is the kernel of the homomorphism $H^2(\mathscr{C}, \mathbb{Z}) \to H^0(|\mathscr{L}^d|, R^2 p_{2_*}\mathbb{Z})$. The latter kernel is equal to $p_1^* \Sigma^\perp \oplus p_2^* H^2(|\mathscr{L}^d|, \mathbb{Z})$, by Lemma 7.3 (3). We conclude that $F^1 H^2(\mathscr{C}, \mathbb{Z}) / p_2^* H^2(|\mathscr{L}^d|, \mathbb{Z})$ is isomorphic to both $H^1(|\mathscr{L}^d|, R^1 p_{2_*}\mathbb{Z})$ and $p_1^* \Sigma^\perp$.

(3) The composition $H^2(\mathscr{C}, \mathbb{Z}) \to H^0(R^2 p_{2_*}\mathbb{Z}) \hookrightarrow \Sigma^*$ factors through $H^2(S, \mathbb{Z})$. If Σ is saturated, then the composition is surjective, since $H^2(S, \mathbb{Z})$ is unimodular. Thus, $d_2^{0,2} : H^0(R^2 p_{2_*}\mathbb{Z}) \to H^2(R^1 p_{2_*}\mathbb{Z})$ vanishes. The sheaf $p_{2_*}\mathbb{Z}$ is the trivial local system and the homomorphism $H^4(|\mathscr{L}^d|, p_{2_*}\mathbb{Z}) \cong H^4(|\mathscr{L}^d|, \mathbb{Z}) \to H^4(\mathscr{C}, \mathbb{Z})$ is the injective pull-back homomorphism p_2^*. Thus the differential $d_2^{2,1} : H^2(R^1 p_{2_*}\mathbb{Z}) \to H^4(p_{2_*}\mathbb{Z})$ vanishes. We conclude that

$E_2^{2,1} := H^2(R^1 p_{2_*}\mathbb{Z})$ is isomorphic to $E_\infty^{2,1}$. Now $E_\infty^{2,1}$ vanishes, since $H^3(\mathscr{C},\mathbb{Z})$ vanishes. □

Let $\mathscr{A}^0$ be the kernel of the homomorphism deg, given in (7.1). Then $\mathscr{A}^0$ is a subsheaf of $R^1 p_{2_*}\mathscr{O}^*_{\mathscr{C}}$ and we get the short exact sequences

$$0 \longrightarrow \mathscr{A}^0 \longrightarrow R^1 p_{2_*}\mathscr{O}^*_{\mathscr{C}} \stackrel{\deg}{\longrightarrow} R^2 p_{2_*}\mathbb{Z} \longrightarrow 0, \tag{7.3}$$

$$0 \longrightarrow R^1 p_{2_*}\mathbb{Z} \longrightarrow R^1 p_{2_*}\mathscr{O}_{\mathscr{C}} \longrightarrow \mathscr{A}^0 \longrightarrow 0, \tag{7.4}$$

and the long exact

$$\cdots \to H^1(|\mathscr{L}^d|, R^1 p_{2_*}\mathbb{Z}) \to H^1(|\mathscr{L}^d|, R^1 p_{2_*}\mathscr{O}_{\mathscr{C}}) \to H^1(|\mathscr{L}^d|, \mathscr{A}^0) \to \cdots$$

Lemma 7.5. *The group $H^0(|\mathscr{L}^d|, \mathscr{A}^0)$ is isomorphic to $NS(S) \cap \Sigma^\perp$. The composite homomorphism*

$$H^2(S,\mathbb{Z}) \to H^{0,2}(S) \stackrel{p_1^*}{\to} H^{0,2}(\mathscr{C}) \cong \widetilde{\text{Ш}} \to H^1(|\mathscr{L}^d|, \mathscr{A}^0)$$

factors through an injective homomorphism from $H^2(S,\mathbb{Z})/[\Sigma^\perp + NS(S)]$ into the kernel of the homomorphism $H^1(|\mathscr{L}^d|, \mathscr{A}^0) \to \text{Ш}$.

Proof. The space $H^0(|\mathscr{L}^d|, R^1 p_{2_*}\mathscr{O}_{\mathscr{C}})$ vanishes, by Lemma 7.3 (1). Hence, $H^0(|\mathscr{L}^d|, \mathscr{A}^0)$ is the kernel of the homomorphism $H^1(|\mathscr{L}^d|, R^1 p_{2_*}\mathbb{Z}) \to \widetilde{\text{Ш}} \cong H^{0,2}(S)$. Compose the above homomorphism with the isomorphism $\Sigma^\perp \cong H^1(|\mathscr{L}^d|, R^1 p_{2_*}\mathbb{Z})$ of Lemma 7.4 in order to get the isomorphism $H^0(|\mathscr{L}^d|, \mathscr{A}^0) \cong NS(S) \cap \Sigma^\perp$.

We have a commutative diagram with short exact rows

$$\begin{array}{ccccccccc}
0 & \longrightarrow & \frac{\Sigma^\perp}{NS(S)\cap\Sigma^\perp} & \longrightarrow & \widetilde{\text{Ш}} & \stackrel{j}{\longrightarrow} & \ker[H^1(\mathscr{A}^0) \to H^2(R^1 p_{2_*}\mathbb{Z})] & \longrightarrow & 0 \\
 & & \downarrow & & \downarrow \cong & & \downarrow & & \\
0 & \longrightarrow & \frac{H^2(S,\mathbb{Z})}{NS(S)} & \longrightarrow & H^2(S,\mathscr{O}_S) & \longrightarrow & H^2(S,\mathscr{O}_S^*) & \longrightarrow & 0.
\end{array} \tag{7.5}$$

The top row is obtained from the long exact sequence of sheaf cohomologies associated to the short exact sequence (7.4). The left vertical homomorphism is injective and the right vertical homomorphism is surjective. The co-kernel of the former is isomorphic to the kernel of the latter and both are isomorphic to $H^2(S,\mathbb{Z})/[\Sigma^\perp + NS(S)]$. Setting

$$\text{Ш}^0 := \ker[H^1(\mathscr{A}^0) \to H^2(R^1 p_{2_*}\mathbb{Z})], \tag{7.6}$$

we see that the right vertical homomorphism fits in the short exact sequence

$$0 \longrightarrow \frac{H^2(S,\mathbb{Z})}{\Sigma^\perp + NS(S)} \longrightarrow Ш^0 \longrightarrow Ш \longrightarrow 0. \tag{7.7}$$

The statement of the Lemma follows. □

Let $Ш^0$ be the group given in Eq. (7.6). Classes of Ш represent torsors for the relative Picard group scheme, while classes of $Ш^0$ represent torsors for the relative Pic^0 group scheme. This comment will be illustrated in Example 7.8 below.

7.2 A Universal Family of Tate-Shafarevich Twists

Let S be the marked $K3$ surface in Diagram (5.4) and $M_H(u)$ the moduli space of H-stable sheaves of pure one-dimensional support on S in that Diagram. Recall that $c_1(u)$ is the first Chern class of $\mathcal{L}^d$, for a nef line-bundle $\mathcal{L}$ on S, and the support map $\pi : M_H(u) \to |\mathcal{L}^d|$ is a Lagrangian fibration.

Let σ be a section of $R^1p_{2_*}(\mathcal{O}^*_{\mathscr{C}})$ over an open subset U of $|\mathcal{L}^d|$. Assume that σ is the image of a section $\tilde{\sigma}$ of $R^1p_{2_*}(\mathcal{O}_{\mathscr{C}})$ over U. Then σ lifts to an automorphism of the open subset $\pi^{-1}(U)$ of $M_H(u)$. This is seen as follows. Fix a point $t \in |\mathcal{L}^d|$ and denote by C_t the corresponding divisor in S. Denote by $\sigma(t)$ the image of σ in $H^1(C_t, \mathcal{O}^*_{C_t})$ and by $L_{\sigma(t)}$ the line-bundle over C_t with class $\sigma(t)$. A sheaf F over C_t is H-stable, if and only if $F \otimes L_{\sigma(t)}$ is H-stable, since tensorization by $L_{\sigma(t)}$ induces a one-to-one correspondence between the set of subsheaves, which is slope-preserving, since $L_{\sigma(t)}$ belongs to the identity component of the Picard group of C_t.

Let s be an element of $Ш^0$. We can choose a Čech 1-co-cycle $\sigma := \{\sigma_{ij}\}$ for the sheaf $\mathscr{A}^0$ representing s in $Ш^0$, with respect to an open covering $\{U_i\}$ of $|\mathcal{L}^d|$, such that each σ_{ij} is the image of a section $\tilde{\sigma}_{ij}$ of $R^1p_{2_*}(\mathcal{O}_{\mathscr{C}})$, since the homomorphism $R^1p_{2_*}(\mathcal{O}_{\mathscr{C}}) \to \mathscr{A}^0$ is surjective. The co-cycle $\{\sigma_{ij}\}$ may be used to re-glue the open covering $\pi^{-1}(U_i)$ of $M_H(u)$ to obtain a separated complex manifold M_σ together with a proper map $\pi_\sigma : M_\sigma \to |\mathcal{L}^d|$. The latter is independent of the choice of the co-cycle, by the following Lemma, so we denote it by

$$\pi_s : M_s \to |\mathcal{L}^d|. \tag{7.8}$$

Lemma 7.6. *Let $\sigma := \{\sigma_{ij}\}$ and $\sigma' := \{\sigma'_{ij}\}$ be two co-cycles representing the same class in $Ш^0$. Then there exists an isomorphism $h : M_\sigma \to M_{\sigma'}$ satisfying the equation $\pi_{\sigma'} \circ h = \pi_\sigma$. If the lattice Σ of Lemma 7.4 has finite index in $NS(S)$, then h depends canonically on σ and σ'.*

Proof. There exists a co-chain $h := \{h_i\}$ in $C^0(\{U_i\}, \mathscr{A}^0)$, such that $h_i\sigma_{ij} = \sigma'_{ij}h_j$, possibly after refining the covering and restricting the co-cycles σ and σ' to the refinement. Each h_i is the image of a section $\tilde{h}_i$ of $R^1p_{2_*}\mathcal{O}_{\mathscr{C}}$, possibly after further refinement of the covering, since the sheaf homomorphism $R^1p_{2_*}\mathcal{O}_{\mathscr{C}} \to \mathscr{A}^0$ is surjective. Hence, h_i lifts canonically to an automorphism of $\pi^{-1}(U_i)$. The co-chain $\{h_i\}$ of automorphisms glues to a global isomorphism from $M_{\sigma'}$ to M_σ, by the equality $h_i\sigma_{ij} = \sigma'_{ij}h_j$.

If $h' := \{h'_i\}$ is another co-chain satisfying the equality $\delta(h) = \sigma(\sigma')^{-1}$, then $h^{-1}h'$ is a global section of $\mathscr{A}^0$. The assumption that Σ has finite index in $NS(S)$ implies that $H^0(\mathscr{A}^0)$ vanishes, by Lemma 7.5. Hence $h = h'$ and the above refinements are not needed. □

In the relative setting the above construction gives rise to a natural proper family

$$\tilde{\pi} : \mathscr{M} \to \widetilde{Ш} \times |\mathscr{L}^d|,$$

which restricts over $\{0\} \times |\mathscr{L}^d|$ to $\pi : M_H(u) \to |\mathscr{L}^d|$, and over $\tilde{s} \in \widetilde{Ш}$ to $\pi_{j(\tilde{s})} : M_{j(\tilde{s})} \to |\mathscr{L}^d|$. Indeed, let $(\{U_i\}, \tilde{\sigma}_{ij})$ be a Čech co-cycle representing a non-zero class $\tilde{\sigma}$ in $H^1(|\mathscr{L}^d|, R^1p_*\mathscr{O}_{\mathscr{C}})$. Let

$$\tau : \widetilde{Ш} \to \mathbb{C} \tag{7.9}$$

be the function satisfying $\tau(x)\tilde{\sigma} = x$. Then $(\{\widetilde{Ш} \times U_i\}, \exp(\tau\tilde{\sigma}_{ij}))$ is a global co-cycle representing the desired family. Let

$$f : \mathscr{M} \to \widetilde{Ш}$$

be the composition of $\tilde{\pi}$ with the projection to $\widetilde{Ш}$.

Proposition 7.7. *If the weight 2 Hodge structure of S is non-special, then M_s is Kähler, for all $s \in Ш^0$.*

Proof. There is an open neighborhood of the origin in $\widetilde{Ш}$, over which the fibers of f are Kähler, by the stability of Kähler manifolds [42, Theorem 9.3.3]. Let $j : \widetilde{Ш} \to Ш^0$ be the homomorphism given in Eq. (7.5) The kernel $\ker(j)$ is isomorphic to the group $[\Sigma^\perp + NS(S)]/NS(S)$, by Lemma 7.5. As a subgroup of the base $\widetilde{Ш}$ of the family f, the kernel $\ker(j)$ acts on the base. Let z be an element of $\ker(j)$ and $\tilde{s}$ an element of $\widetilde{Ш}$. The fibers $M_{\tilde{s}}$ and $M_{\tilde{s}+z}$ of f are both isomorphic to $M_{j(\tilde{s})}$ by Lemma 7.6. Let $V \subset \widetilde{Ш}$ be the subset consisting of points over which the fiber of f is Kähler. Then V is an open and $\ker(j)$-invariant subset of $\widetilde{Ш}$. Note that $\ker(j)$ is a finite index subgroup of $H^2(S,\mathbb{Z})/NS(S)$. The kernel $\ker(j)$ is a dense subgroup of $\widetilde{Ш}$, if and only if the image of $H^2(S,\mathbb{Z})/NS(S)$ is dense in $H^{0,2}(S)$, by Lemma 7.3 (4). This is indeed the case, by the assumption that the weight 2 Hodge structure of S is non-special, and Lemmas 5.4 and 5.5. The complement V^c of V in $\widetilde{Ш}$ is $\ker(j)$ invariant. If non-empty, then V^c is dense and closed and so equal to $\widetilde{Ш}$. But we know that V is non-empty. Hence, $V = \widetilde{Ш}$. □

Example 7.8. Consider the case where $d = 1$ and $\mathrm{Pic}(S)$ is cyclic generated by the line bundle $\mathscr{L}$ of degree $2n-2$, $n \geq 2$. Then $H^2(|\mathscr{L}^d|, R^1{p_2}_*\mathbb{Z})$ vanishes, by Lemma 7.4 (3), and $Ш^0 = H^1(\mathscr{A}^0)$. The linear system $|\mathscr{L}|$ consists of integral curves, and so we can find an open covering $\{U_i\}$ of $|\mathscr{L}|$, and sections $\zeta_i : U_i \to \mathscr{C}$, such that $p_2 \circ \zeta_i$ is the identity. Set $D_i := \zeta_i(U_i)$. We get the line bundle $\mathscr{O}_{p_2^{-1}(U_i)}(D_i)$, which restricts to a line bundle of degree 1 on fibers of p_2 over points of U_i. Let h_i be the section of $R^1 p_{2_*}\mathscr{O}^*_{\mathscr{C}}$ over U_i corresponding to $\mathscr{O}_{p_2^{-1}(U_i)}(D_i)$ and denote by $h := \{h_i\}$ the corresponding co-chain in $C^0(\{U_i\}, R^1 p_{2_*}\mathscr{O}^*_{\mathscr{C}})$.

Consider the Lagrangian fibrations $\pi_0 : M_{\mathscr{L}}(0,\mathscr{L},\chi) \to |\mathscr{L}|$ and $\pi_1 : M_{\mathscr{L}}(0,\mathscr{L},\chi+1) \to |\mathscr{L}|$, for some integer χ. The push-forward of every rank 1 torsion free sheaf on a curve in the linear system $|\mathscr{L}|$ is an $\mathscr{L}$-stable sheaf on S, since the curve is integral. Hence, the section h_i induces an isomorphism $h_i : \pi_0^{-1}(U_i) \to \pi_1^{-1}(U_i)$. The co-boundary $(\delta h)_{ij} := h_j h_i^{-1}$ is a co-cycle in $Z^1(\{U_i\}, \mathscr{A}^0)$ representing a class $s \in \text{Ш}^0$ mapping to the identity in Ш. The Lagrangian fibration $\pi_s : M_s \to |\mathscr{L}|$, associated to the class s in Eq. (7.8) with $u = (0,\mathscr{L},\chi)$, coincides with $\pi_1 : M_{\mathscr{L}}(0,\mathscr{L},\chi+1) \to |\mathscr{L}|$, by the commutativity of the following diagram.

$$\begin{array}{ccccccc}
\pi_0^{-1}(U_j) & \overset{\supset}{\longleftarrow} & \pi_0^{-1}(U_{ij}) & \xrightarrow{h_i^{-1}h_j} & \pi_0^{-1}(U_{ij}) & \overset{\subset}{\longrightarrow} & \pi_0^{-1}(U_i) \\
\downarrow{\scriptstyle h_j} & & & & & & \downarrow{\scriptstyle h_i} \\
\pi_1^{-1}(U_j) & \overset{\supset}{\longleftarrow} & \pi_1^{-1}(U_{ij}) & \xrightarrow{id} & \pi_1^{-1}(U_{ij}) & \overset{\subset}{\longrightarrow} & \pi_1^{-1}(U_i).
\end{array}$$

The moduli spaces $M_{\mathscr{L}}(0,\mathscr{L},\chi)$ and $M_{\mathscr{L}}(0,\mathscr{L},\chi+1)$ are not isomorphic for generic $(S,\mathscr{L})$, since their weight 2 Hodge structures are not Hodge isometric.

The kernel of $\text{Ш}^0 \to \text{Ш}$ is cyclic of order $2n-2$, by the exactness of the sequence (7.7). The class s constructed above generates the kernel. This is seen as follows. The sheaf $R^2 p_{2_*}\mathbb{Z}$ is trivial, in our case, and the homomorphism deg, given in (7.3), maps the 0-co-chain h to a global section of $R^2 p_{2_*}\mathbb{Z}$, which generates $H^0(R^2 p_{2_*}\mathbb{Z})$. Hence, δh generates the image of the connecting homomorphism $H^0(R^2 p_{2_*}\mathbb{Z}) \to H^1(\mathscr{A}^0)$ associated to the short exact sequence (7.3). The latter image is precisely the kernel of $\text{Ш}^0 \to \text{Ш}$.

7.3 *The Period Map of the Universal Family is Étale*

Denote by $T_{\pi_s} := \ker\left[d\pi_s : TM_s \to \pi_s^* T|\mathscr{L}^d|\right]$ the relative tangent sheaf of $\pi_s : M_s \to |\mathscr{L}^d|$.

Lemma 7.9. *The vertical tangent sheaf T_{π_s} is isomorphic to $\pi_s^* T^*|\mathscr{L}^d|$.*

Proof. Let $\text{sing}(\pi_s)$ be the support of the co-kernel of the differential $d\pi_s : TM_s \to \pi_s^* T|\mathscr{L}^d|$. We use Assumption 7.1 to prove that the co-dimension of $\text{sing}(\pi_s)$ in M_s is ≥ 2. The generic fiber of π_s is smooth, since M_s is smooth. All fibers of π_s have pure dimension n [29]. Hence, the only way $\text{sing}(\pi_s)$ could contain a divisor is if π_s

has fibers with a non-reduced irreducible component over some divisor in $|\mathcal{L}^d|$. The generic divisor in the linear system $|\mathcal{L}^d|$ is a smooth curve, by Assumption 7.1 (1) and [32, Prop. 1]. The fiber of π_s, over a reduced divisor $C \in |\mathcal{L}^d|$, is isomorphic to the compactified Picard of C, consisting of $\mathcal{L}$-stable sheaves of Euler characteristic χ with pure one-dimensional support C, which are the push forward of rank 1 torsion free sheaves over C. If C is an integral curve, then the moduli space of rank 1 torsion free sheaves over C with a fixed Euler characteristic is irreducible and reduced [1]. If C is reduced (possibly reducible) with at worst ordinary double point singularities, then the compactified Picard is reduced, by a result of Oda and Seshadri [37]. Assumption 7.1 (2) thus implies that $\mathrm{sing}(\pi_s)$ has co-dimension ≥ 2 in M_s.

Let U be the complement of $\mathrm{sing}(\pi_s)$ in M_s. The isomorphism $TM_s \to T^*M_s$, induced by a non-degenerate global holomorphic 2-form, maps the restriction of T_{π_s} to U isomorphically onto the restriction of $\pi_s^*T^*|\mathcal{L}^d|$. The isomorphism $TM_s \to T^*M_s$ must map T_{π_s} as a subsheaf of the locally free $\pi_s^*T|\mathcal{L}^d|$, by the fact that $\mathrm{sing}(\pi_s)$ has codimension ≥ 2. But T_{π_s} is a saturated subsheaf of TM_s. Hence, the image of T_{π_s} is also saturated in T^*M_s, and is thus equal to $\pi_s^*T^*|\mathcal{L}^d|$. □

When the $K3$ surface S is non-special, the fibers of the family f are irreducible holomorphic symplectic manifolds, by Proposition 7.7 and the fact that Kähler deformations of an irreducible holomorphic symplectic manifold remain such [4]. Denote by

$$\eta : R^2 f_*\mathbb{Z} \to (\Lambda)_{\widetilde{\text{III}}} \tag{7.10}$$

the trivialization, which restricts to the marking η_1 in Diagram (5.4) over the point $0 \in \widetilde{\text{III}}$. Let $P_f : \widetilde{\text{III}} \to \Omega^+_{\alpha^\perp}$ be the period map of the family f and the marking η. Let $dP_f : T_{\tilde{s}}\widetilde{\text{III}} \to H^{2,0}(M_s)^* \otimes H^{1,1}(M_s)$ be the differential at $\tilde{s}$ of the period map.

Lemma 7.10. *The differential dP_f is injective, for all $\tilde{s}$ in $\widetilde{\text{III}}$, and its image is equal to $H^{2,0}(M_s)^* \otimes \pi_s^* H^{1,1}(|\mathcal{L}^d|)$.*

Proof. Let $\psi : H^{2,0}(M_s)^* \otimes H^1(|\mathcal{L}^d|, T^*|\mathcal{L}^d|) \to H^1(M_s, T_{\pi_s})$ be the composition of

$$1\otimes\pi_s^*: H^{2,0}(M_s)^*\otimes H^1(|\mathcal{L}^d|, T^*|\mathcal{L}^d|)\to H^0(M_s, \wedge^2 TM_s)\otimes H^1(M_s, \pi_s^*T^*|\mathcal{L}^d|)$$

with the contraction homomorphism $H^0(M_s, \wedge^2 TM_s) \otimes H^1(M_s, \pi_s^*T^*|\mathcal{L}^d|) \to H^1(M_s, T_{\pi_s})$. Let $\kappa_{\tilde{s}} : T_{\tilde{s}}\widetilde{\text{III}} \to H^1(M_s, TM_s)$ be the Kodaira-Spencer map. We have the commutative diagram.

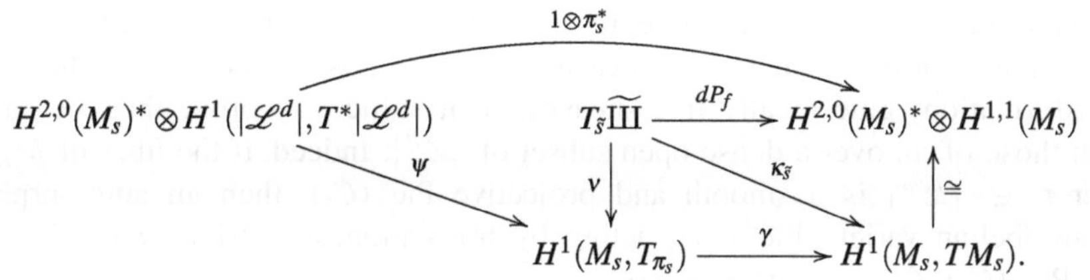

Above, the right vertical homomorphism is induced by the sheaf homomorphism $TM_s \to T^*M_s$, associated to a holomorphic 2-form, and γ is induced by the inclusion of the relative tangent sheaf T_{π_s} as a subsheaf of TM_s. The homomorphism ν is defined as follows. A tangent vector ξ at a class $\tilde{s}$ of $\widetilde{\text{Ш}}$ is represented by a co-cycle of infinitesimal automorphisms – tangent vector fields – which are vertical, being a limit of translations by local sections of the image of $R^1p_{2_*}\mathscr{O}_{\mathscr{C}}$ in $R^1p_{2_*}\mathscr{O}^*_{\mathscr{C}}$. So ξ corresponds to an element $\nu(\xi)$ in $H^1(M_s,T_{\pi_s})$.

The top right triangle commutes, by Griffiths' identification of the differential of the period map [9]. The middle triangle commutes, by definition of the family f. The commutativity of the outer polygon is easily verified. The top horizontal homomorphism $1\otimes\pi_s^*$ is injective, with image equal to the tangent line to the fiber of q. Hence, it suffices to prove that ψ and ν have the same image in $H^1(M_s,T_{\pi_s})$. The latter statement would follow once we prove that ν is an isomorphism.

The homomorphism ν is induced by the pullback

$$\pi_s^*H^1(|\mathscr{L}^d|,R^1p_{2_*}\mathscr{O}_{\mathscr{C}}) \to H^1(M_s,\pi_s^*R^1p_{2_*}\mathscr{O}_{\mathscr{C}}),$$

followed by the homomorphism of sheaf cohomologies induced by an injective sheaf homomorphism

$$\tilde{\nu}\ :\ \pi_s^*R^1p_{2_*}\mathscr{O}_{\mathscr{C}} \to T_{\pi_s}.$$

The domain of $\tilde{\nu}$ is isomorphic to $\pi_s^*T^*|\mathscr{L}^d|$, by Lemma 7.3, and its target is isomorphic to $\pi_s^*T^*|\mathscr{L}^d|$, by Lemma 7.9. Hence, $\tilde{\nu}$ is an isomorphism. It remains to prove that $H^1(M_s,\pi_s^*T^*|\mathscr{L}^d|)$ is one dimensional. We have the exact sequence

$$\begin{aligned}0 &\to H^1(|\mathscr{L}^d|,\pi_{s_*}\pi_s^*T^*|\mathscr{L}^d|) \to H^1(M_s,\pi_s^*T^*|\mathscr{L}^d|)\\ &\to H^0(|\mathscr{L}^d|,T^*|\mathscr{L}^d|\otimes R^1\pi_{s_*}\mathscr{O}_{M_s}).\end{aligned}$$

The left hand space is one-dimensional. It remains to prove that the right hand one vanishes. It suffices to prove that $R^1\pi_{s_*}\mathscr{O}_{M_s}$ is isomorphic to $T^*|\mathscr{L}^d|$, since $T^*|\mathscr{L}^d|\otimes T^*|\mathscr{L}^d|$ does not have any non-zero global sections.

When $s = 0$ and $M_0 = M_H(u)$, then M_0 is projective and $R^1\pi_{0_*}\mathscr{O}_{M_0}$ is isomorphic to $T^*|\mathscr{L}^d|$, by [30, Theorem 1.3]. Let us show that the sheaves $R^1\pi_{s_*}\mathscr{O}_{M_s}$ are naturally isomorphic to $R^1\pi_{0_*}\mathscr{O}_{M_0}$, for all s in Ш. The fibrations π_s agree, by definition, over the open sets in a Čech covering of $|\mathscr{L}^d|$, and the gluing transformations for the co-cycle representing the class s do not change the induced sheaf transition functions for the sheaves $R^1\pi_{s_*}\mathcal{O}_{M_s}$, as we show next. The gluing transformations glue locally free sheaves, so it suffices to prove that they agree with those of π_0 over a dense open subset of $|\mathcal{L}^d|$. Indeed, if the fiber of $M_H(u)$ over $t \in |\mathcal{L}^d|$ is a smooth and projective $\mathrm{Pic}^d(C_t)$, then an automorphism of an abelian variety $\mathrm{Pic}^d(C_t)$, acting by translation, acts trivially on the fiber $H^1(\mathrm{Pic}^d(C_t), \mathcal{O}_{\mathrm{Pic}^d(C_t)})$ of $R^1\pi_*\mathcal{O}_{M_H(u)}$. □

7.4 The Tate-Shafarevich Line as the Base of the Universal Family

Let $q : \Omega^+_{\alpha^\perp} \to \Omega^+_{Q_\alpha}$ be the morphism given in Eq. (4.3).

Theorem 7.11. *Assume that the weight 2 Hodge structure of S is non-special and Assumption 7.1 holds. Then the period map P_f of the family f maps $\widetilde{Ш}$ isomorphically onto the fiber of the morphism q through the period of $M_H(u)$.*

Proof. We already know that P_f is non-constant, by Lemma 7.10. The statement implies that P_f is an affine linear isomorphism of one-dimensional complex affine spaces. It suffices to prove the statement for a dense subset in moduli, since the condition of being affine linear is closed. We may thus assume that $\mathrm{Pic}(S)$ is cyclic generated by $\mathcal{L}$. Then $H^0(|\mathcal{L}^d|, \mathcal{A}^0)$ is trivial, by Lemma 7.5.

Set $\Gamma := c_1(\mathcal{L})^\perp$. Note that $NS(S) = \mathbb{Z}c_1(\mathcal{L})$ and Γ has finite index in $H^2(S,\mathbb{Z})/NS(S)$. Let

$$e : \Gamma \to \widetilde{Ш}$$

be the composition of the projection $\Gamma \to H^{0,2}(S)$ with the isomorphisms $H^{0,2}(S) \cong H^{0,2}(\mathcal{C}) \cong \widetilde{Ш}$ of Lemma 7.3. Then e is injective and its image is dense in $\widetilde{Ш}$, by Lemma 5.4.

Given an element $x \in \widetilde{Ш}$, we get a marked pair (M_x, η_x), as above. $M_H(u)$ will be denoted by M_0, it being the fiber of f over the origin in $\widetilde{Ш}$. We associate next to an element $\gamma \in \Gamma$ a canonical isomorphism

$$h_\gamma : M_0 \to M_{e(\gamma)}.$$

Let $\tau : \widetilde{Ш} \to \mathbb{C}$ be the function given in (7.9), which was used in the construction of the family f. Let $\tilde{\sigma} := \{\tilde{\sigma}_{ij}\}$ be the co-cycle used in that construction. Let a be the 1-co-cycle given by $a_{ij} := \exp(\tau(e(\gamma))\tilde{\sigma}_{ij})$. Then $M_{e(\gamma)}$ is the Tate-Shafarevich twist of M_0 with respect to the co-cycle a. The 1-co-cycle a is a co-boundary in $Z^1(\{U_i\}, \mathcal{A}^0)$, by Lemma 7.5 and the definition of Γ. Thus, there exists a 0-co-chain $h := \{h_i\}$ in $C^0(\{U_i\}, \mathcal{A}^0)$, satisfying $\delta h = a$. The co-chain h is unique, since $H^0(\mathscr{A}^0)$ is trivial, by our assumption on S. The co-chain h determines the isomorphism $h_\gamma : M_0 \to M_{e(\gamma)}$ (Lemma 7.6).

We define next a monodromy representation associated to the family f. Denote by $h_{\gamma *} : H^2(M_0, \mathbb{Z}) \to H^2(M_{e(\gamma)}, \mathbb{Z})$ the isomorphism induced by h_γ. Let

$$\mu : \Gamma \to Mon^2(M_0)$$

be given by the composition $\mu_\gamma := \eta_0^{-1} \circ \eta_{e(\gamma)} \circ h_{\gamma *}$ of the parallel-transport operator $\eta_0^{-1} \circ \eta_{e(\gamma)}$ and the isomorphism $h_{\gamma *}$.

Claim 7.12. The map μ is a group homomorphism.

Proof. Let γ_1, γ_2 be elements of Γ and set $\gamma_3 := \gamma_1 + \gamma_2$. Let the topological space B be the quotient of $\widetilde{\mathrm{III}}$ obtained by identifying the four points 0, $e(\gamma_1)$, $e(\gamma_2)$, $e(\gamma_3)$. The family f descends to a family $\bar{f} : \overline{\mathscr{M}} \to B$ by identifying the fiber $M_{e(\gamma_i)}$ with M_0 via the isomorphisms h_{γ_i}, $1 \leq i \leq 3$. Then μ_{γ_i} is the monodromy operator corresponding to any loop in B, which is the image of some continuous path from 0 to $e(\gamma_i)$ in $\widetilde{\mathrm{III}}$. Let $\bar{0} \in B$ be the image of $0 \in \widetilde{\mathrm{III}}$. The statement now follows from the fact that the monodromy representation of $\pi_1(B, \bar{0})$ in $H^2(M_0, \mathbb{Z})$ is a group homomorphism. □

The image of $\widetilde{\mathrm{III}}$ via the period map is contained in the fiber of q, since the differential of the morphism $q \circ P_f$ vanishes, by Lemma 7.10. It follows that the variation of Hodge structures of the local system $R^2 f_* \mathbb{Z}$ over $\widetilde{\mathrm{III}}$ is the pullback of the one over the fiber of q via the period map $P_{\underline{f}}$. Let η be the trivialization of $R^2 f_* \mathbb{Z}$ given in Eq. (7.10). Given a point $x \in \widetilde{\mathrm{III}}$, set $\alpha_x := \eta_x^{-1}(\alpha)$. Then $\alpha_x = \pi_x^*(c_1(\mathscr{O}_{|\mathscr{L}^d|}(1)))$ and the sub-quotient variation of Hodge structures $\alpha_x^\perp / \mathbb{Z}\alpha_x$ is trivial.

The vertical tangent sheaf T_{π_x} is naturally isomorphic to T_{π_0}, as we saw in the last paragraph of the proof of Lemma 7.10. The 2-form w_x induces an isomorphism $\pi_{x*} T_{\pi_x} \stackrel{w_x}{\to} T^*|\mathscr{L}^d|$, by Lemma 7.9. We get the composite isomorphism $\pi_{0*} T_{\pi_0} \cong \pi_{x*} T_{\pi_x} \stackrel{w_x}{\to} T^*|\mathscr{L}^d|$. Let w_x be the unique holomorphic 2-form, for which the composite isomorphism is equal to $\pi_{0*} T_{\pi_0} \stackrel{w_0}{\to} T^*|\mathscr{L}^d|$. Such a form w_x exists, since the endomorphism algebra of $T^*|\mathscr{L}^d|$ is one dimensional.

We show next that the class of w_x is the $(2,0)$ part of the flat deformation of the class of w_0 in the local system $R^2 f_* \mathbb{C}$. It suffices to prove the local version of that statement. Let x_0 be a point of $\widetilde{\mathrm{III}}$. There is a differentiable trivialization of $f : \mathscr{M} \to \widetilde{\mathrm{III}}$, over an open analytic neighborhood U of x_0, and a C^∞ family of complex structures J_x, $x \in U$, such that (M_{x_0}, J_x) is biholomorphic to M_x. Furthermore, the complex structures J_{x_0} and J_x restrict to the same complex structure on each fiber of π_{x_0} and π_{x_0} is holomorphic with respect to both. Both complex structures induce the same complex structure on $\mathrm{Hom}\left(T_{\pi_{x_0}}, \pi_{x_0}^* T_{\mathbb{R}}^* |\mathscr{L}^d|\right)$ and the two forms w_{x_0} and w_x induce the same section in the complexification of that bundle. Hence, the difference $w_{x_0} - w_x$ is a closed 2-form in $\pi_{x_0}^* \wedge^2 T_{\mathbb{R}}^* |\mathscr{L}^d| \otimes_{\mathbb{R}} \mathbb{C}$. Being closed, the latter 2-form must be the pull-back of a closed 2-form θ on $|\mathscr{L}^d|$, since fibers of π_{x_0} are connected. Now the cohomology class of $\pi_{x_0}^*\theta$ is of type $(1,1)$ with respect to all complex structures, since $H^{1,1}(|\mathcal{L}^d|) = H^2(|\mathcal{L}^d|, \mathbb{C})$. Hence, the class of w_x is the $(2,0)$ part of the class of w_{x_0} with respect to the complex structure J_x.

There exists a constant $c_x \in \mathbb{C}$, such that the equality

$$\eta_x(w_x) = \eta_0(w_0) + c_x\alpha$$

holds in $\Lambda_{\mathbb{C}}$, by the characterization of ω_x in the above paragraph. The function $c : \widetilde{\text{Ш}} \to \mathbb{C}$ defined above is equivalent to the period map P_f and is thus holomorphic and its derivative is no-where vanishing, by Lemma 7.10. If $x = e(\gamma)$, we get $\eta_0^{-1}\eta_{e(\gamma)}(w_{e(\gamma)}) = w_0 + c_{e(\gamma)}\alpha_0$, Now $h_\gamma(w_0) = w_{e(\gamma)}$, by definition of w_x, $x \in \widetilde{\text{Ш}}$, and the construction of h_γ. We get the equality

$$\mu_\gamma(w_0) = w_0 + c_{e(\gamma)}\alpha_0. \tag{7.11}$$

The composition $c \circ e : \Gamma \to \mathbb{C}$ is a group homomorphism,

$$c(e(\gamma_1) + e(\gamma_2)) = c(e(\gamma_1)) + c(e(\gamma_2)),$$

by Eq. (7.11) and Claim 7.12. The image $e(\Gamma)$ is dense in $\widetilde{\text{Ш}}$ and so $e(\Gamma) \times e(\Gamma)$ is dense in $\widetilde{\text{Ш}} \times \widetilde{\text{Ш}}$. We conclude that c is a group homomorphism, $c(x_1 + x_2) = c(x_1) + c(x_2)$, for all $(x_1, x_2) \in \widetilde{\text{Ш}} \times \widetilde{\text{Ш}}$. Continuity of c implies that it is a linear transformation of real vector spaces. Indeed, given x_1, x_2 in $\widetilde{\text{Ш}}$, $c(ax_1 + bx_2) = ac(x_1) + bc(x_2)$, for all $a, b \in \mathbb{Z}$, hence also for all $a, b \in \mathbb{Q}$, and continuity implies that the equality holds also for all $a, b \in \mathbb{R}$. The map c is holomorphic, hence it is a linear transformation of one-dimensional complex vector spaces, which is an isomorphism, since c is non-constant. This completes the proof of Theorem 7.11.

□

Let X be an irreducible holomorphic symplectic manifold of $K3^{[n]}$-type and $\pi : X \to \mathbb{P}^n$ a Lagrangian fibration. Set $\alpha := \pi^* c_1(\mathcal{O}_{\mathbb{P}^n}(1))$. Let d be the divisibility of $(\alpha, \bullet)$. Let $(S, \mathcal{L})$ be the semi-polarized $K3$ surface associated to (X, α) in Diagram (5.4) and χ the Euler characteristic of the Mukai vector u in that diagram. Choose a u-generic polarization H on S.

Theorem 7.13. *Assume that X is non-special and $(S, \mathcal{L})$ satisfies Assumption 7.1. Then X is bimeromorphic to a Tate-Shafarevich twist of the Lagrangian fibration $M_H(0, \mathcal{L}^d, \chi) \to |\mathcal{L}^d|$.*

Proof. Fix a marking $\eta : H^2(X, \mathbb{Z}) \to \Lambda$. Starting with the period of (X, η), Theorem 7.11 exhibits a marked triple (X', α', η'), with $\eta'(\alpha') = \eta(\alpha)$, in the same connected component $\mathfrak{M}^+_{\eta(\alpha)^\perp}$ as the triple (X, α, η), such that the class α' is semi-ample as well and the periods $P(X, \eta)$ and $P(X', \eta')$ are equal. Furthermore, the Lagrangian fibration $\pi' : X' \to |\mathcal{L}^d|$ induced by α' is a Tate-Shafarevich twist of $\pi_0 : M_H(0, \mathcal{L}^d, \chi) \to |\mathcal{L}^d|$. Step 1 of the proof of Theorem 1.3 yields a bimeromorphic map $f : X \dashrightarrow X'$, which is shown in Step 2 of that proof to satisfy $f^*(\alpha') = \alpha$ (see Eq. (6.1)). □

Proof (of Theorem 1.5). The condition that $NS(X) \cap \alpha^{\perp}$ is cyclic generated by α implies that the semi-polarized $K3$ surface $(S, \mathscr{L})$, associated to (X, α), has a cyclic Picard group generated by $\mathscr{L}$. Assumption 7.1 thus holds, by Remark 7.2. Theorem 1.5 thus follows from Theorem 7.13. □

Acknowledgements I would like thank Yujiro Kawamata, Christian Lehn, Daisuke Matsushita, Keiji Oguiso, Osamu Fujino, Thomas Peternell, Sönke Rollenske, Justin Sawon, and Kota Yoshioka for helpful communications. I would like to thank the two referees for their careful reading and insightful comments and suggestions.

References

1. A. Altman, A. Iarrobino, S. Kleiman, Irreducibility of the compactified Jacobian, in *Real and Complex Singularities* (Sijtho and Noordho, Alphen aan den Rijn). Proceedings of Ninth Nordic Summer School/NAVF Symposium in Mathematics, Oslo, 1976 (1997), pp. 1–12
2. W. Barth, K. Hulek, C. Peters, A. Van de Ven, *Compact Complex Surfaces*, 2nd edn. Ergebnisse der Mathematik und ihrer Grenzgebiete. 3. Folge. A Series of Modern Surveys in Mathematics, 4 (Springer, Berlin, 2004)
3. A. Bayer, E. Macri, MMP for moduli of sheaves on $K3$s via wall-crossing: nef and movable cones, Lagrangian fibrations (2013, Electronic preprint). arXiv:1301.6968
4. A. Beauville, Varietes Kähleriennes dont la premiere classe de Chern est nulle. J. Differ. Geom. **18**, 755–782 (1983)
5. A. Beauville, Holomorphic symplectic geometry: a problem list, in *Complex and Differential Geometry*, ed. by W. Ebeling et al. Springer Proceedings in Mathematics, vol. 8 (Springer, Dordrecht, 2011), pp.49–63
6. S. Boucksom, Divisorial Zariski decompositions on compact complex manifolds. Ann. Sci. École Norm. Sup. (4) **37**(1), 45–76 (2004)
7. F. Campana, Isotrivialité de certaines familles hyperkählérian de variété non-projective. Math. Z. **252**(1), 147–156 (2006)
8. F. Campana, K. Oguiso, T. Peternell, Non-algebraic hyperkähler manifolds. J. Diff. Geom. **85**, 397–424 (2010)
9. J. Carlson, M. Green, P. Griffiths, J. Harris, Infinitesimal variations of Hodge structure I. Compos. Math. **50**(2–3), 109–205 (1983)
10. I. Dolgachev, M. Gross, Elliptic three-folds I: Ogg-Shafarevich theory. J. Algebraic Geom. **3**(1) 39–80 (1994)
11. O. Fujino, On Kawamata's theorem, in *Classification of Algebraic Varieties*, ed. by C. Faber, G. van der Geer, E. Looijenga. EMS Series of Congress Reports (European Mathematical Society, Zurich, 2011), pp. 305–315
12. R. Howlett, Introduction to Coxeter groups. Lectures delivered at the A.N.U. Geometric Group Theory Workshop, Jan/Feb 1996. http://www.maths.usyd.edu.au/u/bobh/bobhres.html#infCoxpubs
13. D. Huybrechts, Compact hyperkähler manifolds: basic results. Invent. Math. **135**(1), 63–113 (1999); Erratum: Invent. Math. **152**(1), 209–212 (2003)
14. D. Huybrechts, A global Torelli theorem for hyperkähler manifolds (after Verbitsky). Séminaire Bourbaki, 63 année, 2010–2011, no. 1040

15. D. Huybrechts, M. Lehn, *The Geometry of Moduli Spaces of Sheaves*, 2nd edn. (Cambridge University Press, Cambridge, 2010)
16. Y. Kawamata, Pluricanonical systems on minimal algebraic varieties. Invent. Math. **79**, 567–588 (1985)
17. K. Kodaira, On compact analytic surfaces, II. Ann. Math. (2) **77**, 563–626 (1963); III. Ann. Math. **78**, 1–40 (1963)
18. C. Lai, Varieties fibered by good minimal models (Electronic preprint). Math. Ann. **350**(3), 533–547 (2011). arXiv:0912.3012v2
19. E. Looijenga, C. Peters, Torelli theorems for Kähler $K3$ surfaces. Compos. Math. **42**(2), 145–186 (1980)
20. E. Markman, Integral generators for the cohomology ring of moduli spaces of sheaves over Poisson surfaces. Adv. Math. **208**, 622–646 (2007)
21. E. Markman, On the monodromy of moduli spaces of sheaves on K3 surfaces. J. Algebra. Geom. **17**, 29–99 (2008)
22. E. Markman, Integral constraints on the monodromy group of the hyperkähler resolution of a symmetric product of a $K3$ surface. Internat. J. Math. **21**(2), 169–223 (2010)
23. E. Markman, A survey of Torelli and monodromy results for hyperkahler manifolds, in *Complex and Differential Geometry*, ed. by W. Ebeling et al. Springer Proceedings in Mathematics, vol. 8 (Springer, Dordrecht, 2011), pp. 257–323
24. E. Markman, Prime exceptional divisors on holomorphic symplectic varieties and monodromy-reflections. Kyoto J. Math. **53**(2), 345–403 (2013)
25. E. Markman, S. Mehrotra, Hilbert schemes of $K3$ surfaces are dense in moduli (Electronic preprint). arXiv:1201.0031
26. D. Markushevich, Rational Lagrangian fibrations on punctual Hilbert schemes of $K3$ surfaces. Manuscr. Math. **120**(2), 131–150 (2006)
27. D. Matsushita, On deformations of Lagrangian fibrations (Preprint). arXiv:0903.2098
28. D. Matsushita, On isotropic divisors on irreducible symplectic manifolds (Electronic preprint). arXiv:1310.0896
29. D. Matsushita, Equidimensionality of Lagrangian fibrations on holomorphic symplectic manifolds. Math. Res. Lett. **7**(4), 389–392 (2000)
30. D. Matsushita, Higher direct images of dualizing sheaves of Lagrangian fibrations. Am. J. Math. **127**(2), 243–259 (2005)
31. D. Matsushita, A talk given at *Hyper-Kähler Geometry*, Simons center for Geometry and Physics, Oct 2012. A video is available at http://scgp.stonybrook.edu/archives/3434
32. A.L. Mayer, Families of $K3$ surfaces. Nagoya Math. J. **48**, 1–17 (1972)
33. D. Morrison, Some remarks on the moduli of $K3$ surfaces, in *Classification of Algebraic and Analytic Manifolds*, ed. by K. Ueno. Progress in Mathematics, vol. 39 (Birkhäuser, Boston, 1983), pp. 303–332
34. S. Mukai, Symplectic structure of the moduli space of sheaves on an abelian or K3 surface. Invent. Math. **77**, 101–116 (1984)
35. S. Mukai, On the moduli space of bundles on K3 surfaces I, in *Vector Bundles on Algebraic Varieties*. Proceedings of Bombay Conference, 1984, Tata Institute of Fundamental Research Studies, no. 11, Bombay (Oxford University Press, 1987), pp. 341–413
36. V.V. Nikulin, Integral symmetric bilinear forms and some of their applications. Math. USSR Izvestija **14**(1), 103–167 (1980)
37. T. Oda, C.S. Seshadri, Compactifications of the generalized Jacobian variety. Trans. Am. Math. Soc. **253**, 1–90 (1979)
38. K. O'Grady, The weight-two Hodge structure of moduli spaces of sheaves on a K3 surface. J. Algebra. Geom. **6**(4), 599–644 (1997)
39. J. Sawon, Lagrangian fibrations on Hilbert schemes of points on $K3$ surfaces. J. Algebra. Geom. **16**(3), 477–497 (2007)
40. M. Verbitsky, A global Torelli theorem for hyperkähler manifolds. Duke Math. J. **162**(15), 2929–2986 (2013). arXiv:0908.4121

41. M. Verbitsky, Cohomology of compact hyper-Kähler manifolds and its applications. Geom. Funct. Anal. **6**, 601–611 (1996)
42. C. Voisin, *Hodge Theory and Complex Algebraic Geometry I*. Cambridge Studies in Advanced Mathematics, vol. 76 (Cambridge University Press, Cambridge, 2002)
43. K. Yoshioka, Bridgeland's stability and the positive cone of the moduli spaces of stable objects on an abelian surface (Electronic preprint). arXiv:1206.4838
44. K. Yoshioka, Moduli spaces of stable sheaves on abelian surfaces. Math. Ann. **321**(4), 817–884 (2001)

Contact Kähler Manifolds: Symmetries and Deformations

Thomas Peternell and Florian Schrack

Dedicated to Klaus Hulek on his 60th birthday.

Abstract We study complex compact Kähler manifolds X carrying a contact structure. If X is almost homogeneous and $b_2(X) \geq 2$, then X is a projectivised tangent bundle (this was known in the projective case even without assumption on the existence of vector fields). We further show that a global projective deformation of the projectivised tangent bundle over a projective space is again of this type unless it is the projectivisation of a special unstable bundle over a projective space. Examples for these bundles are given in any dimension.

1 Introduction

A contact structure on a complex manifold X is in some sense the opposite of a foliation: there is a vector bundle sequence

$$0 \to F \to T_X \to L \to 0,$$

where T_X is the tangent bundle and L a line bundle, with the additional property that the bilinear map, induced by the Lie bracket,

$$F \times F \to L, \ (v, w) \mapsto [v, w]/F$$

is everywhere non-degenerate.

T. Peternell (✉) • F. Schrack
Mathematisches Institut, Universität Bayreuth, 95440 Bayreuth, Germany
e-mail: thomas.peternell@uni-bayreuth.de; florian.schrack@uni-bayreuth.de

A. Frühbis-Krüger et al. (eds.), *Algebraic and Complex Geometry*, Springer Proceedings in Mathematics & Statistics 71, DOI 10.1007/978-3-319-05404-9_11,

Suppose now that X is compact and Kähler or projective. If $b_2(X) = 1$, then at least conjecturally the structure is well-understood: X should arise as minimal orbit in the projectivised Lie algebra of contact automorphisms. Beauville [5] proved this conjecture under the additional assumption that the group of contact automorphisms is reductive and that the contact line bundle L has "enough" sections.

If $b_2(X) \geq 2$ and X is projective, then, due to [20] and [9], X is a projectivized tangent bundle $\mathbb{P}(T_Y)$ (in the sense of Grothendieck, taking hyperplanes) over a projective manifold Y (and conversely every such projectivised tangent bundle carries a contact structure). If X is only Kähler, the analogous conclusion is unknown. By [9], the canonical bundle K_X is still not pseudo-effective in the Kähler setting, but—unlike in the projective case—it is not known whether this implies uniruledness of X.

If however X has enough symmetries, then we are able to deal with this situation:

Theorem 1. *Let X be an almost homogeneous compact Kähler manifold carrying a contact structure. If $b_2(X) \geq 2$, then there is a compact Kähler manifold Y such that $X \simeq \mathbb{P}(T_Y)$.*

Here a manifold is said to be almost homogeneous, if the group of holomorphic automorphisms acts with an open orbit. Equivalently, the holomorphic vector fields generate the tangent bundle T_X at some (hence at the general) point.

In this setting it might be interesting to try to classify all compact almost homogeneous Kähler manifolds X of the form $X = \mathbb{P}(T_Y)$. Section 4 studies this question in dimension 3.

In the second part of the paper we treat the deformation problem for projective contact manifolds. We consider a family

$$\pi\colon \mathscr{X} \to \Delta$$

of projective manifolds over the 1-dimensional disc $\Delta \subset \mathbb{C}$. Suppose that all $X_t = \pi^{-1}(t)$ are contact for $t \neq 0$. Is then X_0 also a contact manifold?

Suppose first that $b_2(X_t) = 1$. Then—as discussed above—X_t should be homogeneous for $t \neq 0$. Assuming homogeneity, the situation is well-understood by the work of Hwang and Mok. In fact, then X_0 is again homogeneous with one surprising 7-dimensional exception, discovered by Pasquier and Perrin [26] and elaborated further by Hwang [18]. Therefore one has rigidity and the contact structure survives unless the Pasquier–Perrin case happens, where the contact structure does not survive. We refer to [18] and the references given at the beginning of Sect. 5. Therefore—up to the homogeneity conjecture—the situation is well-understood.

If $b_2(X_t) \geq 2$, the situation gets even more difficult; so we will assume that X_t is homogeneous for $t \neq 0$. We give a short argument in Sect. 2, showing that then X_t is either $\mathbb{P}(T_{\mathbb{P}_n})$ or a product of a torus and $\mathbb{P}_n$. Then we investigate the *global projective rigidity* of $\mathbb{P}(T_{\mathbb{P}_n})$:

Theorem 2. *Let $\pi: \mathscr{X} \to \Delta$ be a projective family of compact manifolds. If $X_t \simeq \mathbb{P}(T_{\mathbb{P}_n})$ for $t \neq 0$, then either $X_0 \simeq \mathbb{P}(T_{\mathbb{P}_n})$ or $X_0 \simeq \mathbb{P}(V)$ with some unstable vector bundle V on $\mathbb{P}_n$.*

The assumption that X_0 is projective is indispensable for our proof. In general, X_0 is only Moishezon, and in particular methods from Mori theory fail. In case X_0 is even assumed to be Fano, the theorem was proved by Wiśniewski [31]; in this case $X_0 \simeq \mathbb{P}(T_{\mathbb{P}_n})$. The case $X_0 \simeq \mathbb{P}(V)$ with an unstable bundle really occurs; we provide examples in all dimensions in Sect. 6. In this case X_0 is no longer a contact manifold.

In general, without homogeneity assumption, X_t is the projectivisation of the tangent bundle of some projective variety Y_t; here we have only some partial results, see Proposition 3. If however X_t is again homogeneous ($t \neq 0$) and not the projectivization of the tangent bundle of a projective space, then X_t is a product of a torus A_t and a projective space, and we obtain a rather clear picture, described in Sect. 7.

The work on the project was started in collaboration with Kristina Frantzen. We would like to heartily thank her for her contributions to Sects. 2–4. We also thank Alan Huckleberry and the referee for very valuable comments.

2 Homogeneous Kähler Contact Manifolds

We first study homogeneous manifolds which are projectivized tangent bundles.

Proposition 1. *Let Y be compact Kähler. Then $X = \mathbb{P}(T_Y)$ is homogeneous if and only if Y is a torus or $Y = \mathbb{P}_n$.*

Proof. One direction being clear, assume that X is homogeneous; thus Y is homogeneous, too. The theorem of Borel and Remmert [8] says that

$$Y \cong A \times G/P$$

where G/P is a rational homogeneous manifold (G a semi-simple complex Lie group and P a parabolic subgroup) and A a torus, one factor possibly of dimension 0. Let $d = \dim A \geq 0$.

We first assume that $d > 0$. If we denote by π_1 and π_2 the two projections from Y to A and G/P, then

$$T_Y = \pi_1^* \mathscr{O}_A^d \oplus \pi_2^* T_{G/P} \simeq \mathscr{O}_Y^d \oplus \pi_2^* T_{G/P}.$$

This leads to an inclusion

$$Z := \mathbb{P}(\mathscr{O}_Y^d) \subset X$$

with normal bundle

$$N_{Z/X} = \mathcal{O}_Z(1) \otimes \pi^* q^*(\Omega^1_{G/P}) = p^*(\mathcal{O}(1)) \otimes \pi^* q^*(\Omega^1_{G/P})$$

(use the formula on p. 12, before (0.5) in [3] to compute the normal bundle). Here $\pi: X \to Y$, $p: Z = \mathbb{P}_{d-1} \times Y \to \mathbb{P}_{d-1}$ and $q: Y \to G/P$ are the projections. Now, X being homogeneous, $N_{Z/X}$ is spanned. This is only possible when $\dim G/P = 0$ so that $Y = A$. If $d = 0$, then X is rational homogeneous, hence Fano. This is to say that T_Y is ample, hence $Y = \mathbb{P}_n$ (we do not need Mori's theorem here because Y is already homogeneous).

Proposition 1 is now applied to obtain

Proposition 2. *Let X be a homogeneous compact Kähler manifold with contact structure and* $\dim X = 2n - 1$. *Then either X is a Fano manifold (and therefore $X \simeq \mathbb{P}(T_{\mathbb{P}_n})$, by Proposition 1, unless $b_2(X) = 1$) or*

$$X \cong A \times \mathbb{P}_{n-1} = \mathbb{P}(T_A),$$

where A denotes a complex torus of dimension n and T_A its holomorphic tangent bundle.

Proof. Again by the theorem of Borel–Remmert, $X \cong A \times G/P$ where G/P is rational-homogeneous and A a torus, one factor possibly of dimension 0. If A has dimension 0, then X is Fano. Therefore in the case $b_2(X) \geq 2$, the variety X is of the form $X = \mathbb{P}(T_Y)$ by [20]. Then we conclude by Proposition 1.

So we may assume $\dim A > 0$. Since a torus does not admit a contact structure, it follows that the factor G/P is nontrivial, i.e. $\dim G/P \geq 1$. We consider the projection $\pi: X \cong A \times G/P \to A$. Every fiber is G/P and in particular a Fano manifold. We may therefore use the arguments of [20], Proposition 2.11, to conclude that every fiber is $\mathbb{P}_{n-1}$. Note that the arguments used in [20], Proposition 2.11 do not use the assumption that X is projective. This completes the proof.

3 The Almost Homogeneous Case

The aim of this section is to generalize the previous section to almost homogeneous contact manifolds.

3.1 Almost Homogeneous Projectivized Tangent Bundles

We begin with the following general observation.

Lemma 1. *Let Y be a compact complex manifold and let $X = \mathbb{P}(T_Y)$ be its projectivised tangent bundle. If X is almost homogeneous, then Y is almost homogeneous.*

We already mentioned that if X is homogeneous, so is Y.

Proof. Let $\pi: X \to Y$ be the bundle projection and consider the relative tangent sequence

$$0 \to T_{X/Y} \to T_X \to \pi^* T_Y \to 0.$$

Since at a general point of X the tangent bundle T_X is spanned by global sections, so is $\pi^* T_Y$. So if $y \in Y$ is general, if $x \in \pi^{-1}(y)$ is general and $v \in (\pi^* T_Y)_x$, then there exists

$$s \in H^0(X, \pi^*(T_Y))$$

such that $s(x) = v$. Since $s = \pi^*(t)$ with $t \in H^0(Y, T_Y)$, we obtain $t(y) = v \in T_{Y,y}$. Thus Y is almost homogeneous.

Remark 1. Note that, conversely, the projectivized tangent bundle $X = \mathbb{P}(T_Y)$ of an almost homogeneous manifold Y is in general **not** almost homogeneous. This is illustrated by the following examples.

Example 1. We start in a quite general setting with a projective manifold Y of dimension n. We assume that Y is almost homogeneous with $h^0(Y, T_Y) = n$. Furthermore we assume

$$h^0(Y, \Omega^1_Y \otimes T_Y) = h^0(Y, \mathrm{End}(T_Y)) = 1, \tag{1}$$

an assumption which is e.g. satisfied if T_Y is stable for some polarization. We let $X = \mathbb{P}(T_Y)$ be the projectivized tangent bundle with projection $\pi : X = \mathbb{P}(T_Y) \to Y$ and hyperplane bundle $\mathscr{O}_X(1)$. Pushing forward the relative Euler sequence to Y yields

$$0 \to \mathscr{O}_Y \to \Omega^1_Y \otimes \pi_*(\mathscr{O}_X(1)) \to \pi_* T_{X/Y} \to 0.$$

Since $\pi_*(\mathscr{O}_X(1)) = T_Y$, we obtain

$$0 \to \mathscr{O}_Y \to \Omega^1_Y \otimes T_Y \to \pi_* T_{X/Y} \to 0.$$

This sequence splits via the trace map $\Omega^1_Y \otimes T_Y \simeq \mathrm{End}(T_Y) \to \mathscr{O}_Y$, so we obtain the exact sequence

$$0 \to H^0(Y, \mathscr{O}_Y) \to H^0(Y, \Omega^1_Y \otimes T_Y) \to H^0(Y, \pi_* T_{X/Y}) \to 0.$$

Using assumption (1) we find

$$H^0(X, T_{X/Y}) = H^0(Y, \pi_* T_{X/Y}) = 0.$$

Now the relative tangent sequence with respect to $\pi: X \to Y$ yields an exact sequence

$$0 \to H^0(X, T_{X/Y}) \to H^0(X, T_X) \to H^0(X, \pi^*(T_Y)) \simeq H^0(Y, T_Y)$$

and therefore

$$h^0(T_X) \leq h^0(T_Y).$$

Hence $h^0(T_X) \leq n$, and X cannot be almost homogeneous.

Notice that an inequality $h^0(T_X) \leq 2n - 2$ suffices to conclude that X is not almost homogeneous. Therefore we could weaken the assumptions $h^0(T_Y) = n$ and $h^0(\mathrm{End}(T_Y)) = 1$ to

$$h^0(T_Y) + h^0(\mathrm{End}(T_Y)) \leq 2n - 2.$$

We give two specific examples.

First, let Y be a del Pezzo surface of degree six, i.e., a three-point blow-up of $\mathbb{P}_2$. Its automorphisms group is $(\mathbb{C}^*)^2 \rtimes S_3$. In particular, Y is almost homogeneous and $h^0(T_Y) = 2$. Since $h^0(\mathrm{End}(T_{\mathbb{P}_2})) = 1$ and Y is a blow up of $\mathbb{P}_2$, each endomorphism of T_Y induces an endomorphism of $T_{\mathbb{P}_2}$ and it follows that

$$h^0(T_Y \otimes \Omega^1_Y) = h^0(\mathrm{End}(T_Y)) = 1. \tag{2}$$

Hence the assumptions of our previous considerations are fulfilled and $X = \mathbb{P}(T_Y)$ is not almost homogeneous.

Here is an example with $b_2(Y) = 1$. We let Y be the Mukai–Umemura Fano threefold of type V_{22}, [23]. Here $h^0(T_Y) = 3$ and Y is almost homogeneous with $\mathrm{Aut}^0(Y) = \mathrm{SL}_2(\mathbb{C})$. Since T_Y is known to be stable (see e.g. [27]), again all assumptions are satisfied and $X = \mathbb{P}(T_Y)$ is not almost homogeneous.

3.2 *The Albanese Map for Almost Homogeneous Manifolds*

A well-known theorem of Barth–Oeljeklaus determines the structure of the Albanese map of an almost homogeneous Kähler manifold.

Theorem 3 ([4]). *Let X be an almost homogeneous compact Kähler manifold. Then the Albanese map $\alpha: X \to A$ is a fiber bundle. The fibers are connected, simply-connected and projective.*

Remark 2. All fibers X_a of α are almost homogeneous.

Proof. Let $x, y \in X_a$ be two general points of a general fiber X_a. Then there exists $f \in \mathrm{Aut}(X)$ with $f(x) = y$. Since the automorphism f is fiber preserving, we obtain an automorphism of X_a mapping x to y. Hence the general fiber X_a is almost homogeneous. Therefore, since α is a fiber bundle, all fibers are almost homogeneous.

3.3 The Case $q(X) = 0$

If the irregularity of X is $q(X) = 0$, the Albanese map is trivial, and it follows that X itself is simply-connected and projective.

Lemma 2. *Let X be an almost homogeneous compact Kähler manifold with contact structure. If $q(X) = 0$ and $b_2(X) \geq 2$, then $X \cong \mathbb{P}(T_Y)$ is a projectivised tangent bundle.*

Proof. X being projective, the results of [20] apply. Combining them with [9] (cf. Corollary 4) yields the desired result.

Remark 3. The case where $q(X) = 0$ and $b_2(X) = 1$ remains to be studied. Here X is an almost homogeneous Fano manifold. It would be interesting to find out whether the results of [5] apply. That is, one has to check whether Aut(X) is reductive and whether the map associated with the contact line bundle L is generically finite.

In order to study the second property, consider the long exact sequence

$$0 \to H^0(X, F) \to H^0(X, T_X) \to H^0(X, L) \to \ldots$$

If $H^0(X, F) \neq 0$ then X has more than one contact structure [22], Proposition 2.2, hence Corollary 4.5 of [19] implies that $X \cong \mathbb{P}_{2n+1}$ or $X \cong \mathbb{P}(T_Y)$.

If $H^0(X, F) = 0$ then L has "many sections" and the map associated with L is expected to be generically finite.

3.4 The Case $q(X) \geq 1$

If the irregularity of X is positive, then the Albanese map $\alpha: X \to A$ is a fiber bundle. We denote its fiber by X_a.

Lemma 3. *Let X be an almost homogeneous compact Kähler manifold with contact structure and $q(X) \geq 1$. If the fiber X_a of the Albanese map fulfills $b_2(X_a) = 1$, then $X \cong \mathbb{P}(T_A) = \mathbb{P}_n \times A$, where A is the Albanese torus of X.*

Proof. Since $b_2(X_a) = 1$, then X_a (being uniruled) is a Fano manifold. We may therefore apply Proposition 2.11 of [20] (which works perfectly in our situation) to conclude that $\alpha: X \to A$ is a $\mathbb{P}_n$-bundle. The proof of Theorem 2.12 in [20] can now be adapted to conclude that $X \cong \mathbb{P}(T_A)$. To be more specific, we already know in our situation that $X = \mathbb{P}(\mathscr{E})$ with $\mathscr{E} = \alpha_*(L)$. The only thing to be verified is the isomorphism $\mathscr{E} \simeq T_A$. But this is seen as in the last part of the proof of Theorem 2.12 in [20], since Sect. 2.1 of [20] works on any manifold.

So $X \simeq \mathbb{P}(T_A)$ and $X \cong \mathbb{P}_n \times A$.

It remains to study the case where the fiber X_a fulfills $b_2(X_a) \geq 2$. In this case we consider a relative Mori contraction (over A; the projection is a projective morphism, [24], (4.12))

$$\varphi: X \to Y.$$

Lemma 4. *We have* $\dim X > \dim Y$.

Proof. The lemma follows from the fact that the restriction map $\varphi_a = \varphi|X_a$ is not birational. This can be shown by the same arguments as in Lemma 2.10 of [20] using the length of the contraction and the restriction of the contact line bundle to the fiber X_a. Again the projectivity of X is not needed in Lemma 2.10.

As above, we may now apply Proposition 2.11 of [20] and conclude that the general fiber of φ is $\mathbb{P}_n$. It remains to check that φ is a $\mathbb{P}_n$-bundle and $X \cong \mathbb{P}(T_Y)$. This is done again as in Theorem 2.12 of [20] with Fujita's result generalized to the Kähler setting by Lemma 5. Also the compactness assumption in [11] is not necessary, this will be important later.

Lemma 5. *Let X be a complex manifold, $f: X \to S$ a proper surjective map to a normal complex space S. Let L be a relatively ample line bundle on X such that $(F, L_F) \simeq (\mathbb{P}_r, \mathscr{O}(1))$ for a general fiber F of f. If f is equidimensional, then f is a $\mathbb{P}_r$-bundle.*

Proof. Since the statement is local in S, we may assume S to be Stein. Then we can simply copy the proof of Lemma 2.12 in [11].

In total, we obtain

Theorem 4. *Let X be a compact almost homogeneous Kähler contact manifold, $b_2(X) \geq 2$. Then $X = \mathbb{P}(T_Y)$ with a compact Kähler manifold Y.*

The arguments above actually also show the following.

Theorem 5. *Let X be a compact Kähler contact manifold. Let $\phi: X \to Y$ be a surjective map with connected fibers such that $-K_X$ is ϕ-ample and such that $\rho(X/Y) = 1$ (we do not require the normal variety Y to be Kähler). Then Y is smooth and $X = \mathbb{P}(T_Y)$.*

One might wonder whether this is still true when X is Moishezon or bimeromorphic to a Kähler manifold. Although there is no apparent reason why the theorem should not hold in this context, at least the methods of proof completely fail. More generally, also the assumption that X is almost homogeneous should be unnecessary. If X is still Kähler, a Mori theory in the non-algebraic case seems unavoidable. Already the question whether X is uniruled is hard.

3.5 Conclusion and Open Questions

1. In all but one case we find that a compact almost homogeneous Kähler contact manifold X has the structure of a projectivised tangent bundle. The remaining case where $q(X) = 0$ and $b_2(X) = 1$ is discussed in Remark 3.
2. Can one classify all Y (necessarily almost homogeneous) such that $\mathbb{P}(T_Y)$ is almost homogeneous? The case where $\dim Y = 2$ will be treated in the next section. One might also expect that if $Y = G/P$, then X should be almost homogeneous. In case Y is a Grassmannian or a quadric, this has been checked by Goldstein [13]. Of course, if $Y = \mathbb{P}_n$, then X is even homogeneous.

4 Almost Homogeneous Contact Threefolds

In this section we specialize to almost homogeneous contact manifolds in dimension 3.

Theorem 6. *Let X be a smooth compact Kähler threefold which is of the form $X = \mathbb{P}(T_Y)$ for some compact (Kähler) surface Y.*

1. If X is almost homogeneous, then Y is a minimal surface or a blow-up of $\mathbb{P}_2$ or $Y = \mathbb{F}_n = \mathbb{P}(\mathcal{O}_{\mathbb{P}_1} \oplus \mathcal{O}_{\mathbb{P}_1}(-n))$ for some $n \geq 0$, $n \neq 1$.
2. If Y is minimal, then X is almost homogeneous if and only if Y is one of the following surfaces.
   - *$Y = \mathbb{P}_2$*
   - *$Y = \mathbb{F}_n$ for some $n \geq 0$, $n \neq 1$*
   - *Y is a torus*
   - *$Y = \mathbb{P}(\mathcal{E})$ with $\mathcal{E}$ a vector bundle of rank 2 over an elliptic curve which is either a direct sum of two topologically trivial line bundles or the non-split extension of two trivial line bundles.*

Proof. Suppose X is almost homogeneous. Then Y is almost homogeneous, too (Lemma 1). By Potters' classification [29], Y is one of the following.

1. $Y = \mathbb{P}_2$
2. $Y = \mathbb{F}_n = \mathbb{P}(\mathcal{O}_{\mathbb{P}_1} \oplus \mathcal{O}_{\mathbb{P}_1}(-n))$ for some $n \geq 0, n \neq 1$
3. Y is a torus

4. $Y = \mathbb{P}(\mathscr{E})$ with $\mathscr{E}$ a vector bundle of rank 2 over an elliptic curve which is either a direct sum of two topologically trivial line bundles or the non-split extension of two trivial line bundles
5. Y is a certain blow-up of $\mathbb{P}_2$ or of $\mathbb{F}_n$.

This already shows the first claim of the theorem, and it suffices to assume Y to be a minimal surface of the list and to check whether $X = \mathbb{P}(T_Y)$ is almost homogeneous. In cases (1) and (3) this is clear; X is even homogeneous.

To proceed further, consider the tangent bundle sequence

$$0 \to T_{X/Y} \to T_X \to \pi^*(T_Y) \to 0.$$

Notice

$$h^0(T_{X/Y}) = h^0(-K_{X/Y}) = h^0(S^2 T_Y \otimes K_Y).$$

Applying π_* and observing that the connecting morphism

$$T_Y \to R^1\pi_*(T_{X/Y})$$

(induced by the Kodaira-Spencer maps) vanishes since π is locally trivial, it follows that

$$H^0(X, T_X) \to H^0(X, \pi^*(T_Y)) = H^0(Y, T_Y)$$

is surjective. If therefore

$$H^0(X, T_{X/Y}) \simeq H^0(Y, S^2 T_Y \otimes K_Y) \neq 0, \tag{$*$}$$

the tangent bundle T_X is obviously generically spanned and therefore X is almost homogeneous.

In case (4), ($*$) is now easily verified: Let $p: \mathbb{P}(\mathscr{E}) \to C$ be the $\mathbb{P}_1$-fibration over the elliptic curve C. The tangent bundle sequence reads

$$0 \to -K_Y \to T_Y \to \mathscr{O}_Y \to 0.$$

Since T_Y is generically spanned, the map $H^0(\mathscr{O}_Y) \to H^1(-K_Y)$ must vanish, so that the sequence splits:

$$T_Y \simeq -K_Y \oplus \mathscr{O}_Y.$$

Thus $S^2 T_Y \otimes K_Y \simeq -K_Y \oplus \mathscr{O}_Y \oplus K_Y$ and ($*$) follows.

Now if $Y = \mathbb{F}_n$ as in (2), let $p: Y \to \mathbb{P}_1$ be the natural projection. The relative tangent sequence then reads

$$0 \to T_{Y/\mathbb{P}_1} \to T_Y \to p^*\mathscr{O}_{\mathbb{P}_1}(2) \to 0. \tag{$**$}$$

Taking the second symmetric power and tensorizing with K_Y yields

$$0 \to T_Y \otimes T_{Y/\mathbb{P}_1} \otimes K_Y \to S^2 T_Y \otimes K_Y \to p^* \mathcal{O}_{\mathbb{P}_1}(4) \otimes K_Y \to 0,$$

so, by $(**)$, we obtain an inclusion

$$H^0(T_{Y/\mathbb{P}_1}^{\otimes 2} \otimes K_Y) \subset H^0(S^2 T_Y \otimes K_Y).$$

Now by the relative Euler sequence, $T_{Y/\mathbb{P}_1} \simeq \mathcal{O}_Y(2) \otimes p^* \mathcal{O}_{\mathbb{P}_1}(n)$, and thus

$$H^0(T_{Y/\mathbb{P}_1}^{\otimes 2} \otimes K_Y) \simeq H^0(\mathcal{O}_Y(2) \otimes p^* \mathcal{O}_{\mathbb{P}_1}(n-2)).$$

Now since

$$p_*(\mathcal{O}_Y(2) \otimes p^* \mathcal{O}_{\mathbb{P}_1}(n-2)) \simeq \mathcal{O}_{\mathbb{P}_1}(n-2) \oplus \mathcal{O}_{\mathbb{P}_1}(-2) \oplus \mathcal{O}_{\mathbb{P}_1}(-n-2),$$

we have shown $(*)$ to be true for $n \geq 2$. If $n = 0$, i.e., $Y \simeq \mathbb{P}_1 \times \mathbb{P}_1$, the sequence $(**)$ splits and an easy calculation shows that $(*)$ is satisfied also in this case.

Remark 4. The case that Y is a non-minimal rational surface in Theorem 6 could be further studied, but this is a rather tedious task.

5 Deformations I: The Rational Case

We consider a family $\pi: \mathcal{X} \to \Delta$ of compact manifolds over the unit disc $\Delta \subset \mathbb{C}$. As usual, we let $X_t = \pi^{-1}(t)$. We shall assume X_t to be a projective manifold for all t, so we are only interested in projective families here. If now X_t is a contact manifold for $t \neq 0$, when is X_0 still a contact manifold?

If $b_2(X_t) = 1$, there is a counterexample due to [26], see also [18]. Here the X_t are 7-dimensional rational-homogeneous contact manifolds and X_0 is a non-homogeneous non-contact manifold. If one believes that any Fano contact manifold with $b_2 = 1$ is rational-homogeneous, then due to the results of Hwang and Mok, this is the only example where a limit of contact manifolds with $b_2 = 1$ is not contact.

If $b_2(X_t) \geq 2$, it is no longer true that the limit X_0 is always a contact manifold, as can be seen from the following example: We let $\mathcal{Y} \to \Delta$ be a family of compact manifolds such that $Y_t \simeq \mathbb{P}_1 \times \mathbb{P}_1$ for $t \neq 0$ and $Y_0 \simeq \mathbb{F}_2$. Then there exist line bundles $\mathcal{L}_1$ and $\mathcal{L}_2$ on $\mathcal{Y}$ such that $\mathcal{L}_1|Y_t \simeq \mathcal{O}_{\mathbb{P}_1 \times \mathbb{P}_1}(2,0)$ and $\mathcal{L}_2|Y_t \simeq \mathcal{O}_{\mathbb{P}_1 \times \mathbb{P}_1}(0,2)$ for every $t \neq 0$. If we let $\mathcal{X} := \mathbb{P}(\mathcal{L}_1 \oplus \mathcal{L}_2)$, then $X_t \simeq \mathbb{P}(T_{Y_t})$ for $t \neq 0$, but $X_0 \not\simeq \mathbb{P}(T_{Y_0})$.

However $\mathbb{P}(T_{\mathbb{P}_1 \times \mathbb{P}_1})$ is not homogeneous; in fact by Proposition 1, $\mathbb{P}(T_{\mathbb{P}_n})$ is the only homogeneous rational contact manifold with $b_2 \geq 2$. In this prominent case we prove global projective rigidity, i.e., $X_0 = \mathbb{P}(T_{\mathbb{P}_n})$, unless X_0 is the projectivization of some unstable bundle, so that both contact structures coming from the two projections $\mathbb{P}(T_{\mathbb{P}_n}) \to \mathbb{P}_n$ survive in the limit. In the "unstable case", the contact structure does not survive. The special case where X_0 is Fano is due to Wiśniewski [31]; here global rigidity always holds.

There is a slightly different point of view, asking whether projective limits of rational-homogeneous manifolds are again rational-homogeneous. As before, if $b_2(X_t) = 1$, this is true by the results of Hwang and Mok with the 7-dimensional exception. In case $b_2(X_t) \geq 2$, this is false in general (e.g. for $\mathbb{P}_1 \times \mathbb{P}_1$), but the picture under which circumstances global rigidity is still true is completely open.

Theorem 7. *Let $\pi\colon \mathscr{X} \to \Delta$ be a family of compact manifolds. Assume $X_t \simeq \mathbb{P}(T_{\mathbb{P}_n})$ for $t \neq 0$. If X_0 is projective, then either $X_0 \simeq \mathbb{P}(T_{\mathbb{P}_n})$ or $X_0 \simeq \mathbb{P}(V)$ with some unstable vector bundle V on $\mathbb{P}_n$.*

Proof. Since $K_{\mathscr{X}}$ is not π-nef, there exists a relative Mori contraction (see [24], (4.12), we may shrink Δ)

$$\Phi\colon \mathscr{X} \to \mathscr{Y}$$

over Δ. Put $\Delta^* = \Delta \setminus \{0\}$ and $\mathscr{X}^* = \mathscr{X} \setminus X_0$; $\mathscr{Y}^* = \mathscr{Y} \setminus Y_0$. Now $\phi_t = \Phi|X_t$ is a Mori contraction for any t (cf. [21], (12.3.4), but this is pretty clear in our simple situation), *unless possibly ϕ_t is biholomorphic* for $t \neq 0$.

Now since $\mathscr{X}$, Δ and π are smooth, the latter case cannot occur by [32], (1.3), so ϕ_t is the contraction of an extremal ray for any $t \in \Delta$. Let $\tau\colon \mathscr{Y} \to \Delta$ be the induced projection and set $Y_t = \tau^{-1}(t)$, so that $Y_t \simeq \mathbb{P}_n$ for $t \neq 0$. Recall that $X_t \simeq \mathbb{P}(T_{\mathbb{P}_n})$ for $t \neq 0$ and that $\mathbb{P}(T_{\mathbb{P}_n})$ carries two projections to $\mathbb{P}_n$. Since $\mathscr{Y}$ is normal, the normal variety Y_0 must also have dimension n.

From the exponential sequence, Hodge decomposition and the topological triviality of the family $\mathscr{X}$, it follows that

$$\mathrm{Pic}(\mathscr{X}) \simeq H^2(\mathscr{X}, \mathbb{Z}) \simeq \mathbb{Z}^2$$

and that

$$\mathrm{Pic}(X_0) \simeq H^2(X_0, \mathbb{Z}^2) \simeq \mathbb{Z}^2.$$

Furthermore, the restriction $\mathrm{Pic}(\mathscr{X}) \to \mathrm{Pic}(X_0)$ is bijective. As an immediate consequence, we can write

$$-K_{\mathscr{X}} = n\mathscr{H}$$

with a line bundle $\mathscr{H}$ on $\mathscr{X}$. Let $\mathscr{H}_t = \mathscr{H}|X_t$ so that $\mathscr{H}_t \simeq \mathscr{O}_{\mathbb{P}(T_{\mathbb{P}_n})}(1)$ for $t \neq 0$.

Claim. $Y_0 \simeq \mathbb{P}_n$.

In fact, by our previous considerations, there is a unique line bundle $\mathscr{L}$ on $\mathscr{X}$ such that

$$\mathscr{L}|X_t = \phi_t^*(\mathscr{O}_{\mathbb{P}_n}(1))$$

for $t \neq 0$. Moreover $\mathscr{L}|X_0 = \phi_0^*(\mathscr{L}')$ with some ample line bundle $\mathscr{L}'$ on Y_0. Therefore by semi-continuity,

$$h^0(\mathscr{L}') = h^0(\mathscr{L}|X_0) \geq n+1$$

and

$$c_1(\mathscr{L}')^n = 1.$$

Hence by results of Fujita [12], (I.1.1), see also [6], (III.3.1), we have $(Y_0, \mathscr{L}') \simeq (\mathbb{P}_n, \mathscr{O}(1))$.

In particular we obtain

Sub-Corollary 1. *$\mathscr{Y}$ is smooth and $\mathscr{Y} \simeq \mathbb{P}_n \times \Delta$.*

Next we notice that the general fiber of ϕ_0 must be $\mathbb{P}_{n-1}$, since it is a smooth degeneration of fibers of ϕ_t (by the classical theorem of Hirzebruch and Kodaira [14]).

One main difficulty is that ϕ_0 might not be equidimensional. If we know equidimensionality, we may apply ([11], 2.12) to conclude that $X_0 = \mathbb{P}(\mathscr{E}_0)$ with a locally free sheaf $\mathscr{E}_0$ on Y_0.

We introduce the torsion free sheaf

$$\mathscr{F} = \Phi_*(\mathscr{H}) \otimes \mathscr{O}_{\mathscr{Y}}(-1).$$

Since

$$\operatorname{codim} \Phi^{-1}(\operatorname{Sing}(\mathscr{F})) \geq 2,$$

the sheaf $\mathscr{F}$ is actually reflexive and of course locally free outside Y_0. In the following Sublemma we will prove that $\mathscr{F}$ is actually locally free.

Sub-Lemma 1. *$\mathscr{F}$ is locally free and therefore $\mathscr{X} = \mathbb{P}(\mathscr{F})$.*

Proof. As explained above, it is sufficient to show that

$$\phi_0 : X_0 \to \mathbb{P}_n$$

is equidimensional. So let F_0 be an irreducible component of a fiber of ϕ_0. Then F_0 gives rise to a class

$$[F_0] \in H^{2k}(X_0, \mathbb{Q}),$$

where we denote by k the codimension of F_0 in X_0. Obviously $k \leq n$, and we must exclude the case that $k < n$.

So we assume in the following that $k < n$. Then, since X_0 is homeomorphic to $\mathbb{P}(T_{\mathbb{P}_n})$, the Leray–Hirsch theorem (cf. [16], Theorem 17.1.1) gives

$$\dim H^{2k}(X_0, \mathbb{Q}) = k + 1.$$

Now if we denote by H the class of a hyperplane in $\mathbb{P}_n$, and by L the class of an ample divisor on X_0, then the classes

$$L^k, L^{k-1}.(\phi_0^* H), \ldots, L.(\phi_0^* H)^{k-1}, (\phi_0^* H)^k \tag{3}$$

form a basis of $H^{2k}(X_0, \mathbb{Q})$, which can be seen as follows: By the dimension formula given above, it is sufficient to show linear independency, so assume that we are given $\lambda_0, \ldots, \lambda_k \in \mathbb{Q}$ such that

$$\sum_{\ell=0}^{k} \lambda_\ell L^{k-\ell}.(\phi_0^* H)^\ell = 0. \tag{4}$$

Now let $\ell_0 \in \{0, \ldots, k\}$. By induction, we assume that $\lambda_\ell = 0$ for all $\ell < \ell_0$. Then intersecting (4) with $L^{n-k-1+\ell_0}.(\phi_0^* H)^{n-\ell_0}$ yields

$$\lambda_{\ell_0} L^{n-1}.(\phi_0^* H)^n = 0,$$

thus $\lambda_{\ell_0} = 0$ since $L^{n-1}.(\phi_0^* H)^n > 0$.

So (3) is indeed a basis of $H^2(X_0, \mathbb{Q})$ and we can write

$$[F_0] = \sum_{\ell=0}^{k} \alpha_\ell L^{k-\ell}.(\phi_0^* H)^\ell \tag{5}$$

for some $\alpha_0, \ldots, \alpha_k \in \mathbb{Q}$. We now let $\ell_0 \in \{0, \ldots, k\}$ and assume that $\alpha_\ell = 0$ for $\ell < \ell_0$. We observe that $[F_0].(\phi_0^* H)^{n-\ell_0} = 0$ since F_0 is contained in a fiber of ϕ_0 and $\ell_0 \leq k < n$. Hence, intersecting (5) with $L^{n-k-1+\ell_0}.(\phi_0^* H)^{n-\ell_0}$ yields

$$0 = \alpha_{\ell_0} L^{n-1}.(\phi_0^* H)^n,$$

so we deduce $\alpha_{\ell_0} = 0$ as before. Therefore by induction, we have $[F_0] = 0$, which is impossible, X_0 being projective.

Now we set $V = \mathscr{F}|X_0$. If the bundle V is semi-stable, then $V \simeq T_{\mathbb{P}_n}$ and the theorem is settled.

Suppose in Theorem 7 that $X_0 \simeq \mathbb{P}(V)$ with an unstable bundle V (we will show in Sect. 6 that this can indeed occur). Then X_0 does not carry a contact structure.

In fact, otherwise $X_0 \simeq \mathbb{P}(T_S)$ with some projective variety S, [20]. Hence X_0 has two extremal contractions, and therefore X_0 is Fano. Hence T_S is ample and thus $S \simeq \mathbb{P}_n$ (or apply Wiśniewski's theorem). Therefore we may state the following

Corollary 1. *Let $\pi: \mathscr{X} \to \Delta$ be a family of compact manifolds. Assume $X_t \simeq \mathbb{P}(T_{\mathbb{P}_n})$ for $t \neq 0$. If X_0 is a projective contact manifold, then $X_0 \simeq \mathbb{P}(T_{\mathbb{P}_n})$.*

In the situation of Theorem 7, we had two contact structures on $X_t \simeq \mathbb{P}(T_{\mathbb{P}_n})$, given by the two projections to $\mathbb{P}_n$. This phenomenon is quite unique because of the following result [20], Proposition 2.13.

Theorem 8. *Let X be a projective contact manifold of dimension $2n-1$ admitting two extremal rays in the cone of curves $\overline{NE}(X)$. Then $X \simeq \mathbb{P}(T_{\mathbb{P}_n})$.*

Here is an extension of Theorem 8 to the non-algebraic case.

Theorem 9. *Let X be a compact contact Kähler manifold admitting two contractions $\phi_i: X \to Y_i$ to normal compact Kähler spaces Y_i. This is to say that $-K_X$ is ϕ_i-ample and that $\rho(X/Y_i) = 1$. Then X is projective and therefore $X = \mathbb{P}(T_{\mathbb{P}_n})$.*

Proof. We already know by Theorem 5 that $X = \mathbb{P}(T_{Y_i})$. Let $F \simeq \mathbb{P}_{n-1}$ be a fiber of ϕ_2. Then the restriction $\phi_1|F$ is finite. We claim that Y_1 must be projective. In fact, consider the rational quotient, say $f: Y_1 \dashrightarrow Z$, which is an almost holomorphic map to a compact Kähler manifold Z. By construction, the map f contracts the images $\phi_1(F)$, hence $\dim Z \leq 1$. But then Z is projective and therefore Y_1 is projective, too (e.g. by arguing that Y_1 cannot carry a holomorphic 2-form and applying Kodaira's theorem that Kähler manifolds without 2-forms are projective).

By symmetry, Y_2 is projective, too. Since the morphisms ϕ_i induce a finite map $X \to Y_1 \times Y_2$ (onto the image of X), the variety X is also projective.

Any projective contact manifold X with $b_2(X) \geq 2$ is of the form $X = \mathbb{P}(T_Y)$. Therefore it is natural ask for generalizations of Theorem 7, substituting the projective space by other projective varieties.

Proposition 3. *Let $\pi: \mathscr{X} \to \Delta$ be a projective family of compact manifolds X_t of dimension $2n-1$. Assume that $X_t \simeq \mathbb{P}(T_{Y_t})$ for $t \neq 0$ with (necessarily projective) manifolds $Y_t \neq \mathbb{P}_n$. Assume that $H^q(X_t, \mathscr{O}_{X_t}) = 0$ for $q = 1, 2$ for some (hence all) t. Then the following statements hold.*

1. *There exists a relative contraction $\Phi: \mathscr{X} \to \mathscr{Y}$ over Δ such that $\Phi|X_t$ is the given $\mathbb{P}_{n-1}$-bundle structure for $t \neq 0$.*
2. *If $\phi_0 := \Phi|X_0$ is equidimensional, then $X_0 \simeq \mathbb{P}(\mathscr{E}_0)$ with a rank-n bundle $\mathscr{E}$ over the projective manifold Y_0; and Y_0 is the limit manifold of a family $\tau: \mathscr{Y} \to \Delta$ such that $Y_t \simeq \tau^{-1}(t)$ for $t \neq 0$. In other words, $\mathscr{X} \simeq \mathbb{P}(\mathscr{E})$ such that $\mathscr{E} = T_{\mathscr{Y}/\Delta}$ over $\Delta \setminus \{0\}$.*

Proof. Since $Y_t \neq \mathbb{P}_n$ by assumption, every X_t, $t \neq 0$, has a unique Mori contraction, the projection $\psi_t: X_t \to Y_t$, by Theorem 8. Notice that since $H^q(X_t, \mathscr{O}_{X_t}) = 0$ for $q = 1, 2$, we have $\mathrm{Pic}(\mathscr{X}) \simeq H^2(\mathscr{X}, \mathbb{Z})$ and the restriction map

$\mathrm{Pic}(\mathscr{X}) \to \mathrm{Pic}(X_0)$ is bijective. Therefore, as in the proof of Theorem 7, we obtain a relative Mori contraction

$$\Phi: \mathscr{X} \to \mathscr{Y}$$

over Δ, and necessarily $\Phi|X_t = \phi_t$ for all $t \neq 0$ (we use again [32], (1.3)). This already shows Claim (1).

If ϕ_0 is equidimensional, we apply—as in the proof of Theorem 7—[6], (III.3.2.1), to conclude that there exists a locally free sheaf $\mathscr{E}_0$ of rank n on Y_0 such that $X_0 \simeq \mathbb{P}(\mathscr{E}_0)$, proving (2).

Theorem 10. *Let $\pi: \mathscr{X} \to \Delta$ be a projective family of compact manifolds X_t of dimension $2n-1$. Assume that $X_t \simeq \mathbb{P}(T_{Y_t})$ for $t \neq 0$ with (necessarily projective) manifolds $Y_t (\neq \mathbb{P}_n)$. Assume that $H^q(X_t, \mathscr{O}_{X_t}) = 0$ for $q = 1, 2$ for some (hence all) t. Assume moreover that*

1. *$\dim X_0 \leq 5$, or*
2. *$b_{2j}(Y_t) = 1$ for some $t \neq 0$ and all $1 \leq j < \frac{n}{2}$.*

Then there exists a relative contraction $\Phi: \mathscr{X} \to \mathscr{Y}$ over Δ such that $\Phi|X_t$ is the given $\mathbb{P}_{n-1}$-bundle structure for $t \neq 0$. Moreover there is a locally free sheaf $\mathscr{E}$ on $\mathscr{Y}$ such that $\mathscr{X} \simeq \mathbb{P}(\mathscr{E})$ and $\mathscr{E}|Y_t \simeq T_{Y_t}$ for all $t \neq 0$.

Proof. By the previous proposition it suffices to show that $\phi_0 = \Phi|X_0$ is equidimensional.

1. First suppose that $\dim X_0 \leq 5$. Then $1 \leq \dim Y_0 \leq 3$. The case $\dim Y_0 = 1$ is trivial. If $\dim Y_0 = 2$, then all fibers must have codimension 2, because ϕ_0 does not contract a divisor (the relative Picard number being 1). If $\dim Y_0 = 3$, then by [1], (5.1), we cannot have a 3-dimensional fiber. Since again there is no 4-dimensional fiber, ϕ_0 must be equidimensional also in this case.
2. If $b_{2j}(Y_t) = 1$ for some t and all $1 \leq j \leq \frac{n}{2}$, then $b_{2k}(X_t) = k+1$ for $k < n$ and we may simply argue as in Sublemma 1 to conclude that ϕ_0 is equidimensional (the smoothness of Y_0 is not essential in the reasoning of Sublemma 1).

6 Degenerations of $T_{\mathbb{P}_n}$

In view of Theorem 7, we can ask the question which bundles can occur as degenerations of $T_{\mathbb{P}_n}$, i.e., for which rank-n bundles V on $\mathbb{P}_n$ there exists a rank-n bundle $\mathscr{V}$ on $\mathbb{P}_n \times \Delta$ such that

$$\mathscr{V}_t := \mathscr{V}|\mathbb{P}_n \times \{t\} \simeq \begin{cases} T_{\mathbb{P}_n}, & \text{for } t \neq 0, \\ V, & \text{for } t = 0. \end{cases}$$

In the case that $n \geq 3$ is odd, it was already observed by Hwang in [17] that one can easily construct a nontrivial degeneration of $T_{\mathbb{P}_n}$ as follows: We consider the null correlation bundle on $\mathbb{P}_n$, which is a rank-$(n-1)$ bundle N on $\mathbb{P}_n$ given by a short exact sequence

$$0 \longrightarrow N \longrightarrow T_{\mathbb{P}_n}(-1) \longrightarrow \mathscr{O}_{\mathbb{P}_n}(1) \longrightarrow 0.$$

(cf. [25], (I.4.2)). The existence of this sequence now implies that $T_{\mathbb{P}_n}$ can be degenerated to $N(1) \oplus \mathscr{O}_{\mathbb{P}_n}(2)$.

When n is even, matters become more complicated, but we can still obtain nontrivial degenerations:

Proposition 4. *Let $n \geq 2$. Then there exists a rank-n bundle $\mathscr{V}$ on $\mathbb{P}_n \times \Delta$ such that $\mathscr{V}_t \simeq T_{\mathbb{P}_n}$ for $t \neq 0$ and $h^0(\mathscr{V}_0(-2)) = 1$, so in particular $\mathscr{V}_0 \not\simeq T_{\mathbb{P}_n}$.*

Proof. We construct an inclusion of vector bundles

$$A\colon \Omega^1_{\mathbb{P}_n \times \Delta/\Delta}(2) \oplus \mathscr{O}_{\mathbb{P}_n \times \Delta} \hookrightarrow \mathscr{O}_{\mathbb{P}_n \times \Delta}(1)^{\oplus(n+1)} \oplus \Omega^1_{\mathbb{P}_n \times \Delta/\Delta}(2)$$

via a family $A = (A_t)_{t \in \Delta}$ of matrices

$$A_t = \begin{pmatrix} \alpha_t & \beta_t \\ \gamma_t & \delta_t \end{pmatrix}$$

of sheaf homomorphisms

$$\alpha_t\colon \Omega^1_{\mathbb{P}_n}(2) \to \mathscr{O}_{\mathbb{P}_n}(1)^{\oplus(n+1)}, \qquad \beta_t\colon \mathscr{O}_{\mathbb{P}_n} \to \mathscr{O}_{\mathbb{P}_n}(1)^{\oplus(n+1)},$$

$$\gamma_t\colon \Omega^1_{\mathbb{P}_n}(2) \to \Omega^1_{\mathbb{P}_n}(2), \qquad \delta_t\colon \mathscr{O}_{\mathbb{P}_n} \to \Omega^1_{\mathbb{P}_n}(2),$$

which we define as follows: We take α_t and β_t to be the natural inclusions coming from the Euler sequence and its dual, where we choose the coordinates on $\mathbb{P}_n$ such that

$$\beta_t(\mathscr{O}_{\mathbb{P}_n}) \not\subset \alpha_t(\Omega^1_{\mathbb{P}_n}(2)).$$

This implies that the map

$$\alpha_t \oplus \beta_t\colon \Omega^1_{\mathbb{P}_n}(2) \oplus \mathscr{O}_{\mathbb{P}_n} \to \mathscr{O}_{\mathbb{P}_n}(1)^{\oplus(n+1)}$$

is generically surjective. Since $\Omega^1_{\mathbb{P}_n}(2) \simeq \Lambda^{n-1}(T_{\mathbb{P}_n}(-1))$ is globally generated, a general section in $H^0(\Omega^1_{\mathbb{P}_n}(2))$ has only finitely many zeroes. Since $\Omega^1_{\mathbb{P}_n}(2)$ is homogeneous, we can thus choose the map δ_t in such a way that its zeroes are disjoint from the locus where $\alpha_t \oplus \beta_t$ is not surjective. Finally we let $\gamma_t = t \cdot \mathrm{id}$.

Now in order to show that A is an inclusion of vector bundles, we need to show that for any point $(p,t) \in \mathbb{P}_n \times \Delta$, the matrix

$$A_t(p) = \begin{pmatrix} \alpha_t(p) & \beta_t(p) \\ \gamma_t(p) & \delta_t(p) \end{pmatrix} \in \mathbb{C}^{(2n+1)\times(n+1)}$$

has rank $n+1$. For semicontinuity reasons, shrinking Δ if necessary, we can assume $t = 0$, then the rank condition follows easily from the choice of α_0, β_0, γ_0, δ_0.

We now let

$$\mathscr{V} := \operatorname{coker} A.$$

It remains to investigate the properties of the bundles $\mathscr{V}_t := \mathscr{V}|\mathbb{P}_n \times \{t\}$. For each $t \in \Delta$, we have an exact sequence of vector bundles

$$0 \longrightarrow \Omega^1_{\mathbb{P}_n}(2) \oplus \mathscr{O}_{\mathbb{P}_n} \longrightarrow \mathscr{O}_{\mathbb{P}_n}(1)^{\oplus(n+1)} \oplus \Omega^1_{\mathbb{P}_n}(2) \longrightarrow \mathscr{V}_t \longrightarrow 0. \tag{6}$$

We want to calculate $H^q(\mathscr{V}_t(-1-k))$ for $k = 0, \dots, n$. From the Bott formula we obtain for $(k,q) \in \{0,\dots,n\}^2$:

$$h^q(\Omega^1_{\mathbb{P}_n}(1-k)) = \begin{cases} 1, & \text{for } (k,q) = (1,1), \\ 0, & \text{otherwise.} \end{cases}$$

Now if we tensorize (6) with $\mathscr{O}_{\mathbb{P}_n}(-1-k)$, take the long exact cohomology sequence and observe that $H^q(\delta_0) = 0$ for every q, we get for $(k,q) \in \{0,\dots,n\}^2$:

$$h^q(\mathscr{V}_0(-1-k)) = \begin{cases} n+1, & \text{for } (k,q) = (0,0), \\ 1, & \text{for } (k,q) \in \{(1,0),(1,1),(n,n-1)\}, \\ 0, & \text{otherwise.} \end{cases}$$

Similarly, if we observe that $H^q(\delta_t) = \mathrm{id}$ for $t \neq 0$, we obtain for $t \neq 0$, $(k,q) \in \{0,\dots,n\}^2$:

$$h^q(\mathscr{V}_0(-1-k)) = \begin{cases} n+1, & \text{for } (k,q) = (0,0), \\ 1, & \text{for } (k,q) \in \{(n,n-1)\}, \\ 0, & \text{otherwise.} \end{cases}$$

The proposition now follows from Lemma 6.

Lemma 6. *Let V be a vector bundle on $\mathbb{P}_n$ such that for any $(k,q) \in \{0,\dots,n\}^2$, we have*

$$h^q(V(-1-k)) = \begin{cases} n+1, & \textit{for } (k,q) = (0,0), \\ 1, & \textit{for } (k,q) = (n,n-1), \\ 0, & \textit{otherwise}. \end{cases}$$

Then $V \simeq T_{\mathbb{P}_n}$.

Proof. We consider the Beilinson spectral sequence for the bundle $V(-1)$, which has E_1-term

$$E_1^{pq} = H^q(V(-1+p)) \otimes \Omega_{\mathbb{P}_n}^{-p}(-p)$$

(cf. [25], (II.3.1.3)).

By assumption, $E_1^{pq} = 0$ for $(p,q) \notin \{(0,0), (-n, n-1)\}$ and

$$E_1^{0,0} = \mathcal{O}_{\mathbb{P}_n}^{\oplus(n+1)}, \quad E_1^{-n,n-1} = \mathcal{O}_{\mathbb{P}_n}(-1).$$

In particular, the only nonzero differential occurs at the E_n-term, namely

$$d_n^{-n,n-1}\colon E_n^{-n,n-1} \to E_n^{0,0}.$$

Since $E_\infty^{pq} = 0$ for $p+q \neq 0$ and $E_\infty^{-p,p}$ are the quotients of a filtration of $V(-1)$, the differential $d_n^{-n,n-1}$ induces a short exact sequence

$$0 \longrightarrow \mathcal{O}_{\mathbb{P}_n}(-1) \overset{d_n^{-n,n-1}}{\longrightarrow} \mathcal{O}_{\mathbb{P}_n}^{\oplus(n+1)} \longrightarrow V(-1) \longrightarrow 0. \tag{7}$$

Now since V is locally free, the map $d_n^{-n,n-1}$ cannot have zeroes, so (7) must be an Euler sequence, whence $V(-1) \simeq T_{\mathbb{P}_n}(-1)$.

7 Deformations II: Positive Irregularity

A homogeneous compact contact Kähler manifold X of dimension $2n+1$ with $b_2(X) \geq 2$ is either $\mathbb{P}(T_{\mathbb{P}_{n+1}})$ or a product $A \times \mathbb{P}_n$ with a torus A of dimension $n+1$. Here we study in general the Kähler deformations of $A \times \mathbb{P}_n$, where A is an m-dimensional torus.

Theorem 11. *Let* $\pi\colon \mathcal{X} \to \Delta$ *be a family of compact manifolds over the unit disc* $\Delta \subset \mathbb{C}$. *Assume* $X_t \simeq A_t \times \mathbb{P}_n$ *for* $t \neq 0$, *where* A_t *is a torus of dimension* m. *If* X_0 *is Kähler, then the relative Albanese morphism realises* $\mathcal{X}$ *as a submersion* $\alpha\colon \mathcal{X} \to \mathcal{A}$, *where* $p\colon \mathcal{A} \to \Delta$ *is torus bundle such that* $p^{-1}(t) \simeq A_t$ *for* $t \neq 0$. *Moreover there is a locally free sheaf* $\mathcal{E}$ *over* $\mathcal{A}$ *such that* $\mathcal{X} = \mathbb{P}(\mathcal{E})$, $\mathcal{X}_t \simeq \mathbb{P}(\mathcal{E}_t)$ *for all* t *and* $\mathcal{E}|A_t \simeq \mathcal{O}_{A_t}^{n+1}$ *for* $t \neq 0$.

Proof. Let $m = \dim A_t = q(X_t)$ for $t \neq 0$. Hodge decomposition on X_0 gives $q(X_0) = m$. Let

$$\alpha\colon \mathscr{X} \to \mathscr{A}$$

be the relative Albanese map. Then $\mathscr{A} \to \Delta$ is a torus bundle and

$$\alpha_t = \alpha|X_t\colon X_t \to A_t$$

is the Albanese map for all t. Since α_t is surjective for all $t \neq 0$, the map α is surjective, too, and so is α_0.

For $t \neq 0$, $X_t \simeq A_t \times \mathbb{P}_n$, so $b_2(X_t) - b_2(A_t) = 1$. Since the families $\pi\colon \mathscr{X} \to \Delta$ and $p\colon \mathscr{A} \to \Delta$ are topologically trivial, we also have $b_2(X_0) - b_2(A_0) = 1$. Thus $h^{1,1}(X_0) - h^{1,1}(A_0) = 1$. Choose a Kähler form ω on X_0. Then we find a positive number λ and a closed $(1,1)$-form u on A_0 such that

$$c_1(-K_{X_0}) = [\lambda\omega] + [\alpha^*(u)].$$

Now we apply [7], Theorem 1.1, to conclude that α_0 is projective. The proof of Theorem 1.1 also shows that $-K_{X_0}$ is α_0-ample, which, however, is anyway clear in our situation since

$$\rho(X_0/A_0) = 1.$$

Furthermore, we have

$$(-K_{X_0})^{n+1} = (-K_{X_t})^{n+1} = 0. \tag{8}$$

From this we conclude that α_0 is equidimensional. In fact, let F be an irreducible component of a fiber of α_0 and assume that $d := \dim F \geq n+1$. Then, since $-K_{X_0}$ is α_0-ample, we obtain

$$(-K_{X_0})^d.F > 0,$$

which contradicts (8).

We have thus shown that α is equidimensional and therefore flat. Since $-K_{X_t}$ is divisible by $n+1$ in $\operatorname{Pic}(X_t)$ for $t \neq 0$, so is $-K_{X_0}$ (first argue topologically, then use the fact that $R^2\pi_*(\mathscr{O}_{X_0})$ is locally free, hence cannot have a non-zero section which vanishes on $\Delta \setminus 0$). Now write $-K_{X_0} = \mathscr{L}_0^{n+1}$ and apply Lemma 3.10 to $\mathscr{L}_0$.

Alternatively, argue as follows. Since X_0 is smooth, a general fiber of α_0 is smooth and hence isomorphic to $\mathbb{P}_n$. This means that the analytic set

$$\{\, s \in \mathscr{A} \mid \alpha^{-1}(s) \not\simeq \mathbb{P}_n \,\}$$

has codimension ≥ 2 in $\mathscr{A}$. We can now apply [2], Theorem 2, to conclude that α is a $\mathbb{P}_n$-bundle.

The existence of $\mathscr{E}$ follows from [10], (4.3).

Remark 5. One might hope to weaken the assumption that X_0 is Kähler and just assume X_0 to be in class $\mathscr{C}$. In this context, it should be noticed that Popovici [28] has shown that any global deformation of projective manifolds is automatically in class $\mathscr{C}$. The Kähler version of Popovici's theorem is still open.

Example 2. We cannot conclude in Theorem 11 that $X_0 \simeq A_0 \times \mathbb{P}_n$, even if $m = n = 1$. Take e.g. a rank-2 vector bundle $\mathscr{F}$ over $\mathbb{P}_1 \times \Delta$ such that $\mathscr{F}|\mathbb{P}_1 \times \{t\} = \mathscr{O}^2$ for $t \neq 0$ and $\mathscr{F}|\mathbb{P}_1 \times \{0\} = \mathscr{O}(1) \oplus \mathscr{O}(-1)$. Let $\eta\colon A \to \mathbb{P}_1$ be a two-sheeted covering from an elliptic curve A and set $\mathscr{E} = (\eta \times \mathrm{id})^*(\mathscr{F})$. Then $\mathscr{X} = \mathbb{P}(\mathscr{E})$ is a family of compact surfaces X_t such that $X_t = A \times \mathbb{P}_1$ for $t \neq 0$ but X_0 is not a product. Notice also that X_0 is not almost homogeneous.

It is a trivial matter to modify this example to obtain a map to a 2-dimensional torus which is a product of elliptic curves. Therefore the limit of a Kähler contact manifold with positive irregularity might not be a contact manifold again.

Corollary 2. *Assume the situation of Theorem 11. Then the following assertions are equivalent.*

1. $X_0 \simeq A_0 \times \mathbb{P}_n$.
2. $\mathscr{E}_0$ *is semi-stable for some Kähler class* ω.
3. X_0 *is homogeneous.*
4. X_0 *is almost homogeneous.*

Proof. (1) implies (2). Under the assumption of (1), there is a line bundle L on A_0 such that $\mathscr{E}_0 \simeq L^{\oplus n+1}$. Hence $\mathscr{E}$ is semi-stable for actually any choice of ω.

(2) implies (3). From the semi-stability of $\mathscr{E}_0$ and $h^0(\mathscr{E}_0) \geq n+1$, it follows easily that $\mathscr{E}_0$ is trivial and that X_0 is homogeneous as product $A_0 \times \mathbb{P}_n$. In fact, choose $n+1$ sections of $\mathscr{E}_0$ and consider the induced map $\mu\colon \mathscr{O}_{A_0}^{n+1} \to \mathscr{E}_0$. By the stability of $\mathscr{E}_0$, the map μ is generically surjective. Hence $\det \mu \neq 0$, hence an isomorphism, so that μ itself is an isomorphism.

The implication "(3) implies (4)" is obvious.

(4) implies (1). Consider the tangent bundle sequence

$$0 \to T_{X_0/A_0} \to T_{X_0} \to \alpha_0^*(T_{A_0}) \to 0.$$

Since X_0 is almost homogeneous, all vector fields on A_0 must lift to X_0. Consequently the connecting map

$$H^0(X_0, \pi^*(T_{A_0})) \to H^1(X_0, T_{X_0/A_0})$$

vanishes, and therefore the tangent bundle sequence splits. Let $\mathscr{F} = \alpha_0^*(T_{A_0})$. As a limit of the foliations $\mathscr{F}_t := \alpha_t^*(T_{A_t}) = T_{X_t/\mathbb{P}_n}$, also $\mathscr{F} \subset T_{X_0}$ is a foliation and it has compact leaves (the limits of tori in $A_t \times \mathbb{P}_n$). By [15], 2.4.3, there exists an equi-dimensional holomorphic map $f: X_0 \to Z_0$ to a compact variety Z_0 such that the set-theoretical fibers F of f are leaves of $\mathcal{F}$. Since the fibers F have an étale map to A_0, they must be tori again. It is now immediate that $Z_0 = \mathbb{P}_n$ and that $X_0 = A_0 \times \mathbb{P}_n$.

Corollary 3. *Assume in Theorem 11 that* $m = 2$ *and* $n = 1$. *Then either* $X_0 \simeq A_0 \times \mathbb{P}_1$, *or* $X_0 = \mathbb{P}(\mathcal{E}_0)$ *and one of the following holds:*

1. *There is a torus bundle* $p: A_0 \to B_0$ *to an elliptic curve* B_0 *and the rank-2 bundle* $\mathcal{E}_0$ *on* A_0 *sits in an extension*

$$0 \to p^*(\mathcal{L}_0) \to \mathcal{E}_0 \to p^*(\mathcal{L}_0^*) \to 0$$

with $\deg \mathcal{L}_0 > 0$.

2. *The rank 2-bundle* $\mathcal{E}_0$ *sits in an extension*

$$0 \to \mathcal{S} \to \mathcal{E}_0 \to \mathcal{I}_Z \otimes \mathcal{S}^* \to 0$$

with an ample line bundle $\mathcal{S}$ *and a finite non-empty set* Z *of degree* $\deg Z = c_1(\mathcal{S})^2$.

Proof. By Corollary 2 we may assume that $\mathcal{E}_0$ is not semi-stable for some (or any) Kähler class ω. Let $\mathcal{S}$ be a maximal destabilising subsheaf, which is actually a line bundle, leading to an exact sequence

$$0 \to \mathcal{S} \to \mathcal{E}_0 \to Q \to 0.$$

Notice that $Q \simeq \mathcal{I}_Z \otimes \mathcal{S}^*$, where Z is a finite set or empty. Taking c_2 and observing that $c_2(\mathcal{E}_0) = 0$ gives

$$c_1(\mathcal{S})^2 = \deg Z.$$

The destabilisation property reads

$$c_1(\mathcal{S}) \cdot \omega > 0.$$

Since $h^0(\mathcal{E}_0) \geq 2$, we deduce that $h^0(\mathcal{S}) \geq 2$, in particular, $\mathcal{S}$ is nef, $\mathcal{S}$ being maximal destabilizing.

If $\mathcal{S}$ is ample, there is nothing more to prove, hence we may assume that $\mathcal{S}$ is not ample. $\mathcal{S}$ being nef, $c_1(\mathcal{S})^2 = 0$ and $\mathcal{S}$ defines a submersion $p: A_0 \to B_0$ to an elliptic curve B_0 such that there exists an ample line bundle $\mathcal{L}_0$ with $\mathcal{S} = p^*(\mathcal{L}_0)$. Therefore we obtain an extension

$$0 \to p^*(\mathcal{L}_0) \to \mathcal{E}_0 \to p^*(\mathcal{L}_0^*) \to 0,$$

as required.

Remark 6. The second case in Corollary 3 really occurs. Take a finite map $f: \mathscr{A} \to \mathbb{P}_2 \times \Delta$ over Δ and a rank-2 bundle $\mathscr{F}$ on $\mathbb{P}_2 \times \Delta$ such that $\mathscr{F}|\mathbb{P}_2 \times \{t\} \simeq \mathscr{O}^2$ for $t \neq 0$ and such that $\mathscr{F}_0$ is not trivial. For examples see e.g. [30]. Now $\mathscr{E} = f^*(\mathscr{F})$ gives an example we are looking for.

Corollary 4. *Assume in Theorem 11 that $m = 2$ and $n = 1$. Let $\Phi: T_{\mathscr{X}/\Delta} \to \frac{-K_{\mathscr{X}}}{2}$ be a morphism such that $\Phi|X_t = \phi_t$ is a contact morphism (i.e., defines a contact structure) for $t \neq 0$. Suppose that*

$$\phi_0: T_{X_0} \to \frac{-K_{X_0}}{2}$$

does not vanish identically. Then the kernel $\mathscr{F}_0$ of ϕ_0 is integrable (in contrast to the maximally non-integrable bundle $\mathscr{F}_t$).

Proof. We consider a family (ϕ_t) of morphisms

$$\phi_t: T_{X_t} \to \mathscr{H}_t$$

such that ϕ_t is a contact form for $t \neq 0$ and $-K_{X_t} = 2\mathscr{H}_t$. Consider the (torsion free) kernel $\mathscr{F}_0$ of ϕ_0. We need to show that the induced map

$$\mu: (\bigwedge^2 \mathscr{F}_0)^{**} = \det \mathscr{F}_0 \to \mathscr{H}_0.$$

vanishes. Since the determinant of the kernel $\mathscr{F}_t$ of ϕ_t is isomorphic to $\mathscr{H}_t$, we conclude that

$$\det \mathscr{F}_0 \simeq \mathscr{H}_0 \otimes \mathscr{O}_{X_0}(E) \qquad (*)$$

with an effective (possibly vanishing) divisor E on X_0. Now the induced map

$$\mu: \det \mathscr{F}_0 \to \mathscr{H}_0$$

must have zeroes, otherwise X_0 would be a contact manifold, hence $X_0 \simeq A_0 \times \mathbb{P}_1$. Thus $\mu = 0$ by $(*)$, and $\mathscr{F}_0$ is integrable.

References

1. M. Andreatta, J. Wiśniewski, A view on contractions of higher-dimensional varieties, in *Algebraic Geometry–Santa Cruz 1995*. Proceedings of Symposia in Pure Mathematics, vol. 62, Part 1 (American Mathematical Society, Providence, 1997), pp. 153–183
2. C. Araujo, J. Ramón-Marí, Flat deformations of $\mathbb{P}^n$. arXiv.AG/1212.3593v1
3. D. Barlet, Th. Peternell, M. Schneider, On two conjectures of Hartshorne's. Math. Ann. **286**, 13–25 (1990)
4. W. Barth, E. Oeljeklaus, Über die Albanese-Abbildung einer fast-homogenen Kählermannigfaltigkeit. Math. Ann. **211**, 47–62 (1974)

5. A. Beauville, Fano contact manifolds and nilpotent orbits. Commun. Math. Helv. **73**, 566–583 (1998)
6. M. Beltrametti, A.J. Sommese, *The Adjunction Theory of Complex Projective Varieties* (de Gruyter, Berlin, 1995)
7. J. Bingener, On deformations of Kähler spaces II. Arch. d. Math. **41**, 517–530 (1983)
8. A. Borel, R. Remmert, Über kompakte homogene Kählersche Mannigfaltigkeiten. Math. Ann. **145**, 429–439 (1961/1962)
9. J.P. Demailly, On the Frobenius integrability of certain holomorphic p-forms, in *Complex Geometry, Göttingen, 2000*, ed. by I. Bauer et al. (Springer, Berlin, 2002), pp. 93–98
10. G. Elencwajg, *The Brauer Groups in Complex Geometry*. Lecture Notes in Mathematics, vol. 917 (Springer, Berlin/New York, 1982), pp. 222–230
11. T. Fujita, On polarized manifolds whose adjoint bundles are not semipositive. Adv. Stud. Pure Math. **10**, 167–178 (1985)
12. T. Fujita, *Classification Theories of Polarized Varieties*. London Mathematical Society Lecture Notes Series, vol. 155 (Cambridge University Press, Cambridge/New York, 1990)
13. N. Goldstein, Ampleness in complex homogeneous spaces and a second Lefschetz theorem. Pac. J. Math. **106**(2), 271–291 (1983)
14. F. Hirzebruch, K. Kodaira, On the complex projective spaces. J. Math. Pures Appl. **36**, 201–216 (1957)
15. A. Höring, Uniruled varieties with split tangent bundle. Math. Z. **256**(3), 465–479 (2007)
16. D. Husemoller, *Fibre Bundles*, 3rd edn. (Springer, New York, 1994)
17. J.M. Hwang, Rigidity of rational homogeneous spaces, in *International Congress of Mathematicians*, Madrid, ed. by M. Sanz Solé, vol. II (European Mathematical Society, Zürich, 2006), pp. 613–626
18. J.M. Hwang, Deformations of the space of lines on the 5-dimensional hyperquadric (2010, Preprint)
19. S. Kebekus, Lines on contact manifolds. J. Reine Angew. Math. **539**, 167–177 (2001)
20. S. Kebekus, Th. Peternell, A.J. Sommese, J. Wisniewski, Projective contact manifolds. Invent. Math. **142**, 1–15 (2000)
21. J. Kollár, S. Mori, Classification of three-dimensional flips. J. Am. Math. Soc. **5**(3), 533–703 (1992)
22. C. LeBrun, Fano manifolds, contact structures, and quaternionic geometry. Int. J. Math. **6**(3), 419–437 (1995)
23. S. Mukai, H. Umemura, Minimal rational threefolds, in *Algebraic Geometry*, Tokyo/Kyoto, 1982. Lecture Notes in Mathematics, vol. 1016 (Springer, Berlin, 1983), pp. 490–518
24. N. Nakayama, The lower semicontinuity of the plurigenera of complex varieties, in *Algebraic Geometry*, Sendai, 1985, ed. by T. Oda. Advanced Studies in Pure Mathematics, vol. 10 (North-Holland, Amsterdam, 1987), pp. 551–590
25. C. Okonek, M. Schneider, H. Spindler, *Vector Bundles on Complex Projective Spaces*. Progress in Mathematics, vol. 3 (Birkhäuser, Boston, 1980)
26. B. Pasquier, N. Perrin, Local rigidity of quasi-regular varieties. Math. Z. **265**(3), 589–600 (2010)
27. Th. Peternell, J. Wiśniewski, On stability of tangent bundles of Fano manifolds with $b_2 = 1$. J. Algebr. Geom. **4**(2), 363–384 (1995)
28. D. Popovici, Limits of projective manifolds under holomorphic deformations. arXiv.AG/09102032v1
29. J. Potters, On almost homogeneous compact complex surfaces. Invent. math. **8**, 244–268 (1968)
30. U. Schafft, Nichtsepariertheit instabiler Rang-2-Vektorbündel auf $\mathbb{P}_2$. J. Reine Angew. Math. **338**, 136–143 (1983)
31. J. Wiśniewski, On Fano manifolds of large index. Manuscr. Math. **70**, 145–152 (1991)
32. J. Wiśniewski, On deformation of nef values. Duke Math. J. **64**(2), 325–332 (1991)

List of Participants

Apostol Apostolov (Leibniz Universität Hannover)
Miguel A. Barja (Universitat Politècnica de Catalunya)
Alexey Basalaev (Hannover University)
Arnaud Beauville (University of Nice)
Gilberto Bini (Università degli Studi di Milano)
Mark Blume (Universität Münster)
Pawel Borowka (University of Bath)
Nathan Broomhead (Leibniz University Hannover)
Gregor Bruns (University of Leipzig)
Chiara Camere (Leibniz Universität Hannover)
Lucia Caporaso (Università Roma Tre)
Sebastian Casalaina-Martin (University of Colorado)
Fabrizio Catanese (Universität Bayreuth)
François Charles (Rennes)
Ciro Ciliberto (Università Roma 2)
Fabien Cléry (University of Amsterdam)
Elisabetta Colombo (Università Statale di Milano)
Slawomir Cynk (Jagiellonian University)
Ananyo Dan (Humboldt Universität zu Berlin)
Mohammad Dashtpeymad (Universität Regensburg)
Matthew Dawes (University of Bath)
Maria Donten-Bury (Mathematics Institute, University of Warsaw)
Wolfgang Ebeling (Leibniz Universität Hannover)
Ron Erez (Tel-Aviv University)
Hélène Esnault (Universität Duisburg-Essen)
Andrea Fanelli (Imperial College London)
Gavril Farkas (Humboldt Universität zu Berlin)
Lucja Farnik (Jagiellonian University)
Sara Angela Filippini (Università dell' Insubria)
Anne Frühbis-Krüger (Leibniz Universität Hannover)

A. Frühbis-Krüger et al. (eds.), *Algebraic and Complex Geometry*, Springer Proceedings in Mathematics & Statistics 71, DOI 10.1007/978-3-319-05404-9,

Irene Galstian (University of London)
Andreas Gathmann (Technische Universität Kaiserslautern)
Yoshinori Gongyo (University of Tokyo)
Rosario González Dorrego (UAM)
Hans-Christian Graf v. Bothmer (Universität Hamburg)
Gert-Martin Greuel (Kaiserslautern)
Valery Gritsenko (Lille University)
Samuel Grushevsky (Stony Brook University)
Jack Hall (KTH)
Lars Halvard Halle (University of Oslo)
Kangjin Han (KIAS)
Thomas Hawes (University of Oxford)
Bas Heijne (Leibniz Universität Hannover)
Nigel Hitchin (University of Oxford)
Andrea Hofmann (University of Stavanger)
Victoria Hoskins (Leibniz Universität Hannover)
Klaus Hulek (Leibniz Universität Hannover)
Yoonsuk Hyun (KIAS)
Anton Isopoussu (University of Cambridge)
Atsushi Ito (University of Tokyo)
Piotr Jedrzejewicz (Nicolaus Copernicus University, Torun)
Malek Joumaah (Leibniz Universität Hannover)
Seung-Jo Jung (University of Warwick)
Jesse Kass (University of Michigan)
Inder Kaur (Humboldt Universitaet, Berlin)
Masayuki Kawakita (RIMS, Kyoto University)
Yujiro Kawamata (University of Tokyo)
Remke Kloosterman (Humboldt Universitaet zu Berlin)
Andreas Leopold Knutsen (University of Bergen)
Herbert Kurke (Humboldt-Universität zu Berlin)
Radu Laza (Stony Brook University)
Margherita Lelli-Chiesa (Humboldt Universität Berlin)
Christian Liedtke (Universität Bonn)
Niels Lindner (HU Berlin)
Wenfei Liu (Universität Bielefeld)
Michael Lönne (Leibniz Universität Hannover)
Eyal Markman (University of Massachusetts Amherst)
Marcio Jose Martins (Physics Dept., Universidade Federal de Sao Ca)
Emilia Mezzetti (University of Trieste)
Michael Miesener (Universität Hannover)
Andreas Mihatsch (WWU Münster)
Giovanni Mongardi (Università Roma Tre)
Shigeru Mukai (RIMS Kyoto)
Fabian Müller (Humboldt-Universität zu Berlin)
Stefan Müller-Stach (Universität Mainz)

Juan Carlos Naranjo (U. Barcelona)
Howard Nuer (Rutgers University)
Yuji Odaka (Imperial College)
Keiji Oguiso (Osaka University)
Hisanori Ohashi (Tokyo University of Science)
Angela Ortega (Humboldt Universität)
John Ottem (University of Cambridge)
Nicola Pagani (Leibniz University Hannover)
Jinhyung Park (KAIST)
Stefano Pascolutti (Riemann Center Hannover)
Thomas Peternell (Universität Bayreuth)
Chris Peters (Université Grenoble I)
Nathan Pflueger (Harvard)
Anh Pham Ngoc (TU Kaiserslautern)
David Ploog (Leibniz University Hannover)
Piotr Pragacz (IMPAN)
Sławomir Rams (IMUJ Cracow/ IAG LU Hannover)
Kristian Ranestad (Univ of Oslo)
Joaquim Roé (Universitat Autònoma de Barcelona)
Sönke Rollenske (Universität Bielefeld)
Gregory Sankaran (University of Bath)
Taro Sano (University of Warwick)
Frithjof Schulze (Institut fuer Algebraische Geometrie)
Georg Schumacher (Universität Marburg)
Matthias Schütt (Leibniz Universität Hannover)
Mingmin Shen (University of Cambridge)
Alejandro Soto (Universität Tübingen)
Tomasz Szemberg (UP Krakow)
Nicola Tarasca (Leibniz Universität Hannover)
Mehdi Tavakol (University of Amsterdam)
Orsola Tommasi (Leibniz Universität Hannover)
Peter Toth (HU Berlin)
Stefano Urbinati (University of Utah)
Ravi Vakil (Stanford University)
Gerard van der Geer (Universiteit van Amsterdam)
Wim Veys (KU Leuven)
Claire Voisin (Université Paris VI)
Malte Wandel (Leibniz Universität Hannover)
Benjamin Wieneck (Leibniz Universität Hannover)
Kimiko Yamada (Okayama University of Science)

Complete List of Talks

Beauville, Arnaud (Nice): Abelian Varieties Associated to Gaussian Lattices

Let Γ be a self-dual lattice, endowed with an automorphism of square -1. Then $A_\Gamma = \Gamma_\mathbb{R}/\Gamma$ is a principally polarized abelian variety, with an automorphism ι of square -1. I will show that the configuration of ι-invariant theta divisors of A_Γ follows a pattern very similar to the classical theory of theta characteristics; as a consequence A_Γ has a high number of vanishing thetanulls. When $\Gamma = E_8$ we recover the 10 vanishing thetanulls of the abelian fourfold discovered by R. Varley.

Caporaso, Lucia (Rome): Tropical Methods for the Geometry of Algebraic Curves and Their Moduli Spaces

The talk will be a survey on the interplay between tropical/combinatorial and algebro geometric techniques, with focus on the case of tropical and algebraic curves and their moduli spaces.

Catanese, Fabrizio (Bayreuth): Topological Methods in Moduli Theory

The structure of the moduli spaces of surfaces and higher dimensional varieties is sometimes rather elusive if one only uses algebraic methods. But, over the complex numbers, sometimes the homotopy type of an algebraic variety determines the structure of the moduli spaces. I will explain this via examples of rigid and weakly

A. Frühbis-Krüger et al. (eds.), *Algebraic and Complex Geometry*, Springer Proceedings in Mathematics & Statistics 71, DOI 10.1007/978-3-319-05404-9, rigid varieties (connected moduli space up to complex conjugation), such as curves, Abelian varieties, Kodaira surfaces, varieties isogenous to a product, Beauville varieties. I will then present some main results on the Inoue type varieties recently introduced in joint work with Ingrid Bauer. A dominant role is played also by the study of moduli spaces of curves with a group G of automorphisms. Here, I will present some recent works, in particular joint work with Michael Loenne and Fabio Perroni, concerning the irreducibe components for some special groups, as Abelian and Dihedral, and the irreducible stable components of the latter moduli spaces for a general group G (extending work of Dunfield and Thurston in the free action case).

Ciliberto, Ciro (Rome): Construction and Properties of Some Irregular Surfaces

In this talk I will explain a few constructions of some non-trivial irregular surfaces with no irrational pencils (among them a recent interesting example by Schoen which I will look from a slightly different viewpoint than the original one). Some interesting properties of these surfaces will also be discussed. This is (experimental) work in progress with M. Mendes Lopes and X. Roulleau.

Esnault, Hélène (Essen): Index and Euler Characteristic over Henselian Fields with Algebraically Closed Residue Fields

Over a henselian field with algebraically closed residue field of residue characteristic 0 or p large, the index of a smooth projective variety divides the Euler characteristic of any coherent sheaf. (Joint with Marc Levine and Olivier Wittenberg)

Farkas, Gavril (Berlin): Syzygies of Torsion Bundles and the Geometry of the Level l Modular Variety over M_g

In joint work with Chiodo, Eisenbud and Schreyer, we formulate, and in some cases prove, three statements concerning the purity of the resolution of various rings one can attach to a generic curve of genus g and a torsion point of order l in its Jacobian. These statements can be viewed an analogues of Green's Conjecture and we verify them computationally for bounded genus. We then compute the cohomology class of the corresponding non-vanishing locus in the moduli space $R_{g,l}$ of twisted level l curves of genus g and use this to derive results about the birational geometry of $R_{g,l}$. For instance, we prove that $R_{g,3}$ is a variety of general type when $g > 11$.

I will also discuss the surprising failure of the Prym-Green Conjecture for genera which are powers of 2.

Gathmann, Andreas (Kaiserslautern): The Relative Tropical Inverse Problem for Curves in a Tropical Plane

The idea of tropical geometry is to associate to an algebraic variety X a polyhedral complex, the so-called tropicalization $trop(X)$ of X. One can then study $trop(X)$ by purely combinatorial means and try to translate the results back to the original variety X. One of the central problems in this process is the "tropical inverse problem", i.e. the question which polyhedral complexes can be realized as tropicalizations of an algebraic variety. We study this question in a relative setting: given a tropical curve C contained in the tropicalization of a plane X, is there an algebraic curve contained in X that tropicalizes to C? We give a complete algorithmic answer to this problem and use this to reprove certain known and some new general criteria for realizability.

Geer, Gerard van der (Amsterdam): Vector-Valued Picard Modular Forms and Curves of Genus Three

The talk deals with vector-valued Picard modular forms on a unitary group of signature $(2, 1)$ over the ring of Eisenstein numbers. We construct modular forms and discuss the structure of modules of such modular forms. It is related to the cohomology of local systems on a moduli space of curves of genus three. This is joint work with Fabien Clery and Jonas Bergstroem.

Gritsenko, Valery (Lille): Exceptional Arnold Singularities and Their Automorphic Discriminants

The semi-universal deformations of exceptional Arnold singularities can be presented as modular varieties of orthogonal type. We construct their automorphic discriminants and we show that three of them determine Lorentzian Kac-Moody Lie algebras. We consider two types of generalized automorphic forms on the full space of deformations (a non-classical homogeneous domain defined by E. Looijenga and K. Saito) and we give an answer of some old problems of K. Saito. At the end we formulate open research questions in this area.

Grushevsky, Sam (Stony Brook): Towards the Stable Cohomology of $\overline{A_g}$

The stable (for g much larger than the degree) cohomology of the moduli space A_g of principally polarized abelian varieties was computed by Borel, as the group cohomology of the symplectic group. Using topological methods, Charney and Lee computed the cohomology of Satake-Baily-Borel compactification. In this talk we will discuss the question of computing the stable cohomology of toroidal compactifications, and in particular will discuss the stabilization of suitable cohomology for the perfect cone toroidal compactification, and the computation of some these stable cohomology groups. Joint work in progress with Klaus Hulek and Orsola Tommasi.

Halle, Lars (Oslo): Néron Component Groups and Base Change

Let K be a complete discretely valued field and let A be an abelian K-variety. In this talk I will discuss the Néron component series of A. This is a formal power series in $\mathbb{Z}[[T]]$ which measures how the number of connected components of the special fiber of the Néron model of A varies under tame extensions of K. In case A is wildly ramified, it is particularly challenging to describe the properties of this series. I will present some results for Jacobians and abelian varieties with potential multiplicative reduction, and discuss a few open problems in this setting. If time permits, I will also mention generalizations to semi-abelian varieties. This is joint work with Johannes Nicaise (Leuven).

Hitchin, Nigel (Oxford): The Hyperholomorphic Line Bundle

On a hyperkähler manifold with a circle action preserving just one complex structure there is a natural hyperholomorphic line bundle. This is a constituent of the physicists' hyperkähler/quaternionic Kähler correspondence and was treated in a differential geometric fashion by A Haydys. We show how to construct this via a holomorphic bundle on the twistor space and consider examples including the moduli space of Higgs bundles.

Kawamata, Yujiro (Tokyo): Derived Categories from the Viewpoint of the Minimal Model Program

I would like to consider some problems concerning the minimal model program and the derived categories; MMP and semi-orthogonal decompositions, K-equivalence and the Fourier-Mukai partners, finiteness of models, and the cone conjecture.

Laza, Radu (Stony Brook): The KSBA Compactification for the Moduli Space of Degree Two K3 Pairs

A classical (and still open) problem in algebraic geometry is the search for a geometric compactification for the moduli of polarized K3 surfaces (X, L). If one considers instead K3 pairs (X, H) with H a divisor in the linear system $|L|$, the resulting moduli space has a natural geometric compactification given by the general MMP framework (pioneered by Kollár, Shepherd-Barron, and Alexeev). In this talk, I will discuss the existence of a good compactification for the moduli of K3 pairs in all degrees, and then discuss in detail the degree 2 case.

Liedtke, Christian (Stanford): On the Birational Nature of Lifting

Whenever a variety X lifts from characteristic p to characteristic zero, say over the Witt ring, then many classical results over the complex numbers hold for X, and certain "characteristic p pathologies" cannot occur, simply because one can reduce modulo p. But then, lifting results are difficult, and in general, varieties do not lift. However, in many situations, it is possible or easier to lift a birational model of X, maybe even one that has "mild" singularities. Thus, a natural question is whether the liftability of such a birational model implies that of our original X. We will show that this completely fails in dimension at least 3, that this question is surprisingly subtle in dimension 2, and that it is trivial in dimension 1. This is joint work with Matthew Satriano.

Markman, Eyal (Amherst): Lagrangian Fibrations on Holomorphic Symplectic Varieties of $K3^{[n]}$ Deformation Type

Let X be an irreducible holomorphic symplectic manifold deformation equivalent to the Hilbert scheme of n points on a K3 surface. Let L be a nef line-bundle, which is isotropic with respect to the Beauville-Bogomolov pairing. Assume that the two-dimensional subspace spanned by cohomology classes of type $(2, 0)$ and $(0, 2)$ on X does not contain non-zero integral classes. We prove that L is base point free and it induces a Lagrangian fibration from X onto a projective space, whose general fiber is a Jacobian.

Mukai, Shigeru (RIMS): Enriques Surfaces and Abelian Surfaces

Many families of Enriques surfaces are constructed from abelian surfaces via Kummer surfaces. Such Enriques surfaces promote and shed new light on the study of Abelian surfaces. In this talk I review known results in this direction and reconstruct such Enriques surfaces canonically from their periods in several cases.

Keiji Oguiso (Osaka): Automorphism Groups of Calabi-Yau Manifolds of Picard Number Two

We prove that the automorphism group of an odd dimensional Calabi-Yau manifold of Picard number two is always a finite group. This makes a sharp contrast to the automorphism groups of K3 surfaces and hyperkähler manifolds and birational automorphism groups, as we shall see. We also clarify the relation between finiteness of the automorphism group (resp. birational automorphism group) and the rationality of the nef cone (resp. movable cone) for a hyperkähler manifold of Picard number two. We will also discuss a similar conjectual relation for a Calabi-Yau threefold of Picard number two, together with existence of rational curve, expected by the cone conjecture.

Peternell, Thomas (Bayreuth): Differential Forms, Foliations and Ricci Flat Varieties

I will discuss possible generalizations of the Beauville-Bogomolov decomposition theorem to singular varieties and present recent results on the decomposition of the tangent sheaf of singular varieties with trivial canonical classes (joint work with D.Greb and S.Kebekus).

Sankaran, Gregory (Bath): Stable Homology for Partial Compactifications of A_g

I shall describe joint work (in progress) with Jeff Giansiracusa that aims to prove stability results for homology or cohomology of suitable toroidal partial compactifications of A_g. The methods come partly from homotopy theory, using stability results for homology of $GL(\mathbb{Z})$ due to Charney, Dwyer and van der Kallen.

Tommasi, Orsola (Hannover): Cohomology of Local Systems of Low Weight on M_2

In this talk we discuss different techniques for the computation of the cohomology of the moduli space of non-singular curves of genus 2 with marked points. We concentrate on the case of genus 2 curves with 4 marked points and we explain how such research is motivated by the study of the structure of the tautological ring of the moduli space of stable curves.

Ravi Vakil (Stanford): Stabilization of Discriminants in the Grothendieck Ring

We consider the "limiting behavior" of *discriminants*, by which we mean informally the closure of the locus in some parameter space of some type of object where the objects have certain singularities. We focus on the space of partially labeled points on a variety X, and linear systems on X. These are connected—we use the first to understand the second. We describe their classes in the "ring of motives", as the number of points gets large, or as the line bundle gets very positive. They stabilize in an appropriate sense, and their stabilization can be described in terms of the motivic zeta values. The results extend parallel results in both arithmetic and topology. I will also present a conjecture (on "motivic stabilization of symmetric powers") suggested by our work. Although it is true in important cases, Daniel Litt has shown that it contradicts other hoped-for statements. This is joint work with Melanie Wood.

Voisin, Claire (Paris): Symplectic Involutions of K3 Surfaces Act Trivially on Zero-Cycles

A symplectic involution of a K3 surface acts trivially on the space of holomorphic 2-forms, hence Bloch's conjecture predicts that it acts trivially on 0-cycles of degree 0 modulo rational equivalence. This statement has been proved by Huybrechts and Kemeny for the pairs (K3, involution) in one of the three series described by Sarti and Van Geemen. We will show how to prove the result in all cases.